The Whitefly, *Bemisia tabaci* (Homoptera: Aleyrodidae) Interaction with Geminivirus-Infected Host Plants

Winston M.O. Thompson
Editor

The Whitefly, *Bemisia tabaci* (Homoptera: Aleyrodidae) Interaction with Geminivirus-Infected Host Plants

Bemisia tabaci, Host Plants and Geminiviruses

Springer

Editor
Dr. Winston M.O. Thompson
Plant Protection
National Agricultural Research Institute
Mon Repos, East Coast Demerara, Guyana, South America
15216 NE 8th St., H3
98007 Bellevue, Washington
USA
winstont335@gmail.com

ISBN 978-94-007-1523-3 e-ISBN 978-94-007-1524-0
DOI 10.1007/978-94-007-1524-0
Springer Dordrecht Heidelberg London New York

Library of Congress Control Number: 2011930863

Cover design: SPi Publisher Services

Printed on acid-free paper

Springer is part of Springer Science+Business Media (www.springer.com)

Dedicated to Iris and Christina

Preface

Whiteflies cause significant problems to agricultural production worldwide. There are various biotypes, but B-biotype is of particular importance because of its polyphagous feeding habit, high fecundity and resistance to a wide range of insecticides. It causes direct feeding damage such as the silverleaf condition in squash, but its efficacy in successfully transmitting several geminiviruses is responsible for a number of disease epidemics around the world. Examples include *Cotton leaf curl virus* (CLCuV) in Pakistan and India, *Tomato yellow leaf curl China virus* (TYLCCNV) in China and *Tomato yellow leaf curl virus* in various parts of the world. In Africa and India, the cassava biotypes pose similar problems. *East African cassava mosaic virus* and *African cassava mosaic virus* are effectively transmitted by the cassava biotype *B. tabaci.* In India, the *Indian cassava mosaic virus* is also transmitted by a cassava biotype that is genetically incompatible with the biotype transmitting *East African cassava mosaic virus* and *African cassava mosaic virus*.

The pathosystems involving B-biotype and crops such as cotton and tomatoes, and the respective geminiviruses: CLCuV, TYLCCNV present similar consequences as the pathosystems involving the cassava biotype, cassava and the geminiviruses affecting cassava. The interaction of vector, virus and host plant in some pathosystems, results in high population levels of the vector, which is responsible for several disease epidemics. In more complex situations, mixed infections and recombinant viruses involved in mixed infections contribute to the interplay of host plant, vector and viruses. Effects of infected host plants on population increase of the vector have been related to improved nutritional status of the host plant and/or suppressed plant defense mechanisms towards the vector. It is worthy to note that not all interactions are favorable to the vector, suggesting that pathosystems vary in the outcome of disease epidemics.

The objective of this E-Book is to introduce the different pathosystems along with the most recent findings and research endeavors. The various systems, each with its own challenge and complexity will unequivocally contribute to existing knowledge. With evolving geminiviruses and the appearance of new *B. tabaci* biotypes, new interaction events and disease epidemics can be anticipated. To this end, chapters are included to deal with emerging geminiviruses, and the distinction between *B. tabaci* biotypes using advanced molecular techniques. This E-Book will be a good reference source, comprising related chapters devoted to an improved understanding of the intricacies underlying geminivirus disease epidemics in various parts of the world. Since the ultimate goal is to advance such understanding into sustainable management practices against *B. tabaci* and the geminiviruses they transmit, concluding chapters deal with management, and possible applications of Remote Sensing and Geographic Information Systems (GIS) technology.

This book will be of value to researchers in the biological and agricultural sciences, graduate students and corporations linked to the agricultural industry.

Acknowledgements

Sincere thanks to all the contributors for their commitment to this project. It was truly a team effort. Hearty thanks to Dr. Claude Fauquet for his kind support during this project.

To my technical team: Konrad Thomas, Juneann Lynch, Jean Barker, Shirley Rodrigues, Michelle Theresa and Joan Thomas.

And to the following individuals who have been of tremendous help and support: Barclay, Ronald, Leonard, Richard and Maurice.

Contents

Contributors

Yehezkel Antignus Agricultural Research Organization (ARO), The Volcani Center, Institute of Plant Protection, P.O.B. 6, Bet Dagan 50250, Israel
antignus@volcani.agri.gov.il

Henryk Czosnek The Otto Warburg Minerva Center for Agricultural Biotechnology & the Robert H. Smith Institute for Plant Science and Genetics in Agriculture, The Hebrew University of Jerusalem, Rehovot 76100, Israel
czosnek@agri.huji.ac.il

P.J. De Barro CSIRO Ecosystem Sciences, 120 Meiers Road, Indooroopilly 4068, Australia
Paul.debarro@csiro.au

James H. Everitt United States Department of Agriculture, Agricultural Research Service, Kika de la Garza Subtropical Agricultural Research Center, Weslaco, TX 78596, USA
james.everitt@ars.usda.gov

Dan Gerling Department of Zoology, Tel Aviv University, Ramat Aviv 69978, Israel
dange@tauex.tau.ac.il

Murad Ghanim Department of Entomology, Agricultural Research Organization, Volcani Center, Bet Dagan 50250, Israel
ghanim@agri.gov.il

Robert L. Gilbertson Department of Plant Pathology, University of California-Davis, Davis, CA 95616, USA
rlgilbertson@ucdavis.edu

T. J. Henneberry Western Cotton Research Laboratory, Phoenix, AZ 85040–8803, USA

A. Rami Horowitz Department of Entomology, Agricultural Research Organization (ARO), Gilat Research Center, M.P. Negev 85280, Israel
hrami@volcani.agri.gov.il

Bikash Mandal Advanced Centre for Plant Virology, Indian Agricultural Research Institute, New Delhi 110 012, India

R. S. Mann Department of Entomology and Nematology, University of Florida, Bldg. 970, Natural Area Drive, Gainesville, FL 32611-0620, USA
mannrs@ufl.edu

Francisco J. Morales Virology Research Unit, International Center for Tropical Agriculture, A.A. 6713, Cali, Colombia
f.morales@cgiar.org

Eric Natwick UC Desert Research and Extension Center, University of California Cooperative Extension, 1050 E. Holton Road, Holtville, CA 92250, USA
etnatwick@ucdavis.edu

M.S. Palaniswami Division of Crop Protection, Central Tuber Crops Research Institute, Trivandrum 695017, India
palaniswamims@yahoo.com

Maria Rojas Department of Plant Pathology, University of California-Davis, Davis, CA 95616, USA
mjrojas@ucdavis.edu

Manoj K. Singh Advanced Centre for Plant Virology, Indian Agricultural Research Institute, New Delhi 110 012, India

Winston M.O. Thompson Plant Protection, National Agricultural Research Institute, Mon Repos, East Coast Demerara, Guyana, South America

15216 NE 8th St., H3 98007 Bellevue, Washington, USA
winstont335@gmail.com

Anupam Varma Advanced Centre for Plant Virology, Indian Agricultural Research Institute, New Delhi 110 012, India
anupamvarma@vsnl.net

Chenghai Yang United States Department of Agriculture, Agricultural Research Service, Kika de la Garza Subtropical Agricultural Research Center, Weslaco, TX 78596, USA
chenghai.yang@ars.usda.gov

Abbreviations

AISA	Airborne Imaging Spectroradiometer for Applications
ANOVA	Analysis of variance
AVIRIS	Airborne Visible/Infrared Imaging Spectrometer
BCIP	5-Bromo-4-chloro-3-indolyl phosphate
CASI	Compact Airborne Spectrographic Imager
CIR	Color Infrared
DNA	Deoxyribonucleic acid
dNTP	deoxyribonucleoside triphosphate
GIS	Geographic Information System
HYDICE	Hyperspectral Digital Imagery Collection Experiment
HyMap	Hyperspectral Mapper
IPM	Integrated Pest Management
IRM	Insecticide Resistance Management
ISEM	Immunosorbent Electron Microscopy
kDa	Kilodalton
NBT	Nitro blue tetrazolium
NCBI	National Center for Biotechnology Information
NIR	Near Infrared
ORF	Open Reading Frame
PCR	Polymerase Chain Reaction
PNPP	*p*-Nitrophenyl Phosphate
RAPD	Random Amplified Polymorphic DNA
RH	Relative Humidity
RNA	Ribonucleic acid
SDS-PAGE	Sodium Dodecyl Sulfate Polyacrylamide Gel Electrophoresis
TAS-ELISA	Triple Antibody Sandwich Enzyme Linked Immunosorbent Assay
TBS	Tris Buffered Saline
UV	Ultraviolet
WTGs	Whitefly Transmitted Geminiviruses

Chapter 1
Introduction: Whiteflies, Geminiviruses and Recent Events

Winston M.O. Thompson

Abstract The first part of this chapter introduces the whitefly as an important economic pest affecting agricultural crops worldwide. It deals with whitefly development and classification, whitefly biotypes and whiteflies as important vectors of plant viruses. Among such viruses are the geminiviruses which are discussed in terms of genetic constitution, host plants and insect vector, of which the whitefly *Bemisia tabaci* is among the most destructive of the vectors. The developments over the past two decades as these relate to *B. tabaci* and transmitted geminiviruses are highlighted. The second part of the chapter introduces the forthcoming chapters of the book.

1.1 The Whitefly

Whiteflies (Homoptera/Hemiptera: Aleyrodidae) are insect pests of significant economic importance affecting agricultural crops such as tomatoes, cotton, cassava and beans, as well as ornamentals. Of importance is the fact that they have worldwide distribution and as such are commonly known insect pests and vectors to entomologists, virologists, agriculturists and growers. They are about 2–3 mm in length, and wings are present in the adult stage of both sexes. The wings are generally opaque and covered with a whitish powder or wax. Abdomen lacks cornicles (tubular structures located dorsally towards the posterior end of the abdomen), and the hind wings are nearly as long as the forewings (Borror et al. 1989).

W.M.O. Thompson (✉)
Plant Protection, National Agricultural Research Institute, Mon Repos,
East Coast Demerara, Guyana, South America
15216 NE 8th St., H3 98007 Bellevue, Washington, USA
e-mail: winstont335@gmail.com

W.M.O. Thompson (ed.), *The Whitefly, Bemisia tabaci (Homoptera: Aleyrodidae) Interaction with Geminivirus-Infected Host Plants*,
DOI 10.1007/978-94-007-1524-0_1, © Springer Science+Business Media B.V. 2011

Unlike many Homopterans that undergo paurometabolous development (gradual metamorphosis), the metamorphosis of whiteflies is different showing a pattern more towards complete metamorphosis (holometabolous development). Borror et al. (1989) describes the metamorphosis as “Intermediate”. There are five instars including the adult. The first instar is active, while the following three are inactive or sessile. During metamorphosis wing development is internal and the wing pads are everted at the end of the third instar, appearing in the fourth instar which resembles a pupa.

Most whitefly species are oligophagous, but most whitefly pest species are polyphagous. There are however some oligophagous whitefly pest species such as *Aleurocybotus* spp. and *Aleurolobus* spp. that affect plants in the Family: Gramineae, and *Asterochiton* spp. affecting plants of *Acer* spp. (Byrne et al. 1990).

Many whitefly pest species are multivoltine, producing several generations a year. They also tend to develop resistance rather quickly to a large number of pesticides (Byrne et al. 1990). As vectors of pathogens, although bacteria and fungi can be transmitted by whiteflies (Costa 1976) whitefly pest species are of greater economic importance as vectors of plant viruses. Some Geminiviruses, Carlaviruses, Nepoviruses, Potyviruses and Closteroviruses can be transmitted by this group of insects (Byrne et al. 1990).

Brown (1994) reported 1,100 species of whiteflies worldwide, and only three are recognised as vectors of plant viruses. Of this number *Bemisia tabaci* (Gennadius) is considered the most important of the whitefly vectors of plant viruses, and the only whitefly species transmitting geminiviruses (Duffus 1987; Harrison 1985). Additionally, *B. tabaci* can also cause direct damage through sucking phloem sap and secretion of honeydew, which in particular, has caused serious problems in the cotton industry (Pollard 1955).

The Indian subcontinent is the suspected centre of origin of *B. tabaci* based on the presence of a large number of its natural enemies in that region. It has been suggested that the spread of the species probably occurred from the Indian subcontinent to Africa, Europe and the Americas, through the movement of plant material by man (Cock 1986).

Under favourable conditions *B. tabaci* can undergo 11–15 generations a year (Avidov 1956), and a female can lay between 100 and 300 eggs in her lifetime, which varies from 3 to 6 weeks (Azab et al. 1971; Bethke et al. 1991).

Early taxonomic separation of whitefly species was for the most part dependent on morphological characteristics of the pupal case (Gill 1992; Mound and Halsey 1978). This approach however, had some weaknesses because morphological characteristics can be altered based on species adaptation to a specific host. For example Russel (1958) reported that the early literature identified several genera and species of whiteflies that are now grouped under the single species, *B. tabaci;* some 18 previously described *Bemisia* species are now classified as *B. tabaci.* International collections of *B. tabaci* were shown to be genetically variable (Costa et al. 1993; Brown 1994). Wool et al. (1993) reported that such populations differed in the ability to utilise specific host plants for feeding and reproductive purposes. These populations also showed differences in virus transmission characteristics (Bedford et al. 1992, 1994). Although these differences occurred, the morphological characteristics of the pupal case were indistinguishable (Mound and Halsey 1978; Russel 1958).

From these earlier studies, it was recognized that a more reliable classification system depended on genetic and biochemical properties, and host plant adaptation, in addition to the morphological characteristics. Subsequently, the application of molecular techniques such as Polymerase Chain Reaction (PCR) and the use of DNA probes, as well as biochemical tools utilised for determining esterase banding patterns, had made possible the identification of different biotypes of *B. tabaci*.

Through the application of DNA sequencing, Fauquet et al. (1998) were able to identify a distinctive population of *B. tabaci* suspected as the driving factor for the cassava mosaic disease (CMD) epidemic in Uganda. In addition, a general esterase marker had been developed for identification of the B-biotype of *B. tabaci* (Brown et al. 1992; Bedford et al. 1992).

Characteristics of the B-biotype of *B. tabaci* include: Their highly polyphagous nature, feeding on a wide range of host plants of distinctive Families (Bedford et al. 1994); their high resistance to a wide range of insecticides including DDT, endosulfan and methyl parathion (Byrne and Devonshire 1993; Perrings et al. 1993); their ability to induce physiological disorders in several plant species within Cucurbita and Brassica, and *Solanum lycopersicum* L. (Bedford et al. 1992; Schuster et al. 1990). Some of these symptoms include the "silverleaf" condition in squash, tomato irregular ripening and broccoli light stalk (Toscano et al. 1994).

Some authorities were of the opinion that the B-biotype had sufficient distinguishing characteristics to warrant its placement under a new species. As a result the new species name, ***Bemisia argentifolii*** Perrings and Bellows, was sometimes seen in the literature. However, experimental findings of Byrne et al. (1998) questioned the species status of *B. argentifolii* since they reported interbreeding between the B-biotype and other biotypes of *B. tabaci*. In their work, they observed two insensitive forms of acetylcholinesterase (AchE) in the species responsible for resistance against carbamates and organophosphorus insecticides. According to their findings, AchE transcended the biotype boundaries established by the esterase binding pattern, and individuals of different biotypes showed the heterozygous form of AchE. They concluded that this was evidence of interbreeding having occurred. Thus, at that time, against a background of varying taxonomic opinion in relation to this particular group of whiteflies, it was not surprising that it was referred to in the literature as *B. argentifolii* or B biotype *B. tabaci*. Interestingly, with the appearance of other biotypes the more recent question has been whether *B. tabaci* is a species of many biotypes or rather a cryptic species complex.

While the effects of B biotype were seen from the mid1980s, another whitefly of great significance, *B. tabaci* Q biotype emerged on the scene *c.*10 years later. The Q biotype was observed in the Mediterranean basin (Guirao et al. 1997). It showed some of the early traits of B-biotype; exhibiting resistance to many insecticides, high fecundity and the capability to displace its competitors. In Spain during 2001–2002 B-biotype was displaced by the indigenous Q-biotype (Brown 2007). The Q biotype is closely monitored by researchers because of its potential to cause significant crop damage and losses through its ability to rapidly expand its population, transmit geminiviruses and overcome the effects of insecticides.

1.2 Geminiviruses

Geminiviruses are plant viruses with one or two single stranded circular DNA molecules. The single genome is *c.* 2.5–3.0 kb (Bisaro 1996; Gutierrez 2000). The capsid consists of two icosahedral particles of 18–30 nm in size (Gutierrez 1999).

As pathogens of significant economic importance, geminiviruses were responsible for tremendous yield losses. *Tomato yellow leaf curl virus* (TYLCV) in some fields caused up to 100% crop loss in the Dominican Republic and losses were estimated at more than US $10 million (Gilbertson et al. 2007). *Cotton leaf curl virus* has affected 50% of the crop in Pakistan (Ali et al. 1993) and severe cassava mosaic disease had caused a famine situation in parts of Uganda (Otim-Nape et al. 1994).

In an earlier classification in the 1980s, the family *Geminiviridae* showed four subgroups of geminiviruses transmitted by Homopterous insects. Subgroups I and II were transmitted by leafhoppers and planthoppers, whereas subgroups III and IV contain viruses transmitted by the whitefly, *B. tabaci* (Brown 1994).

Geminiviruses of subgroups I and II contain two icosahedrons (subunits) fused together giving a twin-like appearance, with a single circular DNA strand holding the units together (Goodman 1981; Goodman et al. 1980; Stanley and Davies 1985). In subgroup III two circular DNA strands are present, each in a separate coat protein (capsid). Members of subgroup IV are monopartite and contain a single DNA strand (Brown 1994).

A more recent classification, place geminiviruses into three genera: monogeminiviruses, hybrigeminiviruses and bigeminiviruses. Hybrigeminiviruses and monogeminiviruses are transmitted by different species of planthoppers and leafhoppers, and bigeminiviruses are transmitted by whiteflies (Gray and Banerjee 1999).

The two component geminiviruses show a conserved 200 nucleotide (nt) noncoding intergenic region (Matthews 1991). This region is capable of forming a hairpin loop and within this loop is a conserved sequence: TAATATTAC, seen in all geminiviruses (Lazarowitz 1987; Bisaro 1996).

Both DNA A and DNA B are required for infectivity of *African cassava mosaic virus* (ACMV) (Stanley 1983) as well as for *Tomato golden mosaic virus* (TGMV) and *Bean golden mosaic virus* (BGMV) (Morinaga et al. 1988). Sequences of the two DNAs are different except for the 200 nt conserved region (Matthews 1991). Comparisons of sequences of ACMV (Matthews 1991) with sequences of TGMV (Hamilton et al. 1984) and BGMV (Howarth et al. 1985) reveal that six of the open reading frames (ORFs) are conserved. The two ORFs in DNA B of ACMV are required for infectivity (Etessami et al. 1988).

In ACMV the coat protein gene is located in DNA A (Townsend et al. 1985). This is also the case for TGMV (Kallender et al. 1988). Based on the findings of Briddon et al. (1990), coat protein was observed to be a function of vector specificity. Coat protein is not essential for infectivity of ACMV (Etessami et al. 1989) or TGMV (Gardiner et al. 1988).

Bipartite whitefly transmissible geminiviruses can occur in mixtures within the host especially in Solanaceous plants (Garzon-Tiznado et al. 1993) and in Cucurbits

(Lazarowitz 1992) thereby allowing for genetic recombination and transencapsulation between viruses (Brown 1994). Padidam et al. (1999) have identified up to 420 significant recombinant fragments, and recombination events could occur both within and between genera suggesting the high versatility or plasticity of this group of viruses.

1.2.1 Genera of Geminiviruses

In the most recent classification, Geminiviruses belong to the Family: *Geminiviridae*. The genera within this family of viruses are described by several researchers (Lazarowitz 1992; Bisaro 1996; Fauquet et al. 2000; Briddon and Markham 2001).

Mastrevirus of which *Maize streak virus* is the type species, is the oldest and most diverse group. The viruses in this group are monopartite, transmitted by leafhoppers and they infect mainly monocotyledonous plants. These viruses encode the movement protein (MP) and the capsid protein (CP) on the virus sense strand. The complementary strand encodes the Rep protein and the Rep A protein; that is exclusive to Mastreviruses.

Curtovirus of which *Beet curly top virus* is the type species. The viruses in this group are transmitted by leafhoppers and infect dicotyledonous plants. These viruses are also monopartite and are thought to have originated from an ancient recombination event between a *Mastrevirus* and a *Begomovirus* (Rybicki 1994). The virus sense strand encodes the MP, CP and a V2 protein. The complementary strand has four open reading frames (ORFs) namely; Rep, C2, REn and C4.

Begomovirus of which *Bean golden mosaic virus* is the type species. These are transmitted only by whiteflies of the species *B. tabaci*. They include a large group of economically important viruses affecting dicotyledonous plants. Members in this group are mainly bipartite.

In genome A there are four ORFs on the complementary strand: AC1 (Rep), AC2 (TrAP), AC3 (REn) and AC4. The virus sense strand encodes the CP. On the B genome, two proteins encoded on the ORFs: BV1 and BC1 are involved in movement.

Topocuvirus of which *Tomato pseudo-curly top virus* is the type species. These are thought to have originated from Begomoviruses interacting with another virus. These viruses are transmitted by tree hoppers and they affect dicotyledonous plants (Gray and Banerjee 1999).

1.3 Significant Events over the Last Two Decades

The first evidence in the displacement of Biotype A by Biotype B in the USA was seen in 1990 when Arizona fields showed respectively compositions of 70% and 30% for the B and A biotypes (Brown 2007). Not surprisingly this newer more abundant biotype was found to be resistant to several insecticides. This 'abundance' characteristic of a *B. tabaci* biotype, in later years became an important driving force in geminivirus disease epidemics. The early 1990s were quite eventful with severe cassava mosaic virus disease seriously affecting cassava in Uganda (Otim-Nape et al. 1994; Gibson et al. 1996), Tomato yellow leaf curl being introduced into the Dominican Republic (Salati et al. 2002) and Cotton leaf curl disease from Pakistan spreading rapidly, moving into Northern India (Varma et al. 1993). Incidentally during the 1990s farmers and researchers in Indonesia observed symptoms of pepper yellow leaf curl on chilli peppers, but at that time the disease was not as yet causing serious problems. There was a complete different scenario however at the turn of the century when a condition referred to as Penyakit Kuning was of serious concern to chili pepper farmers in Indonesia (Chap. 9).

In 1991–1992, host associated biotypes were described on cassava and Okra (Burban et al. 1992) and Legg et al. (1994) reported cassava and non-cassava (sweet potato) biotypes in Uganda. There were also reports on the biological characterization of *B. tabaci* biotypes from various locations (Brown et al. 1992; Bedford et al. 1994). During this time B-biotype was rapidly spreading in Latin America and the Caribbean (Costa et al. 1993; Brown 1994) and had also moved into Brazil (Lourencao and Nagai 1994). The Silverleaf condition on squash was observed in the Southwestern USA (Liu et al. 1992; Cohen et al. 1992). This period had also seen the introduction of TYLCV into the USA (Polston et al. 1994) and the Caribbean region (Rojas and Gilbertson 2008). In 1994 B-Biotype was introduced into Australia (Gunning et al. 1997). Between 1994 and 1996, B-biotype displaced the Jatropha and Sida biotypes in Puerto Rico (Brown 2007). In 1996, Q-biotype was recognized as an important native pest in the Mediterranean basin (Guirao et al. 1997). At this time recombination of geminiviruses was observed (Zhou et al. 1997). In the late 1990s, displacement did not only occur among the vectors but among the viruses as well. In Spain, *Tomato yellow leaf curl Sardinia virus* (TYLCSV) was displaced by TYLCV (Sanchez-Campos et al. 1999). There was also the first report of TYLCV in Japan (Kato et al. 1998) and TYLCV had also spread into Puerto Rico (Bird et al. 2001). Interestingly B-biotype was not always the predominant biotype for competitive space and resources. In the period 2001–2002, B-biotype was displaced by the indigenous Q-biotype in Spain (Brown 2007). Around this time B biotype was first reported in Argentina (Viscarret et al. 2003). In the period 2003–2005, it was found that the silverleaf condition was also produced by a non-B biotype from Uganda (Sseruwagi et al. 2005), and that some *B. tabaci* populations infesting cassava could infest other host plants (Thompson 2003; Sseruwagi et al. 2006). Silverleaf condition was also induced by the Ms biotype of *B. tabaci*, indigenous to the Islands Southwest of the Indian Ocean. This latter biotype was found to be closely related to biotypes B and Q (Delatte et al. 2005). Years 2005–2006 had seen the introduction of Q-biotype into the USA, China, Japan and Mexico (Dennehy et al. 2005; Dong et al. 2006; Ueda and Brown 2006; Martinez-Carrillo and Brown 2007). In the 2005–2006 period severe cassava mosaic virus disease continued to spread and evidence indicated the involvement of an invasive biotype (Legg et al. 2002). In the USA, the Q biotype was found to be restricted to greenhouse grown plants (Brown 2007). In 2007, *B. tabaci* Q biotype was observed in Syria (Fujiie et al. 2009). During this period it was once again displacing the B biotype in the Shandong Province of China (Chu et al. 2010).

The last two decades of events that occurred simultaneously or in tandem project a pattern of new emerging *B. tabaci* biotypes, movement of biotypes into other geographic locations, spread of begomoviruses into other areas, and in some cases displacement of one biotype by another. The displacement mechanism has also been seen among the begomoviruses, but the biological phenomenon quite common with begomoviruses is their propensity to undergo recombination; the consequences of which have presented challenges not only in terms of management but in the development and application of a robust dependable classification system. The classification of geminiviruses has been revised on a number of occasions (Fauquet et al. 2000, 2003), and some of the difficulties were related to the appropriate placement of new geminiviruses and/or recombinant viruses that were incongruous with the existing system. Although there is presently a more updated classification system of the geminiviruses even at levels below the species taxon (Fauquet et al. 2008), it is anticipated that further updates will become necessary as more geminiviruses begin to emerge.

1.4 Objectives and Outline

The objectives of this book is to present the different pathosystems to examine the consequences of the interaction of *B. tabaci* with begomoviruses: *East African cassava mosaic virus*-Uganda (EACMV-UG), *East African cassava mosaic virus* (EACMV), TYLCV and *Cotton leaf curl virus* (CLCuV). Also since geminiviruses continue to evolve and new *B. tabaci* biotypes emerge frequently, it is important to devote attention to whitefly biotypes, and evolving geminiviruses. The ultimate objective is to explore and present ecologically sound management practices for *B. tabaci* whiteflies and the geminiviruses they transmit.

The following chapter by Morales, presents a comprehensive handling of the subject on geminiviruses affecting Latin America and the Caribbean. It identifies problems of *B. tabaci* being associated with intensive cultivation, abusive use of pesticides, favorable conditions for the vector and disturbed ecosystems. The introduction of B-biotype, its adaptation to indigenous and exotic spp. and its efficacy in transmitting geminiviruses are also addressed. Mention is made of the original wild type plant spp. of *B. tabaci*, the appearance of *Bean golden mosaic virus* and *Bean golden yellow mosaic virus* and the distinction between these. One of the highlights of this chapter is the historical naming of Tomato yellow vein streak virus (Syn: Potato deforming mosaic virus) and the scientific debate regarding one name versus the other.

Chapter 3 by Czosnek and Ghanim, deals with TYLCV, vector transmission characteristics of this virus and in detail, outlines the intriguing events of virus travel through the vector's body. The important role of the GroEL homologue is emphasized along with the interaction of virus with vector proteins. The chapter also reports on the inimical effects of the virus on the vector and introduces the whitefly functional genomics project and its benefits in elucidating for example, cellular determinants involved in transmission and the interactions involved during translocation of the virus within the vector.

Chapter 4 by Mann, reports the complex etiology of Cotton Leaf curl disease and the implicated begomoviruses. Details are provided on whitefly morphometrics, development and feeding behavior. This chapter deals with the factors influencing transmission, from acquisition to inoculation, and provides evidence of the non-mutualistic or pathogenic consequences resulting from the interaction of *B. tabaci* with *Cotton leaf curl virus*-infected plants.

Chapter 5 by Thompson addresses the severe cassava mosaic disease in Africa. The components of the pandemic and the driving ecological factors are considered. In this chapter the disease is also discussed along physiological and biochemical perspectives with the view of enhancing an understanding of the disease dynamics. Chapter 5 progresses into Chap. 6 that examines the consequences of the interaction between the vector and one of the parents of the causative agent responsible for the Uganda pandemic, *East African cassava mosaic virus* (EACMV). It deals with the important research question regarding the effects on the vector caused by the virus as a non-recombinant as opposed to it in recombinant form.

Chapter 7 by Palaniswami and Henneberry presents the interaction of *Indian cassava mosaic virus* (ICMV) with its vector at different trophic levels with detection of the virus within the vector and infected plants through molecular tools and serology. The characteristics of ICMV transmission is explored using various acquisition and inoculation feeding schedules and a distinction is made between the cassava and sweet potato biotypes based on biological assays and isozyme banding pattern. The influence of infected plants on *B. tabaci* population development is highlighted along with the effects of insect infestation on plant pathogenesis proteins within plants.

Chapter 8 by Thompson examines the role of amino acids in the interaction of *B. tabaci* with host plants, either infected or uninfected. The chapter also sheds light on the probable involvement of other factors that may be beneficial or detrimental to the vector during its association with the host plant.

Chapter 9 by De Barro, emphasizes the importance of a standardized system in whitefly identification and classification. It provides details on the various molecular tools utilized in whitefly identification with guidelines on the appropriate application of these. It points out the shortcomings that could result from failure to judiciously and meticulously apply the technology. The capacity of *B. tabaci* biotypes to invade other territories and become established is discussed along with the contributing factors. One of the highlights of this chapter is the author's discussion of the contradictory findings of two different groups of researchers on the question of the Uganda pandemic and the involvement of an invasive biotype.

Chapter 10 by Varma et al. is a comprehensive handling of the emerging geminiviruses around the world. This chapter deals with the associated WTGs of cassava, cucurbits, legumes, Malvaceae and solanaceous crops. Spread dynamics of these are presented along with propelling factors for geminivirus evolution. The versatility and adaptation of geminiviruses are clearly underscored with cassava WTGs now infecting legumes and Jatropha, and non-leguminous WTGs affecting legumes. With emerging geminiviruses, new challenges and research endeavors can be expected.

Chapter 11 by Horowitz et al., covers management of whiteflies and outlines the principles of ecologically sound control practices. The main areas of management: chemical, biological, physical and cultural are discussed. The newer insecticides are introduced and the prospects of these for implementation are illuminated. The significant argument is that chemical control is not the panacea to insect pest and disease problems. Successful control is more based on Integrated Pest Management approaches.

In Chap. 12 by Gilbertson et al.: IPM strategies for WTGs, a schematic outlay of the strategy is presented. The approach is versatile and pragmatic, it hinges on a combination of practices employed from pre to post cultivation, and it emphasizes the need for regional coordination. The strategy is demonstrated with case studies on two crops (one annual and one perennial). Importantly, the authors deliberated on considerations of logistics; which is consistent with the general success of such approaches.

Remote Sensing technologies by Yang and Everitt, are presented in Chap. 13. The application of Remote Sensing along with GPS and GIS for detecting and mapping whiteflies is illustrated and the merits and importance of these approaches could be appreciated for forecasting purposes that impinge on strategies for both whitefly and geminivirus management.

Chapters are self-contained. As such coverage of common ground in some instances is inevitable, and construed as a way of allowing the contributors to fully express and emphasize the essential core principles. This book is made possible through the collaborative work of some of the leading researchers in the field.

References

Ali M, Ahmad Z, Hussain T, Tanveer M (1993) Cotton leaf curl virus situation in the Punjab, 1991–1992. Department of Agricultural Extension, Government of Punjab, Lahore

Avidov Z (1956) Bionomics of the tobacco whitefly (Bemisia tabaci Gennad) in Israel. Ktavim 7:25–41

Azab AK, Megahed MM, El-Mirsawi DE (1971) On the biology of Bemisia tabaci (Genn). Bull Soc Entomol Egypte 55:305–315

Bedford ID, Briddon RW, Markham PG, Brown JK, Rosell RC (1992) A new species of Bemisia or biotype of Bemisia tabaci (Genn) as a future pest of European agriculture. Proc Plant Health Eur Single Market BCPC Monogr 54:381–386

Bedford ID, Briddon RW, Brown JK, Rosell RC, Markham PG (1994) Geminivirus transmission and biological characterisation of Bemisia tabaci (Gennadius) biotypes from different geographic regions. Ann Appl Biol 125:311–325

Bethke JA, Paine TD, Nuessly GS (1991) Comparative biology, morphogenics and development of two populations of Bemisia tabaci (Homoptera Aleyrodidae) on cotton and poinsettia. Ann Entomol Soc Am 84:407–411

Bird J, Idris AM, Rogan D, Brown JK (2001) Introduction of the exotic Tomato yellow leaf curl virus-Israel in tomato to Puerto Rico. Plant Dis 85:1028

Bisaro DM (1996) Geminivirus replication. In: De Pamphilis M (ed) Eukaryotic DNA replication. Cold Spring Harbor Laboratory, Cold Spring Harbor

Borror DJ, Triplehorn CA, Johnson NE (1989) An introduction to the study of insects, 6th edn. Harcourt Brace Jovanovich College Publishers, New York

Briddon RW, Markham PG (2001) Complementation of bipartite begomovirus movement functions by topocuviruses and curtoviruses. Arch Virol 146:1811–1819

Briddon RW, Pinner MS, Stanley J, Markham PG (1990) Geminivirus coat protein gene replacement alters insect specificity. Virology 177:85–94

Brown JK (1994) Current status of Bemisia tabaci as a plant pest and virus vector in agro-ecosystems worldwide. FAO Plant Prot Bull 42:3–32

Brown JK (2007) The Bemisia tabaci complex: genetic and phenotypic variation and relevance to TYLCV-vector interactions. In: Czosnek H (ed.) Tomato yellow leaf curl virus disease: management, molecular biology, breeding for resistance. Springer, Dordrecht, pp 25–56

Brown JK, Coats S, Bedford ID, Markham PG, Bird J (1992) Biotype characterisation of Bemisia tabaci populations based on esterase profiles, DNA fingerprinting, virus transmission and bioassay to key host plant species. Phytopathology 82:1104

Burban C, Fishpool LDC, Fauquet C, Fargette D, Thouvenel JC (1992) Host- associated biotypes within West African populations of the whitefly Bemisia tabaci (Genn.) (Hom., Aleyrodidae). J Appl Entomol 113:416–423

Byrne FJ, Devonshire AL (1993) Insensitive acetyl cholinesterase and esterase polymorphism in susceptible and resistant populations of the tobacco whitefly Bemisia tabaci (Genn). Pestic Biochem Phys 45:34–42

Byrne DN, Bellows TS Jr, Parrella MP (1990) Whiteflies in agricultural systems. In: Gerling D (ed.) Whiteflies: their bionomics, pest status and management. Intercept, Andover

Byrne FJ, Gunning RV, Cahill M, Denholm I, Devonshire AL (1998) Laboratory and field evidence of interbreeding between biotypes of Bemisia tabaci (Gennadius) (Homoptera: Aleyrodidae). In: Second International Workshop on Bemisia and Geminiviral Diseases (Program and abstracts L 16), San Juan

Chasen R (1995) Geminiviruses: a twin approach to replication. Plant Cell 7:659–661

Chu D, Wan FH, Zhang YJ, Brown JK (2010) Change in the biotype composition of Bemisia tabaci in shandong province of China from 2005 to 2008. Environ Entomol 39:1028–1036

Cock MJW (ed.) (1986) Bemisia tabaci – a literature review on the cotton whitefly with an annotated bibliography. FAO/IIBC, Ascot

Cohen S, Duffus JE, Liu HY (1992) A new Bemisia tabaci biotype in the Southwestern United States and its role in silverleaf of squash and transmission of lettuce infectious yellows virus. Phytopathology 82:86–90

Costa AS (1976) Whitefly-transmitted plant diseases. Annu Rev Phytopathol 14:429–449

Costa HS, Brown JK, Sivasupramaniam S, Bird J (1993) Regional distribution, insecticide resistance and reciprocal crosses between the A and B biotypes of Bemisia tabaci. Insect Sci Appl 14:255–266

Delatte H, Reynaud B, Granier M, Thornary L, Lett JM, Goldbach R, Peterschmitt M (2005) A new silverleaf-inducing biotype Ms of Bemisia tabaci (Hemiptera: Aleyrodidae) indigenous to the islands of the south-west Indian Ocean. Bull Entomol Res 95:29–35

Dennehy TJ, DeGain BA, Harpold VS, Brown JK, Morin S, Jeff A, Fabrick JA, Byrne FJ, Nichols RL (2005) New challenges to management of whitefly resistance to insecticides in Arizona. Vegetable report, College of Agriculture and Life Sciences, University of Arizona, Tucson

Dong C, Zhang YJ, Brown JK, Cong B, Xu BY, Wu QJ, Zhu GR (2006) The introduction of the exotic Q biotype of B. tabaci from the Mediterranean region into China on ornamental crops. Fla Entomol 89:168–174

Duffus JE (1987) Whitefly transmission of plant viruses. In: Harris KF (ed) Current topics in vector research, vol 4. Springer, New York

Etessami P, Callis R, Ellwood S, Stanley J (1988) Delimitation of essential genes of cassava latent virus DNA 2. Nucleic Acids Res 16:4811–4829

Etessami P, Watts J, Stanley J (1989) Size reversion of African cassava mosaic virus coat protein gene deletion mutants during infection of Nicotiana benthamiana. J Gen Virol 70:277–289

Fauquet CM, Pita J, Deng D, Tores-Jerez I, Otim-Nape WG, Ogwal S, Sangare A, Beachy RN, Brown JK (1998) The East African cassava mosaic virus epidemic in Uganda. In: Second International Workshop on Bemisia and Geminivirus Diseases. San Juan, Abstract L-4

Fauquet CM, Maxwell DP, Gronenborn B, Stanley J (2000) Revised proposal for naming geminiviruses. Arch Virol 145:1743–1761

Fauquet CM, Bisaro DM, Briddon RW, Brown J, Harrison BD, Rybicki EP, Stenger DC, Stanley J (2003) Revision of taxonomic criteria for species demarcation in the family Geminiviridae, and an updated list of begomovirus species. Arch Virol 148:405–421

Fauquet CM, Briddon RW, Brown JK, Moriones E, Stanley J, Zerbini M, Zhou X (2008) Geminivirus strain demarcation and nomenclature. Arch Virol 153:783–821

Fujiie A, Omar AMS, Sawas AB et al (2009) Geographic distribution of Bemisia tabaci biotypes collected from Autumn- cultured potato fields in Syria. J ISSASS 15:12–20

Gardiner WE, Sunter G, Brand L, Elmer JS, Roger SG, Bisaro DM (1988) Genetic analysis of tomato golden mosaic virus: the coat protein is not required for systemic spread or symptom development. EMBO J 7:899–904

Garzon-Tiznado JA, Torres-Pacheco I, Ascencio-Ibanez JT, Herrera-Estrella L, Rivera-Bustamante RF (1993) Inoculation of peppers with infectious clones of a new geminivirus by a biolistic procedure. Phytopathology 83:514–521

Gibson RW, Legg JP, Otim-Nape GW (1996) Unusually severe symptoms are a characteristic of the current epidemic of mosaic virus disease of cassava in Uganda. Ann Appl Biol 128:479–490

Gilbertson R, Rojas M, Kon T, Jaquez J (2007) Introduction of Tomato yellow leaf Curl virus into the Dominican Republic: the development of a successful integrated pest management strategy. In: Czosnek H (ed.) Tomato yellow leaf curl virus disease: management, molecular biology, breeding for management. Springer, Dordrecht, pp 279–304

Gill RJ (1992) A review of the sweetpotato whitefly in Southern California. Pan Pac Entomol 68:144–152

Goodman RM (1981) Geminiviruses. J Gen Virol 54:9–21

Goodman RM, Shock TL, Haber S, Browning KS, Bowers GR Jr (1980) The composition of bean golden mosaic virus and its single stranded DNA genome. Virology 106:168–171

Gray SM, Banerjee N (1999) Mechanisms of Arthropod transmission of plant and animal viruses. Microbiol Mol Biol R 63:128–148

Guirao P, Beitia F, Cenis JL (1997) Biotype determination of Spanish populations of Bemisia tabaci (Hemiptera: Aleyrodidae). Bull Entomol Res 87:587–593

Gunning RV, Byrne FJ, Devonshire AL (1997) Electrophoretic analysis of non-B and B-Biotype Bemisia tabaci (Gennaduis) (Hemiptera: Aleyrodidae) in Australia. Aust J Entomol 36:245–250

Gutierrez C (1999) Geminivirus DNA replication. Cell Mol Life Sci 56:313–329

Gutierrez C (2000) DNA replication and cell cycle in plants: learning from geminiviruses. EMBO J 19:792–799

Hamilton WDO, Stein VE, Coutts RHA, Buck KW (1984) Complete nucleotide sequence of the infectious cloned DNA components of tomato golden mosaic virus: Potential coding regions and regulatory sequences. EMBO J 3:2197–2205

Harrison BD (1985) Advances in geminivirus research. Annu Rev Phytopathol 23:55–82

Howarth AJ, Caton J, Bossert M, Goodmam RM (1985) Nucleotide sequence of bean golden mosaic virus and a model for gene regulation in geminiviruses. Proc Natl Acad Sci USA 82:3572–3576

Kallender H, Petty IDT, Stein VE, Panico M, Blench IP, Etienne AT, Morris HR, Coutts RHA, Bucks KW (1988) Identification of the coat protein gene of tomato golden mosaic virus. J Gen Virol 69:1351–1357

Kato K, Onuki M, Fuji S, Hanada K (1998) The first occurrence of tomato yellow leaf curl virus in tomato (Lycopersicon esculentum Mill). Ann Phytopathol Soc Jpn 64:552–559
Lazarowitz SG (1987) The molecular characterisation of Geminiviruses. Plant Mol Biol Rep 4:177–192
Lazarowitz SG (1992) Geminiviruses: genome structure and gene function. Crit Rev Plant Sci 11:327–349
Legg JP, Gibson RW, Otim-Nape GW (1994) Genetic polymorphism amongst Ugandan populations of Bemisia tabaci (Gennadius) (Homoptera: Aleyrodidae), vector of African cassava mosaic geminivirus. Trop Sci 34:73–81
Legg JP, French R, Rogan D, Okao-Okuja G, Brown JK (2002) A distinct Bemisia tabaci (Gennadius) (Homoptera: Sternorrhyncha: Aleyrodidae) genotype cluster is associated with the epidemic of severe cassava mosaic virus disease in Uganda. Mol Ecol 11:1219–1229
Liu HY, Cohen S, Duffus JE (1992) The use of isozyme patterns to distinguish sweet potato whitefly (Bemisia tabaci) biotypes. Phytoparasitica 20:187–194
Lourenção AL, Nagai H (1994) Surtos populacionais de Bemisia tabaci, no estado de São Paulo. Bragantia 53:53–59
Martinez-Carrillo JL, Brown JK (2007) First report of the Q biotype of Bemisia tabaci in Southern Sonora, Mexico. Phytoparasitica 35:282–284
Matthews REF (1991) Plant virology, 4th edn. Academic, New York
Morinaga T, Ikegami M, Arai T, Yazaki K, Miura K (1988) Infectivity of cloned tandem dimer DNAs of bean golden mosaic virus. J Gen Virol 69:897–902
Mound LA, Halsey SH. (eds.) (1978) Whitefly of the world: a systematic catalogue of the aleyrodidae (Homoptera) with host plant and natural enemy data. British Museum (Natural History)/ Wiley, London/Chichester
Otim-Nape GW, Bua A, Baguma Y (1994) Accelerating the transfer of improved production technologies: controlling African cassava mosaic virus disease epidemics in Uganda. Afr Crop Sci J 2:479–495
Padidam M, Sawyer S, Fauquet CM (1999) Possible emergence of new geminiviruses by frequent recombination. Virology 265:218–225
Perrings TM, Cooper AD, Rodrigues RJ, Farrar CA, Bellows TSJ (1993) Identification of a whitefly species by genomic and behavioural studies. Science 259:74–77
Pollard DG (1955) Feeding habits of the cotton whitefly, Bemisia tabaci Genn. (Homoptera: Aleyrodidae). Ann Appl Biol 43:664–671
Polston JE, Bois D, Serra CA, Concepcion S (1994) First report of a tomato yellow leaf curl-like geminivirus in the Western Hemisphere. Plant Dis 78:831
Rojas MR, Gilbertson RL (2008) Emerging Pplant viruses: Aa diversity of mechanisms and opportunities. In: Roossinck MJ (ed.) Plant virus evolution. Springer, Berlin/Heidelberg, pp 27–52
Russel LM (1958) Synonyms of Bemisia tabaci (Gennadius) (Homoptera: Aleyrodidae). Bull Brooklyn Entomol Soc 52:122–123
Rybicki EP (1994) A phylogenetic and evolutionary justification for three genera of Geminiviridae. Arch Virol 139:49–77
Salati R, Nahkla MK, Rojas MR, Guzman P, Jaquez J, Maxwell DP, Gilbertson RL (2002) Tomato yellow leaf curl virus in the Dominican Republic: characterisation of an infectious clone, virus monitoring in whiteflies and identification of reservoir hosts. Phytopathology 92:487–496
Sánchez-Campos S, Navas-Castillo J, Camero R, Saria C, Díaz JA, Moriones E (1999) Displacement of Tomato yellow leaf curl virus (TYLCV)-Sr by TYLCV-Is in tomato epidemics in Spain. Phytopathology 89:1038–1043
Schuster DJ, Mueller TF, Kring JB, Price JF (1990) Relationship of the sweetpotato whitefly to a new tomato first disorder in Florida. Hortscience 25:1618–1620
Sseruwagi P, Legg JP, Maruthi MN, Colvin J, Rey MEC, Brown JK (2005) Genetic diversity of Bemisia tabaci (Gennadius) (Hemiptera: Aleyrodidae) populations and presence of the B biotype and a non-B biotype that can induce silverleaf symptoms in squash, in Uganda. Ann Appl Biol 147:253–265

Sseruwagi P, Maruthi MN, Colvin J, Rey MEC, Brown JK, Legg JP (2006) Colonization of non- cassava plant species by cassava whiteflies (Bemisia tabaci) in Uganda. Entomol Exp Appl 19:145–153

Stanley J (1983) Infectivity of the cloned geminivirus genome requires sequences from both DNAs. Nature 305:643–645

Stanley J, Davies JW (1985) Structure and function of the DNA genome of Geminiviruses. In: Davies JW (ed.) Molecular plant virology, vol 2, Replication and Gene Expression. CRC Press, Boca Raton

Thompson WMO (2003) A new host plant species for the cassava biotype of Bemisia tabaci (Gennadius) (Hom., Aleyrodidae). J Appl Entomol 127:374–376

Toscano N, Henneberry T, Castle S (1994) Population dynamics and pest status of silverleaf whitefly in the USA. Arab J Pl Prot 12(2):137–142

Townsend R, Stanley J, Curson SJ, Short MN (1985) Major polyadenylated transcripts of cassava latent virus and location of the gene encoding the coat protein. EMBO J 4:33–37

Ueda S, Brown JK (2006) First report of the Q biotype of Bemisia tabaci in Japan bymitochondrial cytochrome oxidase I sequence analysis. Phytoparasitica 34:405–411

Varma A, Malathi VG, Handa A, Aiton M, Harrison BD, Varma JP, Singh RP, Singh M, Srivastava M, Singh J (1993) Occurrence of leaf-curl of cotton and okra in Northern India. In: Abstracts of the 6th international congress of plant pathology, Montreal

Viscarret MM, Torres-Jerez Agostini De, Manero E, Lopez SN, Botto EE, Brown JK (2003) Mitochondrial DNA evidence for a distinct new world group of Bemisia tabaci (Gennadius) (Hemiptera: Aleyrodidae) indigenous to Argentina and Bolivia, and presence of the Old World B biotype in Argentina. Ann Entomol Soc Am 96:65–72

Wool D, Gerling D, Bellotti AC, Morales FJ (1993) Esterase electrophoretic variation in Bemisia tabaci (Genn) (Homo: Aleyrodidae) among host plants and localities in Israel. J Appl Entomol 115:185–196

Zhou X, Liu Y, Calvert L, Munoz C, Otim-Nape GW, Robinson DJ, Harrison BD (1997) Evidence that DNA of a geminivirus associated with severe cassava mosaic disease in Uganda has arisen by interspecific recombination. J Gen Virol 78:2101–2111

Chapter 2
Interaction Between *Bemisia tabaci*, Begomoviruses, and Plant Species in Latin America and the Caribbean

Francisco J. Morales

Abstract The tropical and sub-tropical agricultural regions of Latin America and the Caribbean have suffered a high incidence of viruses transmitted by the whitefly *Bemisia tabaci* in several food and industrial crops of socio-economic importance. These crop production problems have been particularly severe in the past four decades, due to the adoption of modern agricultural practices, such as the extensive and intensive cultivation of a larger number of susceptible crops, a marked increase in the use of agro-chemicals, new cultural practices, and the accelerated evolution of pests under these conditions. The damage caused by these pests is also closely associated with the occurrence of suitable environmental conditions for the reproduction of *B. tabaci*. This chapter describes the interaction of *B. tabaci* with wild and cultivated plant species in Latin America and the Caribbean, and describes the most effective methods of integrated plant pest and disease management available.

Keywords Whitefly • Geminiviridae

2.1 Introduction

Latin America and the Caribbean (LAC) have the highest incidence and genetic diversity of begomoviruses transmitted by *Bemisia tabaci* worldwide, despite the Old World origin of this whitefly species (Brown 1994). Begomoviruses have small but highly efficient replication mechanisms with an unlimited capacity to generate genetic variants (Gutierrez 1999; Howarth and Vandemark 1989). Although *B. tabaci*

F.J. Morales (✉)
Virology Research Unit, International Center for Tropical Agriculture,
A.A. 6713, Cali, Colombia
e-mail: f.morales@cgiar.org

W.M.O. Thompson (ed.), *The Whitefly, Bemisia tabaci (Homoptera: Aleyrodidae) Interaction with Geminivirus-Infected Host Plants*,
DOI 10.1007/978-94-007-1524-0_2,

was probably introduced from Asia or Africa into the Americas during the colonization of this continent by the Spanish and Portuguese, over 500 years ago, historical records indicate that *B. tabaci* was not an important pest or virus vector in the New World until the 1940s. It is important to point out that LAC region was inhabited by numerous pre-Columbian societies that had a very diversified agriculture, which was gradually replaced by a few native and exotic crops of interest to the European colonists. These imported agricultural systems, characterized by the monoculture of food and industrial commodities, not only persist until the present time, but have experienced a significant increase in the area planted, mainly for export purposes. The intensification of cropping systems in LAC has been accompanied by a noticeable increase in the application of agro-chemicals, and the creation of irrigation systems that change the micro-climate of regions previously undisturbed.

2.1.1 The Whitefly Species Bemisia tabaci

The original biotype introduced into tropical America must have found different environments where it could easily adapt and reproduce thanks to the abundance of vegetation, warm temperatures the year round, and adequate humidity to complete its life cycle without seasonal interruptions. Even in sub-tropical regions of LAC, *B. tabaci* could survive thanks to the mild winters of these regions of the Americas.

B. tabaci is currently considered a species complex of races showing different behavior regarding their host preference, fecundity, environmental adaptation, and efficiency of virus transmission (De Barro et al. 2005; Perring 2001). Phylogenetic analyses of *B. tabaci*'s mitochondrial 16S ribosomal subunit and cytochrome oxidase I (COI) separates Old World from New World populations of *B. tabaci* (Frohlich et al. 1999), which suggests an early arrival of this whitefly species in LAC. The original 'A biotype' predominated in the Americas until the introduction of the 'B biotype' in the late 1980s (Bellows et al. 1994), from the Old World. This investigation showed that there was only one major race in the Americas, the 'New World' race, until the B and, more recently, the Q biotypes were introduced into the Caribbean region (Brown 2007; Segarra et al. 1990). However, a different concept of 'races' of *B. tabaci*, had been proposed for the A biotype in Puerto Rico in the late 1950s (Bird and Maramorosch 1978; Bird and Sanchez 1971). The New World 'race' concept was based on the existence of *B. tabaci* populations that differed in their ability to feed and/or breed on specific host plants, and in their efficiency to transmit native begomoviruses. A similar proposal had been advanced in Brazil (Flores and Silberschmidt 1958), where these behavioral differences were associated to the probable existence of 'ecological biotypes'. Some skeptical scientists quickly pointed out that host-specific whitefly populations may arise naturally in response to the selection pressure imposed by the availability of suitable reproductive plant hosts in a particular agroecosystem, without any concomitant changes in the genetic constitution of *B. tabaci* populations. That is to say that *B. tabaci* populations can eventually adapt to non-preferred hosts in the absence of their primary reproductive hosts (Costa 1965, 1975). The arrival of the B biotype of *B. tabaci* in the Americas, characterized by its extraordinary ability to feed on a significantly larger number of host plants and adapt to different environments, when compared to the A biotype (De Barro 1995), finally made the possible role of the 'physiological races' or 'ecological biotypes' in the Americas, irrelevant (Bird and Maramorosch 1978; Bird and Sanchez 1971).

One of the unexpected results of the introduction of the B biotype of *B. tabaci* in the Americas, was the gradual displacement of the A biotype in most agricultural regions of LAC (Morales 2006a), to the point of extinction in many regions. This observation raised the issue of its comparative ability to transmit the native viruses, and preliminary evidence suggested that the B biotype was not as efficient as the original A biotype of *B. tabaci*, as a vector of New World viruses of the predominant genus, *Begomovirus,* associated with *B. tabaci* (Duffus et al. 1992). A study conducted at CIAT, Colombia, showed that although the B biotype of *B. tabaci* displaced the original A biotype in the Cauca Valley of Colombia (Rodriguez et al. 2005), and that the biotype B is the main vector of a begomovirus that collapsed the production of snap beans in this region, the original A biotype of *B. tabaci* was a more efficient vector (26.6% versus 4.9% transmission efficiency) of the virus. However, considering the significantly higher reproductive capacity of the B biotype, which can generate over 1,000 nymphs per leaf in the Cauca Valley, the observed differences in transmission efficiency become irrelevant in epidemiological terms, particularly considering that the A biotype of *B. tabaci* did not attack snap beans in this region before the introduction of the B biotype. Moreover, the transmission efficiency of the B biotype gradually increased during the three years of the study, suggesting that the efficiency of virus transmission depends on the gradual adaptation of a new whitefly biotype to a new agro-ecosystem or host plant. The recent introduction of the Q biotype of *B. tabaci* into Mexico and Guatemala (Brown 2007) poses a new challenge to LAC farmers, considering that this biotype has been reported to be resistant to the new insecticides currently used for control of the B biotype of *B. tabaci*.

2.1.2 The Original *Bemisia tabaci* Plant Hosts

Upon its introduction into LAC, *B. tabaci* found many potential reproductive hosts, mainly wild plant species of the Euphorbiaceae, Malvaceae, and Solanaceae, naturally found throughout LAC. The so called 'tobacco' (*tabaci*) or 'sweet potato' whitefly *B. tabaci* also found these (*Nicotiana* spp. and *Ipomoea batatas*) and other important, cultivated plant hosts present in the Old World, such as cotton (*Gossypium* spp.) and cassava (*Manihot esculenta*), but the original biotype of *B. tabaci* preferred its wild reproductive hosts in the Americas. Among the most ubiquitous wild hosts of *B. tabaci* in the Americas, the Malvaceae includes over 1,700 plant species, of which those in the genera *Abutilon, Malva* and *Sida,* account for some 400 species of neo-tropical origin (Fryxell 1997). The interaction between these species, *B. tabaci,* and begomoviruses transmitted by this whitefly vector was demonstrated at an early stage in the study of begomoviruses in this region (Bird 1958; Cook 1931, 1955; Silberschmidt 1943; Silberschmidt and Tommasi 1955), where the diseases resulting from this interaction were collectively referred to as 'infectious chlorosis of the Malvaceae' (Orlando and Silberschmidt 1946). A second group of native plant species suspected of harboring begomoviruses, included the families Euphorbiaceae, Leguminosae and Convolvulaceae (Bird et al. 1975; Costa and Bennett 1950).

Cotton (*Gossypium* spp.) was the first cultivated, native, malvaceous plant species in Latin America to suffer the attack of a whitefly-transmitted virus. The causal virus could be readily transmitted from a malvaceous weed species (*Sida micrantha*) to cotton by *B. tabaci*, but not from infected to healthy cotton (Costa 1937). Cotton was also the first crop affected by a virus transmitted by *B. tabaci* in northwestern Mexico (Brown and Nelson 1987; INIFAP 1995). Whereas cotton is not an important host of begomoviruses in LAC, with the exception of *Cotton leaf crumple virus* in northwestern Mexico (Idris and Brown 2004) and Guatemala in the early 1950s (Spillari 1994), this crop is suspected to have played a major role in the epidemiology of *B. tabaci* and begomoviruses in the region. In fact, cotton was the first crop of economic importance that was associated with *B. tabaci* as a pest in Central and South America, as a result of the intensive use of broad-spectrum insecticides in the 1950s (Costa et al. 1973; Spillari 1994).

2.1.3 The Original Begomovirus Reservoirs

The role of weeds as sources of begomoviruses infecting cultivated plant species has been known to LAC virologists since the 1950s. In Brazil, a *B. tabaci*-transmitted virus in malvaceous weeds of the genera *Malva* and *Sida*, was transmitted to cotton (Costa and Carvalho 1960a). Begomoviruses found in *Jatropha gossypifolia* and *Euphorbia prunifolia* (Euphorbiaceae) in Puerto Rico (Bird and Maramorosch 1978) and Brazil (Costa and Bennett 1950), respectively, were found to be pathogens of common bean (*Phaseolus vulgaris*). The common bean disease originally described in Brazil as 'bean mottle dwarf' (currently known as 'bean dwarf mosaic'), was shown in the mid 1960s to be induced by a virus transmitted by *B. tabaci* from malvaceous weeds (Costa 1975). In the mid 1970s, Costa (Costa 1976) demonstrated that malvaceous species were important reservoirs of viruses transmitted by *B. tabaci* to common bean, cotton, okra, and soybean in Brazil. The important role of wild species in the Malvaceae and Euphorbiaceae in the dissemination of begomoviruses by *B. tabaci* to cultivated plant species, has been confirmed in Brazil (Lima et al. 2002a; Lima et al. 2002b). More contemporary reports of crops affected by begomoviruses originating in wild malvaceous species include: okra (*Abelmoschus esculentus*), infected in Mexico by a begomovirus related to a virus originally detected in *Sida* (Torre-Almaraz et al. 2004); a yellow mosaic of lima bean (*Phaseolus lunatus*) in northern Peru, caused by a begomovirus related to Sida mottle virus from Brazil; and a begomovirus infecting passifloraceae in Colombia (Morales et al. 2002).

The genetic variability of begomoviruses found in wild hosts was also apparent at an early stage in the study of whitefly-borne viruses in LAC. Wild *Euphorbia prunifolia, Leonurus sibiricus, Phaseolus longepedunculatus* and species of *Sida*, were shown to be commonly infected by different viruses transmitted by *B. tabaci* (Costa and Carvalho 1960b). An investigation of different wild species: *Aspillia tenella* (Compositae), *Desmodium uncinatum, Macroptilium lathyroides, Rhynchosia minima* (Leguminosae), *Malva* sp., *Malvastrum* sp., *Sida rhombifolia* (Malvaceae), *Euphorbia prunifolia* (Euphorbiaceae), *Melochia villosa* (Sterculiaceae), and *Pavonia* sp. (Boraginaceae), showed them to be sources of begomoviruses transmitted by *B. tabaci* to common bean. The begomoviruses present in *M. lathyroides* and *R. minima* also infected pigeon pea (*Cajanus cajan*). The remaining begomoviruses tested only infected common bean, demonstrating the presence of different begomoviruses in these weeds (Morales 2006a).

2.1.4 *Original Interactions Between B. tabaci, Begomoviruses and Crops in Latin America and the Caribbean*

AS mentioned above, cotton was the first crop affected by *B. tabaci* and begomoviruses in LAC. A high demand for cotton in the 1940s, resulted in a substantial increase in the area planted with this industrial crop in Mexico, Central America and Brazil. Cotton plantings in the Pacific lowlands of Central America saw an increase from a few 1,000 ha in the 1940s, to over 300,000 ha in the 1970s (Spillari 1994). In Brazil, large *B. tabaci* populations developed on cotton as the area planted with this fiber crop increased in the northern zones of the state of Parana and São Paulo, in 1968 (Costa et al. 1973). Probably, the need to control important pests of cotton, such as the cotton boll weevil, was largely responsible for the introduction and intensive use of broad-spectrum insecticides, including DDT, methyl parathion, toxaphene, and malathion (Gilliand et al. 1971), which probably induced the development of resistance in *B. tabaci* and the gradual elimination of its biological control agents.

The decisive role played by the extensive cultivation of a reproductive host of *B. tabaci* in the epidemiology of begomoviruses, is best illustrated by the widespread epidemics of *Bean golden mosaic virus* (BGMV) in Brazil. This virus was first described in 1961 by Costa (Costa 1965) in Brazil, but it was considered 'not currently of sufficient economic importance' at that time. A decade later, Costa (Costa 1975) reported the presence of bean golden mosaic in another important bean-growing state (Parana) of Brazil, and associated its rapid dissemination with an exponential increase in the area planted to soybean, which increased from 1.3 million hectares in the early 1970s, to over 6 million hectares by 1976 (FAO 1994). By 1990, BGMV was already present in all of the main bean-growing states of Brazil (Morales and Anderson 2001). The use of broad spectrum insecticides to control insect pests in soybean, such as stink bugs and caterpillars (Gazzoni et al. 1999), may have also contributed to the elimination of the natural enemies of

B. tabaci on soybean in Brazil. By 1975, there were over 10 million hectares of soybean in Brazil and Argentina, coinciding with severe outbreaks of BGMV and other begomoviruses of common bean, such as *Bean dwarf mosaic virus* (BDMV) in northwestern Argentina. The latter begomovirus caused the loss of over 100,000 ha of highly prized common bean varieties in the period 1978–1981 (Morales and Anderson 2001).

The large *B. tabaci* populations currently observed in soybean-producing regions of LAC, are primarily the result of the extensive area planted to this prime export crop, rather than the capacity of modern soybean cultivars to support a high rate of reproduction of *B. tabaci* (Lima and Lara 2004).

2.2 Main Crops Affected by Bemisia Tabaci-Borne Viruses

2.2.1 *Common Bean*

After maize, common bean (*Phaseolus vulgaris* L.) is the most important food staple in LAC, where this legume originated and was domesticated in different agro-ecosystems located along the entire continental western region from Mexico to Chile (Singh 1988). LAC is the main producer and consumer of common bean in the world, with plantings covering over 10 million hectares located throughout the region, particularly in Brazil, the number one producer of common beans in the world (http://faostat.fao.org).

Interestingly, despite the severe damage inflicted in this legume by begomoviruses transmitted by *B. tabaci* throughout the LAC region, common bean is not a preferred host of this whitefly species. Here, we discuss some of the main factors associated with the attack of common bean by *B. tabaci* and its interaction with the various begomoviruses known to affect common bean production in the different regions of LAC.

2.2.1.1 South America

As mentioned above, Brazil is the main producer of common beans in LAC, and it was in this country where the first viral diseases associated to the presence of *B. tabaci* were observed. Bean golden mosaic was first observed in 1961 in the state of São Paulo, albeit at a low incidence (Costa 1965). The epidemiological potential of this new disease was revealed, when bean golden mosaic rapidly disseminated into the main common bean-producing states of Brazil: Parana, Minas Gerais, Goiania and Bahia (Costa 1975), severely affecting over two million hectares of common bean (Ferreira et al. 2002). The dissemination of bean golden mosaic was clearly associated with the expansion of the newly discovered export crop, soybean (*Glycine max*), a known breeding host of *B. tabaci*. The dissemination of bean golden mosaic in northern Brazil, was only arrested by the eastern boundaries of the Amazon forest, characterized by its abundant rainfall (average annual rainfall over 3,000 mm), unsuitable for both extensive agriculture and the reproduction of *B. tabaci*. Bean golden mosaic occasionally crossed the southern boundary of the tropical zone into the southernmost state of Brazil, Rio Grande do Sul, but its incidence remains low in this state. The hotspot of bean golden mosaic in Brazil is located within its tropical zone, particularly in the 'Minas Gerais Triangle' (20°S–50°W). Southwest of this hotspot, lies the Gran Chaco, a hot and dry alluvial plain that covers parts of Brazil, Argentina, Paraguay and Bolivia. This zone has been traditionally devoted to cotton production and, thus, has been suspected to be an important natural bridge for *Bean golden mosaic virus* (BGMV) to spread between southern Brazil and northwestern Argentina (Morales and Anderson 2001), where this virus emerged around 1983.

The ensuing epidemics of BGMV in NW Argentina (NOA) were also closely associated with the expansion of soybean plantings in this country, from over 400,000 ha in 1975 to three million hectares in 1985 (Morales and Anderson 2001). In the late 1970s, a major outbreak of a new disease of common bean, named 'achaparramiento' (dwarfing), was also associated with the emergence of *B. tabaci* as a new pest in NOA, completely destroyed over 60,000 ha of common beans in this region. The 'achaparramiento' disease of common bean in NW Argentina was shown by the author to be induced by a begomovirus similar to the one originally described by Costa (Costa 1975) in Brazil, transmitted from malvaceous species of *Sida* to common bean (Morales 2006a). This observation confirms the importance of malvaceous species as reservoirs of potentially important begomoviruses in LAC (Frischmuth et al. 1997; Höfer et al. 1997; Jovel et al. 2004; Lima et al. 2002a; Rampersad and Umaharan 2003).

The unprecedented emergence of *B. tabaci* as a new pest and vector in N.W. Argentina, was associated with the planting of common bean in the region, soon after the harvest of the soybean crop in the same area. Thus, in the absence of other suitable hosts in this recently developed agricultural region, the whitefly populations that reproduced on soybean were forced to move onto the newly planted common bean fields.

The soybean boom in South America spans over 35 million hectares (http://faostat.fao.org) that include the tropical plains of the Bolivian provinces of Santa Cruz de la Sierra and Tarija. In these provinces, the area planted to soybean increased from 6,000 ha in 1973 to over 400,000 ha in 1995 (Morales and Anderson 2001). Bean golden mosaic emerged in this region in 1992, and laboratory assays clearly showed that the Brazilian, Argentine, and Bolivian BGMV were the same virus species (Faria and Maxwell 1999; Morales and Anderson 2001).

2.2.1.2 Mesoamerica

The emergence of a 'golden mosaic' of common bean in the 1960s and its association with the presence of *B. tabaci* on common bean plantings in Mesoamerica (southern Mexico to Costa Rica), lead Latin American virologists to believe they were witnessing a widespread pandemic of BGMV in tropical America. In Mesoamerica and the Caribbean, the golden mosaic of common beans had been observed since the late 1960s (Gamez 1970), from Guatemala to Costa Rica, and in the Dominican Republic (Schieber 1970), Puerto Rico (Bird et al. 1973); Jamaica (Pierre 1975); and Cuba (Blanco and Bencomo 1978), in the Caribbean region. The only difference between the Brazilian and Mesoamerican/Caribbean 'golden mosaics', was that the causal virus in Brazil could not be transmitted mechanically; only with the aid of *B. tabaci* (Costa 1976). The Puerto Rican isolate of a whitefly-borne virus isolated from lima bean (*Phaseolus lunatus*) became the first begomovirus to be cloned and sequenced in the Americas (Howarth et al. 1985; Morinaga et al. 1987). This strain would later become the Type strain of *Bean golden mosaic virus* (Rybicki 1994), but the rapid progress in molecular biology methods, lead to the demonstration that the Puerto Rican and Brazilian bean golden mosaic viruses were distinct species (Bellows et al. 1994; Faria et al. 1994; Gilbertson et al. 1993). In 1998, the name of the *P. lunatus* and Mesoamerican/Caribbean common bean begomoviruses was changed to "bean golden yellow mosaic" (Bird et al. 1973). This proposal was later accepted by the International Committee on Taxonomy of Viruses (Rybicki et al. 2000). The current geographic range of *Bean golden yellow mosaic virus* (BGYMV) extends from southern Mexico to Colombia in northern South America (Morales 2006a; Morales and Jones 2004).

2.2.1.3 Northern Mexico

In 1974, a "yellow mosaic" of common bean was observed in the Valley of Culiacan, Sinaloa (Lopez 1974). The disease rapidly spread in the region affecting common bean plantings in Baja California Sur, Sinaloa, Sonora and Nayarit. This disease was initially attributed to the presence of BGMV, but the causal agent was later characterized as a distinct begomovirus (Brown et al. 1999) named *Bean calico mosaic virus* (BCaMV). This virus first attacked summer and winter squash plantings in southwestern United States (Flock and Mayhew 1981), and was characterized in 1977 (Brown et al. 1999; Loniello et al. 1992) as *Squash leaf curl virus* (SLCV). A survey conducted in common bean plantings in Los Mochis and Culiacan (Sinaloa), and Etchojoa (Sonora) demonstrated the presence of three different strains related to BCaMV and SLCV (Morales et al. 2005c).

2.2.2 *Tomato*

Tomato (*Solanum lycopersicum* Mill.) is native to South America (Pickersgill 1977), but it may have been domesticated in Mexico, where the name 'tomatl' (Nahuatl) probably comes from. The genetic improvement of tomato has been greatly neglected in LAC, and it is only in the last decades that tomato has been re-discovered as a high value export crop in this region. Unfortunately, this is one of the most affected crops by whitefly-transmitted viruses, a problem that has been compounded by the intensive and indiscriminate use of insecticides on this crop, often on a daily basis.

2.2.2.1 South America

The first attacks of tomato by whitefly-transmitted viruses occurred in the 1950s, when an "infectious chlorosis" of tomato was observed in Brazil (Flores and Silberschmidt 1967; Flores et al. 1960). This disease was apparently caused by begomoviruses transmitted by *B. tabaci* from wild malvaceous plants to tomato. By 1975, other viral diseases of tomato associated with *B. tabaci* had also been reported in Brazil (Costa 1974; Costa et al. 1975), including *Tomato golden mosaic virus* (TGMV), the first begomovirus characterized in the Americas (Matys et al. 1975). Currently, there are over a dozen begomoviruses affecting tomato production in seven states of eastern Brazil (Andrade et al. 2006; Lima et al. 2000; Ribeiro et al. 2003) some of which are related to TGMV. Some of the new begomoviruses detected show a distant relationship to BGMV and to a begomovirus previously described in Brazil as 'Tomato yellow vein streak virus' (Faria et al. 1997). Ribeiro et al. (Ribeiro et al. 2003) proposed the names 'Tomato chlorotic mottle virus' and 'Tomato rugose mosaic virus' for two of the tomato begomoviruses detected, and concluded that these begomoviruses 'are indigenous to Brazil, and have not been introduced from other regions'. However, 'Tomato yellow vein streak virus' is in fact 'Potato deforming mosaic virus', a begomovirus first described in 1962 from the southeastern potato-growing regions of the Province of Buenos Aires, Argentina (Calderoni et al. 1962; Delhey et al. 1981). This begomovirus was also isolated in 1995 by the author in northwestern Argentina from severely affected common bean plants before it was detected in Brazil infecting tomato (Morales and Anderson 2001). In a recent paper, Ribeiro et al. (Ribeiro et al. 2006) recognize that Tomato yellow vein streak virus and Potato deforming mosaic virus are the same begomovirus, but, unfortunately, conclude that 'potato deforming mosaic disease is caused by an isolate of Tomato yellow vein streak virus', when the potato denomination should take precedent. This observation shows that begomoviruses of common bean and tomato can infect species of the same or different families, and be transported by *B. tabaci* in the tropical and subtropical agricultural regions of eastern Brazil, northern Argentina and southeastern Bolivia. Ribeiro et al. (Ribeiro et al. 2003) also make an interesting observation linking the emergence of new begomoviruses in tomato plantings in Brazil, to the introduction of the new B biotype of *B. tabaci* in the early 1990s. This aggressive and prolific biotype, unlike the original A biotype, readily colonizes tomato in Brazil (Lima et al. 2002b). More recently, two other tomato begomoviruses have been detected in Brazil: one named 'Tomato yellow spot virus', shown to be closely related to begomoviruses infecting *Sida* species, and 'Tomato crinkle leaf yellows virus'. These begomoviruses are apparently closely related as they were shown to form pseudo-recombinants (Andrade et al. 2006). The important role of wild malvaceous species in the emergence of new begomoviruses affecting cultivated plant species in LAC, is thus confirmed by this report. The other begomovirus of tomato in Brazil is 'Tomato severe rugose virus', which seems to be more adapted to species of *Capsicum* than tomato (Bezerra-Agasie et al. 2006; Nozaki et al. 2006), although a recent report cites it as a pathogen of potato as well (Souza-Dias et al. 2007).

Historically, a "yellowish mosaic" of tomato (Debrot et al. 1963) was the second disease of tomato observed in Latin America (Venezuela), in the early 1960s. This symptom was later named 'tomato yellow mosaic', and by 1975, this disease was already present in the main tomato-producing states of Venezuela: Aragua, Carabobo, Guarico, and Lara (Lastra and Gil 1981; Lastra and Uzcátegui 1975). In 1981 and 1985, Venezuelan virologists had noticed and reported that *Tomato yellow mosaic virus* (ToYMV) was also an occasional pathogen of potato (*Solanum tuberosum*) in the state of Aragua, Venezuela. These reports, presented to the scientific community both in English (Debrot 1981) and Spanish (Debrot and Centeno 1985), were disregarded by Roberts et al. (Roberts et al. 1986) when they obtained samples of potato plants infected with ToYMV from Venezuela and proceeded to publish on 'a new geminivirus infecting potatoes'. Based on their molecular characterization of this geminivirus, they changed the name of ToYMV to 'Potato yellow mosaic virus' (Roberts et al. 1986). Unfortunately, as the Venezuelan virologists had pointed out before, ToYMV is the most important neo-tropical begomovirus of tomato in the Caribbean region, where it affects tomato production in Guadeloupe, Martinique, Trinidad and Tobago, Puerto Rico and the Dominican Republic in the Antilles (Polston et al. 1998; Urbino et al. 2004) as well as in Panama (Engel et al. 1998), Colombia (Morales 2006a), and Venezuela (Nava et al. 2006). Furthermore, potatoes are primarily grown in the highlands of Latin America and are not commonly found in the agricultural areas affected by ToYMV. An investigation conducted in 2000 to characterize the original isolate of ToYMV preserved in Venezuela since the early 1980s, clearly demonstrated that the original tomato virus had over 95% nucleotide and amino acid sequence identities with the so-called 'Potato yellow mosaic virus' isolate from Venezuela (Morales et al. 2001).

Tomato yellow mosaic has been spreading in the Caribbean Basin from Venezuela, through the Lesser Antilles (Trinidad and Tobago, Martinique, Guadeloupe); until it reached Puerto Rico and the Dominican Republic in the Greater Antilles. This series of islands are close enough to act as a natural bridge for *B. tabaci*, particularly in a region affected every year by tropical systems suspected of aiding the dissemination of whiteflies and begomoviruses in the Caribbean Basin. However, one cannot discard the illegal transport of infected seedlings or plant material infested with viruliferous *B. tabaci* individuals in these islands. The illegal transport of tomato seedlings may have been responsible for the arrival of ToYMV in Panama, without affecting the tomato plantings in the northern coast of Colombia. Eventually, ToYMV disseminated into south central Colombia, probably down the Magdalena Valley, either transported by whiteflies or tomato farmers. ToYMV breached the natural barrier of the central Andean range, causing heavy yield losses in the young tomato production areas of the Cauca Valley department in western Colombia (Morales et al. 2002). In Colombia, there is evidence that ToYMV and the B biotype of *B. tabaci* are being transported on commercial tomato seedlings from warm, mid-altitude valleys (under 1,000 m) to tomato-producing areas in the central highlands of Colombia (above 1,500 m), where these pests were not present in the past (A.R. Corrales and F.J. Morales, unpublished data).

ToYMV has continued to disseminate in Venezuela from the original detection site in the state of Aragua to neighboring states, following a predominant western path into the Andean states of Trujillo, Merida and Tachira (Nava et al. 2006). This dissemination pattern is probably associated with the drier and warmer climates in northwestern Venezuela and northeastern Colombia, which favor the reproduction of *B. tabaci*. The entry point for ToYMV into Colombia was probably the neighboring department of Northern Santander, as suggested by the recent detection of ToYMV in this region of Colombia (FJ Morales, unpublished data). In 2001, an apparent recombinant begomovirus possessing a partial genome of ToYMV, was described as Tomato Venezuela virus (F Geraud, personal communication). In the same year, a survey conducted in tomato growing areas of the state of Lara, Venezuela, resulted in the detection of a new begomovirus species related to *Merremia mosaic virus* (Morales 2006a). Unfortunately, it appears that the exotic TYLCV monopartite begomovirus has been recently detected in Venezuela affecting tomato (Zambrano et al. 2007).

Begomoviruses are just beginning to emerge in other tomato-producing areas of South America, namely in Peru (Murayama et al. 2005), and temperate countries of the Southern Cone, such as Uruguay (D. Maeso, INIA-Uruguay, personal communication). As tomato production increases in these countries, due to the development of a promising tomato industry in countries such as Chile and Peru, one should expect the emergence of new tomato begomoviruses further south of the current geographic boundary (about 30°S), particularly under glass- or screen-house conditions.

2.2.2.2 Mexico

Northwestern Mexico is one of the main 'hotspots' for *B. tabaci* and whitefly-transmitted viruses in LAC, due to its characteristic dry and hot agro-ecosystems created in large irrigation districts. As early as 1971, tomato growers in Culiacan, Sinaloa, noticed a foliar malformation of their plants, which they named 'enchinamiento' (Gallegos 1978). The causal agent was eventually characterized (Brown and Nelson 1988) as a whitefly-transmitted virus currently known as *Chino del tomate virus*. This virus is now endemic in several tomato-growing states of Mexico, including Jalisco, San Luis Potosi, Guanajuato, Michoacan, Tamaulipas, Morelos, Chiapas, and Baja California Sur (Garzón-Tiznado et al. 2002; Hernandez 1972; Holguin-Peña et al. 2003; Holguin-Peña et al. 2005; Montes-Belmont et al. 1995; Torres-Pacheco et al. 1996). Another begomovirus originally detected in northwestern Mexico in the early 1990s, was named 'Tomato leaf curl (Sinaloa) virus' (Brown et al. 1993), although this name is not appropriate considering that *Tomato leaf curl virus* is an Old World begomovirus that does not exist in LAC (Fauquet et al. 2005).

The Sinaloa begomovirus probably originated in pepper fields in Texas, United States, where it was detected in the mid 1980s (Idris and Brown 2004). This begomovirus has now disseminated down the Pacific coast of Mexico and Central America, affecting tomato in Nicaragua (Rojas et al. 2000) and Costa Rica (Idris et al. 1999). The Huasteca region of Mexico (southern part of the northeastern state of Tamaulipas, eastern San Luis Potosi, and northern Veracruz) witnessed the emergence of 'Pepper Huasteco virus' in 1988. This virus was re-named *PepperHuasteco yellow vein virus* (PHYVV), although it should be called *Pepper yellow vein Huasteco virus* (PYVHV) according to the rules of the International Committee on Taxonomy of Viruses (name of host plant, symptoms, geographic location, and *virus*, in that order). PYVHV is an important pathogen of tomato in the central Mexican states of Jalisco, Morelos and Hidalgo (Morales et al. 2005c). *Pepper golden mosaic virus* (PepGMV), formerly known as 'Texas pepper virus', was first observed in Texas in 1987 (Stenger et al. 1990). This is now an important begomovirus of tomato in Baja California Sur, Sinaloa, Sonora, Nayarit, Hidalgo, and Oaxaca. PYVHV and PepGMV are commonly found together in mixed infections of tomato (Holguin-Peña et al. 2004; Morales 2006a). The dissemination of these begomoviruses into central Mexico is unexpected given the mountain ranges that separate the northwestern coast of Mexico from the central highlands. However, this natural barrier has numerous topographic depressions located between 500 and 1,000 m, where crops such as cotton are grown under arid conditions (BS = Steppe/arid) and prolonged dry seasons (Aw), which favors the reproduction and dissemination of *B. tabaci* (Cardenas et al. 1996).

2.2.2.3 Central America

The Pacific region of Central America has been affected by several begomoviruses that attack tomato and other non-traditional export crops. In Guatemala, begomoviruses belonging to the *Abutilon mosaic virus* and *Squash leaf curl virus* phylogenetic clusters, and PepGMV, are known to infect tomato plants (Mejia et al. 1998; Morales and Anderson 2001). *Tomato severe leaf curl virus* (ToSLCV) was later described from Guatemala, Honduras (Nakhla et al. 1994), Nicaragua (Rojas 2005; Rojas et al. 2000) , and central Mexico (San Luis Potosi and Morelos), causing severe stunting and leaf curling symptoms in tomato fields invaded by high populations of the whitefly vector *B. tabaci* (Mauricio-Castillo et al. 2006). Another begomovirus, 'Tomato mild mottle virus', originally described from Central America (Maxwell et al. 2002; Rojas et al. 2000), was also found later on in Baja California Sur, Mexico (Holguin-Peña et al. 2005), demonstrating the relatively high probability for begomoviruses to disseminate along the Pacific coast of Central America and Mexico.

In El Salvador, the predominant begomovirus affecting tomato in El Salvador is *Tomato mottle virus* (ToMoV), a begomovirus originally detected in Florida, United States (Abouzid et al. 1992b), followed by ToSLCV. *Tomato mosaic Havana virus* (ToMHV), a begomovirus previously identified in Cuba (Martinez-Zubiaur et al. 1998), has also been isolated from infected tomatoes in El Salvador, which demonstrates the capacity that begomoviruses and *B. tabaci* have for long-distance dissemination. Tomato in El Salvador is also affected by a virus originally isolated by the author in 1998 from pipián (*Cucurbita argyrosperma*), a popular food staple in this country (Morales 2006b). This begomovirus was later found in Costa Rica infecting squash, and was named 'Squash yellow (mild) mottle' (Karkashian et al. 2002). This virus has also been found in Nicaragua, where it infects both cucurbits (Ala-Poikela et al. 2005) and tomato (Rojas 2005; Rojas et al. 2000).

The Valley of Comayagua Honduras has been severely affected by the incidence of viruses transmitted by *B. tabaci* in tomato plantings. A survey of various tomato and pepper-producing regions of central Honduras, undertaken by the author and scientists from the Panamerican School of Agriculture (Zamorano), detected the presence of ToMHV, the Cuban begomovirus previously detected in El Salvador (FJ Morales and MM Roca, unpublished data). During this survey, a sweet potato field in central Honduras was showing severe virus-like symptoms induced by a begomovirus closely related (95%) to a *Sweet potato leaf curl virus* (SPLCV) isolate from the United States (Lotrakul et al. 1998). A few begomoviruses apparently related to SPLCV have been isolated from sweet potato in Latin America, including Brazil, Peru, Costa Rica, Mexico and Puerto Rico (Fuentes and Salazar 2003; Loebenstein et al. 2003; Lotrakul et al. 2002), but there is no sequence data for these viruses. The genomic organization of SPLCV is similar to that of monopartite begomoviruses, and the fact that sweet potato leaf curl was first observed in Japan and Taiwan, may suggest that SPLCV could be the second Old World begomovirus introduced in the Americas.

In Nicaragua, Tomato leaf curl Sinaloa virus; *Euphorbia mosaic virus*; *Squash yellow mild mottle virus* (a begomovirus previously detected in the neighboring country of Costa Rica infecting cucurbits), and PepGMV have been detected in tomato plantings (Karkashian et al. 2002; Rojas 2005; Rojas et al. 2000). A survey conducted by the author in collaboration with the National Agricultural University (UNA) in the locality of Tisma, near the capital Managua, resulted in the detection of a begomovirus associated with severe foliar malformation of tomato plants. This begomovirus had a partial sequence similarity over 90% when compared to the corresponding region of 'Tomato chino virus' from Mexico. Other tomato samples gave identities of 88% with Sida yellow vein virus and Okra yellow mosaic virus from Mexico. ToSLCV was once again detected in tomato during this survey, respectively (FJ Morales and E Jimenez, unpublished data).

The first begomovirus of tomato identified in the central valley of Costa Rica was *Tomato yellow mottle virus*, originally misidentified as 'tomato yellow mosaic' (Castillo 1997). The second begomovirus isolated from tomato in this country was the so-called 'Tomato leaf curl Sinaloa virus' (Idris et al. 1999). The northern begomovirus PepGMV was detected in Costa Rica (Karkashian et al. 1998; Lotrakul et al. 2000), demonstrating that a begomovirus originally identified in southwestern United States could disseminate through Mexico and five countries in Central America within a decade.

Tomato production is an important activity in the dry region of Azuero, Panama. The first begomovirus affecting tomato production in this region was initially described in 1998 (Engel et al. 1998) as 'Tomato leaf curl virus', but it was later recognized as a begomovirus related to ToYMV. This begomovirus is currently considered a distinct species (Potato yellow mosaic Panama virus). A recent survey of the Azuero region confirmed the endemic nature of this begomovirus in tomato plantings.

2.2.2.4 Caribbean Region

Tomato production is a very important agricultural activity in the Caribbean region (Morales 2006a, b; Morales and Anderson 2001; Morales et al. 2005a, b), mainly for industrial purposes. This region has been particularly affected by begomoviruses of both Old and New World origins. The Dominican Republic was apparently the port of entry of one of the most destructive begomoviruses from the Old World, namely, *Tomato yellow leaf curl* (TYLCV) (Polston et al. 1996). This virus rapidly spread in the Hispaniola island, paralyzing the tomato paste industry of the Dominican Republic and Haiti (Morales 2006a; Morales 2006b; Morales and Anderson 2001). TYLCV continued to disseminate in the Caribbean, reaching Cuba, Puerto Rico, Guadeloupe, and Jamaica (Bird et al. 2001; Gonzales and Valdes 1995; McGlashan et al. 1994; Urbino and Tassius 1999). In the late 1990s, TYLCV was detected in Yucatan, Mexico (Ascencio-Ibañez et al. 1999). This virus is not only a major pathogen of tomato, but it can also infect other cultivated species, such as pepper, tobacco, common bean, and cucurbits (Dalmon and Marchoux 2000; Martinez-Zubiaur et al. 2002, 2003, 2004). Regarding neo-tropical begomoviruses affecting tomato production in the Caribbean, besides ToYMV from Venezuela, most of the remaining tropical American begomoviruses in the Caribbean have been detected in Cuba. In 1997, *Tomato mottle Taino virus* (ToMoTV) and *Tomato mosaic Havana virus* (ToMHV) were identified in this island (Martinez-Zubiaur 1998; Ramos et al. 1997). ToMoTV is very similar to ToYMV, and these two Caribbean begomoviruses can form pseudo-recombinants (Ramos et al. 1997, 2003).

In Puerto Rico, four begomoviruses are known to infect tomato: TYLCV, ToYMV, ToMoV, and a begomovirus isolated from the weed *Merremia* sp. (Idris et al. 1998). A begomovirus related to the latter virus was found by the author infecting tomato in Venezuela, as reported above. In Jamaica, TYLCV has been present since the early 1990s (McGlashan et al. 1994), as well as a new begomovirus named *Tomato dwarf leaf curl virus* (Roye et al. 1999).

The Yucatan Peninsula is part of Mexico, but it is eco-geographically integrated into the Caribbean region. Yucatan is classified as an ecosystem belonging to the 'Dry Tropics' although the annual precipitation ranges from 600 to 1,500 mm. Peak *B. tabaci* populations in this zone occur in May and June, after 5 months (December–April) of dry weather (0–18 mm/month). Tomato production has been affected in the states of Yucatan, Campeche and Quintana Roo since 1991 by begomoviruses that include TYLCV, ToMoV, PepGMV and PYVHV (Diaz-Plaza et al. 1996; Garrido-Ramirez and Gilbertson 1998; Morales 2006a).

2.2.3 Sweet and Hot Peppers

Peppers (*Capsicum* spp.) are usually grown in the same regions and agro-ecosystems devoted to tomato production, and therefore, these related solanaceous species are also affected by begomoviruses in LAC, particularly in mixed cropping systems where *B. tabaci* finds diverse hosts the year round. In fact, some of the begomoviruses that affect tomato were originally characterized as pathogens of peppers, for instance, PYVHV and PepGMV. Pepper yellow vein was first reported in the 1980s as a disease of Serrano pepper (*Capsicum annuum*) in the Huasteca region (Leal and Quintero 1989). A survey conducted by Dr. Rafael Rivera-Bustamante for the TWFP in Mexico, detected PYVHV in the states of Campeche, Colima, Nayarit, Guanajuato, Veracruz, Morelos, Hidalgo, Queretaro, and San Luis Potosi (Morales et al. 2005c). This virus has also been reported to infect pepper in the northwestern state of Sonora (Ramirez et al. 1998) and is a common virus in the Yucatan Peninsula (Diaz-Plaza et al. 1996; Morales 2006a). Pepper golden mosaic is an important hot pepper production constraint in the states of Colima, Nayarit, Veracruz, Oaxaca, Aguascalientes, Morelos and San Luis Potosi (Morales et al. 2005c). This begomovirus (PepGMV) also affects sweet pepper in the state of Coahuila (Bravo-Luna et al. 2000). PepGMV has been isolated from Tabasco pepper (*C. frutescens*) and Habanero pepper (*C. chinense*) in Costa Rica (Lotrakul et al. 2000) and, recently, in Nicaragua affecting sweet pepper (FJ Morales and AK Martinez, unpublished data). On the other hand, some tomato begomoviruses, such as 'Tomato leaf curl Sinaloa virus' were originally isolated from pepper in the state of Sinaloa, northwestern Mexico (Brown et al. 1993), and are now present in Nicaragua and Costa Rica, as mentioned above. In the Caribbean, *Tomato dwarf leaf curl virus,* also affects peppers in Jamaica (Roye et al. 1999), and it is commonly found in mixed infections with TYLCV. The latter virus has been isolated from pepper in Cuba (Martinez-Zubiaur et al. 2006; Quiñones et al. 2001) and Yucatan, Mexico (Ascencio-Ibañez et al. 1999). In Trinidad and Tobago, begomoviruses related to PYVHV and ToYMV have been isolated from *C. annuum* and *C. frutescens* (Umaharan et al. 1998). Evidence for the infection of pepper by ToYMV was also provided by a recent investigation conducted in Colombia (Morales 2006a). In South America, sweet peppers are attacked by begomoviruses in northeastern Brazil (Lima et al. 2001) and Colombia (Morales 2006a).

Recently, a yellow mosaic and foliar distortion of chilli pepper (*C. baccatum*) observed in the states of Goias and São Paulo, Brazil, was shown to be induced by 'Tomato severe rugose virus' (Bezerra-Agasie et al. 2006; Nozaki et al. 2006).

2.2.4 Cucurbits

The family Cucurbitaceae includes two genera of American origin: *Cucurbita* and *Sechium* (Saade 1995). The first epidemics of begomoviruses in cucurbits were observed in southwestern United States and northwestern Mexico in 1976 (Flock and Mayhew 1981) and 1981 (Dodds et al. 1984; McCreight and Kishaba 1991). The main disease induced by a begomovirus, referred to as 'squash leaf curl', affected squash (*Cucurbita maxima*) and other cucurbit species (*C. argyrosperma, C. pepo* and *Cucumis melo*), causing severe foliar malformations. Squash leaf curl was shown to be caused by *Squash leaf curl virus* (SLCV), a bipartite begomovirus (Lazarowitz and Lazdins 1991) . SLCV moved east into Arizona and, subsequently to Texas, where it was first observed (Isakeit and Robertson 1994) infecting watermelon (*Citrullus lanatus*). SLCV was later detected in the state of Sonora, northwestern Mexico, in 1990, and, farther south in 1992, affecting 'calabacita' (*Cucurbita pepo*) in the state of Sinaloa (Ramirez et al. 1995). SLC-like symptoms had already been observed in experimental cucurbit plots in Los Mochis and Culiacan, state of Sinaloa, in 1988 (Silva et al. 1994).

Surveys conducted in 1999 by the TWFP in Central America, revealed the presence of begomoviruses in melon in the Valley of Zacapa, Guatemala; and in 'pipian' (*C. argyrosperma*) and 'ayote' (*C. moschata*) in El Salvador. The begomoviruses isolated from melon and 'pipian' had partial sequence identities of 85% between them, and sequence identities above 80% when compared to *Squash leaf curl virus* (SLCV) (Morales 2006a). The begomovirus detected in melon in 1999 was later re-isolated by other scientists from the same location and named 'Melon chlorotic leaf curl virus' (Brown et al. 2001). The begomovirus isolated from *C. argyrosperma* in El Salvador, was also found later on in Costa Rica infecting squash, and was named 'Squash yellow mild mottle' (Karkashian et al. 2002). This virus was recently detected in Nicaragua infecting cucurbits (Ala-Poikela et al. 2005), where it is also known to infect tomato (Rojas 2005; Rojas et al. 2000). However, the name given to this virus in Costa Rica, appears in the GenBank as a synonym of 'Melon chlorotic leaf curl virus'. In 1998, a begomovirus was isolated from melon showing chlorosis and leaf rugosing in the Caribbean region of Colombia (Atlantico). This disease, referred to as 'melon chlorosis' (Morales et al. 2000), was associated with the introduction of biotype B of *B. tabaci* in northern Colombia, and a begomovirus showing a >94% identity with a fragment of the coat protein of SLCV, and Melon chlorotic leaf curl virus. A similar begomovirus named 'Melon chlorotic mosaic virus' was recently reported from Venezuela. This begomovirus is also closely related (>80%) to SLCV (Ramirez et al. 2004). The proliferation of begomoviruses in the Americas is closely associated with the practice of considering probable strains of known begomoviruses as new species, based on the current amino acid and nucleotide identity threshold of <90% set by the Geminiviridae Study Group of the International Committee on Taxonomy of Viruses for distinct begomovirus species (Fauquet et al. 2005). The capacity of SLCV to generate pathogenic variants (currently considered as distinct species) will likely continue, as suggested from the numerous begomoviruses of cucurbits described in past years, including 'Cucurbit leaf curl' and 'Cucurbit leaf crumple' viruses found in southwestern United States. In the Caribbean *Tomato yellow leaf curl virus* was recently isolated from squash (*Cucurbita pepo*) in Cuba (Martinez-Zubiaur et al. 2004).

2.2.5 Potato

The first disease of potato (*Solanum tuberosum*) caused by a begomovirus in Latin America, was 'potato deforming mosaic' (Calderoni 1965; Calderoni et al. 1962; Calderoni and Malamud 1965). This disease was first described in the early 1960s from the southeastern potato-growing region of the province of Buenos Aires, Argentina (Delhey et al. 1981). Two decades later, potato deforming mosaic was observed to occur in the states of Rio Grande do Sul (Daniels and Castro 1985) and São Paulo, Brazil. Researchers in the latter state concluded that Potato deforming mosaic virus (PDMV) was different from the begomovirus (ToYMV) that infects potato in Venezuela, based on sequence data obtained in Brazil for PDMV (Vega et al. 1992). However, a recent report from Brazil (Ribeiro et al. 2006) claims that 'potato deforming mosaic' is caused by a begomovirus described in 1997 as 'Tomato yellow vein streak virus', again, based on comparative genomic analyses (Faria et al. 1997). The authors should have concluded that 'Tomato yellow vein streak virus' is a misnomer of PDMV, considering that this potato begomovirus was described 35 years before the report on the infection of tomato by PDMV in Brazil. Curiously, the author had isolated PDMV from common bean in north western Argentina in 1995, before this begomovirus was isolated from tomato in Brazil (Morales and Anderson 2001). A recent report from Brazil (Souza-Dias et al. 2007) cites yet another begomovirus, Tomato severe rugose virus, as the causal agent of a 'deforming mosaic' of potato in that country. This report remains to be thoroughly investigated regarding its etiology and relationship to PDMV.

The second begomovirus reported to infect potato (*Solanum tuberosum*) in Latin America is *Tomato yellow mosaic virus* (ToYMV) ((Debrot 1981; Debrot and Centeno 1985)). This virus was also erroneously re-named 'Potato yellow mosaic virus' later on (Roberts et al. 1986), and this misnomer persists despite molecular evidence confirming the identity of ToYMV as a major viral pathogen of tomato and, occasionally, of potato in Venezuela (Morales and Anderson 2001). *Tomato mottle Taino virus* (ToMoTV) was detected in 1998 infecting potato in Cuba (FJ Morales and G Gonzalez, unpublished data). This finding was later on confirmed in Cuba (Cordero et al. 2003). ToMoTV is closely related to ToYMV, and these begomoviruses have been shown to form pseudorecombinants (Ramos et al. 1997).

2.2.6 Tobacco

Tobacco (*Nicotiana tabacum.*), another plant species native of the Americas, is supposed to be the main host of the whitefly *B. tabaci*, as its names suggests. However, tobacco is not a preferred host of *B. tabaci* or the begomoviruses transmitted by this insect vector in Latin America. Research conducted in the early 1950s in Brazil (Silberschmidt and Tommasi 1955), showed that the begomoviruses associated with the 'infectious chlorosis of *Malvaceae*' did not readily infect tobacco. Nevertheless, the original A biotype of *B. tabaci* occasionally colonized tobacco in Latin America, albeit in relatively low numbers (Morales 2006a), and transmitted opportunistic begomoviruses to tobacco in Brazil, Venezuela, Puerto Rico, Dominican Republic, Mexico, Guatemala, and Colombia (Bird and Maramorosch 1978; Costa and Forster 1939; Morales and Anderson 2001; Morales et al. 2000; Paximadis et al. 1999; Wolf et al. 1949). Symptoms induced by neotropical begomoviruses in tobacco are usually of the 'leaf curl' type (Costa 1976). Susceptible tobacco plants may also show dwarfing and different types of variegation; and disease incidence may be significant (>30%) in some tobacco plantings (Morales et al. 2000).

The introduction of the B biotype of *B. tabaci* in Latin America, has apparently changed the privileged isolation of tobacco from emerging begomoviruses. Two begomoviruses isolated in the state of Chiapas, southern Mexico (Paximadis et al. 1999) were shown to be closely related to *Cabbage leaf curl virus* (Abouzid et al. 1992a) and PepGMV. The begomovirus found infecting tobacco in the Magdalena valley of Colombia, was closely related to a begomovirus of *Merremia* sp. from Puerto Rico (Brown et al. 1995). The Magdalena valley of Colombia is still under attack from the B biotype and begomoviruses currently in the process of characterization. Coincidentally, the Puerto Rican begomovirus from *M. quinquefolia* had been observed to infect tobacco (Bird et al. 1975), and it had been mentioned above that these begomoviruses can also infect tomato. A begomovirus isolated in Cuba, tentatively named 'Tobacco leaf rugose' was related to *Jatropha mosaic virus* from Puerto Rico (Dominguez et al. 2002). Recently, another begomovirus inducing foliar rugosity of tobacco, was reported from Cuba. The virus was considered a new species named '*Tobacco leaf curl Cuba virus*' (Moran et al. 2006). Again, this is a misnomer considering that *Tobacco leaf curl* is a recognized begomovirus that only exists in the Old World (Fauquet et al. 2005). Two additional begomoviruses have been recently isolated from diseased tobacco plants in Cuba: Tobacco yellow crinkle virus, and Tobacco mottle leaf curl virus. The former begomovirus is more closely related to begomoviruses such as Pepper golden mosaic and those related to the Squash leaf curl clade, and the latter begomovirus seems to belong to the Abutilon mosaic virus clade (Morales 2006a).

Tobacco plantations in northwest Argentina (Jujuy) are currently suffering heavy whitefly infestations, apparently related to the recent invasion of this region by biotype B of *B. tabaci*. However, this report must be confirmed because *Trialeurodes vaporariorum* was also present in this region. These reports demonstrate that begomoviruses may also become important pathogens of tobacco in Latin America.

2.2.7 *Soybean*

Soybean (*Glycine max*) is the most extensively cultivated, reproductive host of *B. tabaci* in Latin America (*c.* 39 million hectares), although there are other reproductive hosts that generate higher whitefly populations per plant (Anderson et al. 2005; Morales and Anderson 2001).

The transmission of begomoviruses to soybean by *B. tabaci,* was first observed in the early 1970s in the state of São Paulo, Brazil. Susceptible soybean plants showed leaf crinkling and plant stunting, associated with the presence of large populations of *B. tabaci* (Costa et al. 1973). Reports on occasional detection of begomoviruses infecting soybean in Brazil continue to appear in the literature (Moreira et al. 2005), but their incidence and economic importance remains low (Faria et al. 2000). Other begomoviruses capable of infecting soybean elsewhere in Latin America, have been reported from Venezuela (Debrot and Ordoisgotti 1975) and Mexico, where a begomovirus related to *Pepper golden mosaic virus* and *Cabbage leaf curl virus* (CaLCuV) has been implicated in occasional disease outbreaks in soybean plantings in the state of Sinaloa (Mendez-Lozano et al. 2006b). Curiously, a begomovirus isolated from few symptomatic soybean plants in Colombia, was also related to CaLCuV (Morales et al. 2000). Recently, a begomovirus related to *Rhynchosia golden mosaic virus* was isolated from stunted and chlorotic soybean plants in Sinaloa, Mexico (Mendez-Lozano et al. 2006a). The presence of begomoviruses affecting soybean in north western Argentina, has also been reported in the provinces of Salta and Tucuman (Laguna et al. 2005; Pardina et al. 1998) with incidences of up to 45% in some fields. A recent survey of several soybean fields conducted by the author in northwestern Argentina, showed average virus incidences under 5%. Attempts to recover a begomovirus from symptomatic soybean plants yielded negative results.

It can be concluded here that soybean is a better whitefly than begomovirus host. Whether this situation will remain the same or begomoviruses might become a serious constraint to soybean production in the future, is not known. The presence of the B biotype of *B. tabaci* might change this situation for the worse, considering the potential damage that high populations of this biotype can inflict to soybean, as observed in Ecuador in the absence of begomoviruses (Mendoza 1996).

2.2.8 Fruit Crops

Fruit crops have not escaped infection by begomoviruses, particularly species in the *Passifloraceae*. Passionfruit (*Passiflora edulis*) has been infected by begomoviruses in Puerto Rico (Brown et al. 1993) and Brazil (Novaes et al. 2003), expressing mottling and 'little leaf mosaic', respectively. In the northern coast of Colombia, a plant and fruit malformation disease of giant granadilla (*Passiflora quadrangularis*) was shown in 2000 to be caused by a begomovirus (Morales 2006a). Recently, passionfruit and giant granadilla have been attacked in the Cauca Valley of Colombia by an apparently new begomovirus related (84–88%) to begomoviruses described elsewhere infecting weeds, namely *Sida* and *Wissadula* species (FJ Morales and AK Martinez, unpublished data). In Mexico, tomatillo (*Physalis ixocarpa*) was shown to be a host of *Pepper Huasteco yellow vein virus* (Torre-Almaraz et al. 2002), and more recently of *Tomato yellow leaf curl virus* (Morales 2006a). The current popularity of the neglected tropical fruit species of Latin America, will probably cause an increase in the number of these species attacked by begomoviruses in the near future.

2.3 The Ecology of *Bemisia Tabaci*-Transmitted Viruses in LAC

The epidemiology of BGMV in South America is closely linked to the distribution of *B. tabaci* in this region. From the ecological point of view, this region belongs to the tropical savannas: grasslands with scattered trees and/or shrubs in regions with alternating wet and prolonged dry seasons. Rainfall is moderate, ranging from 400 to 800 mm in the drier areas, and 1,100–1,600 mm in the central plateau (Archibold 1995; Cole 1986). In this region, the dry season lasts from May to September, and corresponds to the winter time of the southern hemisphere. However, the lower average temperatures in winter time are still favorable for the development of *B. tabaci* populations (12–18°C), albeit at a lower reproductive rate (Zambrano et al. 2007). Relative humidity may be a limiting factor for the reproduction of *B. tabaci* during winter time (lows in the 30% range), but dew often falls at night. Some rainfall may take place in July and August in some regions. Rainfall from October to April provides moisture during the warmer period of the year, when *B. tabaci* populations increase on the newly planted crops. The Minas Gerais Triangle corresponds to the main area of savanna woodlands and grasslands ('cerrados'), in the states of Minas Gerais, Goias and Mato Grosso. Characteristic savanna vegetation also occurs in the states of Bahia, Sergipe, Alagoas and Pernambuco, where begomoviruses have also become important pathogens of several crops in the last couple of decades. These are the low tree and shrub savannas of the Brazilian 'sertão' or 'caatinga' (Cole 1986). The dissemination of *B. tabaci* in this region is limited by the high precipitation in the Amazon forest to the north; the low winter temperatures of the temperate zone south of the Tropic of Capricorn; the low temperatures of the Andean highlands to the west; and the Atlantic Ocean to the east.

A study conducted on the geographic distribution of *B. tabaci*, based on a climate probability model that uses a cluster analysis function tool, produced five different climate clusters in the BGMV region (Morales and Jones 2004). However, all clusters had a well-defined dry season lasting at least four months, with average rainfall under 80 mm/month, consistent with thee climatic conditions of the savanna regions.

In Central America, the Caribbean and southern Mexico, BGYMV and *B. tabaci* generally affect common bean in the lowlands and mid-altitude valleys under 1,000 m of altitude (Morales 2006a; Morales and Anderson 2001; Morales and Jones 2004). Most of the lowlands in this region are also classified as savanna ecosystems, and, as for 55% of the common bean-producing regions of Latin America, the affected areas in these regions belong to the tropical wet/dry (Aw) climate classification of Koeppen (Morales and Jones 2004), with few arid (BS) zones with artificial irrigation. This region has a well distributed annual rainfall, with a distinct winter season (December–April) characterized by low to moderate precipitation (>50 mm/month) and warm temperatures (up to 23.5°C in January). Therefore, common bean plantings can be heavily affected by BGYMV even during the winter season (Morales et al. 2005a, b). In southern Mexico, the main states affected by BGYMV are Veracruz and Tamaulipas (Gulf Coast), and Chiapas (Gulf of Tehuantepec). In this region, the dry season lasts from January through April, which coincides with the winter season of the northern hemisphere, but temperatures for the dry season seldom drop below 20°C, staying above the lower developmental threshold for *B. tabaci* (Zambrano et al. 2007). In Chiapas, Soconusco and Tapachula have been traditionally planted to cotton and tobacco, which are suitable plant hosts for *B. tabaci*. In Mesoamerica, the main common bean production areas and *B. tabaci* populations are found along the Pacific region, which suffers a prolonged dry and warm season from November through April, which favors the rapid growth of *B. tabaci* populations. Cotton remains an important crop in the Pacific region, and to a lesser extent tobacco, both known hosts of *B. tabaci*. The Caribbean region, on the contrary, receives abundant rainfall 9–11 months of the year, but it also presents periods of low rainfall and moderate temperatures conducive to the development of *B. tabaci* populations.

Northwestern Mexico is a hot and arid region (BW=desert) with scant rain from October through June. However, this is one of the most important agricultural areas of Mexico due to the creation of extensive irrigation districts, with a view to produce high value agricultural commodities for the winter US market. In this region, the average temperature may occasionally drop below 10°C, but it usually remains above the developmental threshold (10°C) of *B. tabaci* (Zalom et al. 1985). Melon, soybean and cotton are the main reproductive hosts of *B. tabaci* in northwestern Mexico (Morales et al. 2005c).

2.4 Integrated Whitefly and Begomovirus Management in LAC

Epidemiology is the 'science of disease in populations' (Vanderplank 1963), and plant disease is considered the outcome of the interaction between populations of plants, pathogens, and the environment (Zadoks and Schein 1979). However, in pathosystems where a vector is implicated, as in the case of whitefly-transmitted viruses, the insect vector plays a critical role in the epidemiology of begomoviruses in agro-ecosystems. Considering that a clear understanding of the epidemiology of pathogens is necessary for the development of reliable and effective disease management systems (Jeger 2004; Royle and Ostry 1995; Zadoks and Schein 1979), this review analyzes the different biotic and abiotic factors that influence begomovirus outbreaks in LAC, but within the realm of circumstantial, etiological and biological epidemiology (MacDonald 1957). Circumstantial epidemiology describes the disease and the circumstances under which it occurs. Etiological epidemiology deals with the causal agents, their plant hosts, and mode of transmission. The characterization of begomoviruses has greatly advanced due to the continuous development of molecular techniques, already implemented in developing countries. Unfortunately, virologists have been more interested in the detection and characterization of 'new' begomoviruses, than with the true concept of 'etiological' and 'evolutionary' epidemiology. Currently, we have an increasing number of 'related' but nevertheless 'distinct' begomovirus species that adds all the time to the existing complexity of begomovirus epidemics. Some begomoviruses, such as *Squash leaf curl virus* (SLCV), seems to be quite capable of adapting to different species of cucurbits, such as squash (*Cucurbita maxima*), pipian (*Cucurbita argyrosperma*), ayote (*Cucurbita moschata*), chayote (*Sechium edule*), cucumber (*Cucumis* sativus), melon (*Cucumis melo*), watermelon (*Citrullus lanatus*), and even species in different families, such as common bean (Leguminosae), found in LAC. On the other hand, *Bean golden mosaic virus* (BGMV), which has over 4 million hectares of common bean monoculture in LAC, does not seem to have the necessary selection pressure to generate variants that may result in reduced fitness or deleterious mutations (Garcia-Arenal et al. 2003).

'Biological' or 'ecological' epidemiology generates basic knowledge on the pathogens, hosts and vectors that comprise a pathosystem. Anderson (Anderson 1994) further proposes that this branch should shift the emphasis from the causal agent to etiological pathways, basically, the role of human activity in the determination of diseases. However, in this chapter, we use the term 'ecological' epidemiology to describe the relation of the organisms associated with the pathosystem, namely, begomoviruses and *B. tabaci*, to their physical surroundings (ecosystems).

In LAC, we have circumstantial evidence suggesting that specific cultivated plant species and cultural practices associated with these cropping systems, have been responsible for the emergence of begomoviruses and *B. tabaci* biotypes as major pests of food and industrial crops. Cotton, was the first crop associated with the emergence of *B. tabaci* as a pest; following the introduction and intensive use of insecticides to control the original pests of cotton, such as the cotton ball weevil. These early, non-selective insecticides eliminated the biological control organisms that maintained *B. tabaci* populations below a damage threshold (Morales 2006a; Morales and Anderson 2001; Spillari 1994). From 1919 to 1947, cotton growers dusted this pest with calcium arsenate, but the second World War motivated the development of chemicals that eventually became insecticides, such as DDT (developed in Switzerland in 1939) and other chlorinated hydrocarbons. Even organophosphates originally developed as potential toxic gasses during the war, eventually became insecticides. The carbamates were developed by Swiss scientists in the 1940s (Philip and Rozeboom 1973). As a result of the intensive use of pesticides in cotton in the 1960s, the whitefly *B tabaci* became a pest of cotton in the Pacific lowlands of Central America. The development of a new insecticide, metamidophos, in 1969, made possible cotton production once more in this region until 1975, when resistance to this insecticide was detected in the existing *B. tabaci* populations. The use of biological control agents temporarily reversed the situation until 1977, but the release and abuse of new insecticides created new epidemics of whiteflies and whitefly-borne viruses that eventually reduced in 90% the area planted to cotton between 1975 and 1990 (Stenger et al. 1990). Cotton was also the first crop to be attacked by *B. tabaci* in the state of Parana, Brazil, in 1967 (Costa 1975). The main factor driving the development of high populations of *B. tabaci* in Brazil was apparently the exponential increase in the area planted to soybean, a suitable reproductive host for *B. tabaci*. Whether insecticides played a role in the case of soybean or not, is not known, although *B. tabaci* was initially considered a pest of soybean in the 1970s, particularly in the states of Parana and São Paulo (Costa 1975), and insecticides were used in soybean against other pests in the 1970s.

Thus, circumstantial epidemiology suggests that *B. tabaci* populations up to the 1960s, were under adequate natural biological control by biological control agents. Epidemiological models suggest that high populations of vectors can favor the evolution of higher virulence in virus populations (Escriu et al. 2003; Roossinck 1997; Seal et al. 2006). These models might explain the clear association between the emergence of different begomovirus-induced diseases previously associated with wild hosts, and the exponential increase of *B. tabaci* populations in soybean plantings, as a result of the increasing area planted and the extended sowing period of this crop in Brazil. These cultural practices permit the emergence of several generations of *B. tabaci* per cropping season until the soybean crop is harvested around January. At this time, common bean plantings are being established during the dry and warm season, thus, providing a continuous supply of food and a breeding host for *B. tabaci* (Costa 1975).

The intensification of crop diversification in Latin America in the 1980s, and the introduction of biotype B of *B. tabaci* in the 1990s, further complicated the situation, as this aggressive and polyphagous new biotype extended its plant host range, bringing different begomoviruses in contact with a larger number of potential hosts (Morales 2006a). At the same time, the year round availability of a larger number of feeding and reproductive hosts, as a result of crop diversification, increased whitefly populations and the chances of begomovirus evolution and adaptation to new plant species. For instance, the number of begomoviruses infecting tomato in Latin America has more than quadrupled in the last two decades (Morales 2006a; Polston and Anderson 1997). As discussed in this chapter, many of these tomato begomoviruses can infect not only various solanaceous plant species, such as hot and sweet peppers, but legumes as well (e.g. common and lima bean). If we extend this purely biological (causal organism) focus to the concept of etiological 'pathways', as suggested by Anderson (Anderson 1994), we could try to understand how human activity and intervention lead to the emergence of new begomovirus and whitefly variants.

The concept of Integrated Pest Management (IPM), initially conceptualized to reduce dependence on pesticides and their effect on the environment, had been built into virus control strategies from the beginning of plant virology (Bos 1999), due to the known in vivo insensitivity of viruses to chemical agents. Unfortunately, the presence of insect vectors (whiteflies) that can also cause direct feeding damage to crops, has lead to extreme cases of insecticide abuse. Hence, the extraordinary capacity of *B. tabaci* to develop resistance to insecticides; and the inefficiency of most insecticides to prevent plant infection by begomoviruses in susceptible crops. Consequently, the selection and development of virus-resistant cultivars has been a major endeavor of plant virologists and breeders since the beginning of plant virology (Morales 2001; Morales and Bos 1988; Thresh 1980).

The search for sources of resistance to neotropical begomoviruses has often been a time-consuming and rather frustrating experience, mainly because of the rare occurrence of cases of immunity to begomoviruses in food and industrial crops. For instance, not a single common bean genotype has been identified as immune to BGMV or BGYMV in over 20,000 common bean accessions screened to date. Nevertheless, some common bean genotypes possessing immunity or high levels of resistance have been identified and used to control an economically important common bean begomovirus, namely *Bean dwarf mosaic virus* (BDMV) in Latin America (Morales 2001; Morales et al. 1990). Considerable progress in breeding for resistance to BGYMV and BGMV has also been achieved in the past three decades by pyramiding resistance genes from different races of common bean (Morales 2001; Morales and Singh 1993). Some of the most recent common bean cultivars produced and released in Central America by the Pan American School (Zamorano, Honduras), possess very high levels of BGYMV resistance, to the extent that these new cultivars could be grown under high begomovirus and *B. tabaci* incidence without chemical protection. However, it is known that the deployment of virus-resistant plant genotypes exerts considerable selection pressure on plant viruses, and begomoviruses are no exception (Segarra et al. 1990; Van den Bosch et al. 2006). Most of the early BGYMV-resistant common bean genotypes released in the 1980s and 1990s, have already shown increased susceptibility to BGYMV within the first decade of cultivation. However, both BGMV and BGYMV have shown to be highly stable from the pathogenic point of view, perhaps because only two or three sources of resistance to common bean begomoviruses have been widely used throughout Latin America (Morales et al. 1990). To date, only some minor changes have been detected in the antigenic properties of most geographical isolates of BGYMV in Central America and the Caribbean (Morales 2006a), which coincided with the arrival of the new biotype (B) of *B. tabaci* in the Americas. The behavior of common bean begomoviruses contrasts with that of other begomoviruses of tomato, peppers and cucurbits in Latin America, which have shown considerable pathogenic variability, as mentioned before in this chapter. However, this situation may be changing, as observed in the case of a new begomovirus that practically annihilated snap bean production in the Cauca Valley of Colombia. The emergence of this begomovirus was associated with the invasion of the valley by the B biotype of *B. tabaci* (Rodriguez et al. 2005), but, more important, the begomovirus was shown to be a recombinant between the A component of BGYMV and the B component of a local tomato begomovirus related to ToYMV (FJ Morales and AK Martinez, unpublished data). This is probably a new adaptation strategy for begomoviruses that do not have a wide host range, such as the common bean begomoviruses in Latin America. The cucurbit, tomato and pepper begomoviruses, on the contrary, have a much wider host range in the Cucurbitaceae and Solanaceae.

Unlike the progress made in the development and release of begomovirus-resistant common bean cultivars, breeding for resistance to tomato or pepper begomoviruses has been largely neglected in Latin America, with few exceptions. Most of the crop improvement for these native plant species, has been carried out mainly in the industrialized world, and therefore, the improved tomato or pepper varieties grown in LAC are not well adapted to these tropical conditions or to the different begomoviruses that exist in this region. The introduction of TYLCV into the Americas, created an opportunity for the major tomato seed companies to market tomato hybrids and varieties bred for resistance to this virus in Israel, Europe or the United States (Lapidot and Polston 2006). The dissemination of TYLCV in southern United States and the Mediterranean region of Europe resulted in the emergence of several active tomato breeding programs engaged in introducing different TYLCV resistance genes derived from various wild species of *Lycopersicon,* such as *L. peruvianum, L. chilense, L. pimpinellifolium, L. cheesmanii,* and *L.hirsutum* (Lapidot and Polston 2006). The early TYLCV-resistant materials were introduced in Caribbean countries affected by this exotic begomovirus in the late 1990s with mixed results, suggesting that the genes for resistance to TYLCV, were not as effective against the local begomoviruses. Also, some of the first improved materials introduced, did not have the required market or industrial properties. However, as breeding efforts intensified by using different sources of virus resistance, a number of recent commercial materials have been showing acceptable levels of field resistance to neotropical tomato begomoviruses. These materials may be less likely to induce begomoviruses to mutate as the resistance genes involved are probably not pathogen-specific. The effectiveness and consequences of deploying genetically-modified plants to control begomoviruses, remain to be seen.

The close association of highly aggressive, polyphagous and fecund virus vector populations with the emergence of new begomoviruses, agrees with studies on plant virus evolution that support the use of control measures to reduce vector/virus populations. High vector populations increase transmission rates and, consequently, the possibility of mixed virus infections (Seal et al. 2006). Hence, one of the complementary IPM measures recommended for the control of begomoviruses in the case of BGYMV- and BGMV-resistant common bean cultivars, is the use of the new generation of systemic insecticides (e.g. imidacloprid, thiametoxam, etc.) at sowing time and thereafter every 20 days, until flowering or, better, the initial stages of pod/fruit formation. The effectiveness of systemic insecticides to control begomoviruses has been demonstrated for ToMoV in Florida, United States. The new threat regarding this strategy, is the recent introduction of biotype Q of *B. tabaci* in the Americas, known to be less sensitive to the new chemistries (Morales 2006a).

A traditional strategy to avoid begomoviruses in Latin America, has been the cultivation of susceptible crops at elevations above 900 m, where the original *B. tabaci* biotypes could not thrive. The arrival and subsequent dissemination of the B biotype of *B. tabaci*, has greatly reduced the effectiveness of this cultural control strategy, as this biotype managed to adapt to agricultural areas above 1,500 m. Crop rotation is another cultural practice recommended to reduce *B. tabaci* populations. In the past, this whitefly species was rather selective, breeding on just a few crops or plant species. This selective behavior gave rise to the concept of 'races' of *B. tabaci* (Bird and Sanchez 1971). However, even the moderately aggressive A biotype of *B. tabaci* was known to colonize non-host plant species when there were no preferred plant hosts in the vicinity. The introduction of the highly polyphagous B biotype of *B. tabaci* further complicated this situation due to its extended host range, which is probably the reason why begomoviruses are increasingly jumping species and plant families. For instance, *Potato deforming mosaic virus* was first detected in potato, next in common bean, and last in tomato. ToYMV is known to infect tomato, potato, pepper and common bean. And SLCV jumped from cucurbits to legumes causing major outbreaks in common bean and different cucurbit species, as mentioned above.

Farmers in general know that the rainy season is the best time to escape heavy yield losses caused by whitefly-borne begomoviruses. Rain causes physical damage to whiteflies and humidity favors the activity of entomo-pathogens, which results in increased mortality rates in whitefly populations (Naranjo et al. 2004). Unfortunately, the dry season in Mesoamerica and some Andean countries often lasts five consecutive months, which forces farmers to grow susceptible crops during the dry season. Also, developing countries in LAC have invested a considerable amount of resources in the construction of irrigation districts in regions affected by a prolonged dry season. Unfortunately, the crop microclimate created by the availability of irrigation water in dry areas, favors the development of *B. tabaci* populations. Crop rotation should be a viable alternative in these regions, but the few crops that are not attacked by *B. tabaci*, such as sugarcane, sorghum or maize, are not as profitable as the high-value horticultural crops (e.g. tomato, chilies, sweet peppers) attacked by whitefly-borne viruses. In some cases, the implementation of legal measures, such as 'susceptible crop-free periods' has been a highly unpopular but effective whitefly and begomovirus control measure. The rationalization of cropping systems would be another effective strategy to reduce the incidence of these pests, but the unpredictable behavior of markets in Latin America has accustomed farmers to plant their most valuable crops at different times of the year, often in successive plantings.

The physical exclusion of whitefly vectors is another control strategy that is gaining interest for crops that are transplanted, such as tomato or sweet and hot peppers. The nurseries of these high-value crops must be completely protected from viruliferous vectors in order to avoid the early infection of seedlings. However, the main objective of 'protected agriculture' is to prevent the infection of susceptible plants during the most critical crop stage: from transplanting to fruit formation. Thus, virus-free seedlings are treated with a new generation systemic insecticide before transplanting, and then covered with a suitable insect-proof material until flowering time. Many farmers are also interested in the use of 'macro-tunnels' (at least 2 m high and about six rows wide). The advantage of these large tunnels is that they can be used without the end or side walls during the rainy season, using systemic insecticides only; and completely covered during the dry season, when high humidity is not a problem for enclosed plants exposed to fungal and bacterial problems. High temperatures must be avoided in macro-tunnels by providing suitable ventilation outlets. Physical protection drastically reduces the use of pesticides, and, therefore, creates suitable conditions for the emergence and use of biocontrol agents. Furthermore, physical exclusion of whiteflies and whitefly-borne viruses does not induce any selection pressure on begomoviruses.

The first outbreaks of whitefly-borne viruses were the consequence of pesticide abuse, at a time when the world was not concerned with environmental pollution or food quality issues. The strict food quality requirements of markets in developed countries, and the availability of more selective insecticides in developing countries, are expected to gradually reduce pesticide abuse and, thus, recover the biological stability of the affected agro-ecosystems. This is a must for the implementation of effective IPM practices to manage *B. tabaci* and, indirectly, the numerous begomoviruses that this whitefly transmits in the region.

References

Abouzid AM, Hiebert E, Strandberg J (1992a) Cloning, identification and partial sequencing of the genome components of a geminivirus infecting the *Brassicaceae*. Phytopathology 82:1070

Abouzid AM, Polston JE, Hiebert E (1992b) The nucleotide sequence of tomato mottle virus, a new geminivirus isolated from tomato in Florida. J Gen Virol 73:3225–3229

Ala-Poikela M, Svensson E, Rojas A, Horko T, Paulin T, Valkonen JP, Kvarnheden A (2005) Genetic diversity and mixed infections of begomoviruses infecting tomato, pepper and cucurbit crops in Nicaragua. Plant Pathol 54:448–459

Anderson PK (1994) A conceptual framework of integrated epidemiology for application to emerging diseases. In: Wilson ME, Levins R, Spielman A (eds) Disease in evolution: global changes and emergence of infectious disease. Ann NY Acad Sci 740:439–444

Anderson PK, Hamon A, Hernandez P, Martin, J (2005) Reproductive crop hosts of *Bemisia tabaci* (Gennadius) in Latin America and the Caribbean. In: Anderson PK, Morales FJ(eds) Whitefly and whitefly-borne viruses in the tropics: building a knowledge base. CIAT Publication No. 341, CIAT, Palmira

Andrade EC, Manhani GG, Alfenas PF, Calegario R, Fontes EPB, Zerbini FM (2006) Tomato yellow spot virus, a tomato-infecting begomovirus from Brazil with a closer relationship to viruses from *Sida* sp., forms pseudorecombinants with begomoviruses from tomato but not from *Sida*. J Gen Virol 87:3687–3696

Archibold OW (1995) Ecology of world vegetation. Chapman & Hall, New York

Ascencio-Ibañez JT, Diaz-Plaza R, Mendez-Lozano J, Monsalve-Fonnegra ZI, Arguello-Astorga GR, Rivera-Bustamante RF (1999) First report of tomato yellow leaf curl geminivirus in Yucatan. Mexico Plant Dis 83:1178

Bellows TS, Perring TM, Gill RJ, Headrick DH (1994) Description of a species of *Bemisia* (Homoptera: Aleyrodidae) infesting North American agriculture. Ann Entomol Soc Am 87:195–206

Bezerra-Agasie IC, Ferreira GB, Avila AC, Inoue-Nagata AK (2006) First report of Tomato severe rugose virus in chilli pepper in Brazil. Plant Dis 90:114

Bird J (1958) Infectious chlorosis of Sida carpinifolia in Puerto Rico. Agricultural Experiment Station of the University of Puerto Rico Technical Paper No. 26, Rio Piedras

Bird J, Maramorosch K (1978) Viruses and virus diseases associated with whiteflies. Adv Virus Res 22:55–110

Bird J, Sanchez J (1971) Whitefly-transmitted viruses in Puerto Rico. J Agric Univ P R 55:461–467

Bird J, Sanchez J, Vakili NG (1973) Golden yellow mosaic of beans *(Phaseolus vulgaris)* in Puerto Rico. Phytopathology 63:1435

Bird J, Sánchez J, Rodríguez RL, Juliá FJ (1975) Rugaceous (whitefly-transmitted) viruses in Puerto Rico. In: Bird J, Maramorosch K (eds) Tropical diseases of legumes. Academic, New York

Bird J, Idris AM, Rogan D, Brown JK (2001) Introduction of the exotic *Tomato yellow leaf curl virus*-Israel in tomato to Puerto Rico. Plant Dis 85:1028

Blanco N, Bencomo I (1978) Afluencia de la mosca blanca (*Bemisia tabaci*) vector del virus del mosaico dorado en plantaciones de frijol. Cienc Agric 2:39–46

Bos L (1999) Plant viruses, unique and intriguing pathogens. Backhuys Publishers, Leiden

Bravo-Luna L, Frias-Treviño GA, Sánchez-Valdés V, Garzon-Tiznado JA (2000) Sources of inoculum and vectors of Texas pepper geminivirus of chilli (*Capsicum annuum*) in Ramos Arizpe, Coahuila, Mexico. Rev Mex Fitopatol 18:97–102

Brown JK (1994) Current status of *Bemisia tabaci* as a plant pest and virus vector in agroecosystems worldwide. FAO Plant Prot Bull 42:3–32

Brown JK (2007) The *Bemisia tabaci* complex: genetic and phenotypic variability drives begomovirus spread and virus diversification. APSnetvFeature. www.apsnet.org/online/feature/btabaci/

Brown JK, Nelson MR (1987) Host range and vector and vector relationships of cotton leaf crumple virus. Plant Dis 71:522–524
Brown JK, Nelson MR (1988) Transmission, host range, and virus-vector relationships of chino del tomate virus, a whitefly-transmitted geminivirus from Sinaloa, Mexico. Plant Dis 72:866–869
Brown JK, Bird J, Fletcher DC (1993) First report of passiflora leaf mottle disease caused by a whitefly-transmitted geminivirus in Puerto Rico. Plant Dis 77:1264
Brown JK, Bird J, Banks G, Sosa M Kiesler et al (1995) First report of an epidemic in tomato caused by two whitefly-transmitted geminiviruses in Puerto Rico. Plant Dis 79:1250
Brown JK, Ostrow KM, Idris AM, Stenger DC (1999) Biotic, Molecular, and phylogenetic characterization of Bean calico mosaic virus, a distinct begomovirus species with affiliation in the Squash leaf curl virus cluster. Phytopathology 89:273–280
Brown JK, Idris AM, Rogan D, Hussein MH, Palmieri M (2001) Melon chlorotic leaf curl virus, a new begomovirus associated with *Bemisia tabaci* infestations in Guatemala. Plant Dis 85:1027
Calderoni AV (1965) An unidentified virus of deforming mosaic type in potato varieties in Argentina. Am Potato J 42:257
Calderoni AV, Malamud O (1965) Enfermedades de la papa. Tech Bull No 49, INTA, Balcarce
Calderoni AV, Garay OA, Pasquale DR, Induni C (1962) Los virus de la papa en la región sudeste de la provincia de Buenos Aires. IDIA 10:389–390, Suppl No
Cardenas JA, Perez F, Nieves F (1996) Campaña contra la mosquita blanca en México. Resumen del Taller de Moscas Blancas, Acapulco
Castillo J (1997) Movimiento diario de Bemisia tabaci en parcelas de tomate , diseminación local del mosaico amarillo y fuentes de inóculo del ToYMV-CR en Guayabo, Costa Rica. MSc thesis, CATIE, Turrialba
Cole MM (1986) The savannas: biogeography and geobotany. Academic, New York
Cook MT (1931) New virus diseases of plants in Porto Rico. J Dep Agric 15:193–195
Cordero M, Ramos PL, Hernandez L, Fernández AI, Echemendia AL et al (2003) Identification of Tomato mottle Taino virus strains in Cuban potato fields. Phytoparasitica 31:478–489
Costa AS (1937) Nota sobre o mosaico do algodoeiro. Rev Agric Piracicaba Bras 12:453–470
Costa AS (1955) Studies on Abutilon mosaic in Brazil. Phytopathol Z 24:97–112
Costa AS (1965) Three whitefly-transmitted diseases of beans in the state of São Paulo, Brazil. FAO Plant Prot Bull 13:121–130
Costa AS (1974) Molestias do tomateiro no Brasil transmitidas pela mosca branca Bemisia tabaci. In: VII Congr An Soc Bras Fitopatol, Brasilia
Costa AS (1975) Increase in the populational density of Bemisia tabaci, a threat of widespread virus infection of legume crops in Brazil. In: Bird J, Maramorosch K (eds) Tropical diseases of legumes. Academic, New York
Costa AS (1976) Whitefly-transmitted plant diseases. Annu Rev Phytopathol 14:429–449
Costa AS, Bennett CW (1950) Whitefly-transmitted mosaic of *Euphorbia prunifolia*. Phytopathology 40:266–283
Costa AS, Carvalho AM (1960a) Mechanical transmission and properties of the Abutilon mosaic virus. Phytopathol Z 37:250–272
Costa AS, Carvalho AM (1960b) Comparative studies between the Abutilon and Euphorbia mosaic virus. Phytopathol Z 38:129–152
Costa AS, Forster R (1939) Uma suspeita de virus do fumo (*Nicotiana tabacum* L.) semelhante a 'leaf curl' presente no estado de São Paulo. J Agron Piracicaba 2:295–302
Costa AS, Costa CL, Sauer HF (1973) Outbreak of whiteflies on crops in Parana and São Paulo. Soc Entom Bras 2:20–30
Costa AS, Oliveira AR, Silva DM (1975) Transmisão mecânica do agente causal do mosaico dourado do tomateiro. In: VIII Congr An Soc Bras Fitopatol, Mossoró
Dalmon A, Marchoux G (2000) Which plants host Tomato yellow leaf curl virus? Phytoma 527:14–17
Daniels J, Castro LA (1985) Ocorrencia do virus do mosaico deformante da batata no Rio Grande do Sul. Fitopatol Bras 10:306

De Barro PJ (1995) *Bemisia tabaci* biotype B: a review of its biology distribution and control. CSIRO Australia Division on Entomology Technical Paper No 36

De Barro PJ, Trueman WH, Frohlich DR (2005) *Bemisia argentifolii* is a race of B. tabaci (Hemiptera:Aleyrodidae): the molecular genetic differentiation of *B. tabaci* populations around the world. Bull Entomol Res 95:193–203

Debrot EA (1981) Natural infection of potatoes in Venezuela with the whitefly-transmitted mosaico amarillo del tomate virus. In: Proceedings of International Workshop, Keeble College, Oxford

Debrot EA, Centeno F (1985) Infección natural de la papa en Venezuela con el mosaico amarillo del tomate, un geminivirus transmitido por moscas blancas. Agron Trop 35:125–138

Debrot E, Ordoisgotti A (1975) Estudios sobre un mosaico amarillo de la soya en Venezuela. Agron Trop 25:435–450

Debrot E, Herold F, Dao F (1963) Nota preliminar sobre un "mosaico amarillento Del tomate" en Venezuela. Agron Trop 13:33–41

Delhey R, Kiehr-Delhey M, Heinze K, Calderoni AV (1981) Symptoms and transmission of potato deforming mosaic in Argentina. Potato Res 24:123–133

Diaz-Plaza R, Aviles-Baeza W, Torres-Pacheco I, Rivera-Bustamante R (1996) Geminivirus en la region horticola de Yucatan, Mexico. In: Resumen Taller Moscas Blancas, Acapulco

Dodds JA, Lee JG, Nameth ST, Laemmlen FF (1984) Aphid- and whitefly-transmitted cucurbit viruses in Imperial County, California. Phytopathology 74:221–225

Dominguez M, Ramos PL, Echemendia AL, Peral R, Crespo J, Andino V (2002) Molecular characterisation of tobacco leaf rugose virus, a new begomovirus infecting tobacco in Cuba. Plant Dis 86:1050

Duffus JE, Cohen S, Liu HI (1992) The sweet potato whitefly in western USA biotypes, plant interactions and virus epidemiology. In: Recent advances in vegetable virus research, 7th conference ISHS Vegetable Virus Workshop Group, Athens

Engel M, Fernandez O, Jeske H, Frischmuth T (1998) Molecular characterization of a new whitefly-transmissible bipartite geminivirus infecting tomato in Panama. J Gen Virol 79:2313–2317

Escriu F, Fraile A, Garcia-Arenal F (2003) The evolution of virulence in a plant virus. Evolution 57:755–765

FAO (1994) Anuario FAO de producción. FAO, Roma

Faria JC, Maxwell DP (1999) Variability in geminivirus isolates associated with Phaseolus spp. in Brazil. Phytopathology 89:262–268

Faria JC, Gilbertson RL, Hanson SF, Morales FJ, Ahlquist P et al (1994) Bean golden mosaic geminivirus type II isolates from the Dominican Republic and Guatemala: Nucleotide sequences, infectious pseudorecombinants, and phylogenetic relationships. Phytopathology 84:321–329

Faria JC, Souza-Dias JAC, Slack SA, Maxwell DP (1997) A new geminivirus associated with tomato in the state of São Paulo. Brazil Plant Dis 81:423

Faria JC, Bezerra IC, Zerbini FM, Ribeiro SG, Lima MF (2000) Current status of geminiviruses in Brazil. Fitopatol Bras 25:125–137

Fauquet CM, Mayo MA, Maniloff J, Desselberger U, Ball LA (2005) Virus taxonomy: eigth report of the international committee on taxonomy of viruses. Elsevier, London

Ferreira CM, Peloso MJ, Faria LC (2002) Feijão na economia nacional. EMBRAPA Doc No 135

Flock RA, Mayhew DE (1981) Squash leaf curl, a new disease of cucurbits in California. Plant Dis 65:75–76

Flores E, Silberschmidt K (1958) Relations between insect and host plant in transmission experiments with 'infectious chlorosis' of Malvaceae. Acad Bras Cienc 30:535–560

Flores E, Silberschmidt K (1967) Contribution to the problem of insect and mechanical transmisión of infectious clorosis of Malvaceae and the disease displayed by *Abutilon thompsonii*. Phytopath Z 60:181–195

Flores E, Silberschmidt K, Kramer M (1960) Observacões da clorose infecciosa das malváceas em tomateiros do campo. Biológico 26:65–69

Frischmuth T, Engel M, Lauster S, Jeske H (1997) Nucleotide sequence evidence to the occurrence of three distinct whitefly-transmitted, Sida-infecting bipartite geminiviruses in Central America. J Gen Virol 78:2675–2682

Frohlich DR, Torres-Jerez I, Bedford ID, Markham PG, Brown JK (1999) A phylogeographical analysis of Bemisia tabaci species complex based on mitochondrial DNA markers. Mol Ecol 8:1683–1691

Fryxell PA (1997) The American genera of Malvaceae-II. Brittonia 49:204–269

Fuentes S, Salazar LF (2003) First report of sweet potato leaf curl virus in Peru. Plant Dis 87:98

Gallegos HML (1978) Enchinamiento del tomate. In: HML Gallegos (ed) Enfermedades de Cultivos del Estado de Sinaloa, Secr. Rec Hidr, Sinaloa

Gamez R (1970) Los virus del frijol en Centroamérica. I. Transmisión por moscas blancas (*Bemisia tabaci* Genn.) y plantas hospedantes del virus del mosaico dorado. Turrialba 21:22–27

Garcia-Arenal F, Fraile A, Malpica JM (2003) Variability and genetic structure of plant virus populations. Int Microbiol 6:225–232

Garrido-Ramirez ER, Gilbertson RL (1998) First report of tomato mottle geminivirus infecting tomatoes in Yucatan. Mexico Plant Dis 82:592

Garzón-Tiznado JA, Acosta-García G, Torres-Pacheco I, Gonzales-Chavira M, Rivera-Bustamante Maya-Hernandez V, Guevara-Gonzalez RG (2002) Presence of the geminiviruses pepper huasteco virus (PHV), Texas pepper virus-tamaulipas, and Chino del tomate virus (CdTV) in the states of Guanajuato, Jalisco, and San Luis Potosí, Mexico. Rev Mex Fitopatol 20:45–52

Gazzoni DL, Corso IC, Miguel M (1999) Effect of insecticides on predators and parasitoids of soybean pests. Pesqui Agropecu Gaucha 5:255–264

Gilbertson R, Faria JC, Ahlquist P, Maxwell DP (1993) Genetic diversity in geminiviruses causing bean golden mosaic disease: the nucleotide sequence of the infectious cloned DNA components of a Brazilian isolate of bean golden mosaic virus. Phytopathology 83:709–715

Gilliand FR, Dumas WT, Arant FS, Ivey HW (1971) Cotton insect control with ULV-applied insecticides. Bulletin of Agricultural Experiment Station, Auburn University, No 414

Gonzales AG, Valdes RS (1995) Virus del encrespamiento amarillo de las hojas del tomate (TYLCV) en Cuba. CEIBA 36:103

Gutierrez C (1999) Geminivirus DNA replication. Cell Mol Life Sci 56:313–329

Hernandez RF (1972) Estudio sobre la mosca blanca en el estado de Morelos. Agric Tec Mex 3:165–170

Höfer P, Engel M, Jeske H, Frischmuth T (1997) Sequence of a new bipartite geminivirus isolated from the common weed Sida rhombifolia in Costa Rica. J Gen Virol 78:1785–1790

Holguin-Peña RJ, Vazquez-Juarez R, Rivera-Bustamante R (2003) First report of a geminivirus associated with leaf curl in Baja California Peninsula tomato fields. Plant Dis 87:1397

Holguin-Peña RJ, Vazquez-Juarez R, Rivera-Bustamante RF (2004) Host range, incidence and phylogeny of Pepper golden mosaic virus (PepGMV) in Baja California Sur, Mexico. Rev Mex Fitopatol 22:206–215

Holguin-Peña RJ, Vazquez-Juarez R, Rivera-Bustamante RF (2005) A new begomovirus causes tomato leaf curl disease in Baja California Sur. Mexico Plant Dis 89:341

Howarth AJ, Vandemark GJ (1989) Phylogeny of geminiviruses. J Gen Virol 70:2717–2727

Howarth AJ, Caton J, Bossert M, Goodman RM (1985) Nucleotide sequence of bean golden mosaic virus and a model for gene regulation in geminiviruses. Proc Natl Acad Sci U S A 82:3572–3576

Idris AM, Brown JK (2004) Cotton leaf crumple virus is a distinct western hemisphere begomovirus species with complex evolutionary relationships indicative of recombination and reassortment. Phytopathology 94:1068–1074

Idris AM, Lee SH, Lewis EA, Bird J, Brown JK (1998) Three tomato-infecting begomoviruses from Puerto Rico. Phytopathology 88:42

Idris AM, Rivas-Platero G, Torres-Jerez I, Brown JK (1999) First report of Sinaloa tomato leaf curl geminivirus in Costa Rica. Plant Dis 83:303

INIFAP (1995) Mosquita blanca en el noroeste de Mexico. Memoria Cientifica 3

Isakeit T, Robertson NL (1994) First report of Squash leaf curl virus on watermelon in Texas. Plant Dis 78:1010

Jeger MJ (2004) Analysis of disese progress as a basis for evaluating disease management practices. Ann Rev Phytopathol 42:61–82

Jovel J, Reski G, Rothenstein D, Ringel M, Frischmuth T, Jeske H (2004) *Sida micrantha* mosaic is associated with a complex infection of begomoviruses different from Abutilon mosaic virus. Arch Virol 149:829–841

Karkashian JP, Nakhla MK, Maxwell DP, Ramirez, P (1998) Molecular characterization of tomato-infecting geminiviruses in Costa Rica. In: International workshop on Bemisia and Geminiviruses, San Juan

Karkashian JP, Maxwell DP, Ramirez P (2002) Squash yellow mottle geminivirus: a new cucurbit-infecting geminivirus from Costa Rica. Phytopathology 92:125

Laguna IG, Fiorona MA, Ploper LD, Galvez MR, Rodriguez PE (2005) Prospección de enfermedades virales del cultivo de soja en distintas areas de producción de Argentina. In: XIII Congreso Latinoamericano de Fitopatologia, Córdoba

Lapidot M, Polston JE (2006) Resistance to Tomato yellow leaf curl virus in tomato. In: Lobenstein G, Carr JP (eds.) Natural resistance mechanisms of plants to viruses. Springer, Dordrecht

Lastra JR, Gil F (1981) Ultrastructural host cell changes associated with tomato yellow mosaic. Phytopathology 71:524–528

Lastra JR, Uzcátegui RC (1975) Viruses affecting tomatoes in Venezuela. Phytopathol Z 84:253–258

Lazarowitz SG, Lazdins IB (1991) Infectivity and complete nucleotide sequence of the cloned genomic components of a bipartite squash leaf curl geminivirus with a broad host range phenotype. Virology 180:58–69

Leal RA, Quintero S (1989) Caracterización de una virosis del chile transmisible por mosquita blanca en la planicie Huasteca. Rev Mex Fitopatol 7:147–149

Lima A, Lara F (2004) Resistencia de genotipos de soja a mosca branca *B. tabaci* (Genn.) Biotipo B (Hemiptera: Aleyrodidae). Neotrop Entomol 33:71–75

Lima JA, Goncalves MF, Oliveira VB, Torres FJ, Miranda AC (2000) Serological and PCR detection of a begomovirus infecting tomato fields in Ibiapaba Mountain, Ceara. Fitopatol Bras 25:104–108

Lima MF, Bezerra IC, Ribeiro SG, Avila AC (2001) Distribution of geminiviruses in tomato and sweet pepper crops in twelve counties of the lower basin of the San Francisco Valley. Fitopatol Bras 26:81–85

Lima GS, Assunção IP, Resende LV, Ferreira MA, Viana TH, Gallindo FA, Freitas NS (2002a) Detection of a begomovirus associated to weeds in the state of Pernambuco and partial molecular characterization of an isolate from Sida rhombifolia. Summa Phytopathol 28:353–356

Lima LHC, Campos L, Moretzsohn MC, Navia D, Oliveira MRV (2002b) Genetic diversity of *Bemisia tabaci* (Genn.) populations in Brazil revealed by RAPD markers. Genet Mol Biol 25:217–223

Loebenstein G, Fuentes S, Cohen J, Salazar LF (2003) Sweet potato. In: Loebenstein G, Thottappilly G (eds) Virus and virus-like diseases of major crops in developing countries. Kluwer, London

Loniello AO, Martinez RT, Rojas MR, Gilbertson RL, Brown JK, Maxwell DP (1992) Molecular characterization of bean calico mosaic geminivirus. Phytopathology 82:1149

Lopez GH (1974) Aumente sus rendimientos en frijol en el Valle de Culiacán. CIAS-INIA-SARH Circular No 12

Lotrakul P, Valverde RA, Clark CA, Sim J, de la Torre R (1998) Detection of a geminivirus infecting sweet potato in the United States. Plant Dis 82:1253–1257

Lotrakul P, Valverde RA, De La Torre R, Jeonggu S, Gomez A (2000) Occurrence of a strain of Texas pepper virus in Tabasco and Habanero pepper in Costa Rica. Plant Dis 84:168–172

Lotrakul P, Valverde RA, Clark CA, Hurtt S, Hoy MW (2002) Sweet potato leaf curl virus and related geminiviruses in sweet potato. Acta Hortic 583:135–141

MacDonald G (1957) The epidemiology and control of malaria. Oxford University Press, London

Martinez-Zubiaur Y (1998) Contribución al conocimiento de geminivirus que afectan el cultivo del tomate (Lycopersicon esculentum Mill) en Cuba. Doctoral thesis, Instituto Superior de Ciencias Agropecuarias de la Habana
Martinez-Zubiaur Y, Blas C, Quiñones M, Castellanos C, Peralta E, Romero J (1998) Havana tomato virus, a new bipartite geminivirus infecting tomatoes in Cuba. Arch Virol 143:1757–1772
Martinez-Zubiaur Y, Quiñónez M, Fonseca D, Potter J, Maxwell DP (2002) First report of Tomato yellow leaf curl virus associated with beans, *Phaseolus vulgaris*, in Cuba. Plant Dis 86:814
Martinez-Zubiaur Y, Quiñónez M, Fonseca D, Miranda I (2003) National survey of begomoviruses in tomato crops in Cuba. Rev Prot Veg 18:168–175
Martinez-Zubiaur Y, Fonseca D, Quiñónez M, Palenzuela I (2004) Presence of tomato yellow leaf curl virus infecting squash (Cucurbita pepo) in Cuba. Plant Dis 88:572
Martinez-Zubiaur Y, Muñiz-Martin Y, Quiñones-Pantoja M (2006) New begomovirus infecting pepper plants in Cuba. Plant Pathol 55:817
Matys JC, Silva DM, Oliveira AR, Costa AS (1975) Purificacão e morfologia do virus do mosaico dourado do tomateiro. Summa Phytopathol 1:267–274
Mauricio-Castillo JA, Arguello-Astorga GR, Alpuche-Solis AG, Monreal-Vargas CT, Dias-Gomez O, Torre-Almaraz R (2006) First report of Tomato severe leaf curl virus in Mexico. Plant Dis 90:1116
Maxwell DP, Nakhla MK, Maxwell MD, Ramirez P, Karkashian JP, Roca MM, R M, Faria JC (2002) Diversity of begomoviruses and their management in Latin America. Phytopathology 92:127
McCreight JD, Kishaba AN (1991) Reaction of cucurbit species to squash leaf curl virus and sweetpotato whitefly. J Am Soc Hortic Sci 116:137–141
McGlashan D, Polston JE, Bois D (1994) Tomato yellow leaf curl geminivirus in Jamaica. Plant Dis 78:1219
Mejia L, Teni RE, Nakhla MK, Maxwell DP (1998) Tomato-infecting, whitefly-transmitted geminiviruses in Guatemala. In: International workshop on Bemisia and Geminiviruses, San Juan
Mendez-Lozano J, Perea-Araujo LL, Ruelas-Ayala RD, Leyva-Lopez NE, Mauricio-Castillo JA, Arguello-Astorga GR (2006a) A begomovirus isolated from stunted and chlorotic soybean plants in Mexico is a new strain of Rhynchosia golden mosaic virus. Plant Dis 90:972
Mendez-Lozano J, Quintero-Zamora E, Barbosa-Jasso MP, Leyva-Lopez NE (2006b) A begomovirus associated with leaf curling and chlorosis of soybean in Sinaloa, Mexico, is related to Pepper golden mosaic virus. Plant Dis 90:109
Mendoza J (1996) Que está pasando con la mosca blanca en Ecuador? Rev INIAP 8:8–10
Montes-Belmont R, Espino-García S, Sosa-Hernandez A, Torres-Pacheco R (1995) Evaluación de extractos vegetales para el control de la virosis "chino del tomate" en dos regiones agroecologicas de Mexico. Rev Mex Fitopatol 13:111–116
Morales FJ (2001) Conventional breeding for resistance to Bemisia tabaci-transmitted geminiviruses. Crop Prot 20:825–834
Morales FJ (2006a) History and current distribution of begomoviruses in Latin America. Adv Virus Res 67:127–162
Morales FJ (2006b) Tropical whitefly IPM project. Adv Virus Res 69:249–311
Morales FJ, Anderson PK (2001) The emergence and dissemination of whitefly-transmitted geminiviruses in Latin America. Arch Virol 146:415–441
Morales FJ, Bos L (1988) Bean common mosaic virus. AAB Descriptions of Plant Viruses. No 337
Morales FJ, Jones PG (2004) The ecology and epidemiology of whitefly-transmitted viruses in Latin America. Virus Res 100:57–65
Morales FJ, Singh SP (1993) Breeding for resistance to bean golden mosaic virus in an interracial population of Phaseolus vulgaris L. Euphytica 67:59–63
Morales F, Niessen A, Ramirez B, Castaño M (1990) Isolation and partial characterization of a geminivirus causing bean dwarf mosaic. Phytopathology 80:96–101

Morales FJ, Muñoz C, Castaño M, Velasco AC (2000) Geminivirus transmitidos por mosca blanca en Colombia. Fitopatol Colomb 24:95–98

Morales FJ, Lastra R, Uzcategui RC, Calvert L (2001) Potato yellow mosaic virus: a synonym of Tomato yellow mosaic virus. Arch Virol 146:2249–2253

Morales FJ, Martinez AK, Velasco AC (2002) Nuevos brotes de begomovirus en Colombia. Fitopatol Colomb 26:75–79

Morales FJ, Gomez E, Villar A, Nin J (2005a) Dominican republic. In: Anderson PK, Morales FJ (eds) Whitefly and whitefly-borne viruses in the tropics: building a knowledge base CIAT Publication No 341. Palmira, Colombia

Morales FJ, Gonzales G, Murguido C, Echemendia A, Martinez Y, Hernández Y, Faure B, Chailloux M (2005b) Cuba. In: Anderson PK, Morales FJ (eds) Whitefly and whitefly-borne viruses in the tropics: building a knowledge base CIAT Publication No 341. Palmira, Colombia

Morales FJ, Rivera-Bustamante R, Salinas R, Torres-Pacheco I, Diaz-Plaza R, Aviles W, Ramirez G (2005c) Mexico. In: Anderson PK, Morales FJ (eds) Whitefly and whitefly-borne viruses in the tropics: building a knowledge base. CIAT Publication No 341, CIAT, Palmira

Moran YM, Ramos PL, Dominguez M, Fuentes AD, Sanchez I, Crespo JA (2006) Tobacco leaf curl Cuba virus, a new begomovirus infecting tobacco (*Nicotiana tabacum*) in Cuba. Plant Pathol 55:570

Moreira AG, Pereira CO, Andrade EC, Zerbini FM (2005) Caracterización molecular de dos begomovirus que afectan a la soja (Glycine max) y malezas asociadas, en Brasil. In: Resúmenes XIII Congreso Latinoamericano de Fitopatologia, Cordoba

Morinaga T, Ikegami M, Miura K (1987) Molecular cloning of full-length of bean golden mosaic virus DNA. Ann Phytopathol Soc Jpn 53:549–553

Murayama A, Aragon L, Fernández-Northcote E (2005) Nuevo begomovirus del grupo del Nuevo Mundo asociado al encrespamiento de la hoja del tomate en la costa del Peru. In: Resúmenes XIII Congreso Latinoamericano de Fitopatologia, Cordoba

Nakhla MK, Maxwell MD, Hidayat SH, Lange DR, Loniello AO, Rojas MR et al (1994) Two geminiviruses associated with tomatoes in Central America. Phytopathology 84:1155

Naranjo SE, Cañas L, Ellsworth PC (2004) Mortality factors affecting populations of sweet potato whitefly, Bemisia tabaci, in a multi-crop system. Hortic Int 2:14–21

Nava AR, Patte CP, Hiebert E, Polston JE (2006) Detection and variability of begomoviruses in tomato from the Andean states of Venezuela. Plant Dis 90:61–66

Novaes QS, Freitas-Astua J, Auki VA, Kitajima EW, Camargo LE, Rezende JA (2003) Partial characterization of a bipartite begomovirus infecting yellow passion flower in Brazil. Plant Pathol 52:648–654

Nozaki DN, Krause-Sakate R, Hasegawa JM, Cezar MA, Dziuba PH, Pavan MA (2006) First report of Tomato severe rugose virus infecting pepper plants in Brazil. Fitopatol Bras 31:321

Orlando A, Silberschmidt K (1946) Estudos sobre a disseminacão natural do virus da clorose infecciosa das malvaceas (Abutilon virus 1 Baur) e a sua relacão com o inseto-vetor Bemisia tabaci (Genn.) (Homoptera:Aleyrodidae). Arq Inst Biol São Paulo 17:136

Pardina PE, Ploper LD, Truol GA, Hanada K, Platero GR, Ramirez P, Herrera PS, Laguna IG (1998) Detection of a geminivirus in soybean crops in northwestern Argentina. Rev Ind Agric Tucuman 75:51–56

Paximadis M, Idris AM, Torres-Jerez I, Villarreal A, Rey MEC, Brown JK (1999) Characterisation of tobacco geminiviruses in the Old and New World. Arch Virol 144:703–717

Perring TM (2001) The *Bemisia tabaci* species complex. Crop Prot 20:725–737

Philip CB, Rozeboom LE (1973) Medico-veterinary entomology: a generation of progress. In: Smith RF, Mittler TE, Smith CN (eds.) History of entomology. Annual Review Inc., Palo Alto

Pickersgill B (1977) Taxonomy and the origin and evolution of cultivated plants in the New World. Nature 268:591–595

Pierre RE (1975) Observations on the golden mosaic of bean (*Phaseolus vulgaris* L.). In: Bird J, Maramorosch K (eds.) Tropical diseases of legumes. Academic, New York

Polston JE, Anderson PK (1997) The emergence of whitefly-transmitted geminiviruses in tomato in the Western Hemisphere. Plant Dis 81:1358–1369

Polston JE, Bois D, Serra CA, Concepción S (1996) First report of a tomato yellow leaf curl-like geminivirus from tomato in the Western Hemisphere. Plant Dis 78:831

Polston JE, Bois D, Ano G, Poliakoff N, Urbino C (1998) Occurrence of a new strain of potato yellow mosaic geminivirus infecting tomato in the Eastern Caribbean. Plant Dis 82:126

Quiñones M, Fonseca D, Accotto GP, Martinez Y (2001) Viral infections associated with the presence of begomovirus in pepper plants in Cuba. Rev Prot Veg 16:147–151

Ramirez JA, Armenta I, Delgadillo F, Rivera-Bustamante RF (1995) Geminivirus transmitidos por mosquita blanca (Bemisia tabaci Gennn.) en los cultivos de chile y calabacita en el Valle de Mayo, Sonora, México. Rev Mex Fitopatol 13:100–105

Ramirez JA, Cardenas IA, Sánchez FD, Garzon-Tiznado JA (1998) Virus transmitted by whitefly (Bemisia tabaci Gennadius) in pepper and zucchini squash in the Mayo Valley, Sonora, Mexico. Agric Tec Mex 24:37–43

Ramirez P, Chicas M, Salas J, Maxwell D, Karkashian J (2004) Identificacion de un nuevo begomovirus en melon (Cucumis melo L.) en Lara, Venezuela. Rev MIP Agroecol 72:22–30

Ramos PL, Guerra O, Peral R, Oramas P, Guevara RG, Rivera-Bustamante R (1997) Taino tomato mottle virus, a new bipartite geminivirus from Cuba. Plant Dis 81:1095

Ramos PL, Guevara RG, Peral R, Ascencio-Ibañez JT, Polston JE, Arguello GR, Vega JC, Rivera-Bustamante RF (2003) Tomato mottle Taino virus pseudorecombines with PYMV but not with ToMoV: Implications for the delimitation of cis- and trans-acting replication specificity determinants. Arch Virol 148:1697–1712

Rampersad SN, Umaharan P (2003) Detection of two bipartite geminiviruses infecting dicotyledonous weeds in Trinidad. Plant Dis 87:602

Ribeiro SG, Ambrocevicius LP, Avila AC, Bezerra IC, Calegario RF, Fernandes JJ et al. (2003) Distribution and genetic diversity of tomato-infecting begomoviruses in Brazil. Arch Virol 148:281–295

Ribeiro SG, Inoue-Nagata AK, Daniels J, Avila AC (2006) Potato deforming mosaic disease is caused by an isolate of Tomato yellow vein streak virus. Plant Pathol 55:569

Roberts EJ, Buck KW, Coutts R (1986) A new geminivirus infecting potatoes in Venezuela. Plant Dis 70:603

Rodriguez I, Morales H, Bueno JM, Cardona C (2005) El biotipo B de Bemisia tabaci adquiere mayor importancia en el Valle del Cauca. Rev Col Entomol 31:21–28

Rojas A (2005) A complex of begomoviruses affecting tomato crops in Nicaragua. In: Proceedings IX international plant virus epidemiology symposium, Lima

Rojas A, Kvarnheden A, Valkonen JPT (2000) Geminiviruses infecting tomato crops in Nicaragua. Plant Dis 84:843–846

Roossinck MJ (1997) Mechanisms of plant virus evolution. Ann Rev Phytopathol 35:191–209

Roye ME, Wernecke ME, McLaughlin WA, Nakhla MK, Maxwell DP (1999) Tomato dwarf leaf curl virus, a new bipartite geminivirus associated with tomatoes and peppers in Jamaica and mixed infection with Tomato yellow leaf curl virus. Plant Pathol 48:370–378

Royle DJ, Ostry ME (1995) Disease and pest control in the bioenergy crops poplar and willow. Biomass Bioenergy 9:69–75

Rybicki EP (1994) A phylogenetic and evolutionary justification for three genera of Geminiviridae. Arch Virol 139:49–77

Rybicki EP, Briddon RW, Fauquet CM, Maxwell DP, Stanley J, Harrison BD, Brown JK (2000) Geminiviridae. In: Regenmortel MHV, Fauquet CM, Bishop DHL, Carsten EB, Estes MK, Lemon SM, Maniloff J, Mayo MA, McGeoch DJ, Pringle CR, Wickner RB (eds.) Virus taxonomy: seventh report of the international committee on taxonomy of viruses. Academic, San Diego

Saade RL (1995) Estudios taxonómicos y ecogeográficos de las Cucurbitaceae latinoamericanas de importancia económica. Inst Int Recur Fitogen (IPGRI), Rome

Schieber E (1970) Enfermedades del frijol (*Phaseolus vulgaris*) en la República Dominicana. Turrialba 20:20–23

Seal S, VandenBosch F, Jeger MJ (2006) Factors influencing begomovirus evolution and their increasing global significance: implications for sustainable control. Crit Rev Plant Sci 25:23–46

Segarra AE, Bird J, Escudero J (1990) Silvering of Cucurbita moschata (Duchesne) associated with Bemisia tabaci Genn (Homoptera:Aleyrodidae) in Puerto Rico. J Agric Univ P R 74:477–478

Silberschmidt K (1943) Estudos sobre a transmissão experimental da "clorose infecciosa" das malvaceas. Arq Inst Biol São Paulo 14:105–156

Silberschmidt K, Tommasi CR (1955) Observações e estudos sobre especies de plantas suscetíveis à clorose infecciosa das malváceas. Acad Bras Cienc 27:195–214

Silva S, Rodriguez R, Garzon-Tiznado JA, Delgadillo F, Cardenas E (1994) Efecto de una variante del virus enrollamiento de la hoja de la calabaza (VEHC, Squash leaf curl virus, SqLCV) en genotipos de cucurbitáceas. Rev Mex Fitopatol 12:15–20

Singh SP (1988) Patterns of variation in cultivated common bean (*Phaseolus vulgaris*, Fabaceae). Econ Bot 43:39–57

Souza-Dias JA, Sawazaki HE, Miranda FH, Maluf H, Arikita H (2007) Tomato severe rugose virus (ToSRV): a first report of another emerging begomovirus causing deforming mosaic symptoms in the potato crop region of Campinas, SP, Brazil. In: Abstracts 13th International meeting, European Association for Potato Research, Aviemore

Spillari AG (1994) Problemática del complejo mosca blanca-virus en algodón en Centroamerica. In: Memorias III (ed) Taller Centroamericano y del Caribe sobre Mosca Blanca. Antigua Guatemala, Guatemala, pp 23–38

Stenger DC, Duffus JE, Villalon B (1990) Biological and genomic properties of a geminivirus isolated from pepper. Phytopathology 80:704–709

Thresh JM (1980) The origins and epidemiology of some important plant virus diseases. Appl Biol 5:1–65

Torre-Almaraz R, Valverde R, Mendez J, Ibáñez JT, Rivera-Bustamante RF (2002) Preliminary characterization of a geminivirus in tomatillo (*Physalis ixocarpa* B.) in the central region of Mexico. Agrociencia 36:471–481

Torre-Almaraz R, Monsalvo-Reyes AC, Mendez-Lozano J, Rivera-Bustamante RF (2004) Characterization of a new geminivirus associated with okra (Abelmoschus esculentus) yellow mottle symptoms in Mexico. Agrociencia 38:227–238

Torres-Pacheco I, Garzón-Tiznado JA, Brown JK, Rivera-Bustamante RF (1996) Detection and distribution of geminiviruses in Mexico and the southern United States. Phytopathology 86:1186–1192

Umaharan P, Padidam M, Phelps RH, Beachy RN, Fauquet CM (1998) Distribution and diversity of geminiviruses in Trinidad Tobago. Phytopathology 88:1262–1268

Urbino C, Tassius K (1999) First report of Tomato yellow leaf curl virus in tomato in Guadeloupe. Plant Dis 35:46–53

Urbino C, Polston JE, Patte CP, Caruana ML (2004) Characterization and genetic diversity of Potato yellow mosaic virus from the Caribbean. Arch Virol 149:417–424

Van den Bosch F, Akudibilah G, Seal S, Jeger M (2006) Host resistance and the evolutionary response of plant viruses. J App Eco 43:506–516

Vanderplank JE (1963) Plant diseases: epidemics and control. Academic, New York

Vega J, Winter S, Pural A, Hamilton RI (1992) Partial characterization of a whitefly-transmitted geminivirus from potato in Brazil. Fitopatol Bras 17:167

Wolf FA, Whitecomb WH, Mooney WC (1949) Leaf curl of tobacco in Venezuela. J Elisha Mitchel Sci Soc 65:8–47

Zadoks JC, Schein RD (1979) Epidemiology and plant disease management. Oxford University Press, Oxford

Zalom FG, Natwick ET, Toscano NC (1985) Temperature regulation of Bemisia tabaci (Homoptera:Aleyrodidae) populations in Imperial Valley cotton. J Econ Entomol 78:61–64

Zambrano K, Carballo O, Geraud F, Chirinos D, Fernández C, Marys E (2007) First report of Tomato yellow leaf curl virus in Venezuela. Plant Dis 91:768

Chapter 3
Bemisia tabaci – *Tomato Yellow Leaf Curl Virus* Interaction Causing Worldwide Epidemics

Henryk Czosnek and Murad Ghanim

Abstract *Tomato yellow leaf curl virus* (TYLCV) is a begomovirus that threatens tomato production worldwide. TYLCV is transmitted in a circulative manner by the whitefly *Bemisia tabaci*. Once ingested, TYLCV was detected in the insect midgut after 1 h, in the haemolymph after 1.5 h, and in the salivary glands after 7 h. Whiteflies were able to infect tomato plants after 8 h. TYLCV survival in the haemolymph of *B. tabaci* is ensured by the interaction of virus particles with a GroEL homologue produced by the whitefly endosymbionts. The relation between TYLCV and its vector are intricate. Following a short acquisition period, the virus remains associated with *B. tabaci* for the 4 weeks-long adult life of the insect. During this period, infectivity decreased from 100 to 10–20%. The long-term presence of TYLCV in *B. tabaci* was associated with a decrease in longevity and fertility. Similar results were reported with *Tomato yellow leaf curl China virus* (TYLCCNV). The question of whether TYLCV is expressed and replicates in its vector is not settled. Transcripts of TYLCV genes encoded by the genome strand as well as by the genome complementary strand have been detected in *B. tabaci* following virus acquisition. TYLCV was found to be transmitted transovarially up to the adult stage of the first progeny generation, as did *Tomato yellow leaf curl Sardinia virus* (TYLCSV); however contrary to TYLCV, the whiteflies were not able to transmit

H. Czosnek (✉)
The Otto Warburg Minerva Center for Agricultural Biotechnology & the Robert H. Smith Institute for Plant Science and Genetics in Agriculture, The Hebrew University of Jerusalem, Rehovot 76100, Israel
e-mail: czosnek@agri.huji.ac.il

M. Ghanim
Department of Entomology, Agricultural Research Organization, Volcani Center, Bet Dagan 50250, Israel
e-mail: ghanim@agri.gov.il

W.M.O. Thompson (ed.), *The Whitefly, Bemisia tabaci (Homoptera: Aleyrodidae) Interaction with Geminivirus-Infected Host Plants*,
DOI 10.1007/978-94-007-1524-0_3,

the disease to tomato plants. TYLCV and TYLCCN could be transmitted with various efficiency during mating. A cDNA microarray was constructed to discover the genes involved in TYLCV-*B. tabaci* interactions.

3.1 Tomato Yellow Leaf Curl Virus

Tomato yellow leaf curl virus TYLCV causes one of the most devastating diseases affecting the domesticated tomato, worldwide (Czosnek 2007). Besides tomato, TYLCV is capable of infecting more than 30 species in over 12 plant families, including cultivated vegetables, ornamentals, weeds and wild plant species. The virus is transmitted by the sweet potato whitefly *Bemisia tabaci*. The severity of the viral epidemic correlates with the size of the whitefly population that vectors TYLCV. Hence, the TYLCV disease is usually managed by frequent applications of insecticides to contain the whitefly populations in the field and greenhouse. Breeding tomato cultivars resistant to TYLCV has started in the late 1970s. It has consisted of introgressing resistant traits found in wild tomato species into cultivated varieties (Picó et al. 1999). Several tomato cultivars are commercially available, which present excellent levels of resistance, satisfactory yields and good fruit quality.

The whitefly *B. tabaci* vectors a large number of viruses infecting many important agricultural plants, ornamentals and weeds (Brown and Czosnek 2002). *B. tabaci* is a complex of biotypes that differ in their behavior, plant host, ability to induce disorders in plants, ability to transmit plant viruses, the endosymbionts they contain and in their genetic make-up. In addition to the endogenous species, new invasive and better fit biotypes such as B and Q have invaded crop systems, exacerbating damages (Czosnek and Brown 2010). Geminiviruses constitute the most important class of pathogens transmitted by *B. Tabaci*. They have been assigned to the genus *Begomovirus* within the family *Geminiviridae* (van Regenmortel et al. 2000). They are small plant viruses characterized by a 22×38 nm geminate particle consisting of two joined incomplete icosahedra encapsidating a single-stranded genome of approximately 2,800 nucleotides (Goodman 1977; Harrison et al. 1977; Zhang et al. 2001; Böttcher et al. 2004).

Tomato yellow leaf curl virus (TYLCV) is the generic name given to a begomovirus that devastates tomato cultures worldwide. The TYLCV complex includes ten species and numerous isolates distinguishable by their sequence (Fauquet et al. 2008). The viral type is TYLCV from Israel (Navot et al. 1991). Unlike most begomoviruses, which possess two genomic components denominated DNA-A and DNA-B (bipartite), the TYLCV species have a single DNA-A-like genome component (monopartite). The TYLCV genome possesses two open reading frames (ORF) on the virion strand: V1, which encodes the coat protein (CP), and V2, which encodes a suppressor of gene silencing to overcome the plant defense system (Zrachya et al. 2007). The complementary virus strand possesses four ORFs: C1 encodes a replication associated protein (Rep), C2 a transcription activator protein (TrAP), C3 a replication enhancer protein (REn), and C4 embedded within C1

(Gronenborn 2007; Glick et al. 2009). The CP of TYLCV is multifunctional. Beside encapsidating the genome and transporting it in and out of the nucleus (Kunik et al. 1998), the integrity of the CP is required for capsid assembly (Hallan and Gafni 2001), systemic plant infection and vector transmission (Noris et al. 1998). Rep and REn are required for efficient viral DNA replication. V2 encodes a movement-like protein. V2 and TrAP may function as a viral suppressor of RNA silencing. In addition to their role as transcription activators of late viral genes, the C4 protein has been implicated in symptom expression and virus movement. A non-coding 200–300 nucleotide-long intergenic region (IR) located upstream the V2 and C1 ORFs contains a conserved stem-loop structure embedding the origin of replication and signals necessary for the replication and transcription of the viral genome in host plants.

3.2 Ingestion and Inoculation of TYLCV by *B. tabaci*

3.2.1 Acquisition and Transmission

Following landing on a tomato plant, the stylets of *B. tabaci* find their way between the epidermal and parenchymal cells of infected plants before penetrating the vascular tissues and reaching the phloem the insect feeds on during virus acquisition and transmission (Pollard 1955). The parameters of acquisition and transmission of a begomovirus were first defined for TYLCV and were based on biological tests (Cohen and Harpaz 1964; Cohen and Nitzany 1966). Single insects are able to acquire TYLCV and transmit it to tomato plants. The minimum acquisition access period (AAP) and inoculation access period (IAP) of Middle Eastern TYLCV isolates varied from 15 to 60 min and from 15 to 30 min, respectively (Cohen and Nitzany 1966; Ioannou 1985; Mansour and Al-Musa 1992; Mehta et al. 1994). Similar values were reported for a begomovirus closely related to TYLCV, the *Tomato yellow leaf curl Sardinia virus* (TYLCSV) from Italy (Caciagli et al. 1995).

The development of molecular tools has allowed a refinement of these studies. TYLCV was readily detected by Southern blot hybridization in DNA extracted from a single whitefly of the B biotype. The hybridization signals indicated that insects that had access to the same tissues for the same period of time could acquire variable amounts of viral DNA (Zeidan and Czosnek 1991).

PCR allowed detecting TYLCV DNA amounts in a single insect, below the threshold of infectivity (Navot et al. 1992). TYLCV DNA could be amplified in 20% of the individuals tested 5 min after the start of the AAP from the infected plant and in all insects 5 min thereafter (Atzmon et al. 1998). Analysis of the electronic waveforms produced during insect feeding indicated that, following a short probing period, the minimum phloem contact threshold period observed for successful inoculation of TYLCV was 1.8 min (Jiang et al. 2000).

3.2.2 *Transmission Efficiency of TYLCV by the B Biotype of B. tabaci: The Effect of Gender and Age*

A single insect is able to infect a tomato plant with TYLCV following a 24 h AAP; efficiency of transmission reaches 100% when 5–15 insects are used (Cohen and Nitzany 1966; Mansour and Al-Musa 1992; Mehta et al. 1994). Gender and age affect transmission ability (Czosnek et al. 2001). Nearly all the 1–2 week-old adult females from synchronized populations of adult *B. tabaci* were able to infect tomato plants during a 48 h IAP, following a 48 h AAP.. In comparison, only about 20% of the males of the same age under the same conditions were able to infect plants. Infection capacity decreased with age. While 60% of the 3 week-old females infected plants, the males were totally unable to infect tomato plants. Only 20% of the 6 week-old females were able to infect tomato plants. Aging insects acquire fewer viruses than younger individuals (Rubinstein and Czosnek 1997). Seventeen days after eclosion, the adult insects acquired less than half the virus acquired by 10 day-old insects. At the age of 24 days, this amount was only about 10%. At the age of 28 days and thereafter, the viral DNA associated with the insects was undetectable by Southern blot hybridization although the insects retained about 20% of their initial inoculation capacity.

Transmission efficiency of Q biotype is not essentially different from that of B. Transmission of a TYLCSV isolate from Murcia, Spain (TYLCSV-ES) was studied using the B, Q and S biotypes of *B. tabaci* (Jiang et al. 2004). Both B and Q-biotypes of *B. tabaci* were able to transmit TYLCSV-ES from infected tomato plants to *Solanum nigrum* and *Datura stramonium* and *vice versa.* No significant difference was found in transmission efficiency from infected tomato plants to weed plants between the B- and Q-biotypes. The S-biotype could not survive on tomato long enough to acquire or transmit TYLCSV-ES. In these studies, the age and gender of the whiteflies was not taken into account.

3.3 The Path of TYLCV in the Whitefly Host

3.3.1 *Circulative Transmission of TYLCV*

Once ingested by whiteflies, begomoviruses are not immediately available for infection. They need to translocate in the insect digestive tract, penetrate the gut membranes into the haemolymph, crossing the epithelial cells of the whitefly digestive tract which bridge between the gut lumen and the haemolymph. From there, begomoviral particles reach the salivary systems and finally enter the salivary duct from where they are egested with the saliva. Translocation of begomoviruses from the digestive tract to the haemolymph and from the haemolymph to the salivary gland is thought to be mediated by still un-identified receptors.

TYLCV has been traced in these organs using antibodies raised against the CP. The virus was localized in the stylets, associated mainly with the food canal all along the lumen. Similarly, TYLCV was immunolocalized to the proximal part of the descending midgut, the filter chamber and the distal part of the descending midgut and in the primary salivary glands (Brown and Czosnek 2002; Czosnek et al. 2002). The localization patterns of the TYLCSV were similar to those of TYLCV. TYLCSV has been detected in the midgutits microvilli, and in the cytoplasm of the primary salivary gland cells (Medina et al. 2006; Ghanim and Medina 2007; Caciagli et al. 2009). Although viral DNA fragments have been amplified from ovary tissue of whiteflies that acquired TYLCV (Ghanim et al. 1998) and TYLCSV (Bosco et al. 2004), no specific labeling of the TYLCSV CP in ovaries was detected (Caciagli et al. 2009).

3.3.2 Velocity of Translocation

Once ingested, begomoviruses are not immediately infective. The time it takes from the beginning of the AAP to the moment the whitefly efficiently transmits the virus to plants is called the latent period. It may vary due to the experimental conditions or to changes in virus and/or vector with time. For example the latent period of TYLCV was reported to be 21 h in the early 1960s (Cohen and Nitzany 1966) while it was found to be 8 h thirty five years later (Ghanim et al. 2001).

The velocity of translocation of TYLCV DNA and CP was determined using whitefly stylets, head, midgut, haemolymph and salivary glands dissected from a single insect as substrate for PCR and immunocapture-PCR (Ghanim et al. 2001). TYLCV was detected in the head of whiteflies as early as 10 min after the beginning of the AAP and in the midgut approximately after 40 min. TYLCV crossed the midgut and reached the haemolymph 30 min after it was first detected in the midgut, 90 min after the beginning of the AAP. TYLCV was detected in the salivary glands approximately 5.5 h after it was first detected in the haemolymph, 7 h after the beginning of the AAP. Whiteflies were able to infect tomato plants 1 h after the virus was first detected in the salivary system indicating that the threshold amount of virions necessary to obtain an efficient infection is low. Translocation timing of TYLCV DNA and CP overlapped, suggesting that the TYLCV moves as virions. Velocity of *Squash leaf curl virus* (SLCV) translocation in *B. tabaci* was similar (Rosell et al. 1999).

3.4 Virion and Whitefly Determinants Insuring Efficient Transmission of TYLCV by *B. tabaci*

3.4.1 The Virus Capsid

It is likely that, during the transit of begomoviruses in their whitefly vector, the capsid is the structure that is exposed to the whitefly tissues and interacts with insect receptors and chaperons (Morin et al. 2000). Vector specificity of geminiviruses is determined by the coat protein (Höfer et al. 1997) and there is no evidence for the involvement of other virus-encoded proteins in transmission.

Loss of transmission by *B. tabaci* can be caused by a surprisingly small number of amino acid replacements in the CP of begomoviruses. Natural TYLCSV mutants have been isolated which are acquired but not transmitted by *B. tabaci*. A region between amino acids 129 and 152, including Q129, Q134, and D152 is relevant for virion assembly, systemic infection, and transmission by the vector (Noris et al. 1998). The double mutant Q129P and Q134H and a further D152E were acquired by *B. tabaci*, circulated in the insect for up to 10 days as the wild type virus, was immune-detected in the salivary glands, but were not transmissible (Caciagli et al. 2009). A N130D mutant prevented assembly of a correct geminate particle and was not acquired by *B. tabaci* (Caciagli et al. 2009). Hence virion formation and stability are necessary but not sufficient for begomovirus transmissibility. In addition, crossing the salivary gland barrier may not be sufficient for transmission. The region of the CP between amino acids 129 and 152 is also implicated in transmission ability of the bipartite *Watermelon chlorotic stunt virus* WmCSV (Kheyr-Pour et al. 2000) and *Abutilon mosaic virus* AbMV (Höhnle et al. 2001).

3.4.2 Association of Viral Particles with GroEL Homologue Produced by the Insect Endosymbiotic Bacteria

Virions that cross the gut wall into the haemolymph on their way to the salivary gland face a particularly hostile environment. A GroEL homologue produced by the primary endosymbionts of aphids has been shown to play a crucial role in the transmission of luteoviruses (van den Heuvel et al. 1994). Similarly, endosymbiotic bacteria housed in the whitefly bacteriocytes seem to have a cardinal role in safeguarding begomoviruses in the haemolymph. As demonstrated for TYLCV, the GroEL homologue seems to bind to and protect begomoviruses from degradation in the haemolymph; disturbing the GroEL-TYLCV association leads to the degradation of the virus and to a markedly decrease in transmission efficiency of the virus (Morin et al. 1999, 2000). Interestingly, in the yeast two hybrid system, *B. tabaci* GroEL interacted with the CP of TYLCV as well as with the CP of the non-transmissible AbMV (Morin et al. 2000), indicating that the amino acid residues at position 124, 149 and 174, which prevented AbMV from crossing into the insect haemolymph (Höhnle et al. 2001) did not prevent binding to GroEL.

3.4.3 Interaction with Host Proteins

During begomovirus circulative transmission, it is likely that particles interact with whitefly proteins in order to move from the digestive tract into the haemolymph and from the haemolymph into the salivary system. In order to search for insect proteins interacting with TYLCSV, the virus CP was used as bait in a yeast two-hybrid screen against a cDNA library constructed from *B. tabaci* biotype Q (Ohnesorge and Bejarano 2009). A 16 kDa small heat-shock protein (named BtHSP16) belonging to the HSP20/a-crystallin family was shown to bind to the TYLCSV CP. The gene coding for BtHSP16 has an intron and occurs at a single locus in the *B. tabaci* genome. Full-length BtHSP16 is required for the interaction with the CP of TYLCSV. The TYLCSV CP interaction domain with BtHSP16 was located within the conserved region of the N-terminal part of TYLCSV CP (amino acids 47–66), overlapping almost completely with the nuclear localization signal described for the CP of TYLCV (Kunik et al. 1998). The region necessary for transmission of TYLCSV by *B. tabaci* (amino acids 129–152) is not directly involved in the specific interaction between the CP and the BtHSP16.

3.5 TYLCV Replication and Transcription in the Whitefly Host?

3.5.1 *Replication*

Begomovirus replication in its vector remains a controversial issue. The persistence of TYLCV/TYLCSV in *B. tabaci* as infective entities for longer than the latent period, sometimes for the entire life of the insect, (Caciagli and Bosco 1997; Rubinstein and Czosnek 1997), raises the question of replication of the virus in the insect. Accumulation of viral DNA in *B. tabaci* reared on a TYLCV-non host plant, after first feeding on plants infected with a TYLCV isolate from Egypt, has been interpreted as multiplication of TYLCV in its vector (Mehta et al. 1994). We have found that after a short AAP the amount of TYLCV DNA associated with whiteflies detectable by Southern blot hybridization steadily increased after a lag period of 8 h, reaching maximum levels approximately after 16 h and decreasing thereafter (Czosnek et al. 2001). These results have been confirmed by feeding whiteflies with purified virions through membranes, and measuring the viral DNA by quantitative PCR after the insects were transferred to non-host cotton plants (Mahadav et al. 2009). It has to be noted that following acquisition of the closely related TYLCSV, accumulation of viral DNA was not observed (Caciagli and Bosco 1997).

3.5.2 *Transcription*

Begomovirus transcription in its vector was assessed by quantifying selected gene transcripts of the monopartite TYLCV and the bipartite *Tomato mottle virus* (ToMoV), after feeding on virus-infected tomato plants and after subsequent transfer to cotton, a ToMoV and TYLCV non-host plant (Sinisterra et al. 2005). Real-time

RT-PCR was performed using specific primers for three ToMoV genes (AV1, BC1 and BV1) and three TYLCV genes (V1, V2 and C3). The ToMoV gene transcripts rapidly became undetectable in whiteflies following transfer from tomato to cotton, probably because degradation was not accompanied by new synthesis. On the other hand, TYLCV transcripts increased after transfer of whiteflies to cotton, and were readily detected after 7 days indicating active TYLCV transcription. Recently, RNAase-sensitive transcripts of the TYLCV CP gene were identified by *in situ* hybridization using short DNA oligonucleotides complementary to CP RNA. The transcripts were localized mostly to the filter chamber and the descending midgut of the whitefly digestive tract (Ghanim et al. 2009).

3.6 Effect of TYLCV on Longevity and Fertility of *B. tabaci*

TYLCV can be associated with the whitefly vector (B biotype) for the entire life of the vector. (Rubinstein and Czosnek 1997) while TYLCSV is undetectable after approximately 20 days (Caciagli and Bosco 1997). The long-term association of TYLCV with female *B. tabaci* was correlated with a decrease in longevity compared with non-viruliferous insects (Rubinstein and Czosnek 1997). Following a 48 h AAP on TYLCV-infected tomato plants insects reared on eggplant, a TYLCV non-host, the life span of the viruliferous insects was shorter by 5–7 days compared to that of non-viruliferous whiteflies (out of 28–32 days), depending on the time of the year the experiment was conducted . Similarly the long-term association of TYLCV with female *B. tabaci* was correlated with a decrease in fertility (Rubinstein and Czosnek 1997). Following a 48 h AAP on TYLCV-infected tomato plants, the mean number of eggs laid either on tomato or on eggplant during a 7 or 20 days long period decreased by 25–50% (depending on the age of the adult). The decrease in fertility was not observed during the first 24 h following AAP. The percentage of eggs that developed into instars was similar, whether they were laid by infected or non-infected insects. Therefore TYLCV influenced the number of eggs laid but not the emergence of the instars. These experiments were confirmed from observations from the field where TYLCV infection of tomato plants had a deleterious effect on the reproduction of *B. tabaci* (Lapidot et al. 2001).

In a similar experiment the effect of a TYLCV isolate from China (*Tomato yellow leaf curl China virus* TYLCCNV) on two *B. tabaci* biotypes (invasive B and local ZHJ1) was evaluated (Jiu et al. 2007). Following a 48 h AAP on TYLCCNV-infected tobacco plants longevity and fertility of viruliferous B and ZHJ1 insect biotypes on cotton decreased by 40% and 35% respectively. In the same study, the effect of another monopartite geminivirus, the *Tobacco curly shoot virus* (TobCSV) on the two biotypes was appraised following a 48 h AAP on TobCSV-infected tobacco and transfer to cotton plants. The results were just the opposite of those obtained with TYLCCNV. Viruliferous B biotype whiteflies exhibited higher longevity and fertility than non-viruliferous whiteflies, while the effect of TobCSV on ZHJ1 insects was minor.

The performance of the B and ZHJ1 whitefly biotypes on uninfected, TYLCCNV-infected (together with its DNAβ) and TYLCV-infected tomato plants was studied (Liu et al. 2009). The infection of tomato plants by either of the viruses had no or only marginal effects on the development, survival and fecundity of the B biotype. In contrast, survival and fecundity of the ZHJ1 biotype were significantly reduced on virus-infected plants compared to those on uninfected plants. Populations of the B biotype on uninfected and TYLCCNV-infected plants increased at similar rates, whereas population increase of the ZHJ1 biotype on TYLCCNV-infected plants was affected adversely. These asymmetric responses to virus infection of tomato plants between the B and ZHJ1 biotypes are likely to offer advantages to the B biotype in its invasion and displacement of the indigenous biotype.

The effect of TYLCV from Japan on the biology of the Q biotype has been studied lately (Matsuura and Hoshino 2009). In these experiments, the insects were constantly raised on infected or healthy tomato. There were no differences in the survival rate and fecundity of the Q biotype of *B. tabaci* between TYLCV-viruliferous and non-viruliferous individuals. Moreover, the proportion of third- and fourth-instar nymphs did not differ between infected and healthy tomato plants, suggesting that TYLCV symptom expression probably does not have a deleterious effect on the development of nymphal instars on infected tomato plants.

In contrast to TYLCV, the bipartite begomovirus ToMoV does not affect whitefly fertility (McKenzie 2002). Whiteflies of the B biotype infected with ToMoV deposited significantly more eggs on healthy tomato leaves than non-viruliferous whiteflies. There was no significant difference between viruliferous and non-viruliferous whiteflies for the number of adults emerged or the proportion of those adults surviving from the egg stage. There was no significant correlation between the number of eggs deposited per female and progeny survival rates on healthy tomato for whitefly infected with or without the virus. These observations indicate that some begomoviruses have deleterious effects on their insect host while others do not.

3.7 Horizontal and Vertical Transmission of TYLCV

3.7.1 Transovarial Transmission

Using PCR, Southern blot hybridization and transmission tests, we have found that TYLCV was transmitted to the progeny of viruliferous insects with various efficiency. Moreover the progeny of viruliferous insects was able to infect tomato test plants. Dissection and analysis of the reproductive system of viruliferous whiteflies showed that both the ovaries and the maturing eggs contained TYLCV DNA (Ghanim et al. 1998). The closely related TYLCSV was also found to be transmitted transovarially to the first generation progeny. TYLCSV was detected in eggs and nymphs as well as in adults of the first generation progeny (Bosco et al. 2004). However, in contrast to TYLCV, the adult progeny of viruliferous insects were unable to infect tomato plants. It is interesting to note that the same scientists found that TYLCV was detected neither in instars nor in adult progeny of viruliferous females. These divergent results may be due to intrinsic differences in the highly inbred insect colonies raised in the laboratory and used in these experiments. The way in which TYLCV (Ghanim et al. 1998) and TYLCSV (Bosco et al. 2004) enter the whitefly reproductive system is unknown. It is possible that during the maturation of eggs in the ovaries, geminiviral particles penetrate the egg together with the endosymbionts, via an aperture in the membrane (Costa et al. 1995). Invading TYLCV may affect the development of some of the eggs, causing a decrease in fertility (Rubinstein and Czosnek 1997; Jiu et al. 2007; Liu et al. 2009). The vertical transmission of TYLCV and TYLCCNV by the B and Q biotypes of *B. tabaci* was studied using virus isolates and whitefly colonies established in China (Wang et al. 2010). Virus DNA was detected in eggs and nymphs but not in the adults of the first generation progeny, except in the combination of TYLCV and Q biotype whitefly where only about 3% of the adults contained the virus DNA. The offspring adults produced by viruliferous females did not transmit the viruses to plants. These results differed from those reported previously (Ghanim et al. 1998; Bosco et al. 2004).

3.7.2 *Transmission During Mating*

TYLCV can be transmitted between whiteflies of the B biotype in a sex-dependent manner, in the absence of any other source of the virus (Ghanim and Czosnek 2000). TYLCV was transmitted from viruliferous males to non-viruliferous females and from viruliferous females to non-viruliferous males, but not between insects of the same sex. Transmission took place when insects were caged in groups or in couples, in a feeding chamber or on TYLCV non-host cotton plants. Both viruliferous male and female whiteflies can transmit TYLCV to their counterparts; there was no significant difference in the efficiency of viral transmission between the two sexes. Transmission of TYLCV in a gender-related manner was not exclusive to the *B. tabaci* B biotype, but was also shared with the Q biotype, indicating that this biological feature might be widely shared among whiteflies (Ghanim et al. 2007). The bipartite begomoviruses SLCV and WmCSV were shown also to be transmitted horizontally among whiteflies of the B biotype with an efficacy similar to that of TYLCV. The horizontal transmission of TYLCV and TYLCCNV by the B and Q biotypes of *B. tabaci* was studied (Wang et al. 2010). Both TYLCV DNA and TYLCCNV DNA were shown to be transmitted horizontally by each of the two biotypes of the whitefly, but frequency of transmission was usually low. The overall percentage of horizontal transmission for either TYLCCNV or TYLCV in each of the two whitefly biotypes was below 5%. Neither virus species nor whitefly biotypes had a significant effect on the frequency of transmission.

The haemolymph plays a primordial role in the transmission of TYLCV amongst *B. tabaci* individuals of opposite gender. TYLCV was first detected in the haemolymph of the recipient insects about 1.5 h after caging, but was detected neither in the midgut nor in the head at this time. From there, TYLCV followed the pathway associated with acquisition from infected plants and did not cross the gut membranes back into the digestive system (Ghanim et al. 2001, 2007). Hence TYLCV passes from one insect to another by exchange of fluids accompanying intercourse, and reaches the open blood circulative system of the sexual partners. Mating was obligatory in order for TYLCV to pass from one insect to another. Caging together *B. tabaci* and *Trialeurodes vaporariorum*, two whitefly species that do not mate, confirmed that mating is obligatory for TYLCV transmission. The virus ingested by *B. tabaci* was not detected in *T. vaporariorum*, and the virus ingested by *T. vaporariorum* was not found in *B. tabaci*. It has to be noted that while TYLCV is found in the haemolymph of *B. tabaci* after feeding on infected tomato plants, the virus is ingested by *T. vaporariorum*, but it is unable to cross the gut/haemolymph barrier (Czosnek et al. 2002) probably because the later insect does not possess the begomoviral receptors that allow virus to cross the gut wall. Interestingly, TYLCV was not transmitted when individuals from the B biotypes where caged with individuals from the Q biotype (Ghanim et al. 2007) indicating that B and Q biotypes do not mate (Pascual and Callejas 2004).

3.8 Conclusions

3.8.1 TYLCV-Whitefly Co-adaptation

Independent but converging pieces of information suggest that whiteflies and begomoviruses have interacted for geological times (Czosnek and Brown 2010). It is unavoidable that during this long-lasting virus-vector relationship, on the one hand the virus has developed a configuration that ensures both its survival and its efficient transmission by the whitefly host, and on the other hand the insect has elaborated strategies that ensure its safeguard from possible deleterious effects of the virus. Only one whitefly species, *B. tabaci*, transmits begomoviruses, and only one viral protein is indispensible for efficient transmission, the coat protein. The adaptation of the local vector to the local begomovirus (Brown and Idris 2005) is reflected in the parameters of acquisition and transmission. Heterologous (from different regions) whitefly and TYLCV combinations result in much longer acquisition feeding periods and sometimes in a dramatic decrease in the efficiency of virus transmission (McGrath and Harrison 1995). Evolution of begomoviruses might have been towards a better adaptation of the capsid to putative receptors of the local whitefly to ensure optimization of virus transmission.

The long-lasting interactions between whiteflies and begomoviruses have been disturbed during the last 30 years, due to international transportation of produce and plant smuggling. New invasive whitefly biotypes have first coexisted with endogenous biotypes and have displaced them later on. For example the invasion of the Southern USA by the Middle Eastern B biotype, followed by the Q biotype, has almost totally eliminated the endogenous A biotype (Brown 2007). The Q biotype has appeared in the Middle East about 10 years ago and for the time being co-exists with the B biotype (Horowitz et al. 2003). Similarly to whiteflies, TYLCV species have moved new locations. The Middle Eastern TYLCV has been reported in Spain in 1999 (Sánchez-Campos et al. 1999), in the late 1990s in the USA (Polston and Anderson 1997), and in 2003 in Italy (Accotto et al. 2003). On the other hand, Spanish and Italian TYLCV isolates have been identified in the Middle East (Anfoka et al. 2008). In addition to the dissemination of TYLCV species, begomoviruses have a propensity to recombine (Padidam et al. 1999; Fauquet et al. 2005). The natural recombinant detected between TYLCV and TYLCSV in Spain (Navas-Castillo et al. 2000; Monci et al. 2002) demonstrated that the probability of such an occurrence and that the emergence of new recombinant begomoviruses, possibly with increased virulence, is high. Variations in whitefly populations and TYLCV species have disturbed the long-lasting co-evolution of the endogenous TYLCV-whitefly complex. This reshuffling may be reflected in the diversity of results obtained when studying virus-insect interaction, transmission, vertical and horizontal transmissions.

Like many insects, whiteflies house several bacterial endosymbionts (Bauman et al. 1993), including the obligate primary symbiont *Portiera aleyrodidarum* and the facultative secondary symbionts *Arsenophonus, Cardinium, Fritschea, Hamiltonella, Rickettsia,* and *Wolbachia.* Some of these symbionts are biotype specific. In Israel, *Hamiltonella* has been found only in the B biotype, whereas *Arsenophonus* and *Wolbachia* have been found only in the Q biotype. *Rickettsia* was highly prevalent, but not fixed, in both of those biotypes (Chiel et al. 2007). The roles these symbionts play are still unknown although they may influence fitness (Chiel et al. 2009) and TYLCV transmission (Gottlieb et al. 2010). GroEL produced by some endosymbionts were shown to be necessary for TYLCV transmission (Morin et al. 1999, 2000). It is possible that GroELs encoded by the genome of other endosymbionts also play a role in virus transmission. The composition of the bacterial fauna may underlie the adaptation of exogenous whiteflies to new environment and facilitate expansion and enhance efficacy of TYLCV transmission.

3.8.2 *The Whitefly Functional Genomics Project*

As the begomoviral capsid evolved toward a better adaptation of the virus to the insect, some whitefly species may have developed receptors in their digestive and salivary systems that facilitate and optimize begomovirus translocation. The question remains why whiteflies have developed a system that allows the circulative transmission of potentially harmful begomoviruses, as have many circulative plant viruses, instead of confining the virus to the stylet or destroying the virus in the digestive system. Indeed, available data suggest that members of the TYLCV family are reminiscent of insect pathogens and may be deleterious to their whitefly vector (see above).

A functional genomics approach has been taken to understand the patterns of gene expression during whitefly development and during association of whiteflies with begomoviruses. We have constructed three cDNA libraries for non-viruliferous whiteflies (eggs, immature instars, and adults) and two from adult insects that fed on tomato plants infected by two geminiviruses: the monopartite TYLCV and the bipartite ToMoV. The sequence of approximately 20,000 clones has been determined (Leshkowitz et al. 2006). Comparisons with public databases indicated that the libraries contained genes involved in cellular and developmental processes. A cDNA microarray was constructed which represents about 6,000 contigs and singletons. The microarray has been used to study resistance to insecticides (Ghanim and Kontsedalov 2007), the immune response accompanying parasitization by the wasp *Eretmocerus mundus* (Mahadav et al. 2008) and differential heat response of B and Q whitefly biotypes (Mahadav et al. 2009). Genome-based functional genomics will be instrumental in (1) studying the interactions underlying the circulative transmission of begomoviruses within vector and non-vector whitefly species, (2) identifying the cellular determinants involved in transmission, and (3) deciphering the evolutionary history of begomovirus-whitefly complexes. Recently, Illumina sequencing was used to analyze the whitefly transcriptome during the insect development (Wang et al. 2010), greatly enhancing our knowledge of gene expression in this insect and providing a most useful database for further studies.

References

Accotto GP, Bragaloni M, Luison D, Davino S, Davino M (2003) First report of *Tomato yellow leaf curl virus* (TYLCV) in Italy. New disease reports. http://bspp.org.uk/publications/new-disease-reports/ndr.php?id=007021

Anfoka G, Abhary M, Haj Ahmad F, Hussein AF, Rezk A, Akad F, Abou-Jawdah Y, Lapidot M, Vidavski F, Nakhla MK, Sobh H, Atamian H, Cohen L, Sobol I, Mazyad H, Maxwell DP, Czosnek H (2008) Survey of tomato yellow leaf curl disease – associated viruses in the Eastern Mediterranean basin. J Plant Pathol 90:311–320

Atzmon G, van Hoss H, Czosnek H (1998) PCR-amplification of tomato yellow leaf curl virus (TYLCV) from squashes of plants and insect vectors: application to the study of TYLCV acquisition and transmission. Eur J Plant Pathol 104:189–194

Bauman P, Munson MA, Lai C-Y, Clark MA, Baumann L, Moran NA, Campbell BC (1993) Origin and properties of bacterial endosymbionts of aphids, whiteflies, and mealybugs. ASM News 5:21–24

Bosco D, Mason G, Accotto GP (2004) TYLCSV DNA, but not infectivity, can be transovarially inherited by the progeny of the whitefly vector *Bemisia tabaci* (Gennadius). Virology 323:276–283

Böttcher B, Unseld S, Ceulemans H, Russell RB, Jeske H (2004) Geminate structures of *African cassava mosaic virus*. J Virol 78:6758–6765

Brown JK (2007) The *Bemisia tabaci* complex: genetic and phenotypic variation and relevance to TYLCV-vector interactions. In: Czosnek H (ed.) Tomato yellow leaf curl virus disease: management, molecular biology, breeding for resistance. Springer, Dordrecht

Brown JK, Czosnek H (2002) Whitefly transmission of plant viruses. In: Plumb RT (ed.) Advances in botanical research. Plant virus vector interactions. Academic, New York

Brown JK, Idris AM (2005) Genetic differentiation of whitefly *Bemisia tabaci* mitochondrial cytochrome oxidase I, and phylogeographic concordance with the coat protein of the plant virus genus *Begomovirus*. Ann Entomol Soc Am 98:827–837

Caciagli P, Bosco D (1997) Quantitation over time of tomato yellow leaf curl geminivirus DNA in its whitefly vector. Phytopathology 87:610–613

Caciagli P, Bosco D, Al-Bitar L (1995) Relationships of the Sardinian isolate of tomato yellow leaf curl geminivirus with its whitefly vector *Bemisia tabaci* Gen. Eur J Plant Pathol 101:163–170

Caciagli P, Medina Piles V, Marian D, Vecchiati M, Masenga V, Mason G, Falcioni T, Noris E (2009) Virion stability is important for the circulative transmission of *Tomato yellow leaf curl sardinia virus* by *Bemisia tabaci*, but virion access to salivary glands does not guarantee transmissibility. J Virol 83:5784–5795

Chiel E, Gottlieb Y, Zchori-Fein E, Mozes-Daube N, Katzir N, Inbar M, Ghanim M (2007) Biotype-dependent secondary symbiont communities in sympatric populations *of Bemisia tabaci*. Bull Entomol Res 97:407–413

Chiel E, Inbar M, Mozes-Daube N, White JA, Hunter MS, Zchori-Fein E (2009) Assessments of fitness effects by the facultative symbiont *Rickettsia* in the sweetpotato whitefly (Hemiptera: Aleyrodidae). Ann Entomol Soc Am 102:413–418

Cohen S, Harpaz I (1964) Periodic, rather than continual acquisition of a new tomato virus by its vector, the tobacco whitefly (*Bemisia tabaci* Gennadius). Entomol Exp Appl 7:155–166

Cohen S, Nitzany FE (1966) Transmission and host range of the tomato yellow leaf curl virus. Phytopathology 56:1127–1131

Costa HS, Westcot DM, Ullman DE, Rosell RC, Brown JK, Johnson MW (1995) Morphological variation in *Bemisia* endosymbionts. Protoplasma 189:194–202

Czosnek H (2007) Tomato yellow leaf curl virus disease: management, molecular biology, breeding for resistance. Springer, Dordrecht

Czosnek H, Brown J (2010) The whitefly genome – white paper: proposal to sequence multiple genomes of *Bemisia tabaci*. In: Stansly PA, Naranjo SE (eds.) *Bemisia*: bionomics and management of a global pest. 540 pages. Springer, Dordrecht, The Netherlands, pp 503–532

Czosnek H, Ghanim M, Rubinstein G, Morin S, Fridman V, Zeidan M (2001) Whiteflies: vectors – or victims ? – of geminiviruses. In: Maramorosch K (ed.) Advances in virus research. Academic, New York

Czosnek H, Ghanim M, Ghanim M (2002) Circulative pathway of begomoviruses in the whitefly vector *Bemisia tabaci* – insights from studies with *Tomato yellow leaf curl virus*. Ann Appl Biol 140:215–231

Fauquet CM, Sawyer S, Idris AM, Brown JK (2005) Sequence analysis and classification of apparent recombinant begomoviruses infecting tomato in the Nile and Mediterranean basins. Phytopathology 95:549–555

Fauquet CM, Briddon RW, Brown JK, Moriones E, Stanley J, Zerbini M, Zhou X (2008) Geminivirus strain demarcation and nomenclature. Arch Virol 153:783–821

Ghanim M, Czosnek H (2000) Tomato yellow leaf curl geminivirus (TYLCV-Is) is transmitted among whiteflies (*Bemisia tabaci*) in a sex-related manner. J Virol 74:4738–4745

Ghanim M, Kontsedalov S (2007) Gene expression in pyriproxyfen-resistant *Bemisia tabaci* Q biotype. Pest Manag Sci 63:776–783

Ghanim M, Medina V (2007) Localization of tomato yellow leaf curl virus in its whitefly vector *Bemisia tabaci*. In: Czosnek H (ed.) Tomato yellow leaf curl virus disease: management, molecular biology, breeding for resistance. Springer, Dordrecht, pp 171–183

Ghanim M, Morin S, Zeidan M, Czosnek H (1998) Evidence for transovarial transmission of tomato yellow leaf curl virus by its vector the whitefly *Bemisia tabaci*. Virology 240:295–303

Ghanim M, Morin S, Czosnek H (2001) Rate of *tomato yellow leaf curl virus* (TYLCV) Translocation in the circulative transmission pathway of its vector, the whitefly *Bemisia tabaci*. Phytopathology 91:188–196

Ghanim M, Sobol I, Ghanim M, Czosnek H (2007) Horizontal transmission of begomoviruses between *Bemisia tabaci* biotypes. Arthropod Plant Inter 1:195–204

Ghanim M, Brumin M, Popovski S (2009) A simple, rapid and inexpensive method for localization of *Tomato yellow leaf curl virus* and *Potato leafroll virus* in plant and insect vectors. J Virol Methods 159:311–314

Glick E, Levy Y, Gafni Y (2009) The viral etiology of tomato yellow leaf curl disease – a review. Plant Prot Sci 45:81–97

Goodman RM (1977) Single-stranded DNA genome in a whitefly-transmitted plant virus. Virology 83:171–179

Gottlieb Y, Zchori-Fein E, Mozes-Daube N, Kontsedalov S, Skaljac M, Brumin N, Sobol I, Czosnek H, Vavre F, Fleury F, Ghanim M (2010) The transmission efficiency of *tomato yellow leaf curl virus* is correlated with the presence of a specific symbiotic bacterium species. J Virol 84:9310–9317

Gronenborn B (2007) The tomato yellow leaf curl virus: genome and function of its proteins. In: Czosnek H (ed.) Tomato yellow leaf curl virus disease: management, molecular biology and breeding for resistance. Springer, Dordrecht

Hallan V, Gafni Y (2001) Tomato yellow leaf curl virus (TYLCV) capsid protein (CP) subunit interactions: implications for viral assembly. Arch Virol 146:1765–1773

Harrison BD, Barker H, Bock KR, Guthrie EJ, Meredith G, Atkinson M (1977) Plant viruses with circular single-stranded DNA. Nature 270:760–762

Höfer P, Bedford ID, Markham PG, Jeske H, Frischmuth T (1997) Coat protein gene replacement results in whitefly transmission of an insect nontransmissible geminivirus isolate. Virology 236:288–295

Höhnle M, Höfer P, Bedford ID, Briddon RW, Markham PG, Frischmuth T (2001) Exchange of three amino acids in the coat protein results in efficient whitefly transmission of a nontransmissible *Abutilon mosaic virus* isolate. Virology 290:164–171

Horowitz AR, Denholm I, Gorman K, Cenis JL, Kontsedalov S, Ishaaya I (2003) Biotype Q of *Bemisia tabaci* identified in Israel. Phytoparasitica 31:94–98

Ioannou N (1985) Yellow leaf curl and other diseases of tomato in Cyprus. Plant Pathol 345:428–434

Jiang YX, de Blas C, Barrios L, Fereres A (2000) A Correlation between whitefly (Homoptera: Aleyrodidae) feeding behavior and transmission of tomato yellow leaf curl virus. Ann Entomol Soc Am 93:573–579

Jiang YX, Dde Blas C, Bedford ID, Nombela G, Muñiz M (2004) Effect of *Bemisia tabaci* biotype in the transmission of *tomato yellow leaf curl sardinia virus* (TYLCSV-ES) between tomato and common weeds. Span J Agric Res 2:115–119

Jiu M, Zhou XP, Tong L, Xu J, Yang X, Wan FH, Liu SS (2007) Vector-virus mutualism accelerates population increase of an invasive whitefly. PLoS ONE 2:e182

Kheyr-Pour A, Bananej K, Dafalla GA, Caciagli P, Noris E, Ahoonmanesh A, Lecoq H, Gronenborn B (2000) *Watermelon chlorotic stunt virus* from the Sudan and Iran: Sequence comparisons and identification of a whitefly-transmission determinant. Phytopathology 90:629–635

Kunik T, Palanichelvam K, Czosnek H, Citovsky V, Gafni Y (1998) Nuclear import of a geminivirus capsid protein in plant and insect cells: implications for the viral nuclear entry. Plant J 13:121–129

Lapidot M, Friedmann M, Pilowsky M, Ben-Joseph R, Cohen S (2001) Effect of host plant resistance to *tomato yellow leaf curl virus* (TYLCV) on virus acquisition and transmission by its whitefly vector. Phytopathology 91:1209–1213

Leshkowitz D, Gazit S, Reuveni E, Ghanim M, Czosnek H, McKenzie C, Shatters RG Jr, Brown JK (2006) Whitefly (*Bemisia tabaci*) genome project: analysis of sequenced clones from egg, instar, and adult (viruliferous and non-viruliferous) cDNA libraries. BMC Genomics 7:79

Liu J, Zhao H, Jiang K, Zhou X-P, Liu SS (2009) Differential indirect effects of two plant viruses on an invasive and an indigenous whitefly vector: implications for competitive displacement. Ann Appl Biol 155:39–448

Mahadav A, Gerling D, Gottlieb Y, Czosnek H, Ghanim M (2008) Gene expression in the whitefly *Bemisia tabaci* pupae in response to parasitization by the wasp *Eretmocerus mundus*. BMC Genomics 9:342

Mahadav A, Kontsedalov S, Czosnek H, Ghanim M (2009) Thermotolerance and gene expression following heat stress in the whitefly *Bemisia tabaci* B and Q biotypes. Insect Biochem Mol Biol 39:668–676

Mansour A, Al-Musa A (1992) Tomato yellow leaf curl virus: host range and vector-virus relationships. Plant Pathol 41:122–125

Matsuura S, Hoshino S (2009) Effect of tomato yellow leaf curl disease on reproduction of *Bemisia tabaci* Q biotype (Hemiptera: Aleyrodidae) on tomato plants. Appl Entomol Zool 44:143–148

McGrath PF, Harrison BD (1995) Transmission of tomato leaf curl geminivirus by *Bemisia tabaci* effects of virus isolate and vector biotype. Ann Appl Biol 126:307–316

McKenzie CL (2002) Effect of tomato mottle virus (ToMoV) on *Bemisia tabaci* biotype B (Homoptera: Aleyrodidae) oviposition and adult survivorship on healthy tomato. Fla Entomol 85:367–368

Medina V, Pinner MS, Bedford ID, Achon MA, Gemeno C, Markham PG (2006) Immunolocalization of tomato yellow leaf curl Sardinia virus in natural host plants and its vector *Bemisia tabaci*. J Plant Pathol 88:299–308

Mehta P, Wyman JA, Nakhla MK, Maxwell DP (1994) Transmission of tomato yellow leaf curl geminivirus by *Bemisia tabaci* (Homoptera: Aleyrodidae). J Econ Entomol 87:1291–1297

Monci F, Sanchez-Campos S, Navas-Castillo J, Moriones E (2002) A natural recombinant between the geminiviruses tomato yellow leaf curl Sardinia virus and tomato yellow leaf curl virus exhibits a novel pathogenic phenotype and is becoming prevalent in Spanish populations. Virology 303:317–326

Morin S, Ghanim M, Zeidan M, Czosnek H, Verbeek M, van den Heuvel JFJM (1999) A GroEL homologue from endosymbiotic bacteria of the whitefly *Bemisia tabaci* is implicated in the circulative transmission of *tomato yellow leaf curl virus*. Virology 30:75–84

Morin S, Ghanim M, Sobol I, Czosnek H (2000) The GroEL protein of the whitefly *Bemisia tabaci* interacts with the coat protein of transmissible and non-transmissible begomoviruses in the yeast two-hybrid system. Virology 276:404–416

Navas-Castillo J, Sanchez-Campos S, Noris E, Louro D, Accotto GP, Moriones E (2000) Natural recombination between *tomato yellow leaf curl virus-Is* and *tomato leaf curl virus*. J Gen Virol 81:2797–2801

Navot N, Pichersky E, Zeidan M, Zamir D, Czosnek H (1991) Tomato yellow leaf curl virus: a whitefly-transmitted geminivirus with a single genomic component. Virology 185:151–161

Navot N, Zeidan M, Pichersky E, Zamir D, Czosnek H (1992) Use of polymerase chain reaction to amplify tomato yellow leaf curl virus DNA from infected plants and viruliferous whiteflies. Phytopathology 82:1199–1202

Noris E, Vaira AM, Caciagli P, Masenga V, Gronenborn B, Accotto GP (1998) Amino acids in the capsid protein of tomato yellow leaf curl virus that are crucial for systemic infection, particle formation, and insect transmission. J Virol 72:10050–10057

Ohnesorge S, Bejarano ER (2009) Begomovirus coat protein interacts with a small heatshock protein of its transmission vector (*Bemisia tabaci*). Insect Mol Biol 18:693–703

Padidam M, Sawyer S, Fauquet CM (1999) Possible emergence of new geminiviruses by frequent recombination. Virology 265:218–225

Pascual S, Callejas C (2004) Intra- and interspecific competition between biotypes B and Q of *Bemisia tabaci* (Hemiptera: Aleyrodidae) from Spain. Bull Entomol Res 94:369–375

Picó B, Ferriol M, Diez MJ, Nuez F (1999) Developing tomato breeding lines resistant to tomato yellow leaf curl virus. Plant Breed 118:537–542

Pollard DG (1955) Feeding habits of the cotton whitefly. Ann Appl Biol 43:664–671

Polston JE, Anderson PK (1997) The emergence of whitefly-transmitted geminiviruses in tomato in the western hemisphere. Plant Dis 81:1358–1369

Rosell RC, Torres-Jerez I, Brown JK (1999) Tracing the geminivirus-whitefly transmission pathway by polymerase chain reaction in whitefly extracts, saliva, hemolymph, and honeydew. Phytopathology 89:239–246

Rubinstein G, Czosnek H (1997) Long-term association of tomato yellow leaf curl virus (TYLCV) with its whitefly vector *Bemisia tabaci*: effect on the insect transmission capacity, longevity and fecundity. J Gen Virol 78:2683–2689

Sánchez-Campos S, Navas-Castillo J, Camero R, Soria C, Díaz JA, Moriones E (1999) Displacement of tomato yellow leaf curl virus (TYLCV)-Sr by TYLCV-Is in tomato epidemics in Spain. Phytopathology 89:1038–1043

Sinisterra XH, McKenzie CL, Hunter WB, Powell CA, Shatters RG Jr (2005) Differential transcriptional activity of plant-pathogenic begomoviruses in their whitefly vector (*Bemisia tabaci*, Gennadius: Hemiptera Aleyrodidae). J Gen Virol 86:1525–1532

van den Heuvel JFJM, Verbeek M, van der Wilk F (1994) Endosymbiotic bacteria associated with circulative transmission of potato leafroll virus by *Myzus persicae*. J Gen Virol 75:2559–2565

van Regenmortel MHV, Fauquet CM, Bishop DHL, Carstens EB, Estes MK, Lemon SM, Maniloff J, Mayo MA, McGeoch DJ, Pringle CR, Wickner RB (2000) Virus taxonomy: the classification and nomenclature of viruses. The seventh report of the international committee on taxonomy of viruses. Academic, San Diego, 1167 pp

Wang J, Zhao H, Liu J, Jiu M, Qian Y-J, Liu S-S (2010) Low frequency of horizontal and vertical transmission of two begomoviruses through whiteflies exhibits little relevance to the vector infectivity. Ann Appl Biol 157:125–133

Wang X-W, Luan J-B, Li J0M, Bao Y-Y, Zhang C-X, Liu S-S (2010) *De novo* characterization of a whitefly transcriptome and analysis of its gene expression during development. BMC Genomics 11:400

Zeidan M, Czosnek H (1991) Acquisition of tomato yellow leaf curl virus by the whitefly *Bemisia tabaci*. J Gen Virol 72:2607–2614

Zhang W, Olson NH, Baker TS, Faulkner L, Agbandje-McKenna M, Boulton MI, Davis JW, McKenna R (2001) Structure of the *Maize streak virus* geminate particle. Virology 279:471–477

Zrachya A, Glick E, Levy Y, Arazi T, Citovsky V, Gafni Y (2007) Suppressor of RNA silencing encoded by *tomato yellow leaf curl virus*-Israel. Virology 358:159–165

Chapter 4
Bemisia tabaci Interaction with *Cotton Leaf Curl Virus*

R.S. Mann

Abstract Cotton leaf curl disease is the major biotic threat to cotton production in India and Pakistan. The typical symptoms of the disease on cotton and several other malvaceous and solanaceous plant species include vein swelling and formation of leaf-like outgrowths called enations. Severely infected leaves show rolling with spirally twisted leaf petioles, branches and the main stem. Severely infected plants become stunted in growth resulting in high yield losses and poor fiber quality. In association with essential disease-specific satellite components beta DNA satellite and a nonfunctional alpha satellite the disease is caused by a group of whitefly-transmitted Geminiviruses belonging to the genus *Begomovirus* of the family *Geminiviridae*. Like other Begomoviruses *Cotton leaf curl virus* (CLCuV) is exclusively transmitted by a whitefly vector, *Bemisia tabaci* in a circulative, persistent manner, and can be retained from few days to the entire life period of the whitefly. The virus is not transmitted transovarially by the vector. Whiteflies require an acquisition threshold period ranging from 15 min to 4 h and an inoculation threshold period of 5 min to 1 h to successfully transmit the virus depending upon the virus strain, host plant, vector behavior and the abiotic factors. Up to 8 h latent period is also required between the acquisition and inoculation periods for successful transmission of the virus to new plants. The transmission efficiency is greatly improved with an increase in acquisition and inoculation periods and number of whiteflies. The virus has mixed effects on the biology and behavior of whiteflies. Despite the importance of virus-vector interactions in disease epidemiology and spread the studies on interactions between CLCuV and *B. tabaci* are still in infancy. This chapter reviews the basic interactions between this very important virus-vector system.

R.S. Mann (✉)
Department of Entomology and Nematology, University of Florida,
Bldg. 970, Natural Area Drive, Gainesville, FL 32611-0620, USA
e-mail: mannrs@ufl.edu

W.M.O. Thompson (ed.), *The Whitefly, Bemisia tabaci (Homoptera: Aleyrodidae) Interaction with Geminivirus-Infected Host Plants*,
DOI 10.1007/978-94-007-1524-0_4,

4.1 Overview

Cotton (*Gossypium* spp.) is an important cash crop grown for textile, fiber, and edible oil worldwide. The genus *Gossypium* is composed of 50 species, of which only four are cultivated. Among the cultivated species, *G. hirsutum* L. is by far the most important, accounting for more than 85% of total global lint production. The cotton plant is attacked by a wide range of economically important insect and diseases including viral diseases caused by *Cotton leaf curl virus* (CLCuV), *Cotton leaf crumple virus* (CLCrV) and *Cotton Leaf mosaic virus* (CotLM) (Akhtar et al. 2002). Cotton leaf curl disease (CLCuD) is one of the most destructive diseases of cotton in Pakistan and India. Although CLCuD has been known in Pakistan from 1967 (Mahmood 1999; Thakur 2002) the rapid spread of the disease started in 1988 that reduced cotton production by 7.4 million cotton bales accounting for 4.98 billion US$ loss (Mansoor et al. 1999). The disease has rapidly spread to new areas including India and China (Varma et al. 1993; Mansoor et al. 2006; Cai et al. 2010). In India, CLCuD was first noticed in 1989 in some exotic accessions of germplasm of *Gossypium barbadense* in the experimental fields of the Indian Agricultural Research Institute (IARI), New Delhi (Varma 1990; Varma et al. 1993). However, early detection of disease and regular removal of symptomatic plants coupled with avoidance of cultivation of *G. barbadense* restricted its incidence in only IARI fields (Varma and Malathi 2003). The disease was detected in epidemic forms in Rajasthan in 1993 and then in Punjab in 1994 (Singh et al. 1994). Initially the disease was restricted to the *G. hirsutum* varieties planted close to the Pakistan border areas (Varma and Malathi 2003; Monga et al. 2007), however, by 1998 the disease was widespread in Northern India and was responsible for 75% decline in cotton production in Indian Punjab. Currently, the disease is endemic throughout Pakistan and is widespread over the entire northern cotton-producing zone of India. Outside Asia, the disease has been reported in Cameroon, Central African Republic, Egypt, Ghana, Ivory Coast, Nigeria and Sudan. In Nigeria the disease was reported as early as 1912 (Farquharson 1912) and in Sudan in 1924 (Lambart 1924). Presently, CLCuD is one of the most important diseases of cotton in these countries and has the potential to cause a significant impact to cotton production (Brown and Bird 1992; Mahmood 1999; Akhtar et al. 2002). New introductions of CLCuD have recently been reported in China (Cai et al. 2010).

CLCuD is associated with a complex of viral species, including CLCuV. All the species induce similar symptoms on cotton plants. The first symptoms of CLCuV infection appear within 2–3 weeks of inoculation and are initially characterized by deep downward cupping of the youngest leaves. This is followed by either upward or downward curling of the leaf margins, darkening, swelling, and formation of leaf-like outgrowths on veins called "enations" which frequently develop into cup-shaped structures on the undersides of leaves. Appearance of the disease at the seedling stage seriously hampers flowering, boll formation, and maturation. There is a significant decrease in sympodial and monopodial branches in severely infected cotton plants. The severely infected plants show stunted growth, reduced seed cotton yield, and fiber quality (Kapur et al. 1994; Nateshan et al. 1996; Singh et al. 1994; Briddon 2003; Monga et al. 2004; Geering 2010). The economic losses due to CLCuD depend upon the cultivar, time and severity of infection. The virus infection in cotton plants is known to increase peroxidase activity, catechol, phenols, carotenoids, proteins, total sugars, chlorophyll, oil content, lipase enzyme content, and reduce Ca++ and K+ content (Kaur et al. 1998; Ashraf et al. 2004; Kang et al. 2003).

The host range of CLCuV includes *G. hirsutum* and *G. barbadense, Abutilon theophasti* (Nill), *Althea rosea* (Cav), *A. ficifolia, A. kurdica, A. nudiflora, A. pontica, A. sulpurea, Convolvulus arvensis* (L.), *Hibiscus cannabinus* (L.), *H. esculentus, H. ficulneus, H. huegelii, H. trionum, H. sabdariffa, Lavatera cretica* (L.), *Malva alcea* (L.), *M. silvestris, M. moschata, Malvaviscu arboreus* (Car.), *Pavonia hastate* (L.), *Nicotiana tobacum* L., *Rumex dentatus, Sida acuta* (Burm.), *S. alba, S. cordifolia* and *Cucurbit* spp. (Tarr 1957, 1964; Bink 1975; Cauquil and Follin 1983; Fauquet and Thouvenel 1987; Singh 2000; Akhtar et al. 2002; Kang et al. 2004). In addition the disease has been observed on several weeds (Briddon and Markham 2000; Kang et al. 2004; Akhtar et al. 2002). It has also been found on *Corchorus fascicularis* Lau., *Phyllanthus niruri* L., *Clitoria ternatea* L., and *Phaseolus vulgaris, Petunia* spp. and *Urena lobata* in the African subcontinent. The cotton species, *G. arboreum* and *G. herbaceum* are resistant to CLCuV (Cauquil and Follin 1983; Akhtar et al. 2002).

4.2 The Virus: CLCuV

Cotton leaf curl disease is mainly caused by *Cotton leaf curl virus* belonging to the genus *Begomovirus* of the family *Geminiviridae* (Mansoor et al. 2003). Geminiviruses are characterized by circular single stranded DNA (ssDNA) genomes encapsidated in twinned quasi isometric particles of about 18×30 nm. The family *Geminiviridae* consists of four genera of viruses namely, *Begomovirus, Curtovirus, Topocuvirus*, and *Mastrevirus*, based on genome organization, insect vector, and host range (Padidam et al. 1995). The begomoviruses, the largest and most economically significant group of plant viruses pose a serious threat to sustainable agriculture. During the last two decades, new *Begomovirus* species have emerged worldwide, probably as a consequence of the spread of one or more highly polyphagous biotypes of their insect vector, *Bemisia tabaci* (Gennadius) (Homoptera: Aleyrodidae) (Rybicki and Pietersen 1999). Usually multiple begomovirus species have emerged simultaneously in a given region, with the ensuing species complexes causing diseases in a wide variety of plant species, including many of economic importance(Lefeuvre et al. 2007, 2009). All the begomoviruses are transmitted by *B. tabaci* in a persistent manner (Briddon and Markham 2000; Fauquet and Stanley 2005; Seal et al. 2006) and infect only dicotyledonous plants (Briddon and Markham 2000).

In general the begomoviruses have bipartite genomes (designated as DNA-A and DNA-B), of about 2.5–3.0 kb length, encapsidated within twinned quasi-icosahedral particles (Mansoor et al. 2003). However, a small number of begomoviruses such as CLCuV, *Tomato yellow leaf curl virus* (TYLCV), *Ageratum yellow vein virus* (AYVV) etc. have a monopartite genome. In bipartite viruses, the genes required for virus replication and encapsidation by coat protein are encoded by DNA-A component while those required for movement (both intra and intercellular) occur on DNA-B (Murthi et al. 2007; Kumar et al. 2010). In monopartite begomoviruses, these genes are present in DNA-A. However, in recent years, a single stranded DNA component of 1,350 kb length designated as DNA- β (beta satellite) has been associated with monopartite viruses (Briddon et al. 2003; Mansoor et al. 2003; Kirthi et al. 2004; Radhakrishnan et al. 2004). The DNA-A component is not directly required for disease progression, however, DNA-β requires DNA-A for replication and encapsidation and for encoding pairs of proteins involved in inter- and intracellular movement of viruses and induction of characteristic leaf curl symptoms (Noueiry et al. 1994; Briddon et al. 2003; Mansoor et al. 2003). Another nanovirus like, 1.4 kb, circular single stranded DNA called alphasatellite (DNA-1) has been found associated with CLCuD. Although it is presumed that DNA-1 has no role in symptom expression, its exact function is still unclear. DNA-1 is also transmitted by *B. tabaci* (Saunders et al. 2002, 2005; Rahman et al. 2005). The DNA-A component is invariably found in all begomoviruses including CLCuV irrespective of their monopartite or bipartite genome, origin or prevalence (Nadeem et al. 1997; Zhou et al. 1997, 1998; Briddon et al. 2000; Idris and Brown 2000, 2002; Briddon 2003).

Recent studies from India and Pakistan have demonstrated a complex etiology of CLCuD involving CLCuV and a minimum of four other distinct and closely related, *Begomovirus* species namely: *Cotton leaf curl Alabad virus* (CLCuAV), *Cotton leaf curl Kokhran virus* (CLCuKV), *Cotton leaf curl Multan virus* (CLCuMV), and *Cotton leaf curl Rajasthan virus* (CLCuRV) (Nadeem et al. 1997; Zhou et al. 1998; Briddon and Markham 2000). In addition *Cotton leaf curl Bangalore virus* (CLCuBV), *Papaya leaf curl virus* (PaLCuV) and *Tomato leaf curl Bangalore virus* (ToLCBV) are also associated with the disease (Geering 2010). Furthermore, each virus species has several strains and variants identified on the basis of disease prevalence in specific geographic areas (Fauquet et al. 2008; Kumar et al. 2010). In Africa the disease is caused by *Cotton leaf curl Gezira virus* (CLCuGV) that has great diversity in their sequences from the Asian viruses that cause CLCuD (Briddon 2003; Geering 2010). Although, CLCuV is a distinct virus, it shows considerable identity with other disease causing begomoviruses on other plants such as tomato, okra and some weed spp. The epitope profiles of CLCuV-PK (CLCuMV) were indistinguishable from the profiles of viruses causing yellow vein disease of okra in India or Pakistan, or leaf curl of okra *(Abelmoschus esculentus), Hibiscus tiliaceus,* radish or sunflower in Pakistan (Harrison et al. 1997). The virus causing *Hibiscus leaf curl disease* (HLCuD) from Southern India and Pakistan showed 95–97% DNA-A nucleotide identity with CLCuMV (Rajeshwari et al. 2005; Mao et al. 2008). The virus of Yellow vein disease of *Digera arvensis* showed 98% nucleotide sequence identity with CLCuRV (Mubin et al. 2009). Similarly, *Tomato leaf curl virus* (ToLCV) in Pakistan showed 99% sequence identity with CLCuRV (Shahid et al. 2007, 2009).

4.3 The Vector: *Bemisia tabaci*

B. tabaci is one of the most serious cosmopolitan pests of field, vegetable and ornamental crops worldwide. This insect vector is a highly polyphagous pest that feeds on over 700 species of plants in 86 families (Greathead 1986; McAuslane 2009; Brown 2001; De Barro 1995; Jones 2003; Martin et al. 2000; Mugiira et al. 2008). It causes damage by feeding directly on plant sap, excreting honeydew that causes sooty mold, and transmitting more than 100 plant viruses (Brown 2001; De Barro 1995; Jones 2003; Martin et al. 2000; Mugiira et al. 2008).

Direct feeding damage is caused by piercing and sucking sap from the plant foliage. This causes weakening and early wilting of the plant and reduces plant growth rate and yield. Feeding damage also cause leaf chlorosis, leaf withering, premature dropping of leaves and plant death. Excretion of honeydew and deposition of sooty mold reduces photosynthesis as well as the yield and market value of crops (Byrne and Bellows 1991a, b; Jones 2003). Both *B. tabaci* adults and nymphs cause feeding damage, and transmit diseases (Avidov and Harpaz 1969; Mound and Halsey 1978; Costa and Brown 1991; Costa et al. 1996; Mann et al. 2008, 2009). However, nymphal stages are sedentary and do not contribute significantly to disease transmission.

In the past decade, whitefly-transmitted plant viruses have increased in prevalence and distribution because of agricultural intensification due to extensive monocultures, increased fertilizer application, pesticides, and/or irrigation (Thresh 1982; Matson et al. 1997; Varma and Malathi 2003; Seal et al. 2006; Morales and Anderson 2001; Xie and Zhou 2003; Seal et al. 2006). *B. tabaci* was first reported to be a serious pest of cotton in the late 1920 s and early 1930 s in northern India (presently in Pakistan) (Hussain and Trehan 1933). Subsequently, severe infestations on cotton were recorded on various crops in other parts of the world including Sudan, Iran, El Salvador, Mexico, Brazil , Turkey, Israel, Thailand, and the USA by 1981 (Basu 1995). Infested cotton plants are damaged not only by the direct feeding damage, but also by sooty mold (which reduces photosynthesis) and contamination of lint that result in reduced value of the textile and fiber (Naveed et al. 2007). In addition to transmitting CLCuV, *B. tabaci* transmits two other cotton diseases: cotton mosaic disease (CMD) and cotton leaf crumple disease (CLCrD). These reduce the vigor and growth of plants. Cotton mosaic disease is caused by *Cotton Leaf mosaic virus* (CotLM). The disease is prevalent in Africa, South America (Brazil, Columbia), Central America (Guatemala, Nicaragua and El-Salvador), India and Pakistan (Nour 1960; Nour and Nour 1964; Bink 1975; Watkins 1981; Cauquil and Follin 1983; Nelson et al. 1998; Akhtar et al. 2002). Cotton leaf crumple disease is caused by *Cotton leaf crumple virus* (CLCrV) in the USA, Mexico and Guatemala and is presumed to be distinct from *CLCuV* (Idris and Brown 2004; Geering 2010).

Morphologically, the adult *B. tabaci* are soft and whitish-yellow in color when they first emerge from their nymphal exuvia. Within few hours, their two pairs of wings become iridescent white due to the deposition of a powdery wax. The body remains light yellow with a light dusting of wax. The female measures 0.96 mm from the tip of the vertex (head) to the tip of the abdomen, while the male measures 0.82 mm (Mohanty and Basu 1987; Brar et al. 2005). The egg measures approximately 0.21 mm in length and 0.09 mm in width and is somewhat tapered towards the distal end. Eggs are oval and these are generally attached to the lower leaf surface by a short stalk. The eggs obtain moisture through this stalk. The color of the eggs is initially pearly white, but darkens over time. The number of eggs laid by *B. tabaci* females is highly variable, differing between populations, host plant species and environmental conditions (Gerling 1990). On cotton *B. tabaci* generally lays up to 300 eggs that hatch in 6–7 days depending upon the temperature (Butter and Vir 1991; Lynch and Simmons 1993; Simmons 1994; Simmons et al. 2000; Brar et al. 2005). B. *tabaci* develop from an egg to an adult through four nymphal instars. The first nymphal instar is called the crawler. It is oval in shape and measures approximately 0.27 mm in length and 0.14 mm in width. It is whitish-green in color and has two yellow spots, the mycetomes, visible in the abdomen through the integument. It molts to the second nymphal instar, usually in 2–11 days after eclosion (Mohanty and Basu 1987). The crawler is the only mobile immature form of whiteflies which move few centimeters in search of a feeding site before settling. However, they can move to another leaf on the same plant. The first, second and third instar nymphs are flattened and elliptical in shape, greenish-yellow in color, and range from 0.36 mm (second instar) to 0.66 mm (fourth instar) in length (Sharaf and Batta 1985; Brar et al. 2005). The mycetomes are yellow in color. The second and third nymphal instars each last about 2–11 days (Mohanty and Basu 1987; Brar et al. 2005). The fourth or red-eyed nymphal stage is sometimes called the "pupal stage" and is difficult to delineate from the pupal stage (Hussain 1931; Gill 1990; Byrne and Bellows 1991a, b; Salas and Mendoza 1995; Aneja 2000; Mann et al. 2009). Pupa has two very distinctive characters, namely the eyes and the caudal furrow. Eyes appear as two red spots constricted in the middle while the caudal furrow (the median longitudinal depression along the posterior abdominal dorsum between the distal portion of the vasiform orifice and the tip of the abdomen) appear only in the pupa (Byrne and Bellows 1991a). The fourth and red-eyed nymphal stages combined last for 3–11 days (Mohanty and Basu 1987). The second, third and fourth nymphal instars are immobile hence the immatures are thought to play little role in virus transmission. The adult emerges out of the pupal case through a T-shaped suture (Rao and Reddy 1989; Bellows et al. 1994; Hodges and Evans 2005). Adult longevity of *B. tabaci* can range from 10–15 days during summer to 30–60 days during winter (Gerling 1990). The females are comparatively longer lived than males (Azab et al. 1971). The initiation of courtship by female is allowed to males 10 h after pupal eclosion (Li et al. 1989) and the deposition of eggs occur singly on the underside of top and middle leaves of cotton plants (Hussain and Trehan 1933; Natarajan 1990). Both *B. tabaci* adults and immatures are phloem feeders with piercing and sucking mouthparts. In general whitefly stylets follow an intercellular path to pierce the leaf epidermal cells (Walker 1985). *B. tabaci* feeding behavior usually involves initial walking followed by probing to identify the host vascular bundles. *B. tabaci* takes 15–30 min to reach the phloem. Feeding duration and frequency is directly related to the host suitability. Feeding durations are longer on suitable hosts than on less desirable hosts (Lei et al. 1998). The duration and frequency of successful phloem feedings are strongly correlated to virus transmission efficiency depending on the acquisition and inoculation threshold periods of the viruses (Jiang et al. 2001).

Mouthparts and stylet structure of *B. tabaci* are similar to other phloem feeders with piercing–sucking mouth parts (Harris et al. 1995, 1996; Rosell et al. 1995; Hunter et al. 1996). *B. tabaci* utilize stylet bundle to extract plant fluids. The stylet bundle consists of maxillary and mandibular stylets. The maxillary stylets are interlocked to form a food canal in which plant fluids are transported into the body, and the salivary canal through which saliva is injected into the plant (Rosell et al. 1995, 1999). The maxillary stylets are surrounded by the mandibular stylets for support and cutting through plant tissues during probing. The salivary canal excretes saliva to smoothen the penetrating process into plant tissues. Plant fluids are transported into the insect when the stylet penetrates the plant tissue and reaches the phloem. The plant sap enters the stylets and is transferred to the precibarium, a narrow tube connecting the food canal to the cibarium. The pressure differences between the cavities of the plant cell and the insect's body transport the contents into the insect. The food canal extends into the cibarium and esophagus. The esophagus extends from head to the base of the abdomen. From the esophagus, food is pushed into the midgut. The esophagus and the anterior midgut join the anterior hindgut to form a filter chamber. The midgut then proceeds dorsocaudally to join the hindgut and then to the anus within the vasiform orifice (Harris et al. 1996; Czosnek et al. 2001; Ghanim et al. 2001a, b).

The investigations on *B. tabaci* organs and cells involved in transmission of CLCuV and other viruses is seriously lacking. The information on transmission of begomoviruses by *B. tabaci* comes primarily from TYLCV, *Tomato mottle virus* (ToMoV) and *Cabbage leaf curl virus* (CaLCV) (reviewed by Brown and Czosnek 2002). Molecular studies on TYLCV by several researchers have shown that begomovirus particles present in phloem are ingested by *B. tabaci* through the stylets. The virus particles then enter the esophagus and the digestive tract and penetrate the gut membranes into the haemolymph to reach the salivary glands and finally enter the salivary duct from where they are egested with the saliva (Hunter-Fujita et al. 1998; Rosell et al. 1999; Ghanim et al. 2001a, b; Czosnek et al. 2002; Sinisterra et al. 2005).

B. tabaci is believed to be a species complex, with a number of recognized biotypes. According to Basu (1995), severe field population outbreaks of *B. tabaci* in various parts of the world during the last decade and differences in the range of host plants in different localities suggest a broad range of genetic differences within and between populations. Currently, 24 *B. tabaci* biotypes have been identified clustered in seven groups (Perring 2001; Tsagkarakou et al. 2007). *B. tabaci* biotypes have diverse geographical distribution, chemical resistance, host specialization, degree of fecundity, ability to induce phytotoxic disorders upon feeding, host range, dispersal behaviors, prokaryotic endosymbiont composition, and competency for begomovirus transmission (Bird 1957; Bird and Maramorosch 1978; Bedford et al. 1994; Brown et al. 1995; McGrath and Harrison 1995; Basu 1995; Oliveira et al. 2001; Ahmed et al. 2009). B. *tabaci* complex significantly contributes to the emergence of new begomovirus problems in the world (McGrath and Harrison 1995; Varma and Malathi 2003). Although most biotypes can transmit a range of begomoviruses, they do so with differing efficiencies depending on the virus species, vector biotypes and host plants (Bedford et al. 1994; McGrath and Harrison 1995; Maruthi et al. 2002; Jiu et al. 2007).

4.4 *B. tabaci* CLCuV Interactions

Majority of plant viruses depend upon insect vectors for their survival and spread. Insects are considered to be the most important factors in plant virus disease spread. Approximately 80% of the plant viruses depend on insect vectors for transmission. Although much is known about the movement of viruses within the plant cell, little is known about their requirements and mechanisms in insect vectors. Furthermore, the relationships between plant viruses and insect vectors are complex and unique for every virus and insect species. The mouthparts of sucking insects are ideal for the inoculation of virus into plants (Costa 1976). Virus transmission by the vector starts with ingestion of the virus from infected plants and ends with successful transmission to uninfected plants. However, the transmission process is the result of a complex interaction between the host, vector and the pathogen. The rate at which the vector becomes viruliferous and capable of transmitting the virus to new plants depends on the insect species/biotype, the virus and plant species involved. Furthermore, the transmission process is influenced by the physiological status of both the host and the vector, and abiotic factors such as temperature and relative humidity.

Plant viruses are classified as non-persistent, semi-persistent and persistent depending on the period the vector can harbor infectious particles, which can range from minutes to hours (nonpersistent) to days (semipersistent) and to entire life and even inheritance by the insect progeny (persistent) (Hohn 2007). The persistent viruses are further classified as circulative and propagative. In circulative transmission, the viruses are ingested by the vector, travel through the gut wall into the haemolymph and then pass to the salivary glands where they can potentially be transmitted to other plants. They do not multiply in the vector and require longer acquisition and inoculation access periods. These viruses are retained up to 12 h to the entire lifetime of the vector. Between acquisition and inoculation, a latent period of one to several hours is necessary for persistent viruses to be transmitted (Kennedy et al. 1962; Hohn 2007; Moury et al. 2007). In propagative transmission the viruses replicate within the insect vector and are retained throughout the lifetime of the vector.

B. *tabaci* transmits plant viruses in seven distinct virus groups including: geminiviruses, closteroviruses, carlaviruses, potyviruses, nepoviruses, luteoviruses and a DNA-containing rod-shaped virus (Duffus 1987, 1996; Oliveira et al. 2001, 2003). The most economically significant of these are the geminiviruses. Among the geminiviruses, begomoviruses are the most economically significant group of plant viruses transmitted by *B. tabaci*. In general, begomoviruses are transmitted by *B. tabaci* in a persistent or semi-persistent manner.

B. tabaci is the only known vector of CLCuV. Both, the immobile larval stages of *B. tabaci* have the ability to acquire and transmit the virus. Whitefly transmission of CLCuV was experimentally confirmed by Kirkpatrick (1931) in Sudan, Hameed et al. (1994) in Pakistan, and Kapur et al. (1994) in India. CLCuV is not transmitted by aphids, fleas, beetles or jassids (Kirkpatrick 1931; Watkins 1981) and the disease is not spread through sap, seed, soil or mechanical means (Tarr 1957; Singh et al. 1994). However, the virus can be successfully transmitted through grafting (Tarr 1957; Singh et al. 1994; Akhtar et al. 2002).

4.4.1 Minimum Time Required for Successful Acquisition and Inoculation of CLCuV by B. tabaci

Whitefly mediated transmission of CLCuV to cotton plants showed an acquisition threshold period (minimum time to acquire the virus), of 15 min to 4 h depending upon the virus strain involved. The inoculation threshold period (minimum time required to inoculate the plant) ranges from 5 min to 1 h. In addition, up to 8 h latent period (period between acquisition and inoculation period) is required for successful transmission of the virus from infected plants to healthy plants (Nateshan et al. 1996; Mann and Singh 2004a).

Kirkpatrick (1931) reported that the virus is only transmitted by whiteflies and the whole process of acquiring the virus and infecting healthy plants was accomplished in 6.5 h. This included 3, 0.5 and 3 h acquisition, inoculation and latent period respectively. Although, the virus isolates in these studies were unknown, based on current advancements in virus identification in Nigeria and adjoining areas (Tiendrebeogo et al. 2010) it appears that the virus isolate referred to in this study was CLCuGV formerly described as, *Cotton leaf curl Sudan virus* (Stanley et al. 2004). Nateshan et al. (1996) reported that in Southern India CLCuV-K was transmitted between cotton plants by adult *B. tabaci* in a semi persistent manner. *B. tabaci* required a minimum 1 h to acquire the virus from diseased plants. The adults inoculated new plants in 5 min. However, increasing the acquisition access period from 1 h to 24 h, with an inoculation access period of 24 h increased the transmission from 20% to 87%. Similarly, increasing the inoculation access period from 5 min to 24 h increased the proportion of infected cotton seedlings from 13% to 73%. The authors were unable to demonstrate virus transmission with acquisition or inoculation access periods shorter than 1 h or 5 min, respectively. In addition to minimum acquisition and inoculation periods the whiteflies also required a latent period of 8 h between acquisition and inoculation periods, for successful transmission of virus to new plants. Mann and Singh (2004a) reported that in North India *B. tabaci* was able to acquire CLCuV in as low as 20 min and inoculate new plants in 10 min. Like with previous studies, transmission efficiency improved with the increase in acquisition and inoculation periods. The whiteflies required a latent period of 8 h to induce disease symptoms on newly infected cotton plants. Furthermore, percentage virus transmission increased with an increase in acquisition and inoculation access periods. In North India, Khan and Ahmad (2005) detected CLCuRV DNA in *B. tabaci* within 4 h of release on virus infected cotton plants and the whiteflies inoculated the new plants in 1 h. The authors did not explore latent period between the acquisition and the inoculation periods but provided an acquisition period of 24 h in inoculation studies that may have covered the required latent period before inoculation. Mann and Singh (2004a, b) did not verify the virus strain but the collection sites and identification of virus culture at a later stage by other researchers (Kang et al. 2004; Monga et al. 2005, 2007) suggested that the virus strain in those studies could be CLCuRV.

The acquisition and inoculation periods are also known to be influenced by the host plant species and the stage of development of host plants. Whiteflies required less time to acquire the virus from preferred hosts as compared to less preferred hosts. *B. tabaci* acquired the virus more quickly from younger plants as compared to older cotton plants (Mann and Singh 2004a, b). Similarly whiteflies inoculated preferred hosts more quickly as compared to less suitable hosts (Singh 2001). In addition to host suitability and plant age CLCuV acquisition is influenced by the severity of disease symptoms or the virus titer present in plants. *B. tabaci* acquired CLCuV more efficiently from heavily diseased plants than from less severely infected plants (Singh 2000, 2001).

Although, there are notable differences in the acquisition and inoculation periods, in all the above mentioned studies virus transmission efficiency increased with an increase in acquisition and inoculation periods. Khan and Ahmad (2005) attributed the differences in acquisition and inoculation access periods to the differences in vector feeding behavior, specificity of insect vectors on different hosts or the virus titer in different hosts. Seal et al. (2006) suggested that distribution of begomoviruses associated with CLCuD was governed predominantly by a specific vector population, rather than host plant or geographic origin. Furthermore, differences in acquisition and inoculation feeding could be due to variation in virus isolates or the physiology and age of the whiteflies or the host plants. Environmental factors such as temperature could also influence pathogen transmission by insect vectors (Daugherty et al. 2009). However, Czosnek et al. (2001) suggested that these parameters are similar for all begomoviruses and the variations are more likely due to experimental procedures.

4.4.2 CLCuV Persistence in *B. tabaci*

There are very few studies on persistence of CLCuV in *B. tabaci*. CLCuV persistence in *B. tabaci* has been reported from a few days to the entire life period.

Kirkpatrick (1931) reported that the virus was retained for several days to the entire life of *B. tabaci* and was not transmitted transovarially. However, serial transmission studies by Nateshan et al. (1996) of viruliferous *B. tabaci* adults on healthy plants from CLCuV infected plants showed that the virus was retained in *B. tabaci* adults for up to 9 d. Serial transmission studies by Mann and Singh (2004c) showed that *B. tabaci* adults could retain the virus for their entire lifespan. However, both authors showed that the serially transferred viruliferous whiteflies were not consistent in transmitting the virus to new plants i.e. the whiteflies transmitted the virus to new plants on day 1, 2, 5 or 8, but not on day 3, 4, 6 or 7. Reasons for this inconsistency are still unclear. Localization and pathway of CLCuV in cells and organs of *B. tabaci* may shed some light on such transmission irregularities.

4.4.3 Other Parameters Affecting B. tabaci CLCuV Interactions

A single whitefly is capable of acquiring and transmitting the virus to new plants that produce symptoms. Virus transmission is shown to be a direct function of the number of viruliferous whiteflies per plant in several virus transmission studies including CLCuV (Singh et al. 1994; Mann and Singh 2004a). Mann and Singh (2004a) demonstrated that when a single whitefly was allowed an acquisition access period of 24 h, 20% of the experimental whiteflies acquired the virus to induce CLCuV symptoms on healthy cotton plants. The percentage of virus transmission increased to 80, 87, 85, 90 and 88 when numbers of whiteflies were increased to 5, 10, 15, 20 and 25 per plant, respectively. Khan and Ahmad (2005) were unable to detect CLCuV symptoms on *N. benthamiana* plants when a single whitefly was allowed an acquisition access period of 24 h on cotton plants. The authors further reported that a minimum of three whiteflies were required for successful transmission of CLCuV to new uninfected plants and the transmission efficiency increased up to 70% with 10 whiteflies per plant. However, no correlations were observed between whitefly numbers and disease incidence under field conditions (Briddon et al. 2000; Mann et al. 2008).

In general female whiteflies are more prolific vectors of CLCuV than male whiteflies (Nateshan et al. 1996; Mann and Singh 2004a). Female whiteflies transmitted CLCuV 1.5 times more efficiently than male whiteflies. In addition sexual demarcation is also observed in CLCuV retention capacity of male and female whiteflies. Female whiteflies retained the virus up to 7.46 days as compared to 5.08 days by males (Mann and Singh 2004c).

CLCuV transmission is influenced by host plant species and developmental stage. Whitefly mediated transmission by Singh (2000) and Mann and Singh (2004b) on various host plants indicated that whiteflies transmitted CLCuV more efficiently on cotton, tomato and China rose plants than on eggplant or radish. Furthermore, whiteflies inoculated CLCuV to the seedling stage more efficiently than the vegetative, cotyledonary or fruit initiation stages. Pre- and post-acquisition starvation of whitefly up to 4 h did not affect the transmission efficiency of the whiteflies.

4.4.4 CLCuV Affects Fitness of *B. tabaci*

Plant pathogen-vector systems are characterized by complex direct and indirect interactions (Byrne and Bellows 1991a, b; Blair et al. 1995). The interactions between plant-pathogenic viruses and their insect vectors range from the insect functioning as a casual carrier to intimate molecular interactions (Nault 1997). Earlier studies by various researchers suggest that viruses can have positive, neutral or negative effects on their vectors (Costa et al. 1991; Colvin et al. 2006; Jiu et al. 2007). Virus-infected plants undergo changes that affect the biology of insect vectors. Insect vectors feeding on infected plants have been reported to differ in growth rates, longevity, and fecundity compared to those vectors feeding on uninfected plants (Kennedy 1951; Baker 1960). Some vectors preferentially colonize infected plants (Castle and Berger 1993; Castle et al. 1998). Virus infection can improve plant suitability for insect vector survival and reproduction by improving the nutritional quality of the host plant (Kennedy 1951; Baker 1960). However, increased survival may lead to higher virus proliferation in the field (McElhany et al. 1995). Recent studies by Mann et al. (2008) showed that *B. tabaci* deposited significantly fewer eggs on virus infected plants compared to healthy plants. However, CLCuV infection increased percent egg viability of *B. tabaci*. The developmental time of whiteflies from egg to adulthood was significantly reduced on CLCuV infected plants with shorter nymphal and pupal duration. Male and female whiteflies had shorter longevity on CLCuV infected plants compared to healthy plants. The altered *B. tabaci* life history parameters correlated to the altered plant volatiles, morphology or nutrition. Severely diseased infected plants acted as poor host for whiteflies than the uninfected or mildly diseased plants. The egg duration was similar on CLCuV infected and healthy plants regardless of disease severity.

CLCuV is also known to affect behavior of *B. tabaci*. Both viruliferous and healthy whiteflies preferentially settled on healthy cotton plants compared to CLCuV infected plants when given a choice indicating that healthy cotton was a more suitable host, and *B. tabaci* could differentiate between CLCuV infected and healthy plants (Mann et al. 2009). The authors suggested that differential preferences could be due to alterations of nutritional status of virus infected plants because such preferences were missing in newly CLCuV inoculated plants (Mann et al. 2008, 2009). Furthermore, CLCuV infection is known to alter concentrations of peroxidase activity, and levels of catechol, phenols, carotenoids, proteins, total sugars, chlorophyll, oil and lipase enzyme in cotton plants (Kaur et al. 1998; Kang et al. 2003; Ashraf et al. 2004), which are deterrent to whitefly colonization (Ilyas et al. 1991; Butter et al. 1992; Hasan et al. 2000; Selvanarayanan and Muthukumaran 2005; Ravi et al. 2006). In addition to chemical changes the disease causes morphological changes such as pronounced vein and veinlet thickening, roughing of the surface and curling of upper leaves (Kapur et al. 1994; Kang et al. 2003) that may discourage whitefly adults from settling on infected plants. However, Sidhu et al. (2009) showed that viruliferous whiteflies laid fewer eggs and had reduced longevity when reared on CLCuV resistant cotton plants. These results suggest that the interactions between *B. tabaci* and CLCuV are not symbiotic or mutualistic, but pathogenic.

4.5 Conclusion

B. tabaci is the only vector associated with CLCuV that acquires and transmits the virus to healthy plants in a persistent manner. The virus is acquired in 15 min to 4 h and transmitted to new plants in 5 min to 1 h depending on the virus isolates and the experimental conditions. In addition, a latent period of up to 8 h is required for successful transmission of the virus from infected plants to healthy plants. The virus transmission efficiency is greatly boosted with an increase in acquisition and inoculation periods. The virus is retained in adult whiteflies from a few days to the entire life period. Both *B. tabaci* nymphs and adults can successfully acquire and transmit virus to new plants. The transmission efficiency is greatly enhanced with increased numbers of *B. tabaci* adults.

Although the function of geminivirus genes has been thoroughly investigated for the last decade, infection of plant hosts by CLCuV is still not fully understood. The research on CLCuV has focused on molecular characterization of virus species and few basic virus vector relationships. The way CLCuV interact with *B. tabaci* is still unclear. As with other *B. tabaci* transmitted viruses specific interactions should occur between *B. tabaci* and CLCuV for the successful transmission process of virus to new plants (acquisition, multiplication and inoculation). For future research directions, a better understanding is needed on: the differences between CLCuV and other begomoviruses with regard to their interaction with *B. tabaci,* CLCuV effects on whitefly behaviour and physiology, the competency level of biotypes and geographically isolated populations of *B. tabaci* in transmitting CLCuV and the role of endosymbionts in CLCuV transmission. Furthermore, improved knowledge is required on the molecules and/or receptor sites involved in virus-insect host cell interactions and the pathways of CLCuV in *B. tabaci*.

References

Ahmed MZ, Shatters RG, Ren SX, Jin GH, Mandour NS, Qiu BL (2009) Genetic distinctions among the Mediterranean and Chinese populations of *Bemisia tabaci* Q biotype and their endosymbiont Wolbachia populations. J Appl Entomol 133:733–741

Akhtar KP, Haq MA, Hussain M, Khan AI (2002) Whitefly transmitted geminiviruses and associated disorders in cotton: a review. Pak J Phytopathol 14:140–150

Aneja AK (2000) Biology of whitefly *Bemisia tabaci* (Gennadius) on American cotton. Dissertation, Punjab Agricultural University, Ludhiana

Ashraf M, Mahmood YS, Sarwar G, Ashraf M, Naeem M, Zafar S (2004) Physiological and biochemical changes in resistant and susceptible to *Cotton leaf curl virus* (CLCuV) cotton varieties at germination and early seedling stages: changes in lipase, oil content, protein and soluble sugars. Int J Biol Biotechnol 1:217–222

Avidov Z, Harpaz I (1969) Plant pests of Israel. Israel University Press, Jerusalem

Azab AK, Megahed MM, El-Mirswami HD (1971) On the biology of *Bemisia tabaci* (Genn) (Homoptera: Aleyrodidae). Bull Soc Ent Egypte 55:305–15

Baker PF (1960) Aphid behavior on healthy and on yellows virus infected sugar beet. Ann Appl Biol 48:384–391

Basu AN (1995) *Bemisia tabaci* (Gennadius): crop pest principal whitefly vector of plant viruses. Oxford and IBH Publishing, New Delhi
Bedford ID, Briddon RW, Brown JK, Rossel RC, Markham PG (1994) Geminivirus transmission and biological characterisation of *Bemisia tabaci* (Genn) biotypes from different geographic regions. Ann Appl Biol 125:311–325
Bellows TS, Perring TM, Gill RJ, Headrich DH (1994) Description of a species of *Bemisia tabaci* (Homoptera: Aleyrodidae). Ann Entomol Soc Am 87:195–206
Bink FA (1975) Leaf curl and mosaic diseases of cotton in Central Africa. Emp Cott Rev 25:133–141
Bird J (1957) A whitefly transmitted mosaic of Jatropha gossypifolia. Technical Paper, University of Puerto Rico, Agricultural Experiment Station, 22:1–35
Bird J, Maramorosch K (1978) Viruses and virus diseases associated with whiteflies. Adv Virus Res 22:55–110
Blair MW, Basset MJ, Abouzid AM, Hieber E, Polston JE, Mcmillan RT, Graves JRW, Lamberts M (1995) Occurrences of bean golden mosaic virus in Florida. Plant Dis 79:539–533
Brar DS, Aneja AK, Singh J, Mahal MS (2005) Biology of whitefly *Bemisia tabaci* (Gennadius) on American cotton *Gossypium hirsutum* Linnaeus. J Insect Sci 18:48–59
Briddon RW (2003) Cotton leaf curl disease, a multicomponent begomovirus complex. Mol Plant P 4:427–434
Briddon RW, Markham PG (2000) Cotton leaf curl virus disease. Virus Res 71:151–159
Briddon RW, Mansoor S, Bedford ID, Pinner MS, Markham PG (2000) Clones of cotton leaf curl geminivirus induce symptoms atypical of cotton leaf curl disease. Virus Genes 20:19–26
Briddon RW, Bull SE, Imran A, Idris AM, Shahid M, Bedford ID, Poonam D, Narayan R, Siwatch SS, Abdel-Salam AM, Brown JK, Yusuf Z, Markham PG (2003) Diversity of DNA beta, a satellite molecule associated with some monopartite begomoviruses. Virology 312:106–121
Brown JK (2001) The molecular epidemiology of begomovirus spp. In: Khan JA, Dykstra J (eds.) Trends in plant virology. The Haworth Press, New York
Brown JK, Bird I (1992) Whitefly transmission geminivirus in the Americas and the Caribbean basin: past and present. Plant Dis 76:220–228
Brown JK, Czosnek H (2002) Whitefly transmission of plant viruses. In: Plumb RT (ed.) Advances in botanical research, vol 36. Academic, New York
Brown JK, Frohlich DR, Rosell RC (1995) The sweetpotato or silverleaf whiteflies: biotypes of *Bemisia tabaci* or a species complex? Annu Rev Entomol 40:511–534
Butter NS, Vir BK (1991) Response of whitefly, *Bemisia tabaci* (Genn.) to different cotton genotypes under glasshouse conditions. Ind J Entomol 53:115–119
Butter NS, Vir BK, Kaur G, Singh TH, Raheja RK (1992) Biochemical basis of resistance to whitefly *Bemisia tabaci* Genn (Aleyrodidae: Hemiptera) in cotton. Trop Agric 69:119–122
Byrne DN, Bellows TS (1991a) Whiteflies biology. Annu Rev Entomol 36:431–457
Byrne DN, Bellows TS (1991b) Life history traits of the whitefly *Bemisia tabaci* (Homoptera: Aleyrodidae) on six virus infected or healthy plant species. Environ Entomol 20:1102–1107
Cai JH, Xie K, Lin L, Qin BX, Chen BS, Meng JR, Liu YL (2010) Cotton leaf curl Multan virus newly reported to be associated with cotton leaf curl disease in China. New disease reports. http://wwwndrsorguk. Accessed 24 May 2010
Castle SJ, Berger PH (1993) Rates of growth and increase of *Myzus persicae* on virus-infected potatoes according to type of virus-vector relationship. Entomol Exp Appl 69:51–60
Castle SJ, Mowry TM, Berger PH (1998) Differential settling by *Myzus persicae* (Homoptera: Aphididae) on various virus infected host plants. Ann Entomol Soc Am 91:661–667
Cauquil JC, Follin JC (1983) Presumed virus and mycoplasma-like organism diseases in subsaharan Africa and in the rest of the world. Cotton Fibres Trop 38:309–371
Colvin JC, Omongo A, Govindappa MR, Stevenson PC, Maruthi MN (2006) Host plant viral infection effects on arthropod–vector population growth development and behavior: management and epidemiological implications. Adv Virol Res 67:419–452
Costa AS (1976) Whitefly-transmitted plant diseases. Annu Rev Phytopathol 14:429–49

Costa H, Brown JK (1991) Variation in biological characteristics and esterase patterns among populations of *Bemisia tabaci* and the association of one population with silverleaf symptom induction. Entomol Exp Appl 61:211–219

Costa HS, Brown JK, Byrne DN (1991) Life history traits of the whitefly *Bemisia tabaci* (Homoptera Aleyrodidae) on six virus-infected or healthy plant species. Environ Entomol 20:1102–1107

Costa HS, Westcot DM, Ullman DE, Rosell R, Brown JK, Johnson MW (1996) Virus-like particles in the mycetocytes of the sweetpotato whitefly *Bemisia tabaci* (Homoptera Aleyrodidae). J Invert Pathol 67:183–186

Czosnek H, Ghanim M, Morin S, Rubenstein G, Fridman V, Zeidan M (2001) Whiteflies: vectors and victims (?) of geminiviruses. Adv Virus Res 57:291–322

Czosnek H, Ghanim M, Ghanim N (2002) The circulative pathway of begomoviruses in the whitefly vector *Bemisia tabaci*-insights from studies with Tomato yellow leaf curl virus. Ann Appl Biol 140:215–231

Daugherty MP, Bosco D, Almeida RPP (2009) Temperature mediates vector transmission efficiency: inoculum supply and plant infection dynamics. Ann Appl Biol 155:361–369

De Barro PJ (1995) *Bemisia tabaci* biotype B: a review of its biology distribution and control. CSIRO technical paper Australia

Duffus JE (1987) Whitefly transmission of plant viruses. Curr Top Vector Res 4:73–91

Duffus JE (1996) Whitefly borne viruses. In: Gerling D, Mayer RT (eds) *Bemisia*: taxonomy biology damage control and management. Intercept, Andover

Farquharson CO (1912) Report of the mycologist. Report of Agriculture Department, Nigeria. In: Siddique MA, Hughes LC (eds.) Cotton growth in the Gezira environment. Cambridge Publications, London

Fauquet CM, Stanley J (2005) Revising the way we conceive and name viruses below the species level: a review of geminivirus taxonomy calls for new standardized isolate descriptors. Arch Virol 150:2151–2179

Fauquet C, Thouvenel JC (1987) Plant viral diseases in the Ivory Coast Orstorm, Paris. Documentation techniques No. 46

Fauquet CM, Briddon RW, Brown JK, Moriones E, Stanley J, Zerbini M, Zhou X (2008) Geminivirus strain demonstration and nomenclature. Arch Virol 153:783–821

Geering ADW (2010) Cotton leaf curl (Cotton leaf curl Multan virus and others) Pest and diseases image library updated on 2/23/2010 9:49:01 PM Available online: http://wwwpadilgovau Accessed 24 May 2010

Gerling D (1990) Whiteflies: their bionomics pest status and management. Intercept Ltd, Andover

Ghanim M, Morin S, Czosnek H (2001a) Rate of *Tomato yellow leaf curl virus* (*TYLCV*) trans location in the circulative transmission pathway of its vector the whitefly *Bemisia tabaci*. Phytopath 91:188–196. (homoptera: aleyrodidae). Ann Entomol Soc Am 76:310–313

Ghanim M, Rosell RC, Campbell LR, Czosnek H, Brown JK, Ullman DE (2001b) Digestive salivary and reproductive organs of *Bemisia tabaci* (Gennadius) (Hemiptera: Aleyrodidae) biotype B. J Morph 248:22–40

Gill RJ (1990) The morphology of whiteflies. In: Gerling D (ed.) Whitelies: their bionomics pests status and management. Intercept, Andover, pp 13–46

Greathead AH (1986) Host plants. In: Cock MJW (ed.) *Bemisia tabaci,* a literature survey on the cotton whitefly with an annotated bibliography. CAB International, Wallingford

Hameed S, Khalid S, Ul-Haq E, Hashrni AA (1994) Cotton leaf curl disease in Pakistan caused by a whitefly-transmitted geminivirus. Plant Dis 78:529

Harris KF, Pesic-Van Esbroeck Z, Duffus JE (1995) Anatomy of a virus vector. In: Gerling D, Mayer R (eds.) Bemisia 1995: taxonomy biology damage control and management. Intercept, Andover

Harris KF, Pesic-Van EZ, Duffus JE (1996) Morphology of the sweet potato whitefly *Bemisia tabaci* (Homoptera Aleyrodidae) relative to virus transmission. Zoomorphology 116:143–156

Harrison BD, Liu YL, Khalid S, Hameed S, Otim-Nape GW, Robinson DJ (1997) Detection and relationships of cotton leaf curl virus and allied whitefly-transmitted geminiviruses occurring in Pakistan. Ann Appl Biol 130:61–75
Hasan MU, Ahmad F, Wakeel W (2000) Role of biochemical components in varietal resistance of cotton against sucking insect pests. Pak Entomol 22:69–71
Hodges GS, Evans GA (2005) An identification guide to the whiteflies (Hemiptera: Aleyrodidae) of the Southeastern United States. Fla Entomol 88:518–534
Hohn T (2007) Plant virus transmission from the insect point of view. Proc Natl Acad Sci USA 104:17905–17906
Hunter WB, Hiebert E, Webb SE, Polston JE, Tsai J (1996) Precibarial and cibarial chemosensilla in the whiteßy, *Bemisia tabaci* (Gennadius) (Homoptera: Aleyrodidae). Int J Insect Morphol Embrol 25:295–304
Hunter-Fujita FR, Entwistle PF, Evans HF, Crook NE (1998) Insect viruses and pest management. Wiley, New York
Hussain MA (1931) A Preliminary note on the whitefly of cottons in the Punjab. Agric J India 25:508–26
Hussain MA, Trehan KN (1933) Observations on the life history bionomics and control of whitefly of cotton (*Bemisia gossypiperda* M & L). Indian J agric Sci 3:701–06
Idris AM, Brown JK (2000) Identification of a new, monopartite begomovirus associated with leaf curl disease of cotton in Gezira, Sudan. Plant Dis 84:809
Idris AM, Brown JK (2002) Molecular analysis of *Cotton leaf curl virus-Sudan* reveals an evolutionary history of recombination. Virus Genes 24:249–256
Idris AM, Brown JK (2004) *Cotton leaf crumple virus* is a distinct new world begomovirus species with complex evolutionary relationships indicative of recombination and reasortants. Phytopathology 94:1068–1074
Ilyas M, Puri SN, Rote NB (1991) Effects of some morphophysiological characters of leaf on incidence of cotton whitefly. J Maha Agri Univ 16:386–388
Jiang YX, Nombela G, Muñiz M (2001) Analysis by DC-EPG of the resistance to *Bemisia tabaci* on a Mitomato line. Entomol Exp Appl 99:295–302
Jiu M, Zhou XP, Tong L, Xu J, Yang X, Wan FH, Liu SS (2007) Vector-virus mutualism accelerates population increase of an invasive whitefly. PLoS ONE 2:e182
Jones DR (2003) Plant viruses transmitted by whiteflies. Eur J Plant Pathol 109:195–219
Kang SS, Athar M, Cheema SS (2003) Physiological changes in cotton infected with cotton leaf curl virus. Plant Dis Res 18:193–195
Kang SS, Athar M, Cheema SS, Malathi VG, Radhakrishnan G (2004) Quick detection of Cotton leaf curl virus. Ind Phytopathol 57:45–246
Kapur SP, Joginder S, Chopra BL, Sohi AS, Rewal HS, Narang DD (1994) Cotton leaf curl in Punjab. Plant Dis Res 9:86–91
Kaur G, Sohal BS, Singh J, Bajaj KL (1998) Influence of *cotton leaf curl virus* on the polyphenol metabolism of resistant and susceptible cotton leaves. Plant Dis Res 13:23–27
Kennedy JS (1951) Benefits to aphids from feeding on galled and virus-infected leaves. Nature 168:825–826
Kennedy JS, Day MF, Eastop VF (1962) A conspectus of aphids as vectors of plant viruses. Commonwealth Institute of Entomology, London
Khan JA, Ahmad J (2005) Diagnosis monitoring and transmission characteristics of Cotton leaf curl virus. Curr Sci 88:1803–1809
Kirkpatrick TW (1931) Further studies on leaf curl of cotton in the Sudan. Bull Ent Res 22:323–36
Kirthi N, Priyadarshini CGP, Sharma P, Maiya SP, Hemalatha V, Sivaraman P, Dhawan P, Rishi N, Savithri HS (2004) Genetic variability of begomoviruses associated with cotton leaf curl disease originating from India. Arch Virol 149:2047–2057
Kumar A, Kumar J, Khan JA (2010) Sequence characterization of cotton leaf curl virus from Rajasthan: phylogenetic relationship with other members of geminiviruses and detection of recombination. Virus Genes 40:282–289

Lambart AR (1924) Gezira research farm: botanical station annual report-4

Lefeuvre P, Martin DP, Hoareau M, Naze F, Delatte H, Thierry M, Varsani A, Becker N, Reynaud B, Lett JM (2007) *Begomovirus* 'melting pot' in the south-west Indian Ocean islands: molecular diversity and evolution through recombination. J Gen Virol 88:3458–3468

Lefeuvre P, Lett JM, Varsani A, Martin DP (2009) Widely conserved recombination patterns among single-stranded DNA viruses. J Virol 83:2697–2707

Lei H, Tjallingi WF, van Lantern JC (1998) Probing and feeding characteristics of the greenhouse whitefly in association with host plant acceptance and whitefly strains. Entomol Exp Appl 92:299–309

Li T, Vinson BS, Gerling D (1989) Courtship and mating behaviour of *Bemisia tabaci* (Homoptera: Aleyrodidae). Environ Entomol 18:800–06

Lynch RE, Simmons AM (1993) Distribution of immatures and monitoring of adult sweet-potato whitefly *Bemisia tabaci* (Gennadius) (Homoptera Aleyrodidae) in peanut *Arachis hypogaea*. Environ Entomol 22:375–383

Mahmood T (1999) Cotton leaf curl virus disease and its present status in Punjab. Pak Cotton Grower 3:17–18

Mann RS, Singh L (2004a) Retention of *Cotton leaf curl virus* (*CLCUV*) in its vector whitefly *Bemisia tabaci* (Gennadius). Ind J Entomol 66:96–98

Mann RS, Singh L (2004b) Studies on the interaction of *Cotton leaf curl virus* (*CLCuV*) with its vector, *Bemisia tabaci* (Gennadius). J Cotton Res Devel 18:95–98

Mann RS, Singh L (2004c) Studies on the relationship of *Cotton leaf curl virus* (*CLCuV*) with its vector, *Bemisia tabaci* (Gennadius). Ind J Plant Prot 32:140–141

Mann RS, Sidhu JS, Butter NS, Sohi AS, Sekhon PS (2008) Performance of Bemisia tabaci (Hemiptera: Aleyrodidae) on healthy and *Cotton leaf curl virus* infected cotton. Fla Entomol 91:249–255

Mann RS, Sidhu JS, Butter NS (2009) Settling preference of the whitefly *Bemisia tabaci* (Hemiptera: Aleyrodidae) on healthy versus cotton leaf curl virus-infected cotton plants. Int J Trop Insect Sci 29:57–61

Mansoor S, Khan SH, Bashir A, Saeed M, Zafar Y, Malik KA, Briddon R, Stanley J, Makham PG (1999) Identification of a novel circular single stranded DNA associated with cotton leaf curl disease in Pakistan. Virology 259:190–99

Mansoor S, Briddon RW, Bull SE, Bedford ID, Bashir A, Hussain M, SaeedZafar MY, Malik KA, Fauquet C, Markham PG (2003) Cotton leaf curl disease is associated with multiple monopartite begomoviruses supported by a single DNA β. Arch Virol 148:1969–1986

Mansoor S, Amrao L, Amin I, Briddon RW, Malik KA, Zafar Y (2006) First report of cotton leaf curl disease in central and southern Sindh Province in Pakistan. Plant Dis 90:826

Mao M, He Z, Yu H, Li H (2008) Molecular characterization of *Cotton leaf curl Multan virus* and its satellite DNA that infects Hibiscus rosa-sinensis. Chin J Virol 24:64–68

Martin JH, Mifsud D, Rapisarda C (2000) The whiteflies (Hemiptera: Aleyrodidae) of Europe and the Mediterranean Basin. Bull Entomol Res 90:407–448

Maruthi MN, Colvin J, Seal SE, Gibson G, Cooper J (2002) Coadaptation between cassava mosaic geminivirus and their local vector population. Virus Res 86:71–85

Matson PA, Parton WJ, Power AG, Swift MJ (1997) Agricultural intensification and ecosystem properties. Science 277:504–509

Mcauslane HJ (2009) Sweetpotato whitefly B biotype of silverleaf whitefly *Bemisia tabaci* (Gennadius) or *Bemisia argentifolii* Bellows and Perring (Insecta: Hemiptera: Aleyrodidae). EDIS, University of Florida, Gainesville

Mcelhany PL, Real A, Power AG (1995) Vector preference and disease dynamics: a study of barley yellow dwarf virus. Ecology 76:444–457

McGrath PF, Harrison BD (1995) Transmission of tomato leaf curl geminiviruses by *Bemisia tabaci*: effects of virus isolate and vector biotype. Ann Appl Biol 126:307–316

Mohanty AK, Basu AN (1987) Biology of whitefly vector *Bemisia tabaci* (Genn) on four hosts plants throughout the year. J Ent Res 11:15–18

Monga D, Raj S, Mayee CD (2004) Strategies for cotton leaf curl virus disease management. In: National symposium on changing world order-cotton research development and policy in context, ANGRAU, Hyderabad, 10–12 Aug 2004

Monga D, Kumar R, Kumar M (2005) Detection of DNA A and satellite (DNA beta) in cotton leaf curl virus (CLCuV) infected weeds and cotton plants using PCR technique. J Cotton Res Dev 19:105–108

Monga D, Chander S, Kumar M, Singh NP (2007) Distribution pattern of cotton leaf curl virus disease in north India. J Cotton Res Devel 21(2):248–252

Morales FJ, Anderson PK (2001) The emergence and dissemination of whitefly-transmitted geminiviruses in Latin America. Arch Virol 146:415–441

Mound LA, Halsey SH (1978) Whitefly of the World, a systematic Catalogue of the Aleyrodidae (Homoptera) with host plant and natural enemy data British Museum (Natural History). Wiley, London

Moury B, Fabre F, Senoussi R (2007) Estimation of the number of virus particles transmitted by an insect vector. Proc Natl Acad Sci USA 104:17891–17896

Mubin M, Briddon RW, Mansoor S (2009) Diverse and recombinant DNA betasatellites are associated with a begomovirus disease complex of *Digera arvensis*, a weed host. Virus Res 142:208–212

Mugiira RB, Liu SS, Zhou X (2008) Tomato yellow leaf curl virus *and Tomato leaf curl Taiwan virus* invade south-east coast of China. J Phytopathol 156:217–221

Murthi MN, Rekha AR, Mirza SH, Alam SN, Colvin J (2007) PCR-based detection and partial genome sequencing indicate high genetic diversity in Bangladeshi begomoviruses and their whitefly vector *Bemisia tabaci*. Virus Genes 34:373–385

Nadeem A, Weng Z, Nelson MR (1997) Cotton leaf crumple virus and cotton leaf curl virus are two distantly related geminiviruses. Mol Plant Pathol On-line: 0612nadeem. http://www.bspp.org.uk/mppol/ Accessed 24 May 2010

Natarajan K (1990) Natural enemies of *Bemisia tabaci* (Gennadius) and effect of insecticides on their activity. J Biol Control 4:86–88

Nateshan HM, Muniyappa V, Swanson MM, Harrison BD (1996) Host range, vector relations and serological relationships of cotton leaf curl virus from southern India. Ann Appl Biol 128:233–244

Nault LR (1997) Arthropod transmission of plant viruses: a new synthesis. Ann Entomol Soc Am 90:521–541

Naveed M, Salam A, Saleem MA (2007) Contribution of cultivated crops vegetables weeds and ornamental plants in harboring of *Bemisia tabaci* (Homoptera: Aleyrodidae) and associated parasitoids (Hymenoptera: Aphelinidae). J Pestic Sci 80:191–197

Nelson MR, Nadeem A, Ahmad W, Orum TV (1998) Global assessment of cotton viral disease. In: Proceedings of beltwide cotton conference San Diego, Memphis, pp 161–162

Noueiry AO, LucasWJ GRL (1994) Two proteins of a plant DNA virus coordinate nuclear and plasmodesmal transport. Cell 76:925–932

Nour MA (1960) On "leaf curl" of cotton in the Philippines. Plant Prot Bull FAO 8:55–56

Nour MA, Nour JJ (1964) Identification, transmission and host range of leaf curl viruses infecting cotton in the Sudan. Empire Cotton Growing Rev 41:27–37

Oliveira MRV, Henneberry TJ, Anderson P (2001) History current status and collaborative research projects for *Bemisia tabaci*. Crop Prot 20:709–723

Oliveira MRV, Amancia E, Laumann RA, Gomes O (2003) Natural enemies of *Bemisia tabaci* (Gennadius) B biotype and *Trialeurodes vaporariorum* (Westwood) (Hemiptera: Aleyrodidae) in Brasília Brazil. Neotrop Entomol 32:151–154

Padidam M, Beachy RN, Fauquet CM (1995) Tomato leaf curl geminivirus from India has a bipartite genome and coat protein is not essential for infectivity. J Gen Virol 76:25–35

Perring TM (2001) The *Bemisia tabaci* species complex. Crop Prot 20:725–737

Radhakrishnan G, Malathi VG, Varma A (2004) Detection of DNA A and DNA β associated with cotton leaf curl and some other plant diseases caused by whitefly transmitted geminiviruses. Ind Phytopathol 57:53–60

Rahman M, Malik TA, Hussain D, Zafar Y (2005) Inheritance of host plant resistant mechanism against cotton leaf curl virus disease (CLCD) in cotton. Plant Pathol 54:764–772

Rajeshwari R, Reddy RVC, Maruthi MN, Colvin J, Seal SE, Muniyappa V (2005) Host range, vector relationships and sequence comparison of a begomovirus infecting hibiscus in India. Ann Appl Biol 147:15–25

Rao VN, Reddy AS (1989) Seasonal influence on developmental duration of whitefly (*Bemisia tabaci*) in upland cotton (*Gossypium hirsutum*). Ind J Agric Sci 59:283–85

Ravi M, Dhandapani N, Sathiah N, Murugan M (2006) Influence of organic manures and fertilizers on the incidence of sucking pests of sunflower, Helianthus annus L. Ann Plant Prot Sci 14:41–44

Rosell R, Lichty JE, Brown JK (1995) infrastructure of the mouthparts of adult sweetpotato whitefly *Bemisia tabaci* (Gennadius) (Homoptera: Aleyrodidae). Int J Insect Morph Embryol 24:297–306

Rosell RC, Torres-Jerez I, Brown JK (1999) Tracing the geminivirus-whitefly transmission pathway by polymerase chain reaction in whitefly extracts saliva hemolymph and honeydew. Phytopathology 89:239–246

Rybicki EP, Pietersen G (1999) Plant virus problems in the developing world. Adv Virus Res 53:127–175

Salas J, Mendoza O (1995) Biology of the sweetpotato whitefly (Homoptera: Aleyrodidae) on tomato. Fla Entomol 78:154–160

Saunders K, Bedford ID, Stanley J (2002) Adaptation of an autonomously-replicating nanovirus-like DNA component associated with ageratum yellow vein disease from whitefly to leafhopper transmission. J Gen Virol 83:907–913

Saunders K, Bedford ID, Rahman M, Hussain D, Malik T, Zafar Y (2005) Genetics of resistance against cotton leaf curl disease in *Gossypium hirsutum*. Plant Pathol 54:64–772

Seal SE, Van Den Bosch F, Jeger MJ (2006) Factors influencing begomovirus evolution and their increasing global significance: implications for sustainable control. Crit Rev Plant Sci 25:23–46

Selvanarayanan V, Muthukumaran N (2005) Insect resistance in tomato accessions and their hybrid derivatives in Tamil Nadu, India. Comm Agric Appl Biol Sci 70:613–624

Shahid MS, Mansoor S, Briddon RW (2007) Complete nucleotide sequences of *Cotton leaf curl Rajasthan virus* and its associated DNA β molecule infecting tomato. Arch Virol 152:2131–2134

Shahid MS, Liaqat A, Saiqa W (2009) *Cotton leaf curl Rajasthan virus infecting* tomato in Pakistan. Pak J Sci Indus Res 52:319–321

Sharaf N, Batta Y (1985) Effect of some factors on the relationship between the whitefly *Bemisia tabaci* Genn (Homoptera: Aleyrodidae) and the parasitoid *Eretmocerus mundus* Mercet (Hymenoptera: Aphelinidae). Z Angew Entomol 99:267–276

Sidhu JS, Mann RS, Butter NS (2009) Deleterious effects of *Cotton leaf curl virus* (*CLCuV*) on *Bemisia tabaci* (Gennadius). J Entomol 6:62–66

Simmons AM (1994) Oviposition on vegetables by *Bemisia tabaci* (Homoptera Aleyrodidae) – temporal and leaf surface factors. Environ Entomol 22:381–389

Simmons AM, McCutcheon GS, Dufault RJ, Hassell RL, Rushing JW (2000) *Bemisia argentifolii* (Homoptera: Aleyrodidae) attacking species of medicinal herbal plants. Ann Entomol Soc Am 93:856–861

Singh M (2000) Studies on the leaf curl disease of cotton caused by *CLCuV* in Punjab. Dissertation, Punjab Agricultural University, Ludhiana

Singh M (2001) Studies on virus vector relationships of cotton leaf curl virus (*CLCuV*) with its vector *Bemisia tabaci* (Genn.). Dissertation, Punjab Agricultural University, Ludhiana

Singh J, Sohi AS, Mann HS, Kapur SP (1994) Studies on whitefly, *Bemisia tabaci* (Genn.) transmitted cotton leaf curl disease in Punjab. J Insect Sci 7:194–198

Sinisterra XH, McKenzie CL, Hunter WB, Powell CA, Shatters RG Jr (2005) Differential transcriptional activity of plant-pathogenic begomoviruses in their whitefly vector (*Bemisia tabaci* Gennadius: Hemiptera Aleyrodidae). J Gen Virol 86:1525–1532

Stanley J, Bisaro DM, Briddon RW, Brown JK, Fauquet CM, Harrison BD, Rybicki EP, Stenger DC (2004) Geminiviridae. In: Fauquet CM, Mayo MA, Maniloff J, Desselberger U, Ball LA (eds) Virus taxonomy eighth report of the international committee on taxonomy of viruses. Elsevier/Academic, London

Tarr SAJ (1957) Recent observations on diseases of Cotton in the Sudan Gezira. Plant Prot Bull FAO 5:85–88

Tarr SAJ (1964) Virus diseases of cotton. Commonw Mycol Inst Misc Publ 18:26

Thakur PD (2002) Virus diseases of cotton. In: Gupta VK, Paul YS (eds) Diseases of field crops. Indus Publishing Company, New Delhi

Thresh JM (1982) Cropping practices and virus spread. Annu Rev Phytopathol 20:193–218

Tiendrebeogo F, Lefeuvre P, Hoareau M, Villemot J, Traoré AS, Barro N, Traoré VS, Reynaud B, Traoré O, Lett JM (2010) Molecular diversity of *Cotton leaf curl Gezira virus* isolates and their satellite DNAs associated with okra leaf curl disease in Burkina Faso. Virol J 7:48

Tsagkarakou A, Tsigenopoulos CS, Gorman K, Lagnel J, Bedford ID (2007) Biotype status and genetic polymorphism of the whitefly *Bemisia tabaci* (Hemiptera: Aleyrodidae) in Greece: mitochondrial DNA and microsatellites. Bull Entomol Res 97:29–40

Varma A (1990) Changing pattern of plant diseases caused by whitefly transmitted geminiviruses. In: Abstracts of the 8th international congress of virology, Berlin

Varma A, Malathi VG (2003) Emerging geminivirus problems: a serious threat to crop production. Ann Appl Biol 142:145–164

Varma A, Malathi VG, Handa A, Aiton M, Harrison BD, Varma JP, Singh RP, Singh M, Srivastava M, Singh J (1993) Occurrence of leaf-curl of cotton and okra in Northern India. In: Abstracts of the 6th international congress of plant pathology, Montreal

Walker GP (1985) Stylet penetration by the bayberry whitefly as affected by leaf age in lemon Citrus limon. Entomol Exp Appl 39:15–21

Watkins GM (1981) Compendium of cotton diseases. American Phytopathology Society, St Paul

Xie Y, Zhou XP (2003) Molecular characterization of squash leaf curl Yunnan virus a new begomovirus and evidence for recombination. Arch Virol 148:2047–2054

Zhou X, Liu Y, Calvert L, Munoz C, Otim-Nape GW, Robinson DJ, Harrison BD (1997) Evidence that DNA-A of a geminivirus associated with severe cassava mosaic disease in Uganda has arisen by interspecific recombination. J Gen Virol 78:2101–2111

Zhou X, Liu Y, Robinson DJ, Harrison BD (1998) Four DNA-A variants among Pakistani isolates of cotton leaf curl virus and their affinities to DNA-A of geminivirus isolates from okra. J Gen Virol 79:915–925

Chapter 5
Association of *Bemisia tabaci* with the Severe Cassava Mosaic Disease in Uganda

Winston M.O. Thompson

Abstract Severe cassava mosaic disease is associated with *East African cassava mosaic virus*-Uganda variant (EACMV-UG) in association with *African cassava mosaic virus* (ACMV). A pseudorecombinant has also been associated with the pandemic. These begomoviruses occur in mixed infections and through synergism the virus titer increases thus resulting in a rich source of inoculum. The interaction of *Bemisia tabaci* with begomovirus-infected host plants is an important factor implicated in the severe cassava mosaic disease pandemic. Whitefly visual cues to the yellows of infected cassava plants could account for their orientation preference for infected over healthy plants. Based on epidemiological models, such preference with few infected plants could increase the rate of spread of the pathogen and a similar situation could result, if the orientation preference is for healthy plants in the presence of a higher population of infected plants. Disease spread dynamics is exacerbated with high population levels of the vector, which is influenced by an invasive genotype that is particularly fecund, as seen with several other pathosystems around the world and secondly, by plant responses to infection and other stress factors. Plant biochemical reactions to infection result in the availability of nutrients beneficial to the vector. Amino acids for example, have been seen in higher amounts in infected as opposed to healthy plants and with concomitant higher populations of whiteflies on the infected plants. Control measures found to be promising include the use of resistant varieties, use of healthy planting material and roguing. With continued spread of the pandemic these measures represent the epitome of control approaches.

Keywords *Bemisia tabaci* • Cassava plants • *East African cassava mosaic virus*-Uganda variant • Interaction • Vector population

---

W.M.O. Thompson (✉)
Plant Protection, National Agricultural Research Institute, Mon Repos, East Coast Demerara, Guyana, South America
15216 NE 8th St., H3 98007 Bellevue, Washington, USA
e-mail: winstont335@gmail.com

W.M.O. Thompson (ed.), *The Whitefly, Bemisia tabaci (Homoptera: Aleyrodidae) Interaction with Geminivirus-Infected Host Plants*,
DOI 10.1007/978-94-007-1524-0_5, © Springer Science+Business Media B.V. 2011

5.1 Background

Cassava *Manihot esculenta* Crantz (Euphorbiaceae) is an important staple food crop. The tubers can be boiled, providing a dietary source of calories or in other cases used for making chips and flours. In the developed countries the crop is utilised for industrial purposes, in preparing starch flakes and pearls, and in the livestock industry, it forms a component in animal feeds (Cock and Reyes 1985).

Cassava is reportedly the fourth major contributor to a dietary source of calories, superseded only by rice, maize and sugar within the tropics (Cock and Reyes 1985). In the period 1995–1997, world production figures were on average *c.* 165 million metric tons. Of this the African continent accounted for *c.* 84 million metric tons per year (FAO 1998).

An important feature of the crop is its ability to grow and produce well in low fertility soil, withstand attack by locusts and other insect pests, and tolerate drought conditions. In Africa, with increasing population growth and reducing soil fertility, cassava is one of the crops that can still be successfully grown under low soil fertility conditions. In Southern India and Java, with population increase, cassava production increased on low fertility lands, which were unsuitable for rice cultivation (Cock and Reyes 1985).

Cassava mosaic disease (CMD) has been identified as a major biotic constraint to cassava production in 15 out of 18 African countries surveyed, causing between 20% and 90% losses (Fauquet and Fargette 1990). Thresh et al. (1997) estimated total yield losses in Africa at US $44 million per annum. A severe form of the disease has been present in Uganda since 1988 (Pita et al. 1998). The epidemic in Uganda extended southwards from northern Uganda and was responsible for tremendous losses of cassava (Gibson et al. 1996). Since then the epidemic is now considered a pandemic because of the number of countries now affected in various parts of the continent.

5.2 Etiology

5.2.1 *The Begomoviruses Involved and Recombination Events*

The severe cassava mosaic epidemic in Uganda that started in 1988 was associated with the Uganda variant of *East African cassava mosaic virus* (EACMV) referred to as (UgV) (Zhou et al. 1997) and later referred to as EACMV-UG based on a taxonomic review of the family *Geminiviridae* (Fauquet and Stanley 2003). In describing the genetic constitution of the virus Zhou et al. (1997) demonstrated that the virus is the result of recombination between DNA A of EACMV and *African cassava mosaic virus* (ACMV). Samples obtained from within the epidemic area revealed greater severity of symptom types and stunting as compared to those outside of the epidemic area.

Similar type observations were made by Harrison et al. (1997). In determining the cause of cassava mosaic epidemic in Uganda, plants of varying symptom severity were sampled from within the epidemic area, at the periphery and outside of the epidemic area. Cuttings from these were grown in the glasshouse and the leaf extracts were then tested by PCR for ACMV, EACMV and EACMV-UG. Results showed that ACMV was present in all sampled areas and EACMV-UG was found only within the epidemic area. Most ACMV- infected plants exhibited mild to moderate symptom types, most plants infected with EACMV-UG plus ACMV showed severe symptom types and some samples of EACMV-UG alone showed severe symptom types. It was also recognized that a very severely affected plant from ukerewe island, Tanzania showed the presence of both ACMV and EACMV. These results showed the consistent association of EACMV-UG with the cassava mosaic epidemic, concomitant with severe symptom types.

In South Africa a new begomovirus was found affecting cassava plants. Based on complete sequence analysis of its DNA A and DNA B, and comparisons with old and new world begomoviruses, it was found to be most similar to EACMV, but since the percentage sequence similarity between this virus and EACMV was less than 90%, it was designated a distinct virus: *South African cassava mosaic virus* (SACMV) (Berrie et al. 2001). One significant event was evidence of recombination within DNA A of SACMV, providing further evidence of recombination events within the genus *Begomovirus* of the family *Geminiviridae.*

In studying the viruses from stem cuttings obtained from the Cameroon, PCR with primers specific for the coat protein genes of both ACMV and EACMV were used. The analysis revealed the presence of ACMV in all samples and ACMV plus *East African cassava mosaic Cameroon virus* (EACMCV) was present in eight of 50 samples. Following complete nucleotide sequence analysis, EACMCV showed evidence of recombination in both DNA A (AC1, AC3 region) and DNA B (BC1 region). Results also showed greater accumulation of both viruses when they occur in mixed infections as opposed to single infections (Fondong et al. 2000). These findings revealed a synergistic mechanism as well as the capability of EACMCV undergoing double recombination.

Molecular characterization of begomoviruses at various locations within Uganda, including the epidemic area, showed the presence of ACMV-Uganda strain (ACMV-UG), EACMV-UG and a new virus strain designated EACMV-UG3. Most plants infected with ACMV-UG alone showed very mild symptoms, a higher percentage of plants infected with EACMV-UG alone showed severe symptoms and all plants infected with the combination of ACMV-UG and EACMV-UG showed very severe symptom types (Pita et al. 2001). Symptom types were also correlated with virus accumulation, and it was concluded that a synergistic effect resulted from the mixed infection of ACMV-UG and EACMV-UG. In the same study, EACMV-UG DNA-A was widespread and always in association with EACMV-UG3 DNA-B (another recombinant). The researchers demonstrated that the pseudorecombinant virus of EACMV-UG DNA-A plus EACMV UG3 DNA-B was capable of infecting cassava plants and therefore was one of the cassava mosaic viruses responsible for infection in Uganda (Pita et al. 2001). These results show the propensity of begomoviruses to engage in recombination and pseudorecombination, and the complexity of infection with begomoviruses occurring singly, in combination with other begomoviruses or in mixed infections with recombinants and/or pseudorecombinants.

5.3 Symptomology

In studying mosaic diseases of cassava, Storey and Nichols (1938) described symptoms of chlorosis which may vary from pale yellow to white with only a tinge of green, to barely discernible paler than normal areas on the leaf. Chlorotic areas can vary in size from small specks or spots to larger areas on the leaflet. In terms of distribution, leaflets may show a uniform mosaic pattern or the pattern may be localized to a few areas only. These initial symptoms can lead to secondary symptoms such as general stunting, leaf distortion and a reduction in leaflet size. The severity of symptoms may be affected by the variety of the cassava crop, stage of its infection and temperature (Storey and Nichols 1938).

With ACMV, symptom severity is not always directly correlated with accumulation of the virus. In studying the concentration of ACMV in cassava genotypes of varying degrees of resistance, the resistant genotype NR 8083 showed significantly less severity of symptoms than the susceptible TMS 91934, though there were no significant differences in virus concentration. The moderately resistant TMS30572 and NR 8082 showed significantly less symptom severity than the susceptible TMS 91934 and TME 117 but showed significantly higher concentrations of the virus compared to TMS 91934. The resistant and moderately resistant genotypes with high virus concentrations reflected root yield losses. The researchers indicated that symptom severity is not necessarily linked to virus concentration (Ogbe et al. 2003).

It should be noted however, that an association of virus accumulation and symptom severity has been demonstrated with the Uganda pandemic (Pita et al. 2001), and this is one of the factors more than likely responsible for the spread of the pandemic. Typical symptoms types of EACMV-UG infected cassava plants are shown (Fig. 5.1).

5.4 Spread of EACMV-UG

Cassava mosaic geminiviruses are transmitted by the vector *Bemisia tabaci* (Gennadius) (Homoptera: Aleyrodidae), and the diseases could spread through the use of infected cuttings. Grafting has also been effective in transmission type studies (Thompson 2002). In the 1990s the severe cassava mosaic disease was present in almost the whole of Uganda, Southern Sudan, Western Kenya, Eastern Democratic Republic of Congo, and North Western Tanzania (Otim-Nape et al. 1997).

Tadu et al. (2005) reported the results of a virus survey conducted in 2002 in Southern Sudan. EACMV-UG was detected in 13 of 14 samples tested clearly indicating its presence in Sudan. Also, in a survey conducted in Kenya to determine

Fig. 5.1 Typical symptom types of cassava var. Ebwanateraka infected with EACMV-UG

the factors contributing to incidence and severity of CMD as well as the distribution of the associated begomoviruses, results showed that CMD was highest in Western Kenya and lowest in the Coast Province. In Western and Nyanza Provinces 52% of the samples contained EACMV-UG, 17% contained both EACMV-UG and ACMV and 22% contained ACMV (Were et al. 2004). The researchers indicated that in Western Kenya where intensive cultivation occurred, EACMV-UG often exist in combination with ACMV and was responsible for driving the epidemic.

The epidemic had also moved to other areas. In 2000, a survey conducted in Rwanda to determine the presence of EACMV-UG and severe cassava mosaic disease, showed that ACMV incidence was low in the areas surveyed but CMD symptoms were more severe in north-eastern Umukara and in this area infection was more caused by whiteflies as compared to other areas. EACMV-UG was detected only in Umukara (Legg et al. 2001). As with previous observations, it was found that EACMV- UG in co-infection with ACMV resulted in most severe symptom types. The authors concluded that spread of the epidemic to Rwanda was caused by movement of viruliferous whiteflies from neighboring countries of Uganda and/or Tanzania. As expected, presence of EACMV-UG in Burundi has been subsequently reported (Bigirimana et al. 2003).

5.5 EACMV-UG Infected Host Plants Interaction with *B. tabaci*

To further exacerbate the situation of disease spread, infected plants were often found in association with high population levels of *B. tabaci*. This unusually high population level of the vector was attributed to the presence of a new *B. tabaci* genotype cluster. Two genotype clusters were found: Ug2 within the epidemic area and Ug1 at the front or ahead of the epidemic area. Molecular analysis showed that these two genotypes diverged by *c.* 8% and that Ug2 was non-indigenous to Uganda, and recently underwent an expansion of its population range (Legg et al. 2002).

The notion of new biotype was not supported by Maruthi et al. (2001) who demonstrated through experiments that populations within and outside the epidemic area were not reproductively isolated, did not reflect differences in fecundity, nymphal development, sex ratios and numbers of eclosed adults. Furthermore cluster analysis grouped individuals from the pandemic and non-pandemic areas into a single large group.

In studying the mechanisms driving the epidemic in Uganda, a study was designed where healthy cassava cuttings were planted at eight different sites along a North to South direction in Southern Uganda, moving through the epidemic front. Data were collected on virus incidence, severity of symptom types, population levels of *B. tabaci* and the ability of the vector to transmit the begomoviruses. Six months after planting, disease incidence was highest at sites 1–2 and decreased gradually up to site 6 and was lowest at sites 7 and 8. This concurred with highest *B. tabaci* populations at sites 1 and 3. In 1997 the following year, diseased incidence increased at sites 5 and 6 and this was also associated with high population levels of the vector. Notable was the mixed infection of plants within the epidemic area with both ACMV and EACMV-UG causing most severe symptom types (Colvin et al. 2004). The researchers reported whiteflies collected from within the epidemic area were infective unlike those outside the area. Additionally, EACMV-UG was more easily transmitted than ACMV in the mixed infections. Thus from these findings it is clear that factors of begomoviruses in mixed infections implicated with severe symptom types, and associated with high populations of a viruliferous vector, represent a conducive milieu for disease spread.

To have a better understanding of the factors driving the Uganda CMD pandemic a bifurcating approach or perspective is required, taking into account (1) The virus, its enhanced accumulation and (2) the vector, its fecundity, or induced prolificacy.

5.5.1 The Virus

An epidemic could be driven by factors influencing the concentration level of the pathogen. Such effects have been observed in various pathosystems. In a study to determine the effects of nutritional amendments on tomato plants infected with *Fusarium oxysporum* f.sp. *lycopersici* (col) and either *Tomato mosaic virus* or *Tomato bushy stunt virus* it was found that plants infected with the fungus and either virus showed lower fresh weight and smaller leaf areas, compared to plants infected with either pathogen alone. Since the concentration of either virus was higher per unit leaf fresh weight in co-infected plants maintained across different nutrient media it was suggested that the presence of the fungus was an important factor in virus replication (Chant and Gbaja 1985).

In the Uganda pandemic, various synergistic scenarios have been reported reflecting accumulation of the viruses within infected plants. The double recombinant EACMCV in co-infection with ACMV accounted for increased concentration levels of both viruses as compared to single infection by either virus (Fondong et al. 2000).

Other findings showed similar results where plants infected with EACMV-UG plus ACMV reflected synergism and accumulation of both viruses (Pita et al. 2001). These researchers also reported EACMV-UG DNA-A often in association with EACMV-UG3 DNA B, within the epidemic area. EACMV-UG DNA A was important for the replication of DNA B of EACMV-UG3. Thus, in situations of mixed infections and synergism, where virus/viruses accumulate, a rich source of inoculum is made available to prevailing vectors to drive the pandemic.

5.5.2 *The Vector*

5.5.2.1 Attraction to Olfactory and Visual Cues

This interaction will involve the infected host plant undergoing physical and/or biochemical changes that influence the vector. In a study examining the influence of *Cucumber mosaic virus* infected (*Cucurbita pepo* L., cv. Dixie) on foraging cues of *Myzus persica* (Sulzer) and *Aphis gossypii* Glover, results showed that infected plants with significantly reduced quality, deceptively attracted the vectors by emitting a volatile mix of compounds normally emitted from healthy plants. This resulted in the vectors alighting on plants for only a short period before emigrating (Mauck et al. 2010).This mechanism is highly conducive for a virus that is non-persistently transmitted (Mauck et al. 2010), and create conditions favorable for rapid disease spread.

In similar fashion, whiteflies could be attracted to the symptoms of cassava mosaic virus- infected plants. The chlorosis that is commonly associated with infected leaves may serve as a visual cue. Attraction to yellows has been shown with the whitefly *Trialeurodes vaporariorium* (Westwood) (Gerling and Horowitz 1984), and *B. tabaci* attraction to the yellow spectrum of transmitted light (Mound 1962) has been exploited in the use of yellow sticky traps for monitoring population numbers (Palumbo et al. 1995; Chu and Henneberry 1998). When *B. tabaci* colonize the host plant, its duration of stay is dependent on host quality. Whiteflies remain for longer periods of time on better quality host plants and emigrate from these when they become less favorable. This allows sufficient time for the vector to successfully acquire the virus and transmit it to new host plants. Based on transmission studies (Chant 1958; Dubern 1994), provided the vector is allowed an acquisition and inoculative feed of 4–6 h and 10–30 min respectively with a latency period of 4–8 h, the disease could be successfully transmitted. The efficiency of spread will depend on a number of factors that are incorporated into models to predict various outcomes. Some of these include, the proportion of viruliferous vectors, nutritional quality of infected host plants, vector duration on infected host plants, climatic conditions, acquisition efficiency, vector fecundity, vector distance migrated. Such multiple factors present a complex system of interactions that is highly dynamic, the consequences of which are sometimes difficult to predict.

5.5.2.2 Disease Spread Dynamics

In a study of vector preference for healthy or infected plants to gain insight on pathogen spread, Sisterson (2008) describes two modes of preference: orientation preference which is based on olfactory and visual cues, and feeding preference that is related to gustatory responses. This study revealed that feeding preference for healthy or infected plants does not affect the qualitative relationship between percentage infected plants and rate of pathogen spread. With feeding preference for healthy plants, rate of pathogen spread increased unlike that on infected plants where rate of pathogen spread decreased. On the contrary, orientation preference to healthy or infected plants affected the qualitative relationship between percentage infected plants and rate of pathogen spread. With orientation preference to healthy plants, disease spread occurred at a slower rate with few infected plants, but increased as the number of infected plants increased. With orientation preference for infected plants, pathogen spread occurred at a high rate with few infected plants and then decreased as the number of infected plants increased. The results also showed that the magnitude of insect density or movement could have a greater impact than preference on the rate of spread of a pathogen.

In a study examining an epidemiological model when incorporating *B. tabaci* population dynamics in relation to ACMV, investigators found that with the assumption of healthy cassava cuttings being used exclusively, persistent cycles of disease incidence could occur through high transmission efficiency by the vector or large population levels of the vector. When frequency-dependent use of infected cuttings was applied to the model, persistent disease cycles no longer occurred. Instead there were three possible states: disease elimination, persistence of healthy and infected plants or total infection (Holt et al. 1997). It was observed that in some situations, low initial incidence of ACMV presented the risk of attaining total infection more readily than instances of high incidence. The researchers found that in general, incidence of disease reflected a threshold response to changes in the proportion of infected cuttings, with an increase in their use having little impact on incidence. This observation is similar to that made by Sisterson (2008), who reported a higher rate of pathogen spread with initially a smaller number of infected plants but as the number of infected plants increased rate of pathogen spread slowed in a situation where vector orientation is more geared towards infected plants. Holt et al. (1997) also reported, rouging diseased plants had little effect on incidence but had an effect of moving the system away from the threshold of total collapse of healthy plants. On the contrary intensification in the production of healthy plants had the opposite effect of moving the system closer to this threshold. These findings parallel observations by Sisterson (2008), who indicated higher rate of disease spread as number of infected plants increase, and with vector orientation for healthy plants.

Given this scenario it could be anticipated that viruliferous vectors with preference for healthy plants could move the system towards a threshold of collapse of the healthy plant population. These results present important findings that could provide valuable insight on some of the principal mechanisms involved in the spread of the Uganda pandemic.

5.5.2.3 Host Plant Response Effects on the Vector

Effects on the vector could be direct or indirect. Research showed that ACMV-infected leaves have lower cyanide and protein content (Almazan and Theberge 1989) and other results show differences in *B. tabaci* population numbers in relation to the cyanide glucoside content of plants (Dengel 1981). This researcher reported plants with low, intermediate or high levels of cyanide glucoside were associated with high, intermediate or low population levels of the vector.

Additionally, nutrient stress factors can mobilize nitrogenous compounds and herbivores can take advantage of the increase amount of available nitrogen. These effects could result from certain changes in weather patterns, and the physiological response is similar where a variety of unrelated stress factors impinge on plants or plant parts thus affecting metabolism (White 1984). Plants respond by breaking down and mobilizing nitrogen in soluble form away from stressed tissues. When herbivores feed on such parts they benefit from the enriched supplies of nitrogen and this could have a positive effect on herbivore abundance (White 1984). Based on these results it can be expected that other stress type situations will elicit particular type of responses, one of which is the hormoligosis phenomenon ; where sub-lethal doses of a pesticide could elicit behavioral and/or physiological changes of pest populations.

In studying behavioral hormoligosis of *B. tabaci* ovipositional preference on cotton, American cotton (Var. LH 1556) was subjected to various levels of insecticides *viz*., quinalphos, carbaryl, acephate, endosulfan and fenvalerate. These were sprayed repeatedly on potted plants. Results showed that all the insecticides caused an increase in total free amino acids. Plants treated with fenvalerate were most preferred for ovipositing, with 39.3, 37.3 and 36.1 eggs/leaf on concentrations of 38, 50 and 25 g/ha respectively. Ovipositing was second most preferred on acephate (1,500 g/ha) treated plants with 26.9 eggs/leaf. On the untreated control *B. tabaci* oviposited only 14.1 eggs/leaf (Abdullah et al. 2006). The researchers indicated that the findings confirmed behavioral hormoligosis where *B. tabaci* exhibit ovipositional preference on fenvalerate and acephate treated plants, and this may explain the high population numbers of *B. tabaci* when these insecticides are applied repeatedly.

The amino acid asparagine was found in higher concentrations in EACMV-UG infected plants compared to uninfected plants. In the same experiment, *B. tabaci* was used to transmit the virus to healthy plants and the whiteflies were allowed to remain on the plants and their populations monitored. Results showed higher populations of developing nymphs on infected symptomatic plants as compared to uninfected plants.

The researchers indicated that *B. tabaci* and EACMV-UG may be interacting in a mutually beneficial manner and this mechanism may be responsible for driving the epidemic (Holt and Colvin 2001).

The influence of asparagine on *B. tabaci* performance has been observed when whiteflies were feeding on artificial diets (Thompson 2006). There was the trend of increasing oviposition across a concentration range of 0.07, 0.7, 7.0, 70.0 mg/ml in 20%W/V sucrose diets. The effects of various amino acids are reported in Chap. 8 in this book. Based on experimental findings, it was found that the amino acids play a pivotal role in *B. tabaci* survival and reproduction.

5.5.2.4 The Biotype Factor

The involvement of a non-indigenous *B. tabaci* genotype associated with the Uganda epidemic as suggested by Legg et al. (2002) is in stride with the history of evolving new genotypes of *B. tabaci,* and the fact that new genotypes continue to evolve or invade different geographic areas around the world. As a consequence indigenous populations have been outcompeted or displaced by an invasive genotype. The presence of the B-biotype in the USA is a historical example of a successful invasion and adaptation of an exotic biotype. Brown (1994) reported that in the period 1991–1994, the B-biotype was the only whitefly observed in Arizona, Texas, California, Florida, Georgia, Tennessee, New York and Hawaii. Brown (1994) suggested, based on the earliest available data, there was probably a displacement of the indigenous A-biotype by the B-biotype in these states from as early as 1986. The B biotype may have invaded South Western USA as early as 1981, where unusually high populations of whiteflies were observed causing a number of virus-like disease epidemics on important crops in California and the Arizona desert production areas. Crops significantly affected included cotton and *Cucurbita* spp., with an estimated $8 Million loss on Cucurbits alone (Duffus and Flock 1982). The researchers indicated the presence of *B. tabaci* in California since the late 1920s but it had never before attained such high population numbers. The important begomovirus diseases transmitted by these whiteflies were cotton leaf crumple and Squash leaf curl (Duffus and Flock 1982).

In early 2000, high population numbers of whiteflies were seen associated with the silverleaf condition on *Cucurbita* spp. in Southern Turkey. Later test using mitochondrial cytochrome oxidase I gene and sequence comparisons with reference B-biotype sequences, confirmed its presence and it was reportedly infesting 152 host plant species in Turkey (Bayhan et al. 2006). It was concluded that the invasive B biotype was responsible for the displacement of the indigenous cotton haplotype.

In China for example, as with the Uganda pandemic, high population levels of the vector has been associated with infected plants. In the study by Jiu et al. (2007) the performance of the invasive B-biotype and the indigenous ZHJ1 biotype was compared on healthy plants, *Tobacco curly shoot virus* (TbCSV)-infected tobacco plants and *Tomato yellow leaf curl China virus* (TYLCCNV)-infected tobacco plants.

Results showed that B-biotype increased its longevity and fecundity by 6 and 12-fold respectively on TbCSV-infected plants and 7 and 18-fold respectively on TYLCCNV-infected plants. After 56 days the population density on TbCSV-infected plants and TYLCCNV-infected plants was 2 fold and 13-fold that on healthy plants respectively. There were no significant differences in the performance of the indigenous ZHJ1 biotype on healthy or infected plants. The authors indicated that the interaction between the invasive biotype, and the infected plants could contribute to the displacement of the indigenous biotype, invasion of *B. tabaci* B-biotype, and the disease epidemics associated with this vector.

Thus from these findings, the involvement of a new *B. tabaci* biotype in the Uganda pandemic as suggested earlier by Gibson et al. (1996) and later by Legg et al. (2002) may be a plausible proposition, in view of the similar type interactions and consequences between the invasive vector and begomoviruses in different pathosystems around the world.

5.6 Concluding Remarks

5.6.1 *Scenario of the Uganda Pandemic*

In general, the Uganda pandemic involves factors of virus-infected host plants interacting with the vector either directly or indirectly. The begomoviruses found within pandemic areas occur singly or in combination with other begomoviruses, which could be recombinant or pseudorecombinant viruses. In many cases of co-infection, there is accumulation of viruses through synergism, thereby presenting a rich source of inoculum.

Infected plants reveal the typical mosaic type symptoms, with chlorosis of varying patterns, which could serve as a visual cue for whiteflies. At low infection and with orientation preference for infected plants, rate of spread increases, and as the proportion of infected plants increase, with orientation preference for healthy plants, rate of spread also increases (Sisterson 2008). The infected host plays a major role in driving the pandemic through responses to infection and other stress factors. Such responses result in the mobilization of nutrients from which colonizing vectors could benefit. Elevated populations of the vector have been associated with EACMV-UG infected cassava plants that contained elevated levels of asparagine (Holt and Colvin 2001). Also, it has been reported that the Uganda epidemic was associated with high populations of a genotype cluster thought to be non-indigenous to Uganda (Legg et al. 2002).

Irrespective of the basis of the population boost, it can be expected that high populations of a viruliferous vector could be the driving force of the pandemic. It was pointed out that the magnitude of vector density and migration can have a significant impact on disease spread (Sisterson 2008). A depiction of the events of the pandemic is illustrated (Fig. 5.2).

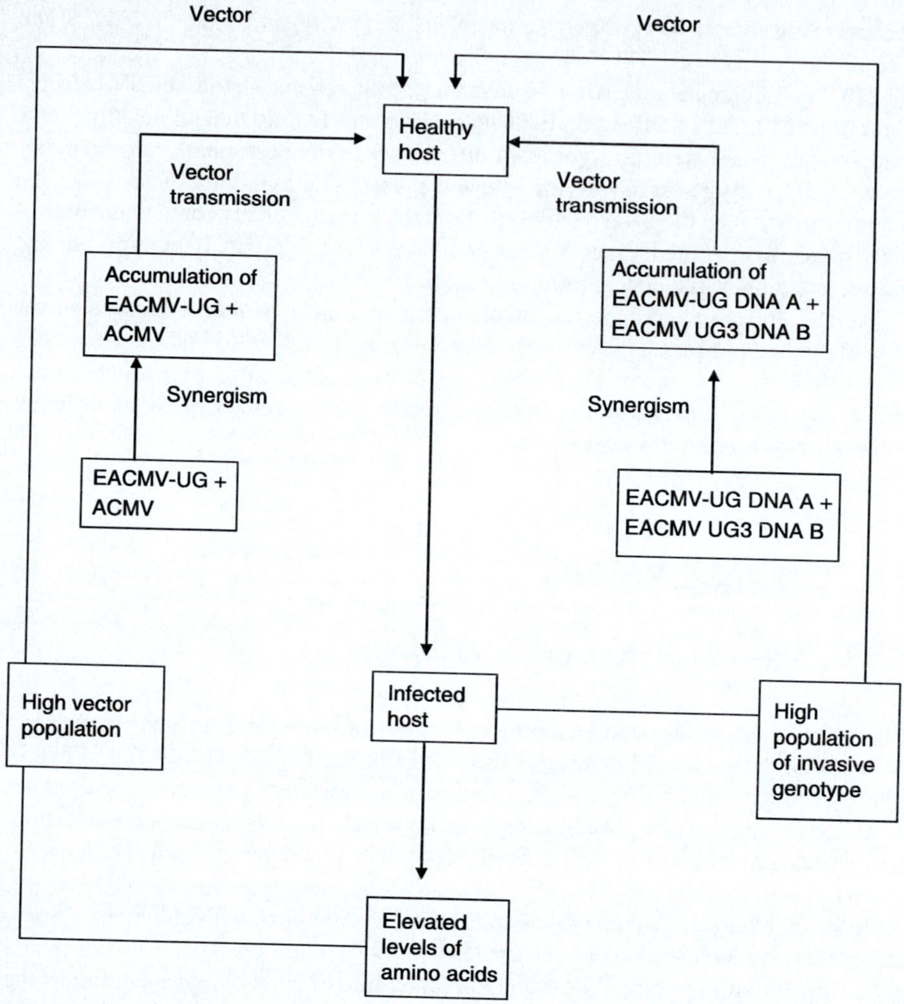

Fig. 5.2 Flow chart of events of the CMD Uganda pandemic

More recently, the co-infection of ACMV with EACMV-UG has been associated with a higher frequency of milder symptom types. Sseruwagi et al. (2004) in studying the begomoviruses in Uganda following the epidemic of the 1990s observed that EACMV-UG was the predominant begomovirus found in the surveyed areas. It was present in 73% of the severely infected plants and 53% in mildly infected plants. ACMV was found mainly in mildly infected plants. Mixed infection of EACMV-UG and ACMV occurred in only 18% of the samples and there was almost a similar distribution between severe and mild symptom forms (16% and 21% respectively).

Unlike previous observations, these results reflected a different situation in Uganda, where the mixed infected plants exhibited in almost equal proportion, the percentage of mildly and severely symptomatic plants.

Legg (1999) had earlier reported progress in managing the Uganda epidemic, through a program of selecting, multiplying and disseminating resistant germplasm material, some of which were obtained from the International Institute of Tropical Agriculture (IITA), Nigeria. Thus the use of resistant varieties is an integral part of the control strategy, and it was suggested that this approach has provided effective control of the pandemic in East Africa (Legg and Thresh 2000). Given the control practices that are promising, and considering the potential threat to other areas, universal awareness and collaborative control measures must be based on sound management principles.

5.6.2 *Principles of Management*

5.6.2.1 Use of Clean Planting Material

Use of virus-free planting material has been recommended (Thresh and Otim-Nape 1994; Bock 1994; Hillocks 1997). The application of virus-free planting material must take serious consideration of the disease pressure in the area as well as the use of resistant varieties. Accordingly such a scheme can be practical and feasible if the disease pressure is low and plants are sufficiently resistant (Bock 1994).

Hillocks (1997) indicated that virus-free cassava plants can be obtained through selection of material within existing stocks, plus subsequent removal (roguing) of infected plants which appear at the sprouting stage. Hillocks (1997) also suggested as an alternative, the generation of virus-free plants for subsequent multiplication for use by farmers, through tissue culture technology coupled with thermotherapy. A similar approach was outlined by Frison (1994) in providing guidelines for the safe movement of cassava planting material. The methodology involved growing cassava cuttings in a growth chamber at temperature 35°C for 1 month. After thermotherapy the apical buds of plants are removed and sterilized. Meristem sticks are then dissected and the explants transferred to glass tubes, which are stored at 29°C (day temperature) and 24°C (night temperature). After about 4–6 weeks, plants developing from the meristem tips are transferred into a soil mixture after which they can be transferred into an insect-free room for virus indexing. Plants free of the virus are multiplied *in vitro*, conserved in medium or long-term storage and safely transported *in vitro*.

5.6.2.2 Roguing

This practice requires farmers' recognition of the disease and willingness to select healthy-looking plants for establishment, then subsequently removing diseased plants (Hillocks 1997). Roguing has been recommended as a cultural control method for cassava in Kenya (Robertson 1987; Guthrie 1990) and Uganda (Otim-Nape et al. 1997). The success of roguing and selection in producing healthy cassava planting material was observed in East Africa (Childs 1957; Bock 1983) and West Africa (Fauquet and Fargette 1990). Roguing however may be inadequate and unacceptable to farmers as a control measure where there is rapid spread of the disease from sources outside the field (Fargette et al. 1990).

Additionally, introduction of clean planting material into infected fields where roguing is not practiced can result in widespread infection of the introduced material. In the early attempts to manage CMD in Uganda, clean planting material of susceptible varieties was introduced into Northern Luwero, Arua, Masindi, Nebbi, Moyo and Kumi districts and distributed to farmers for planting. As a consequence all the fields became totally infected with severe symptoms within 8 months of planting (Otim-Nape et al. 1997). In 1991–1993, a slightly different approach was adopted in the district of Soroti. Before clean material was introduced the extension staff was trained and farmers were advised to destroy all infected plants in the area. This time the results were satisfactory and little re-infection occurred (Otim-Nape et al. 1997).

The importance of rouging has been pointed out by Van den Bosch et al. (2007) who indicated that new and advance control methods for viral diseases of vegetative propagated crops, should take into consideration, the evolutionary responses of the virus in order to reduce the risk of failure. They observed that *in vitro* propagation, breeding and diagnosis have a chance of failure due to selection for virus strains that build-up a high virus titer within plants. The researchers reported the importance of rouging for vegetative propagated plants as having a lower risk of failure since it selects for low virus titer strains and increases the population of healthy plants.

5.6.2.3 Use of Resistant Varieties

Among the most important management approach is the use of resistant varieties. This coupled with healthy planting material could be effective. Morales (2006) reported the use of resistant varieties as a major component in an IPM strategy, followed by phytosanitation and cultural measures.

Based on the experimental findings of Ogbe et al. (2003), cognizance must be given to the basis of resistance. This is an important thought for consideration especially as it relates to management. Resistance in general is based on major principles of tolerance, antixenosis and antibiosis. Where resistance is related to tolerance, in which case obvious effects of infection are not visible, but the pathogen/virus is able to accumulate within the host, there is the danger of a tolerant host serving as a reservoir of inoculum for transmission by vectors. Thus in breeding programs, careful consideration must be given to the basis of resistance most desirably and applicable under field conditions.

5.6.2.4 Avoiding Pesticide Abuse

While pesticides play a pivotal role in control strategies, the abusive use of pesticides deviates from a sustainable pest management philosophy. Morales (2006) pointed out that major contributors to disease epidemics involve the abusive use of pesticides and the utilization of infected plant material. Additionally the injudicious use of pesticides could contribute to disturbed ecosystems, insect pest resistance and the emergence of new insect pests. The situation can be further exacerbated if the principle of hormoligosis becomes operational through pesticide abuse as reported by Abdullah et al. (2006). Abusive use of pesticides should be avoided, and for sustainability, pesticides are best applied within the context of IPM programs. More on management as it pertains to the vector and whitefly transmissible geminiviruses is presented in Chaps. 11 and 12.

References

Abdullah NMM, Singh J, Sohal BS (2006) Behavioral hormoligosis in oviposition preference of *Bemisia tabaci* on cotton. Pestic Biochem Phys 84:1–16

Almazan AM, Theberge RL (1989) Influence of cassava mosaic virus on cassava leaf vegetable quality. Trop Agr (Trinidad) 66:305–308

Bayhan E, Ulusoy MR, Brown JK (2006) Host range, distribution and natural enemies of *Bemisia tabaci* ‘B biotype’ (Hemiptera:Aleyrodidae) in Turkey. J Pest Sci 79:233–240

Berrie LC, Rybicki EP, Rey MEC (2001) Complete nucleotide sequence and host range of *South African cassava mosaic virus*: further evidence for recombination among begomoviruses. J Gen Virol 82:53–58

Bigirimana S, Barumbanze P, Obonyo R, Legg JP (2003) First evidence for the spread of East African cassava mosaic virus-Uganda (EACMV-UG) and the pandemic of severe cassava mosaic disease to Burundi. New Dis Rep 8:Aug–Jan

Bock KR (1983) Epidemiology of cassava mosaic in Kenya. In: Plumb RT, Thresh JM (eds.) Plant virus disease epidemiology. Blackwell, Oxford

Bock KR (1994) Control of African cassava mosaic geminivirus by using virus free planting material. Trop Sci 34:102–109

Brown JK (1994) Current status of *Bemisia tabaci* as a plant pest and virus vector in agro-ecosystems worldwide. FAO Plant Protect B 42:3–32

Chant SR (1958) Studies on the transmission of cassava mosaic virus by *Bemisia* spp (Aleyrodidae). Ann Appl Biol 46:210–215

Chant SR, Gbaja IS (1985) Effect of nutrition on the interaction of viruses and *Fusarium oxysporum* in tomato seedlings. Phytoparasitica 13:47–57

Childs AHB (1957) Trials with virus resistant cassava in Tanga Province, Tanganyika. East Afr Agr J 23:135–137

Chu CC, Henneberry TJ (1998) Development of a silverleaf whitefly (Homoptera: Aleyrodidae) trap. Recent Res Dev Entomol 2:47–54

Cock JH, Reyes JA (eds) (1985) Cassava research production and utilization cassava program. UNDP/CIAT, Cali

Colvin J, Omongo CA, Maruthi MN, Otim-Nape GW, Thresh JM (2004) Dual begomovirus infections and high *Bemisia tabaci* populations: two factors driving the spread of a cassava mosaic disease pandemic. Plant Pathol 53:577–584

Dengel HJ (1981) Untersuchungen über das Auftreten der Imagines von *Bemisia tabaci* (Genn.) auf verschiedenen Manioksorten. Z Pflanzenk Pflanzen 88:355–366
Dubern J (1994) Transmission of Africa cassava mosaic geminivirus by the whitefly *Bemisia tabaci*. Trop Sci 34:82–91
Duffus JE, Flock RA (1982) Whitefly transmitted disease complex of the desert Southwest. Calif Agr 36:4–6
FAO (1998) FAO production yearbook. FAO, Rome
Fargette D, Faucquet C, Grenier E, Thresh JM (1990) The spread of African cassava mosaic virus into and within cassava fields. J Phytopathol 130:289–302
Fauquet C, Fargette D (1990) African cassava mosaic virus: aetiology, epidemiology and control. Plant Dis 74:404–411
Fauquet CM, Stanley J (2003) Geminivirus classification and nomenclature: progress and problems. Ann Appl Biol 142:165–189
Fondong VN, Pita JS, Rey MEC, de Kochko A, Beachy RN, Fauquet CM (2000) Evidence of synergism between African cassava mosaic virus and a new double recombinant geminivirus infecting cassava in Cameroon. J Gen Virol 81:287–297
Frison EA (1994) Sanitation techniques for cassava. Trop Sci 34:146–153
Gerling D, Horowitz AR (1984) Yellow traps for evaluating the population levels and dispersal patterns of *Bemisia tabaci* (Gennadius) (Homoptera: Aleyrodidae). Ann Entomol Soc Am 77(6):753–759
Gibson RW, Legg JP, Otim-Nape GW (1996) Unusually severe symptoms are a characteristic of the current epidemic of mosaic virus disease of cassava in Uganda. Ann Appl Biol 128:479–490
Guthrie J (1990) Controlling African cassava mosaic disease. CTA, Wageningen
Harrison BD, Zhou X, Otim-Nape GW, Liu Y, Robinson DJ (1997) Role of a novel type of double infection in the geminivirus-induced epidemic of severe cassava mosaic in Uganda. Ann Appl Biol 131:437–448
Hillocks RJ (1997) Cassava virus diseases and their control with specific reference to Southern Tanzania. Integr Pest Manag Rev 2:125–138
Holt J, Colvin J (2001) Observation and theory of whitefly-borne virus disease epidemics. In: Jeger MJ, Spence NJ (eds) Biotic interactions in plant-pathogen associations. CAB International, Wallingford
Holt J, Jeger MJ, Thresh JM, Otim-Nape GW (1997) An epidemiological model incorporating vector population dynamics applied to African cassava mosaic virus disease. J Appl Ecol 34:793–806
Jiu M, Zhou X-P, Tong L, Xu J, Yang X, Wan FH, Liu SS (2007) Vector-virus mutualism accelerates population increase of an invasive whitefly. PLoS ONE 2(1):e182. doi:10.1371/journal.pone.0000182
Legg JP (1999) Emergence, spread and strategies for controlling the pandemic of cassava mosaic virus disease in East and Central Africa. Crop Prot 18:627–637
Legg JP, Thresh JM (2000) Cassava mosaic virus disease in East Africa: a dynamic disease in a changing environment. Virus Res 71:135–149
Legg JP, Okao-Okuja G, Mayala R, Muhinyuza JB (2001) Spread into Rwanda of the severe cassava mosaic virus disease pandemic and associated Uganda variant of East African Cassava mosaic virus (EACMV-Ug). New Dis Rep 3: Feb–July
Legg JP, French R, Rogan D, Okao-Okuja G, Brown JK (2002) A distinct *Bemisia tabaci* (Gennadius) (Homoptera: Sternorrhyncha: Aleyrodidae) genotype cluster is associated with the epidemic of severe cassava mosaic virus disease in Uganda. Mol Ecol 11:1219–1229
Maruthi MN, Colvin J, Seal S (2001) Mating compatibility, life history traits and RAPD-PCR variation in *Bemisia tabaci* associated with the cassava mosaic disease pandemic in East Africa. Entomol Exp Appl 99:13–23
Mauck KE, De Moraes CM, Mescher MC (2010) Deceptive chemical signals induced by a plant virus attract insect vectors to inferior hosts. Proc Natl Acad Sci USA 107:3600–3605
Morales FJ (2006) Tropical whitefly IPM project. Adv Virus Res 69:249–311

Mound LA (1962) Studies on the olfaction and colour sensitivity of *Bemisia tabaci* (Genn.) (Homoptera: Aleyrodidae). Entomol Exp Appl 5:99–104

Ogbe FO, Atiri GI, Dixon AGO, Thottappilly G (2003) Symptom severity of cassava mosaic disease in relation to concentration of *African cassava mosaic virus* in different cassava genotypes. Plant Pathol 52:84–91

Otim-Nape GW, Bua A, Thresh JM, Baguma Y, Ogwal S, Semakula GN, Acola G, Byabakama B, Martin A (1997) Cassava mosaic virus disease in Uganda. The current pandemic and approaches to control. Natural Resources Institute, University of Greenwich, Chatham

Palumbo JC, Tonhasca A Jr, Byrne DN (1995) Evaluation of three sampling methods for estimating adult sweet potato whitefly (Homoptera: Aleyrodidae) abundance on cantaloupes. J Econ Entomol 88:1393–1400

Pita JS, Sangare A, Beachy RN, Fauquet CM (1998) Cassava mosaic disease (CMD) in Africa: a parallel study between Uganda (Ug) and Ivory Coast (IC). Phytopathology 88:S 71

Pita JS, Fondong VN, Sangare A, Otim-Nape GW, Ogwal S, Fauquet CM (2001) Recombination, pseudorecombination and synergism of geminiviruses are determinant keys to the epidemic of severe cassava mosaic disease in Uganda. J Gen Virol 82:655–665

Robertson IAD (1987) The whitefly *Bemisia tabaci* (Gennadius) as a vector of African cassava mosaic virus at the Kenyan coast and ways in which the yield losses in cassava Manihot esculenta Crantz caused by the virus can be reduced. Insect Sci Appl 8:797–801

Sisterson MS (2008) Effects of insect vector preference for healthy or infected plants on pathogen spread: insights from a model. J Econ Entomol 101:1–8

Sseruwagi P, Rey MEC, Brown JK, Legg JP (2004) The cassava mosaic geminiviruses occurring in Uganda following the 1990s epidemic of severe cassava mosaic disease. Ann Appl Biol 145:113–121

Storey HH, Nichols RFW (1938) Studies of the mosaic diseases of cassava. Ann Appl Biol 25:790–806

Tadu G, Winter S, Gadelseed AMA, Dafalla DA (2005) Association of East African cassava mosaic virus-Uganda (EACMV-UG) with cassava mosaic disease in Sudan. New Dis Rep 11: Feb–July

Thompson WMO (2002) Comparison of *Bemisia tabaci* (Homoptera: Aleyrodidae) development on uninfected cassava plants and cassava plants infected with East African cassava mosaic virus. Ann Entomol Soc Am 95:387–394

Thompson WMO (2006) Influence of amino acids on cassava biotype *Bemisia tabaci* (Gennadius) (Homoptera: Aleyrodidae) when feeding on an artificial system. J Entomol 3:198–203

Thresh JM, Otim-Nape GW (1994) Strategies for controlling African cassava mosaic geminivirus. Adv Dis Vector Res 10:215–236

Thresh JM, Otim-Nape GW, Legg JP, Fargette D (1997) African cassava mosaic disease: the magnitude of the problem? Afr J Root Tuber Crops 2:13–19

Van den Bosch F, Jeger MJ, Gilligan CA (2007) Disease control and its selection for damaging plant virus strains in vegetatively propagated staple food crops, a theoretical assessment. Proc Royal Soc B 274:11–18

Were HK, Winter S, Maiss E (2004) Occurrence and distribution of cassava begomoviruses in Kenya. Ann Appl Biol 145:175–184

White TCR (1984) The abundance of invertebrate herbivores in relation to the availability of nitrogen in stress food plants. Oecologia 63:90–105

Zhou X, Liu Y, Calvert L, Munoz C, Otim-Nape GW, Robinson DJ, Harrison BD (1997) Evidence that DNA-A of a geminivirus associated with severe cassava mosaic disease in Uganda has arisen by interspecific recombination. J Gen Virol 78:2101–2111

Chapter 6
Interaction of *Bemisia tabaci* with *East African cassava mosaic virus*-Infected Plants

Winston M.O. Thompson

Abstract In examining the interaction of *Bemisia tabaci* with *East African cassava mosaic virus* (EACMV) infected cassava plants, the number of eggs oviposited, developing nymphal instars and eclosed adults were not significantly different between uninfected and EACMV-infected plants. Highest mortality occurred on non-viruliferous first instars and on viruliferous fourth instars on infected plants. Highest mortality was also observed on fourth instars of both viruliferous and non-viruliferous whiteflies developing on uninfected plants. Development of non-viruliferous *B. tabaci* was 62% and 77% on uninfected and infected plants respectively and this was 56% and 59% respectively for viruliferous whiteflies. Developmental period of non-viruliferous whiteflies was 25 days on both uninfected and infected plants. Viruliferous whiteflies reflected a developmental period of 25 days on uninfected plants and 26 days on infected plants. Irrespective of the viruliferous nature of *B. tabaci* and infection status of plants, eclosed adults were always in a 1:1 ratio of females to males. Infected or uninfected plants colonized by *B. tabaci* for at least two generations reflected no significant differences in the number of eggs and subsequent developing juveniles. Subsequent studies revealed no interaction effect between plant health status and the viruliferous status of whiteflies in relation to oviposition and total number of eclosed adults. There were no significant differences in the total number of emerged adults between infected and uninfected plants, and between viruliferous and non-viruliferous whiteflies. The effect of EACMV on *B. tabaci* is neutral, in that the virus has not affected the vector through direct or indirect measures.

W.M.O. Thompson (✉)
Plant Protection, National Agricultural Research Institute, Mon Repos, East Coast Demerara, Guyana, South America
15216 NE 8th St., H3 98007 Bellevue, Washington, USA
e-mail: winstont335@gmail.com

W.M.O. Thompson (ed.), *The Whitefly, Bemisia tabaci (Homoptera: Aleyrodidae) Interaction with Geminivirus-Infected Host Plants*,
DOI 10.1007/978-94-007-1524-0_6, © Springer Science+Business Media B.V. 2011

Keywords *Bemisia tabaci* • Cassava host plants *East African cassava mosaic virus* • Interaction • Population numbers

6.1 Introduction

East African cassava mosaic virus is an important begomovirus affecting cassava in Africa. It is often found in mixed infections within plants and in some cases undergoes recombination with other begomoviruses. Importantly it is one of the parents of the recombinant virus associated with severe cassava mosaic disease in Eastern and Central Africa. In this Chapter, the influence of this virus on *Bemisia tabaci* is examined.

6.1.1 *Taxonomy*

East African cassava mosaic virus (EACMV) is a *Begomovirus* within the family *Geminiviridae*. The cassava mosaic geminiviruses within the genus *Begomovirus* consist of two DNA molecules, each approximately 2.8 kbp. DNA-A is responsible for replication, regulation of genes and encapsidation, and DNA-B codes for two proteins responsible for cell to cell movement, symptom regulation and host range (Stanley et al. 2004).

6.1.2 *Isolates of EACMV from Eastern Africa*

A study by Ndunguru et al. (2005) revealed several isolates in Tanzania that show high sequence similarity. An isolate from Yombo Vituka closely resembled the previously described isolate from Tanzania EACMV-[TZ] and was named EACMV-[TZ/YV]. Another isolate from Tanga bearing close resemblance to an isolate from Kenya (EACMV-[KE/K2B]) was named EACMV-[KE/TZT]. One isolate from Mara region in Lake Victoria showed high sequence identity to EACMV-[KE/K2B] and was named EACMV-[KE/TZM]. Also, an isolate from Kagera region, Northwest Tanzania reflecting very high sequence identity to EACMV-[UG2Svr] was named EACMV-UG2 [TZ10].

In terms of genetic relationships between the isolates, there is high sequence similarity within the A fragment between EACMV-[TZ/YV] and EACMV-[TZ], and between EACMV-[KE/TZT] and EACMV-[KE/K2B]. In the coat protein (CP) sequence, there is high sequence similarity between EACMV-[TZ/YV], EACMV-[TZ] and EACMZV-[ZB] from Zanzibar. There is also sequence similarity between EACMV-[KE/TZT] and EACMV-[KE/K2B]. EACMV-[KE/TZM] showed high CP sequence similarity to EACMZV-[ZB]. Plants infected singly by EACMV-[KE/TZM] show severe symptom types. EACMV-UG2 [TZ10] has high DNA-A sequence, and common region (CR) sequence similarity to EACMV-UG2Svr from Uganda (Ndunguru et al. 2005).

6.1.3 Transmission Characteristics of Cassava Mosaic Geminiviruses (CMGs)

As early as 1957, it was found that *Bemisia* spp. acquire the virus during 4–6 h of feeding (Chant 1958). There is a latency period that ranges from 4–8 h after which the whiteflies transmit the virus with an inoculative feeding period of 10–30 min (Seif 1981; Chant 1958). Viruliferous *B. tabaci* continue to transmit for up to 9 days (Chant 1958; Dubern 1994).

The acquisition feeding is influence by feeding status of whiteflies. Starved whiteflies acquire the virus in 3.5 h while unstarved ones acquire it in 5 h (Seif 1981). Although Cassava mosaic viruses (CMVs) are transmitted by the various nymphal instars, transovarian transmission has not been observed (Dubern 1994).

To a great extent, the host plant plays a pivotal role in successful vector transmission of CMVs. Although optimum transmission has been achieved with ten whiteflies, this number was found to produce a low transmission level on resistant genotypes (Seif 1981).

6.1.4 Vector-Host Plant Interaction

The interaction of *B. tabaci* with EACMV-infected cassava plants warrants attention in the light of the positive influence the Uganda variant of EACMV (EACMV-UG) has on the vector. In general, interactions involving insect vectors and their host plants can produce different results depending on the insect vector, the host plant and the virus involved. In some interactions a higher population of insects has been observed when they colonise uninfected rather than infected plants. A higher population level of *Myus persicae* (Sulzer) occurred on uninfected as compared to *Cucumber mosaic virus* (CMV) infected plants of *Nicotiana tabacum* L., *Gomphrena globosa* L. and *Zinnia elegans* Jacq. (Lowe and Strong 1963). Also, *B. tabaci* has shown preference for uninfected tomato plants as opposed to plants infected with *Tomato yellow leaf curl virus* (TYLCV) (Berlinger 1986).

On the contrary, some insect-host plant interactions produce increased populations of the vector on the infected plants. A higher population of *Thrips tabaci* Lindeman was reported on the weed *Emilia sonchifolia* (L.) infected with the yellow spot virus as compared to healthy plants (Carter 1939). *Aphis fabae* Scop. showed faster reproduction on *Sugar beet mosaic virus*-infected plants than on healthy sugar beet plants (Kennedy 1951) and Ajayi and Dewar (1983) observed higher population levels of *Sitobion avenae* (F.) and *Metopalophium dirhodum* (Walker) on *Barley yellow dwarf virus*-infected wheat plants than on healthy wheat plants.

Higher population levels on infected plants can be attributed to biochemical changes favorable to the vector caused by the infection (Carter 1939; Thresh 1967). Changes in the levels of amino acids in plant tissue are also involved with changing population levels of the vector (Ajayi and Dewar 1983; Colvin et al. 1999).

The interaction of *B. tabaci* with healthy plants versus plants infected with the begomoviruses they transmit has shown different results (Costa et al. 1991; Colvin et al. 1999; Thompson 2002). The proceeding section examines the influence of EACMV-infected plants on *B. tabaci,* based on a series of experiments (Thompson 2002; Thompson unpublished) with the view of having a much better understanding of the interaction of vector, host and EACMV; which represents one of the parents of the recombinant virus (EACMV-UG) associated with severe cassava mosaic virus disease and boosted population levels of the vector.

6.2 *B. tabaci* Interaction with Cassava Host Plants

6.2.1 *Populations of B. tabaci on Uninfected Cassava and Cassava Infected with EACMV Over One Generation*

When 1–4 day-old *B. tabaci* from Uganda were allowed to colonize either EACMV-infected or uninfected cassava plants (*Manihot esculentum* Crantz.), there were no significant differences in the number of eggs/five females/3 days and in the number of subsequent developing nymphal instars between uninfected and infected plants. Similarly there were no significant differences in the total number of individuals that successfully developed into adults. This was the case where either non-viruliferous or viruliferous whiteflies were allowed to colonize the plants (Figs. 6.1 and 6.2) (Thompson 2002). These results show that no significant differences occur in *B. tabaci* population numbers between uninfected cassava plants and those infected with EACMV.

6.2.2 *Mortality Parameters on Developmental Stages*

Mortality could be affected by interaction effects involving host plant health status, and vector virulence. In a laboratory study, highest mortality occurred on non-viruliferous first instars, and on viruliferous fourth instars developing on EACMV-infected plants. On uninfected plants however, mortality was highest on fourth instars of both viruliferous and non-viruliferous whiteflies.

Regression analysis showed that where mortality was highest on the fourth instars (K4), mortality was density-independent. Density-independence is generally more related to environmental conditions (Youdeowei and Service 1983).

Thus, developing fourth instars of either viruliferous or non-viruliferous whiteflies on uninfected plants, and developing fourth instars of viruliferous whiteflies on infected plants were more affected by abiotic rather than biotic factors. Losses of late instar nymphs to abiotic factors has been reported (Fishpool and Burban 1994).

Where highest mortality occurred on the first instars (K1), such mortality was density-dependent. Based on regression analysis, there was an inverse relationship between mortality and population density of the first instars, indicating reduced

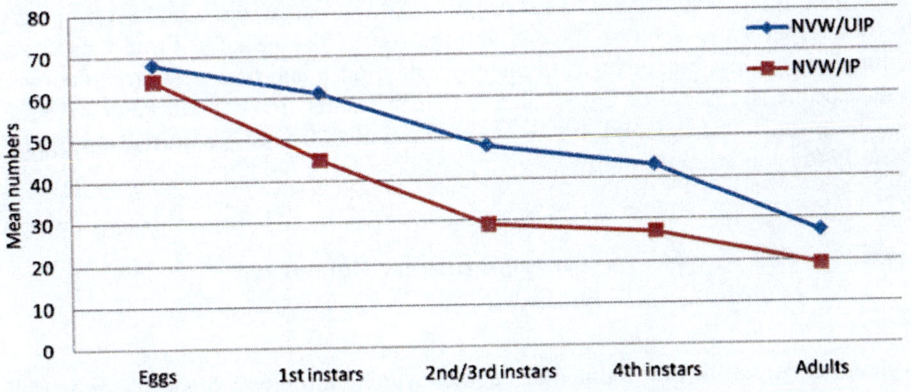

Fig. 6.1 Mean number of eggs, nymphal instars and adults when non viruliferous *B. tabaci* colonize uninfected or infected cassava plants. *NVW/UIP* Non viruliferous whiteflies on uninfected plants, *NVW/IP* Non viruliferous whiteflies on infected plants

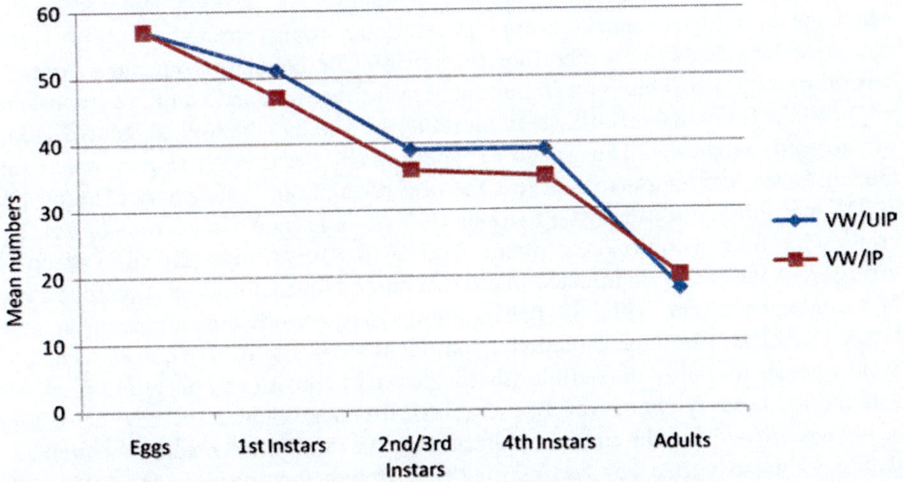

Fig. 6.2 Mean number of eggs, nymphal instars and adults when viruliferous *B. tabaci* colonize uninfected or infected cassava plants. *VW/UIP*, Viruliferous whiteflies on uninfected plants, *VW/IP*, Viruliferous whiteflies on infected plants

mortality of first instar nymphs as the population increased. Other studies have shown high mortality of first instars and that climate often influences this parameter (Horowitz et al. 1984; Mound 1983). Furthermore Riley et al. (1996) revealed highest losses in the small instars of *B. tabaci* on pepper, cotton and lemon and such losses were attributed to host plant effects. Here the findings show that mortality of the first instars can also be affected by population density.

In a study to determine the rates and sources of mortality of *B. tabaci* on field grown cassava in Uganda, highest mean rate of marginal mortality was observed on the fourth instar, followed by mortality on the eggs, first instar, second instar and third instar respectively (Asiimwe et al. 2007). Furthermore highest irreplaceable mortality was seen on the egg and the fourth instar stages. From these findings, it is evident that apart from the eggs, the fourth and first instars are the most susceptible to mortality based on consistent results in both laboratory studies (Thompson 2002) and field studies (Asiimwe et al. 2007), thus contributing significantly to generation mortality.

6.2.3 The Key Factor Concept and its Relevance to Population Dynamics

Identification of the developmental stage on which mortality is highest, is representative of a first step in a life table analysis of *B. tabaci* on cassava. There are few studies on life table analysis of *B. tabaci* (Riley et al. 1996) and these have been attempted mainly on cotton (Horowitz et al. 1984). There have been no published reports on life table analysis of *B. tabaci* on cassava until recently (Asiimwe et al. 2007). In life table studies it is possible to identify the developmental stage on which the mortality accounts for major generation mortality through several generations (the key factor) and therefore responsible for population changes through several generations. This is an important area for research and can have important implications for management, where mortality caused by a biological control agent can be used to replace the key factor (Youdeowei and Service 1983) or complement the key factor in attempts to suppress the pest population. In the study (Thompson 2002), mortality on fourth instar nymphs (K4) of either non-viruliferous or viruliferous whiteflies on uninfected plants, and on first instar nymphs (K2) of non-viruliferous whiteflies on infected plants, accounted for major generation mortality in a single generation. Although these findings may provide indications of the key factor (K4 or K2) in those situations, additional studies will be required to determine whether mortality of fourth or first instars will consistently account for major generation mortality over several generations. It is possible that the key factor can be influenced by the health status of plants and/or the viruliferous nature of whiteflies. If this is the case, then key factors may be different for juveniles developing on infected versus uninfected plants. Possible different effects of different key factors on populations developing on infected or uninfected plants may in part account for population differences between infected and uninfected plants over several generations. This is an area of research, which can be explored in attempts to understand the medium to long terms effects of plant health status on population levels of the vector, when introducing the key-factor concept.

6.2.4 Sex Ratio

Experimental findings consistently show a 1:1 ratio of females to males of *B. tabaci* developing on uninfected plants or infected plants irrespective of the viruliferous nature of ovipositing females.

Table 6.1 Results of Chi Square test of sex ratios for the different whitefly/plant combinations

Whitefly/plant combination	Sex ratio (females: males)	Chi Square value (1 DF) including Yates correction factor	Significance
Non-viruliferous whiteflies on uninfected plants	0.82: 1	2.476	$P>0.05$
Non-viruliferous whiteflies on infected plants	0.96: 1	0.048	$P>0.05$
Viruliferous whiteflies on uninfected plants	1.23: 1	1.370	$P>0.05$
Viruliferous whiteflies on infected plants	1.14: 1	0.393	$P>0.05$

In an experiment where non-viruliferous whiteflies were investigated, the total proportion of the sexes on collectively all uninfected plants was 45% females and 55% males and on the infected plants 49% females and 51% males. Overall, sex ratio (F: M) was 0.82:1 on collectively all uninfected plants and was 0.96:1 on the infected plants (Table 6.1). Based on the Chi Square test and including Yates correction factor adopted from Little and Hills (1978), these ratios did not deviate significantly from a 1:1 ratio (Thompson 2002).

Where viruliferous whiteflies were investigated, the total proportion of the sexes on uninfected plants was 55% females and 45% males and on infected plants 53% females and 47% males. Overall sex ratio on uninfected plants was 1.23:1 and on infected plants was 1.14:1 (Table 6.1). Here again, sex ratios were not significantly different from a 1:1 ratio (Thompson 2002).

6.2.5 *Development and Measures of Central Tendency*

There is no significant difference in the rate of emergence of *B. tabaci* developing on infected or uninfected plants. A study showed colonization of plants by non-viruliferous whiteflies resulted in a mean % emergence of 62 and 77 on uninfected and infected plants respectively. Mean % emergence was 56 and 59 respectively on uninfected and infected plants colonized by viruliferous whiteflies (Thompson 2002).

Developmental period was insignificantly different between various combinations of plant health status and whitefly viruliferous status, with a minimum of 21 days and maximum of 29 days. Where plants were colonized by non-viruliferous whiteflies, mean developmental period to adulthood was 25 days on both uninfected and infected plants, and frequency distribution tables showed that the highest frequency of adult emergence (the mode) occurred at 25 days After Adult Introduction (AAI) on uninfected plants. Highest frequency of emergence occurred at 26 days (AAI) for juveniles developing on the infected plants (Table 6.2) (Thompson 2002). For viruliferous whiteflies, mean developmental period was 25 days and the highest frequency of emergence occurred on day 23 for whiteflies developing on uninfected

Table 6.2 Parameters (days) for the development of *B. tabaci* on cassava plants

Whitefly plant combination	Mean ± SEM	Mode	Range	Minimum	Maximum
Non-viruliferous whiteflies on uninfected plants	25.33 ± 0.40	25	4	23	27
Non-viruliferous whiteflies on infected plants	24.73 ± 0.62	26	7	22	29
Viruliferous whiteflies on uninfected plants	24.77 ± 0.17	23	8	21	29
Viruliferous whiteflies on infected plants	25.81 ± 0.16	26	7	22	29

plants. On infected plants mean development as well as highest frequency of emergence occurred at 26 days AAI (Table 6.2) (Thompson 2002).

6.2.6 Assessment of At Least Two Generations of *B. tabaci* on Uninfected and Infected Plants

In a study to determine the influence of EACMV-infected cassava on *B. tabaci,* non-viruliferous whiteflies were allowed to colonize uninfected or infected plants for 3 days. The number of eggs was counted the day of whitefly removal and a recording was made of the developing juveniles at 11, 17 and 20 days AAI. This schedule of monitoring was based on a study on the development of cassava biotype *B. tabaci* under similar conditions (Thompson 2000). The upper part of the plants containing the developing nymphs was enclosed within a perforated plastic bag 3 days before expected emergence of the adults. Emerged adults were allowed to remain on the plants for a further 4 weeks, when another generation was expected to develop.

In the first 3 weeks after introduction of the whiteflies, there were no significant differences in the number of eggs and in the subsequent nymphal instars developing on uninfected or infected plants. When the emerged adults were allowed to remain on the plants for an additional 4 weeks, the numbers of first and second/third instars were not significantly different between the groups on uninfected or infected plants. There were however, a significantly higher number of fourth instars as well as empty pupal cases on uninfected as compared to infected plants. Considering the total number of nymphs of all instars, and empty pupal cases, no significant differences in total numbers were detected between the groups on uninfected or infected plants (Thompson 2002).

Further studies (Thompson unpublished), examined the interaction effect/significance resulting from the association of host plant health status and virulence or non-virulence of the vector. Experimental plants consisted of uninfected and infected Ebwanateraka cv. cassava plants. The experiment was set up in the form of a 2×2 factorial design: Plant health status (infected and uninfected plants) × Whitefly viruliferous status (viruliferous and non-viruliferous). Five females each from the non-viruliferous and viruliferous *B. tabaci* colonies were caged separately onto a young leaf (position 1–4 from the top of the plant) of the same cassava plant either uninfected or infected with EACMV. The whiteflies were caged on the plants for

48 h after which the leaves enclosed by the clip cages were examined for the number of eggs oviposited using a dissecting microscope. At 22 days AAI, plants were monitored for adult emergence over a 9 day period. Total emerged adults were counted for the various treatment combinations.

In terms of oviposition, there was not an interaction effect between plant health status and whitefly viruliferous status. Also there were no significant differences in the number of eggs oviposited by viruliferous or non-viruliferous whiteflies regardless of the health status of plants colonized. There was however a significantly higher number of eggs oviposited on infected plants than on uninfected plants.

When examining the number of adults that had emerged from eggs, there was not a significant interaction between whitefly viruliferous status and plant health status. The number of adults that had emerged from eggs was not significantly different between non-viruliferous and viruliferous whiteflies, and between uninfected and infected plants. The generation survival was generally higher on the uninfected plants although the difference was not significant.

These findings show that no significant differences exist in *B. tabaci* populations developing on uninfected and EACMV-infected plants. Secondly, viruliferous *B. tabaci* did not produce significantly higher populations than non viruliferous *B. tabaci.* The larger number of eggs oviposited on the infected plants did not result in significantly higher population levels on infected as opposed to uninfected plants because of poorer generation survival on infected as compared to uninfected plants.

6.3 Concluding Remarks

6.3.1 *Influence of EACMV-Infected Cassava Plants on B. tabaci*

The experimental findings from the series of experiments investigating the effects of EACMV-infected cassava plants on *B. tabaci* revealed that the important biological parameters such as sex ratio, development period and percentage emergence were not significantly different between unifected and infected plants regardless of the viruliferous status of the whiteflies. Thus there were no biological advantages of one host type over the other (infected versus uninfected plants) that would trigger a population boost of the vector. The consequence of the interaction of *B. tabaci* with EACMV-infected cassava plants is therefore considered neutral.

6.3.2 *EACMV in Combination with ACMV and Recombination Events*

Infection with EACMV in combination with ACMV could exert differing effects than infection with EACMV alone. A study showed that *Nicotiana benthamiana* plants infected with the Cameroon isolates of EACMV together with ACMV produced more severe symptom types than when infected with either virus alone. Additionally Southern blot analysis of the DNAs revealed higher accumulation of both DNA components of EACMV and ACMV in the mixed infection as compared to single infection by either virus, suggesting a synergistic reaction between EACMV and ACMV (Fondong et al. 2000).

Mixed infection of ACMV from Uganda with the Uganda variant of EACMV (the recombinant EACMV-UG) also produced severe symptom types suggesting a synergistic reaction between ACMV and EACMV-UG (Pita et al. 2001). The severe cassava mosaic virus disease has been characterized by the presence of this recombinant virus in association with high population levels of *B. tabaci* (Legg et al. 2002; Colvin et al. 2004). It is important to note that this vector population explosion in the presence of EACMV-UG-- a virus evolved from the recombination of parent viruses: ACMV and EACMV, has not been seen in the presence of EACMV alone based on the series of experiments (Thompson 2002; Thompson unpublished). It is evident that the consequences of the interaction of *B. tabaci* with EACMV-UG infected plants are quite different to the effects of *B. tabaci* interaction with EACMV-infected plants.

Recombination events of cassava begomoviruses coupled with *B. tabaci,* have caused tremendous losses to cassava growing areas in Eastern and Central Africa in the prevailing pandemic. What is of particular importance is the fact that several recombinant events have resulted in epidemics around the world. A typical example is the severe strain of *Tomato leaf curl New Delhi virus* (tomato leaf curl New Delhi virus-[India: New Delhi: Severe: 1992]) in pseudorecombination with the Varanasi strain of *Tomato leaf curl Gujarat virus* (tomato leaf curl Gujarat virus-[India: Varanasi: 2001]) responsible for severe losses in tomatoes in Northern India (Chakraborty et al. 2008).

6.3.3 Future Trends

6.3.3.1 *B tabaci* Genotypes in Eastern Africa

In a study to examine the vector-virus dynamics related to the spread of cassava mosaic diseases (CMDs) in the post epidemic area in Uganda, sampling of *B. tabaci* from cassava and 22 other plant species revealed the presence of previously reported genotypes; Ugandan 1 (Ug1), Ug2 and Ug 8. *B. tabaci* Ug1 infested cassava as well as non-cassava species such as *Jatropha gossypifolia, Manihot glaziovii, Aspilia aficana, Euphorbia heterophylla* and *Abelmoschus esculentus,* while Ug2 infested only cassava. Ug8 was found to infest *Ipomoea batatas* (sweet potato), *Lycopersicon esculentum* (tomatoes) and *L. nepetifolia* (Sseruwagi et al. 2005). In addition to these, five new genotypes were observed in Uganda and recorded for the first time; Ug3, Ug4, Ug5, Ug6 and Ug7. Ug3 was unique in forming its own cluster and colonized only a single plant species, *Ocimum gratissimum.* The Ug4 genotype had as its closest relative *B. tabaci* on Okra from Ivory Coast, while the Ug5 and Ug6 genotypes had as their closest relatives the North African/Mediterranean/Middle Eastern genotypes, which also contain the B and Q biotypes. *B. tabaci* Ug7 showed close resemblance to *B. tabaci* from Reunion Island, West of the Indian Ocean (Sseruwagi, et al. 2005). This recent observation of new genotypes within this ecosystem is a stark reminder of emerging genotypes of this pernicious vector.

6.3.3.2 The Spread of EACMV

The spread of EACMV across the African continent could be anticipated with the movement of plant material and the vector. A survey conducted in 1997 revealed the presence of the disease in Nigeria (Ogbe et al. 1999). The researchers reported EACMV in association with ACMV in mixed infections with plants showing severe symptom types. More recently, Adjata et al. (2008) reported the presence of EACMV for the first time in Togo. Based on PCR analysis of infected cassava plants EACMV was detected in all five of the regions sampled. Interestingly, it occurred more commonly in mixed infections with ACMV. When examining the incidence of occurrence vis-à-vis single and mixed infection, plants with mixed infection were at least twice as many as those infected with EACMV alone (Adjata et al. 2008).

This scenario is reminiscent of the situation in Uganda, where similar type factors and conditions provided the catalyst for recombination events and the emergence of a hybrid-type begomovirus. A similar type of epidemic could occur in Togo and other parts of Africa if EACMV and ACMV engage in recombination in the presence of a highly fecund vector with the capabilities of a B-biotype, or the Uganda cassava genotype associated with the present pandemic in Eastern and Central Africa.

References

Adjata KD, Muller E, Peterschmitt M, Aziadekey M, Gumedzoe YMD (2008) Incidence of cassava viral diseases and first identification of *East African cassava mosaic virus* and *Indian cassava mosaic virus* by PCR in cassava (*Manihot esculenta* Crantz.) fields in Togo. Am J Plant Physiol 3:73–80

Ajayi O, Dewar AM (1983) The effect of barley yellow dwarf virus on field populations of the cereal aphids *Sitobion avenae* and *Metopolophium dirhodum*. Ann Appl Biol 103:1–11

Asiimwe P, Ecaat JS, Otim M, Gerling D, Kyamanywa S, Legg JP (2007) Life-table analysis of mortality factors affecting populations of *Bemisia tabaci* on cassava in Uganda. Entomol Exp Appl 122:37–44

Berlinger MJ (1986) Host plant resistance to *Bemisia tabaci*. Agric Ecosyst Environ 17:69–82

Carter W (1939) Populations of *Thrips tabaci* with special reference to virus transmission. J Anim Ecol 8:261–271

Chakraborty S, Vanitharani R, Chattopadhyay B, Fauquet CM (2008) Supervirulent pseudorecombination and asymmetric synergism between genomic components of two distinct species of begomovirus associated with severe tomato leaf curl disease in India. J Gen Virol 89:818–828

Chant SR (1958) Studies on the transmission of cassava mosaic virus by *Bemisia* spp (Aleyrodidae). Ann Appl Biol 46:210–215

Colvin J, Otim-Nape GW, Holt J, Omongo C, Seal S, Stevenson P, Gibson G, Cooter RJ, Thresh JM (1999) Factors driving the current epidemic of severe cassava mosaic disease in East Africa. In: VIIth international plant virus epidemiology symposium. Plant virus epidemiology: current status and future prospects. Aguadulce (Almeria), 11–16 Apr 1999, pp 76–77

Colvin J, Omongo CA, Maruthi MN, Otim-Nape GW, Thresh JM (2004) Dual begomovirus infections and high *Bemisia tabaci* populations: two factors driving the spread of a cassava mosaic disease pandemic. Plant Pathol 53:577–584

Costa HS, Brown JK, Byrne DN (1991) Life history traits of the whitefly, *Bemisia tabaci* (Homoptera: Aleyrodidae), on six virus-infected or healthy plant species. Environ Entomol 20:1102–1107

Dubern J (1994) Transmission of African cassava mosaic geminivirus by the whitefly (*Bemisia tabaci*). Trop Sci 34:82–91

Fishpool LDC, Burban C (1994) *Bemisia tabaci*: the whitefly vector of African cassava mosaic geminivirus. Trop Sci 34:55–72

Fondong VN, Pita JS, Rey MEC, de Kochko A, Beachy RN, Fauquet CM (2000) Evidence of synergism between African cassava mosaic virus and a new double-recombinant geminivirus infecting cassava in Cameroon. J Gen Virol 81:287–297

Horowitz AR, Podoler H, Gerling D (1984) Life table analysis of the tobacco whitefly *Bemisia tabaci* (Gennadius) in cotton fields in Israel. Acta Oecol-Oecol Appl 5:221–233

Kennedy JS (1951) Benefits to aphids from feeding on galled and virus-infected leaves. Nature 168:825–826

Legg JP, French R, Rogan D, Okao-Okuja G, Brown JK (2002) A distinct *Bemisia tabaci* (Gennadius) (Hemiptera: Sternorrhyncha: Aleyrodidae) genotype cluster is associated with the epidemic of severe cassava mosaic virus disease in Uganda. Mol Ecol 11:1219–1229

Little TM, Hills FJ (1978) Agricultural experimentation: design and analysis. Wiley, New York

Lowe S, Strong FE (1963) The unsuitability of some viruliferous plants as host for the green peach aphid, *Myzus persicae*. J Econ Entomol 56:307–309

Mound LA (1983) Biology and identity of whitefly vectors of plant pathogens. In: Plumb RT, Thresh JM (eds.) Plant virus epidemiology: the spread and control of insect borne viruses. Blackwell Scientific Publications, Oxford

Ndunguru J, Legg JP, Aveling TAS, Thompson G, Fauquet CM (2005) Molecular biodiversity of cassava begomoviruses in Tanzania: evolution of cassava geminiviruses in Africa and evidence for East Africa being a center of diversity of cassava geminiviruses. Virol J 2:21

Ogbe FO, Atiri GI, Robinson D, Winter S, Dixon AGO, Quin FM, Thottappilly G (1999) First report of East African cassava mosaic begomovirus in Nigeria. Plant Dis 83:398

Pita JS, Fondong VN, Sangaré A, Otim-Nape GW, Ogwal S, Fauquet CM (2001) Recombination, pseudorecombination and synergism of geminiviruses are determinant keys to the epidemic of severe cassava mosaic disease in Uganda. J Gen Virol 82:655–665

Riley D, Nava-Camberos V, Allen J (1996) Population dynamics of *Bemisia* in agricultural systems. In: Gerling D, Mayer RT (eds.) Bemisia 1995: taxonomy, biology, damage, control and management. Intercept, Andover

Seif AA (1981) Seasonal fluctuation of adult population of the whitefly, *Bemisia tabaci* on cassava. Insect Sci Appl 1:363–364

Sseruwagi P, Legg JP, Maruthi MN, Colvin J, Rey MEC, Brown JK (2005) Genetic diversity of *Bemisia tabaci* (Gennadius) (Hemiptera: Aleyrodidae) populations and presence of the B biotype and a non-B biotype that can induce silverleaf symptoms in squash, in Uganda. Ann Appl Biol 147:253–265

Stanley J, Bisaro DM, Briddon RW, Brown JK, Fauquet CM, Harrison BD, Rybicki EP, Stenger DC (2004) Geminiviridae. In: Fauquet CM, Mayo MA, Maniloff J, Desselberger U, Ball LA (eds.) Virus taxonomy, VIIIth report of the ICTV, 8th edn. Elsevier/Academic, London

Thompson WMO (2000) Development, morphometrics and other biological characteristics of the whitefly *Bemisia tabaci* (Gennadius) on cassava. Insect Sci Appl 20:251–258

Thompson WMO (2002) Comparison of *Bemisia tabaci* (Homoptera: Aleyrodidae) development on uninfected cassava plants and cassava plants infected with *East African cassava mosaic virus*. Ann Entomol Soc Am 95:387–394

Thresh JM (1967) Increased susceptibility of black currant bushes to the gall mite vector (*Phytoptus ribis* Nal.) following infection with reversion virus. Ann Appl Biol 60:455–467

Youdeowei A, Service MW (1983) Pest and vector management in the tropics. Longman, London

Chapter 7
Bemisia tabaci (Genn.): Biotypes and Cassava Mosaic Virus in India

M.S. Palaniswami and T.J. Henneberry

Abstract Biology of *Bemisia tabaci* on five different host plants *viz*., cassava, sweet potato, cotton, egg plant and tobacco showed that developmental duration was maximum on cassava and fecundity was maximum on tobacco. Population of cassava whitefly (CWF) and sweet potato whitefly (SPWF) separately maintained on the respective host plants were used in two sets of experiments *viz*., choice and no-choice studies on cassava, sweet potato, egg plant, cotton, tobacco and tomato. Oviposition was recorded on all host plants, but no CWF nymphs emerged beyond the first instars on sweet potato, while there was no pupal formation or adult development of SPWF on cassava. In choice tests, CWF reproduced on cassava, egg plant, tomato and tobacco, but not on cotton and sweet potato. Conversely, SPWF reproduced on sweet potato, cotton, egg plant, tomato and tobacco, but not on cassava, indicating two different biotypes. In isozymes studies five non-shared alleles to CWF and six to SPWF were observed, ascertaining the biotypes. The analysis of amplification products obtained using the single primers revealed that the CWF and SPWF do not share any similarity among them, confirming the biotypes. Secondary endosymbionts were not associated with the two biotypes.

Role of *B. tabaci* CWF and SPWF on *Indian cassava mosaic virus* (ICMV) transmission was elucidated from cassava to cassava and cassava to tobacco, with different number of whiteflies and different acquisition access feeding periods (AAFP) and inoculation access feeding periods (IAFP). Even a single whitefly was capable of ICMV transmission under 10 h/10 h AAFP/IAFP and maximum percentage of transmission was in 48 h/48 h AAFP/IAFP. ICMV transmission in

---

M.S. Palaniswami (✉)
Division of Crop Protection, Central Tuber Crops Research Institute, Trivandrum 695017, India
e-mail: palaniswamims@yahoo.com

T.J. Henneberry
Western Cotton Research Laboratory, Phoenix, AZ 85040–8803, USA

W.M.O. Thompson (ed.), *The Whitefly, Bemisia tabaci (Homoptera: Aleyrodidae) Interaction with Geminivirus-Infected Host Plants*,
DOI 10.1007/978-94-007-1524-0_7, © US Government 2011

cassava plants through *B. tabaci* was ascertained using dot-blot immunoassay, TAS-ELISA and Immuno-scanning electron microscopy. ICMV DNA tests in whitefly stylet, salivary gland and digestive tract showed positive.

The activity of *rhodanese* and *beta-cyanoalanine synthase*; cyanide detoxifying enzymes in *B. tabaci* reared on cassava and sweet potato gave supportive evidence for different biotypes. Whitefly feeding induced pathogenesis related (PR) proteins in both ICMV-free and ICMV-infected cassava plants. Heavy infestation of *B. tabaci* showed increased level of PR proteins. Quantitative measurement and electrophoretic detection of total protein, *peroxidase*, *chitinase* and *ß 1,3 glucanase* in both ICMV-infected and ICMV-free cassava plants under whitefly infested and non-infested conditions revealed that cassava plants with whitefly infestation had an increased level of *peroxidase, chitinase* and *ß 1,3 glucanase,* and decreased level of total proteins.

7.1 Introduction

Insects are by far the most important group of plant virus vectors, both in terms of virus transmission and the economic importance of the transmitted viral diseases. Many insect vectors are also important insect pests. There are more than 14 viruses transmitted by the group of whiteflies belonging to Aleyrodidae. Among 1,300 whitefly species in over 120 genera, only *Bemisia* and *Trialeurodes* are significant in plant virus transmission. In the genus *Bemisia* only *Bemisia tabaci* is found to be an active vector; whereas in *Trialeurodes* genus the species *Trialeurodes vaporariorum, T. abutilonica* and *T. ricini* transmit plant viruses. Among the 37 species of whiteflies identified, *B. tabaci* is the most economic and ubiquitous, distributed worldwide, in tropical, subtropical and temperate zones. Of the known 22 synonyms of *B. tabaci*, seven are from India. There were host associated variations in *B. tabaci* indicating that the Synonyms were in all likehood biotypes of *B. tabaci*. It is further unique in that its host range is wide and varied giving rise to biological types (Palaniswami 2004; Palaniswami et al. 2005).

B. tabaci commonly known as the sweet potato whitefly is an important vector of many plant viruses, causing economic damage to many horticultural crops as well as commercial crops. *B. tabaci* was first observed in India as early as 1905. The center of origin has been suggested as the Indian subcontinent due to the various types of natural enemies observed in the region. By 1919 it had emerged as a menace to cotton farmers and subsequent it posed serious threat to Indian agriculture from Kanyakumari to Kashmir. This insect has been recorded on more than 600 plant species belonging to 74 families and the list is growing. The whitefly outbreaks have been reported during 1985–1987 in southern India and during 1987–1995 in northern India on cotton, tobacco, tomato, egg plant and several horticultural crops. The nymphs and adults of whitefly feed on sap by inserting their sucking mouthparts into the leaf penetrating the phloem region. During feeding plant viruses are consumed along with the plant sap and in turn transmitted to new plants when feeding occurs. Damage caused by whiteflies are many fold *viz*., direct feeding causing stress to plants, virus vectoring, soothy mould due to honey dew, plant growth disorders due to injection of toxins and phoresy (carrying other harmful plant organisms from plant to plant). *B. tabaci* has gained the attention of the world as a pest as well as a vector since the 1980s (Faust 1992). Differences in plant virus transmission capabilities between different populations of *B. tabaci* highlighted the presence of biotypes within *B. tabaci*. Describing so many species of *Bemisia* and later on making them synonyms of *B. tabaci* indicates that *B. tabaci* is a species complex (Perring et al. 1991; Perring 2001) with many biotypes. *B. tabaci* is unique in that its host range is wide and varied giving rise to biological types. In India Palaniswami and Nair (1995) indicated the presence of host associated variation in *B. tabaci*.

B. tabaci species complex transmit a number of viruses belonging to the genus Begomovirus, which are generally bipartite. Geminiviruses form the second largest family of plant viruses, the *Geminiviridae*, represented by four genera: *Mastrevirus, Curtovirus, Topocuvirus* and *Begomovirus*, depending on their vector, host range and genomic characteristics. More than 80% of the known geminiviruses are transmitted by whiteflies and belong to the genus *Begomovirus*. The whitefly-transmitted Geminiviruses produce a wide range of plant symptoms. In some cases localized yellowing or chlorotic spots occur but more often symptoms are yellow veins or bright yellow or golden mosaic. Geminiviruses cause significant yield losses to many crop plants throughout the world (Brown 1994; Brown et al. 1995b). The Genus *Begomovirus* as transmitted by *B. tabaci* has been reported from South and Central America and the African and Indian subcontinents. Among the begomoviruses transmitted by *B. tabaci* cassava mosaic viruses are important. Worldwide many begomoviruses infect cassava *viz.*, *African cassava mosaic virus* (ACMV), *East African cassava mosaic virus* (EACMV), *Indian cassava mosaic virus* (ICMV), *Sri Lankan cassava mosaic virus* (SLCMV) and *South African cassava mosaic virus* (SACMV). These five viruses have a largely non overlapping distribution. There are also recombinant viruses attacking cassava such as EACMV–UG in Uganda resulting from recombination between ACMV and EACMV.

Cassava (*Manihot esculenta* Crantz., Euphorbiaceae) is an important source of food for more than one fifth the world population spread over Africa, Asia and South America. Cassava tubers are a rich source of carbohydrate and leaves are high in proteins, minerals and vitamins. In India cassava was introduced through Kerala by the Portuguese during the eighteenth century. The tubers of cassava saved many lives during famine conditions especially in Kerala. Now the crop has become an important industrial crop for production of sago, starch, animal alcohol and eri silk rearing especially in Tamil Nadu, Kerala, Andhra Pradesh, Maharashtra and Assam states of India. Cassava mosaic disease (CMD) was first observed in Kerala during 1956 and symptoms include leaf chlorotic mottle, distortion of leaves, crinkling and stunting of cassava plant parts.

CMD is an important constraint in the productivity of cassava causing economic loss depending on the degree of infection. It was recognized as a major disease in as early as the 1940s, as farmers were using symptomless or CMD-infected stems for cultivation. Yield loss due to CMD ranges from 17% to 42% (Palaniswami et al. 1996).

The incidence and intensity of CMD varies with varieties and geographic location. In India mean incidence of CMD was about 30% in Tamil Nadu, 23% in Kerala and 20% in Andhra Pradesh. However it was very low (<5%) in Maharashtra, Uttar Pradesh, Gujarat, Chattisgarh and Jharkhand states. The disease is also prevalent in North Eastern states of India. ICMV transmission has been reported from cassava to tobacco, cucurbits and cassava. The field spread of ICMV varied from 5% to 30% depending on varieties. In H 97, H 165 and M4 it was less than 5% while in the indigenous variety Kalikalan it was >50% (Palaniswami et al. 1996).

Under the collaborative project with USIF-USDA detailed investigations were carried out in India on biotypes of *B. tabaci,* ICMV transmission and its parasitoids and the outcome of the project is embodied in this chapter. In order to identity the biotypes of *B. tabaci*, biological assays, cross breeding, isozyme and molecular studies were carried out using the colonies of *B. tabaci* collected separately from sweet potato and cassava in the fields (emerging adult populations from sweet potato and cassava maintained separately for eight generations and same generation individuals were used for various experiments/tests). Cassava (M4/H226), sweet potato (Sree Nandini), egg plant (Pusa Kusum), tobacco (Hema), cotton (Pelta Pine), tomato (Arka Aloka) and bhindi (local culture) were used as host plants/varieties in the different tests (Lisha et al. 2003).

7.2 Biotypes of *Bemisia tabaci*

Biology of *B. tabaci* on cassava (Palaniswami et al. 2001) has been reported in India. In all the host plants tested newly emerged adults required 2–4 days for oviposition. Egg period was shorter on cassava (3.5 days), while nymphal and pupal periods and total life cycle (13.2, 7.7 and 23.5 days respectively) were longer on cassava compared to sweet potato, cotton, egg plant, bhendi and tobacco. The difference in the total developmental duration (from egg to adult emergence) was more or less similar on the different host plants except cassava. The longevity of whitefly on egg plant was the lowest ranging from 5–9 days followed by cotton (7–18 days) compared to the other plants under study. Present study showed that fecundity was maximum on tobacco and minimum on egg plant. The longevity of whitefly on egg plant was the lowest (06.83 ± 1.47) days and maximum on cassava (17.80 ± 9.53) days (Tables 7.1 and 7.2).

Palaniswami et al. (2001) reported 45 eggs on cassava. *B. tabaci* is known to have a host range that is highly variable. Furthermore, biotypes vary with respect to geography, fecundity, dispersal behaviour, insecticide resistance, natural enemy complex, invasive behaviour, plant virus transmission and complement endosymbionts (Brown et al. 1995b). In West Africa and Uganda, differences in host selection occur among different *B. tabaci* populations (Burban et al. 1992; Legg 1996). On the Ivory Coast, the *B. tabaci* "Cassava strain" is restricted to cassava, *Manihot esculenta* Crantz; tree cassava, *M. glaziovii* Muell; and aubergine (egg plant), *Solanum melongena* L. An okra strain has a much wider host range yet does not

Table 7.1 Developmental period of *Bemisia tabaci* on six species of host plants

Host plant	Pre-ovipositional period	Developmental period in days			
		Egg	Nymph	Pupa	Total
Cassava	3.3±0.52	3.5±1.38	13.2±1.94	7.7±2.42	23.5±5.09
Sweet potato	3.5±0.55	4.5±1.38	9.0±0.90	7.0±0.63	19.5±2.17
Cotton	2.8±0.98	5.0±0.90	8.5±1.05	7.0±1.10	20.0±2.37
Egg plant	2.5±0.55	7.5±1.38	9.0±0.90	4.0±0.90	20.5±1.87
Tobacco	2.3±0.52	7.0±0.90	8.0±1.10	5.0±0.90	19.0±1.55
Bhendi	3.2±0.75	5.3±1.21	9.0±1.26	5.0±0.90	17.7±1.21

Table 7.2 Fecundity longevity and sex-ratio of *B. tabaci* on five host plants

Host	Fecundity (no.)	Longevity (days)	Sex-ratio
Cassava	45.00±12.00	17.80±9.53	1:1.80
Sweet Potato	41.67±08.12	16.33±2.73	1:1.59
Cotton	41.30±14.10	13.00±4.98	1:1.20
Tobacco	65.10±36.10	16.00±5.51	1:1.93
Egg Plant	11.83±08.91	06.83±1.47	1:1.65

Mean±SE of six observations ($P<0.01$)

colonize cassava (Burban et al. 1992). The precise taxonomic status of *B. tabaci* biotype B has been the subject of much discussion (Bartlett and Gawel 1993; Brown et al. 1995b; Perring et al. 1993). Electrophoretic pattern differences have been used by numerous authors to clarify *B. tabaci* biotype relationships (Prabhaker et al. 1987; Costa and Brown 1991; Burban et al. 1992; Liu et al. 1992; Brown et al. 1995a). Variation in *B. tabaci* populations in relation to virus transmission has been observed in India (Palaniswami and Nair 1995).

7.2.1 *Biological Assays*

B. tabaci pupae were collected from sweet potato and cassava plants in the field at Trivandrum, Kerala, India. The pupae were held in separate cages in a screen-house. Each colony was maintained for eight generations before experimental use. They were designated as the cassava strain and sweet potato reared whitefly populations.

7.2.1.1 No Choice Host Plant Studies

Five cassava and sweet potato strain whitefly adults were released into leaf clip cages (4.5 cm in diameter) placed onto leaves of cassava, sweet potato, cotton, egg plant and tobacco. After 48 h, adults were removed from the clip cages and the cages were left on the leaves. This experiment was repeated 6–13 times under screenhouse conditions (25–32°C and 65–70% RH).

Cassava strain whitefly oviposited on all host plants. The highest number of eggs were oviposited on tobacco followed by cassava, egg plant which were on par, and were significantly superior to cotton and sweet potato, respectively. The highest percentage of eggs that developed to the adult stage occurred on cassava (94%), followed by egg plant (86%) and tobacco (81%). Although nymphs emerging from the eggs oviposited on cotton developed to the pupal stage, no adult emergence occurred. No cassava reared whitefly nymphs developed beyond the first instar from eggs oviposited on sweet potato. Sweet potato reared whitefly confined in leaf-cages also oviposited eggs on all the host plants. However, the highest numbers were oviposited on sweet potato followed by tobacco, and these were significantly higher than on cotton and cassava. First instar population was significantly lowest on cassava. Although the number of eggs oviposited on cassava and cotton was on par the number of first instars developed was significantly lower on cassava. The adult population was significantly higher on sweet potato followed by tobacco, cotton and egg plant. There was no pupal formation or adult development on cassava. The highest percentage of adult emergence occurred from eggs oviposited on cotton (95%), followed by sweet potato (91%), tobacco (80%) and egg plant (79%).

7.2.1.2 Performance on Host Plants with Choice

Cassava strain whitefly, 30 were collected from the colony and released into each cage containing plant pairs in the screen house (25–32°C and 65–70% RH). Plant pairs in the cages were cassava/cassava, cassava/sweetpotato, cassava/cotton, cassava/egg plant, and cassava/tomato. Plants before placement into cages were trimmed so that each plant had approximately the same leaf area. Leaves on the plants were examined with a hand lens after 3 days and the number of eggs oviposited was counted. For sweet potato reared whitefly, 30 insects from the colony were collected and introduced into each cage containing sweet potato/sweet potato, sweet potato/cassava, sweet potato/cotton, sweet potato/egg plant, sweet potato/tomato. The numbers of eggs per plant were counted after 3 days and plants re-examined daily thereafter as described above. The experiment was repeated on six occasions.

Cassava strain in choice tests oviposited significantly more eggs on cassava compared to sweet potato, cotton, egg plant, and tomato (Table 7.3). The least preferred ovipositional host for cassava strain was tomato followed by sweet potato, egg plant and cotton. The cassava strain when given a choice of cassava and sweet potato, oviposited 484 eggs on cassava, but only 30 eggs on sweet potato. When given a choice of cassava or cotton, oviposited only 96 eggs on cotton with 4% developing to adults compared with 399 eggs on cassava and 87% adult emergence. On tomato, cassava strain laid seven eggs compared to 454 eggs on cassava. None of the first instars from hatched eggs on sweet potato survived to the pupal or adult stage, whereas 87% adult emergence occurred on cassava (Table 7.3). The highest

Table 7.3 Host preference studies of *B. tabaci* reared on cassava

	Population in numbers			
Transfer host	Egg	First instar	Pupa	Adult
Cassava	240(136) 90.00[a]	164(121) 90.00[a]	161(119) 90.00[a]	161(119) 90.00[a]
Sweet potato	30(484) 02.92[a]	22(426) 02.04[a]	00(420) 00.00[a]	00(420) 00.00[a]
Cotton	96(399) 20.78[a]	68(369) 21.78[a]	13(346) 02.81[a]	4(346) 01.10[a]
Egg plant	75(561) 10.01[a]	49(481) 08.49[a]	49(481) 08.49[a]	49(481) 08.49[a]
Tomato	07(454) 01.41[a]	04(410) 00.77[a]	02(410) 00.51[a]	02(410) 00.51[a]
CD (0.05)	08.79	09.96	03.74	03.74

Mean of six observations
Figures in parenthesis are values on cassava
[a]Percentage based on rearing host in arc sin values

percentage of adult emergence from eggs oviposited was 67% on cassava followed by 65% on egg plant, 29% on tomato, 4% on cotton, and no adult emergence from the eggs oviposited on sweet potato.

The least preferred ovipositional host for sweet potato strain was cassava followed by egg plant, tomato and cotton ($P<0.05\%$). The sweet potato-reared population when given a choice, oviposited 777 eggs on sweet potato compared to 293 on cassava. None of the eggs oviposited on cassava developed beyond the first instar. On sweet potato, 69% of the eggs oviposited developed to the adult stage. Sweet potato strain oviposited 391 eggs on cotton with 77% developing to adults and 677 eggs on tomato with 88% developing to adults (Tables 7.3 and 7.4).

The longevity of the cassava whitefly was 18.5 days on cassava, 4.5 days on sweet potato and 8.0 days on cotton. The longevity of the SPWF was 4.8 days on cassava, 16.0 days on sweet potato and 19.5 days on cotton (Fig.7.1a).

The cassava-reared population reproduced on cassava, egg plant, tomato and tobacco, but not on cotton and sweet potato. Conversely, the sweet potato-reared population reproduced on sweet potato, cotton, egg plant, tomato and tobacco, but not on cassava. The comparative biology of CWF and SPWF on acceptable hosts signifies that tobacco and egg plant are common host plants for both strains (Fig. 7.1b, c). ANOVA of percentages of eggs oviposited, nymphs, pupae and adults produced by the cassava strain on alternate hosts (egg plant, cotton, tomato and sweet potato) compared with the cassava rearing host supported the conclusion that cassava was significantly superior to the other host plants. Similarly, ANOVA of percentages of the sweet potato-reared population on transfer host plants confirmed that the preferred host for the sweet potato-reared population was sweet potato followed by cotton, tomato, and egg plant and the least preferred was cassava. Similar distinctions were made between a cassava biotype and an okra/cotton/sweet potato biotype

Table 7.4 Host preference studies of *B. tabaci* reared on sweet potato

	Population in numbers			
Transfer host	Egg	First instar	Pupa	Adult
Sweet potato	186(155) 90.00[a]	140(135) 90.00[a]	129(124) 90.00[a]	129(124) 90.00[a]
Cassava	293(777) 21.71[a]	148(725) 12.31[a]	00(695) 00.00[a]	00(695) 00.00[a]
Cotton	391(717) 32.10[a]	323(578) 32.73[a]	308(536) 34.04[a]	308(536) 34.04
Egg plant	84(287) 13.55[a]	74(275) 15.15[a]	71(271) 14.48[a]	71(271) 14.48[a]
Tomato	677(1403) 27.49[a]	600(1326) 29.91[a]	596(1305) 27.08[a]	596(1305) 27.08[a]
CD (0.05)	05.69	07.44	06.24	06.24

Mean of six observations
Figures in parenthesis are values on sweet potato
[a]Percentage based on rearing host in arc sin values

in Uganda (Legg 1996), and West African *B. tabaci* that were polyphagous and oligophagous biotypes (Burban et al. 1992). This is similar to the polyphagous sweet potato-reared population and the oligophagous cassava-reared population respectively in our study. Reproductive failure of a sweet potato-reared population on cassava in South America was suggested as an indication that *B. tabaci* from the new and old world, respectively, may be distinctive races (Costa and Russel 1975).

The first instar sweet potato-reared whitefly mortality on cassava was also similar to the Ugandan *B. tabaci* biotype and Legg (1996) explained the mortality as a consequence of failure to feed successfully or the possible presence of compounds toxic in non-host leaf tissue. Cassava leaf tissue contains high concentration of the cyanogenic glycoside linamarin localised in the symplast (Mkpong et al. 1990). It is possible that penetration of the phloem tissue by the sweet potato whitefly (SPWF) stylet and salivary secretions might lead to the limited hydrolysis of cyanogenic glycosides and consequent release of toxic cyanide. Survival on cassava would depend on mechanisms either to prevent the release of free cyanide or to effect its detoxification (Legg 1996). The increased rate of oviposition and lower mortality of all stages of cassava-reared and sweet potato reared populations in our study on their preferred hosts appear to fit the host preference criteria of van Lenteren and Noldus (1990). These differences clearly showed the presence of the cassava and sweet potato biotypes in India.

7.2.1.3 Squash Silver Assay

Squash seeds were planted in earthen pots and kept in confined condition in rearing cages. After germination seedlings were transferred into plastic pots and grown in

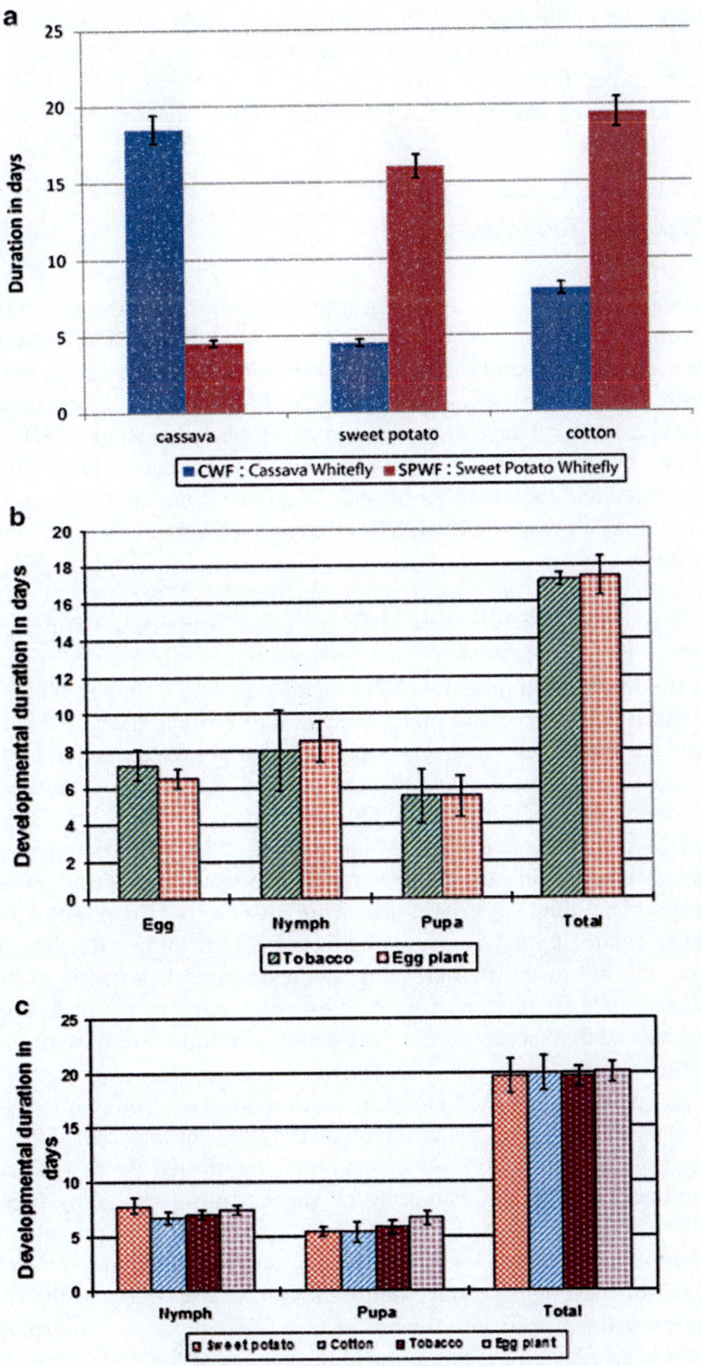

Fig. 7.1 (**a**) Longevity of cassava and sweet potato whiteflies on different host plants; (**b**) Comparative biology of cassava whitefly on tobacco and egg plant; (**c**) Comparative biology of SPWF on sweet potato, cotton, tobacco and egg plant

green house under natural lighting in rearing cages. At the 3-leaf stage each of six plants were exposed to the sweet potato or cassava biotype. After 2 weeks, the seedlings were evaluated for squash silver leaf symptoms. None of the plants exposed either to the cassava or sweet potato biotype developed silverleaf symptoms.

7.2.2 *Crossing Breeding*

The biological consequences of mating interaction between the two biotypes of *B. tabaci* were studied by cross breeding experiments. Three SPWF females were released into a clip cage (a plastic box of 4 cm in diameter and 2 cm in thickness attached with a film clip) that was placed on the brinjal; which is the common host for both biotypes. After 2 days of pre-ovipositional period, a single CWF male was introduced into the same clip cage containing the three female SPWF. Total number of eggs oviposited and total number of male and female emergence were recorded. Similarly female CWF with male SPWF were released into a clip cage for developmental studies.

The results of the cross breeding studies using female SPWF and male CWF, and female CWF and male SPWF showed an increased proportion of male progeny. Cross breeding experiments using female sweet potato biotype and male cassava biotype resulted in a mean progeny of 2.8 females and 17.7 males. While crossing with female cassava biotype and male sweet potato biotype gave 1.33 female and 10.17 males. Breeding within each type yielded higher population of females (13–16) and males (14–18). The male – female ratio was 1:0.75 for cassava biotype and 1:1 for sweet potato biotype (Table 7.5). Cross breeding experiments using female sweet potato biotype and male cassava biotype produced 13.7% females; similarly, female cassava biotype and male sweet potato biotype crossing experiment gave 11.6% females. The offsprings developed from crossing between the two biotypes yielded lower female population compared to normal breeding within the biotypes. The male population in the two crossing studies ranged from 86% to 88%. The percentage of female F1 progeny was very low compared to the male population. The results showed that cassava and sweet potato whiteflies are biotypes not separate species as they do breed with each other.

Several authors have reported crossing experiments between two biotypes (Liu et al. 1992; Byrne et al. 1992). Only Byrne et al. (1992) have successfully reported the interbreeding between two biotypes and have confirmed the results using biochemical markers. Our results also revealed successful interbreeding between the cassava and sweet potato biotype, with the production of female progeny. Because *B. tabaci* is an arrhenotokous insect species i.e. unfertilized eggs produce haploid males, whilst fertilized eggs produce diploid females, assessment of the success of the crossing experiment relied on the production of F1 females. The isolation protocol adopted here allowed for the removal of females at the pupal stage based on the sexual dimorphism and was effective in eliminating intra-strain mating.

Table 7.5 Average number of F1 offspring produced in crossing experiments

Cross (n)	Females ± SD	Males ± SD	Sex ratio M:F
CWF female × CWF male (6)	13.83 ± 8.93	18.5 ± 7.64	1:0.75
SPWF female × SPWF male (7)	16.29 ± 3.63	14.85 ± 3.63	1:1.10
CWF female × SPWF male (6)	1.33 ± 1.97	10.17 ± 8.38	1:0.13
SPWF female × CWF male (10)	2.8 ± 2.1	17.7 ± 7.5	1:0.16

CWF Cassava biotype, *SPWF* sweetpotato biotype, *n* Number of replications, *SD* Standard Deviation

7.2.3 *Isoenzyme Studies*

Specific isozyme patterns have been observed for *B. tabaci* populations collected from different host plant species (Burban et al. 1992; Costa and Brown 1991). Since the adults of the two known whitefly biotypes are morphologically similar, we used electrophoresis to examine variation at enzyme loci ("allozyme or isoenzymes"). We compared the two biotypes with five isozymes, *viz.*, Esterase (EST), Malate dehydrogenase (MDH), Phosphoglucoisomerase (PGI), Alcohol dehydrogenase (ADH) and Phosphoglucomutase (PGM).

7.2.3.1 Gel Electrophoresis

Five enzyme systems, Esterase (EST), Malate dehydrogenase (MDH), Phosphoglucomutase, phosphoglucoisomerase and alcohol dehydrogenase were used in vertical slab native polyacrylamide gel electrophoresis studies. Samples were prepared by homogenizing individual *B. tabaci* in a microfuge tube in 15 μl of 0.05 M Tris-HCl, pH 6.8, containing 10% sucrose and 0.1% Triton-X 100. The homogenates were transferred into wells of the stacking gel (2.5%), superposed on the 10% resolving gel. The reservoir contained Tris 0.06 M Glycine 0.37 M buffer, pH 8.2. Gels were run at 15 mA for 15 min. and then at 19 mA for 120 min. Bromophenol blue was used as the running marker.

Esterase

Cassava strain had three esterase bands, two slow and one fast, whereas the sweetpotato- reared population biotype had four bands, two slow and two fast (Fig. 7.2a). The fourth band was very feeble (Rf value 0.293). The two slow moving bands showed different Rf values for the cassava- reared population (0.150 and 0.204) and sweetpotato- reared population (0.108 and 0.165). A fast moving band with an Rf value of 0.263 occurred with both Cassava and sweet potato strains.

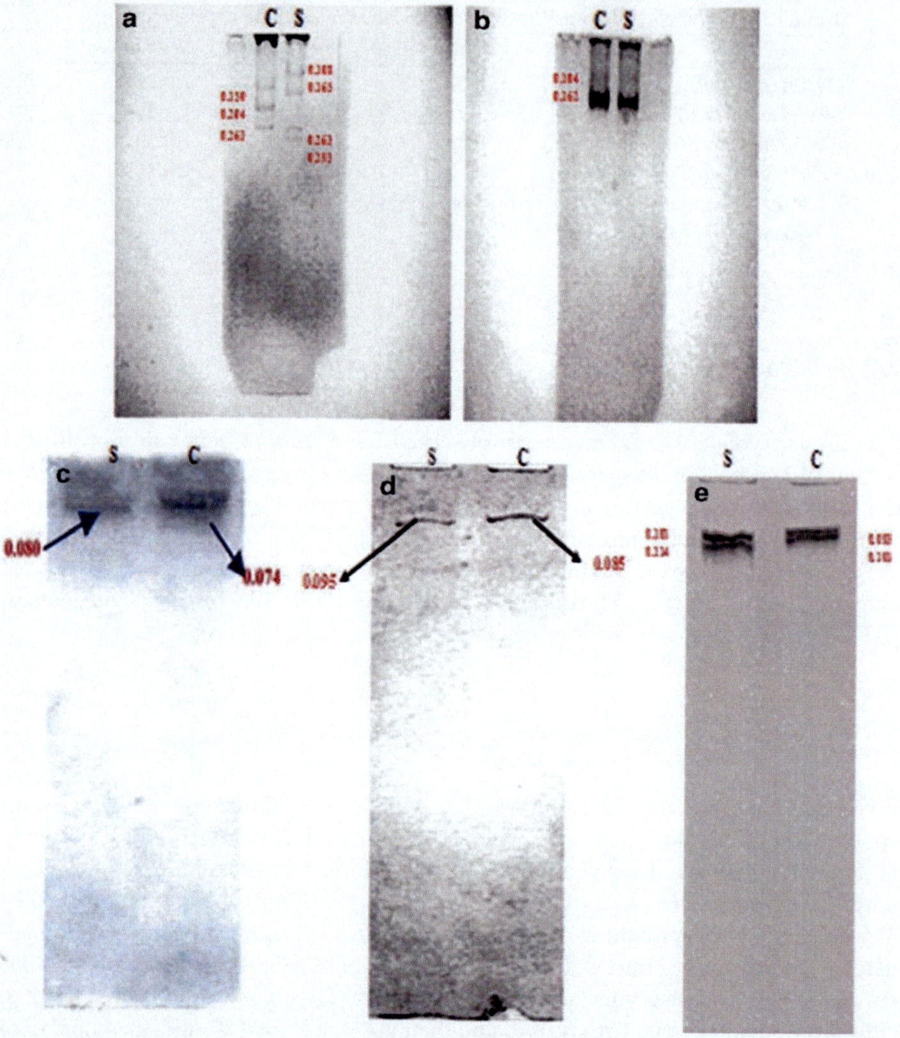

Fig. 7.2 **(a–e)** Isoenzyme studies for confirmation of biotypes (**a** esterase, **b** Malate dehydrogenase, **c** Phosphoglucoisomerase, **d** Alcohol dehydrogenase, **e** Phosphoglucomutase)

Malate Dehydrogenase

The MDH pattern showed no variation in banding pattern between the two biotypes. MDH zymogram showed two bands (Rf values 0.104 and 0.162) for the two biotypes (Fig. 7.2b).

Phosphoglucoisomerase

The banding patterns for the cassava and sweet potato biotypes of the isozyme phosphoglucoisomerase are illustrated in Fig. 7.2c. The bands for the cassava biotype appeared at Rf value 0.074 and that for the sweet potato biotype at 0.08.

Alcohol Dehydrogenase

Figure 7.2d illustrates the isozyme pattern of alcohol dehydrogenase for the cassava and sweet potato whitefly. Both biotypes showed a single banding pattern. The Rf values of these bands are shown. For the cassava biotype the band appeared at 0.085 and for the sweet potato biotype it appeared at 0.095.

Phosphoglucomutase

Figure 7.2e shows the banding pattern for the cassava and sweet potato whitefly when stained for phosphoglucomutase activity. Both biotypes showed two bands for this isozyme. They both shared a common band of Rf value 0.101. The sweet potato biotype had a fast moving band at Rf value 0.114, whereas the cassava biotype had a slow moving band at 0.093.

Prabhaker et al. (1987) described the use of electrophoretic patterns to distinguish three whitefly species. Costa and Brown (1991) utilized esterase-banding pattern to distinguish *Bemisia* populations with different biological characteristics. Brown et al. (1995a) used isozyme techniques to separate biotypes of *B. tabaci*.

The five isozymes in total indicated the presence of 15 allozymic loci and of these both biotypes shared four common loci (26.67%) and differed at eleven loci in the presence or absence of the bands. At five of these loci, the difference between CWF and SPWF was "fixed" that is each of these bands were present in CWF and not in SPWF. The other six loci were present in only SPWF but not in CWF. Since the adults of the two known whitefly biotypes are morphologically similar, we used electrophoresis to examine variation at enzyme loci (allozyme or isoenzymes). Thus 73.33% of the alleles represented fixed differences between CWF and SPWF. The differences in the banding pattern are due to variations in the amino acid content of the molecule, which in turn is dependent on the sequences of the nucleotides in the DNA. Different bands obtained indicate different electrophoretic motilities of the isozymes, which are coded by different alleles or separate genetic loci.

7.2.4 Molecular Evidences

RAPD-PCR markers were explored to differentiate the biotypes of *B. tabaci* within the different Indian populations of *B. tabaci*,the cassava and sweet potato biotypes and to estimate the degree of genetic similarity between the different collections based on the RAPD data. The following oligonucleotide primers (Sigma Genosis, UK) were used for RAPD studies: Four selected primers from the RAPD kits, *viz.*, OPH9, OPH16, OPF2 and OPF12 and the other two primers were single PCR primers $(TCC)_5$, $(GTG)_5$ used by Perring et al. (1993) *viz.*,

1. 5′ GAG GAT CCC T 3′
2. 5′ ACG GTA CCA G 3′
3. 5′ TGT AGC TGG G 3′
4. 5′ TCT CAG CTG G 3′
5. 5′ TCC TCC TCC TCC TCC 3′
6. 5′ GTG GTG GTG GTG GTG 3′

All six primers tested in this study produced RAPD patterns, and the results of F2, F12 and the two single primers are described, as they gave distinct amplification products. Figure 7.3a–f illustrates the amplification products obtained from the six primers (H9, H16, F2, F12, $(GTG)_5$, $(TCC)_5$).

RAPD-PCR performed on template DNA extracted from the 70% alcohol preserved *B. tabaci* showed that the preservation did not have any subsequent effect on RAPD-PCR results. Reproducibility of RAPD-PCR results were tested by conducting replicated reactions (eight for each primer) using identical genotype x primer combinations. In these reactions, RAPD patterns were consistent for any given genotype x primer combination, however the relative intensity of the band was sometimes variable.

Using the primer F2 the CICR (Nagpur, Maharashtra) and Kannur showed 39.9% similarity. The Parbhani collection showed 42.2% similarity with Kannur population (Fig. 7.3c). Using the primer F12 the collections from Parbhani and Nagpur showed 63.5% similarity based on Dice Coefficient value. The Dholi collection showed 55.55 with Nagpur and 53.6% with Parbhani. The SPWF banding pattern differed from other collections and showed 23.8% similarity with Parbhani population. The collection from Kannur showed 44.05 similarity with Nagpur, 40.9% with Dholi and 37.2% with Parbhani (Fig. 7.3d).

Fig. 7.3 (**a**) RAPD-PCR studies using primer H16, Lane 1 – 1 Kb marker, Lane 2 – Cassava whitefly, Lane 3 – Sweet potato whitefly, Lane 4 – Bhubaneswar, Lane 5 – Bangalore, Lane 6-7 – Maharashtra; (**b**) RAPD-PCR studies using primer H9, Lane 1 – 1 Kb marker, Lane 2 – Cassava whitefly, Lane 3 – Sweet potato whitefly, Lane 4 – Bhubaneswar, Lane 5 – Bangalore, Lane 6-7 – Maharashtra; (**c**) RAPD-PCR studies using primer OPF2, Lane 1 – 1 Kb marker, Lane 3 – Sweet potato whitefly, Lane 4 – BBSR, Lane 5 – Bangalore, Lane 6 – Parbhani, Lane 7 – Dholi, Lane 8 – CICR, Lane 9 – Kannur; (**d**) RAPD-PCR studies using primer OPF12, Lane 1 – 1 Kb marker, Lane 3 – Sweet potato whitefly, Lane 4 – BBSR, Lane 5 – Bangalore, Lane 6 – Parbhani, Lane 7 – Dholi, Lane 8 – CICR, Lane 9 – Kannur; (**e**) RAPD-PCR studies using primer $(GTG)_5$, Lane 1 –1 Kb marker, Lane 2 – Cassava whitefly, Lane 3 – Sweet potato whitefly, Lane 4 – Bhubaneswar, Lane 5 – Bangalore, Lane 6–7 – Maharashtra; (**f**) RAPD-PCR studies using primer $(TCC)_5$, Lane 1 – 1 Kb marker, Lane 2 – Cassava whitefly, Lane 3 – Sweet potato whitefly, Lane 4 – Bangalore, Lane 5 – Bhubaneswar, Lane 6 – Dholi

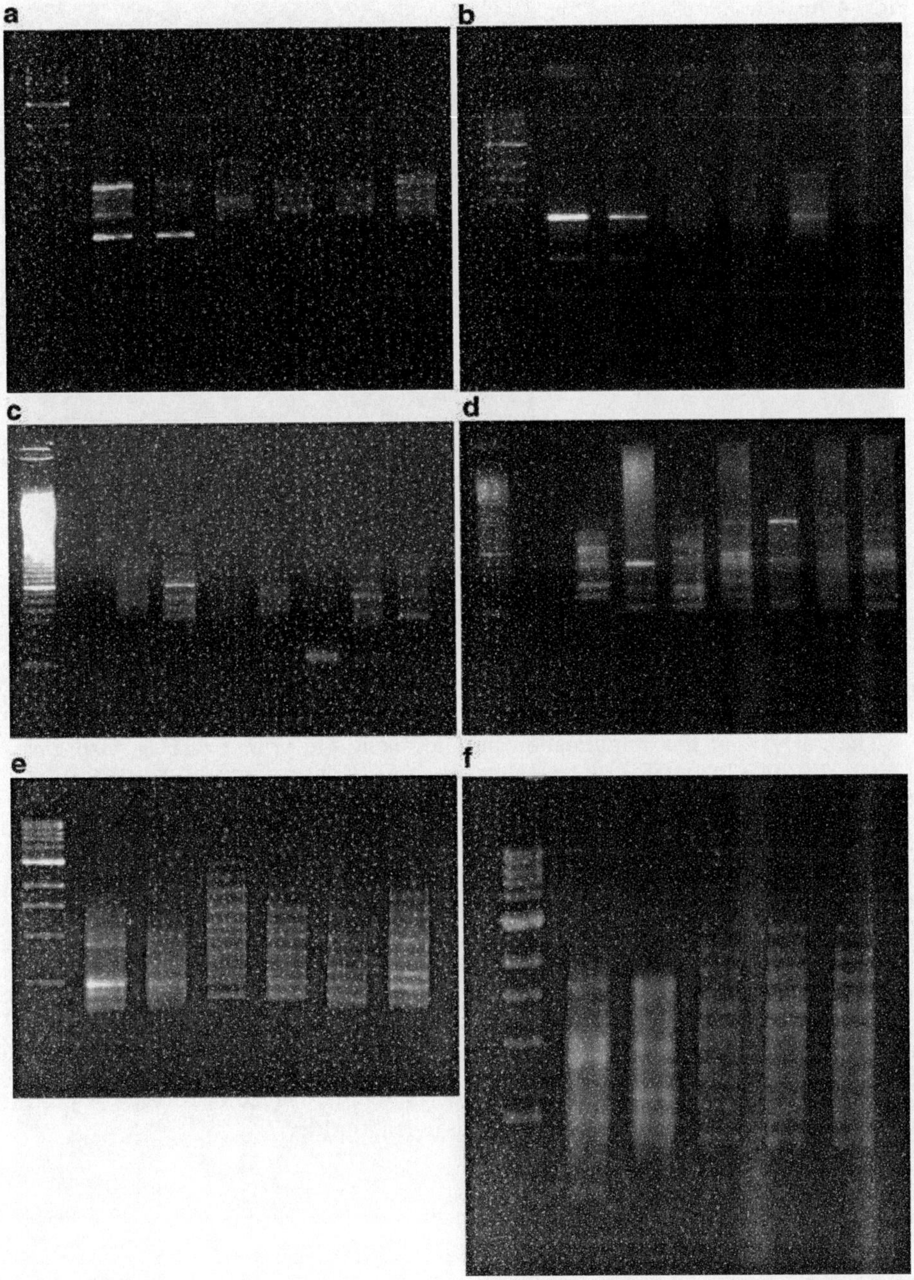
a
b
c
d
e
f

Fig. 7.4 Amplification of D2 region of 28S rRNA of *Bemisia tabaci* (PCR product 600–800 bp). Lane 1 – 1 kb marker, Lane 2 – Cassava whitefly, Lane 3 – Sweet potato whitefly, Lane 4 – Bhubaneswar

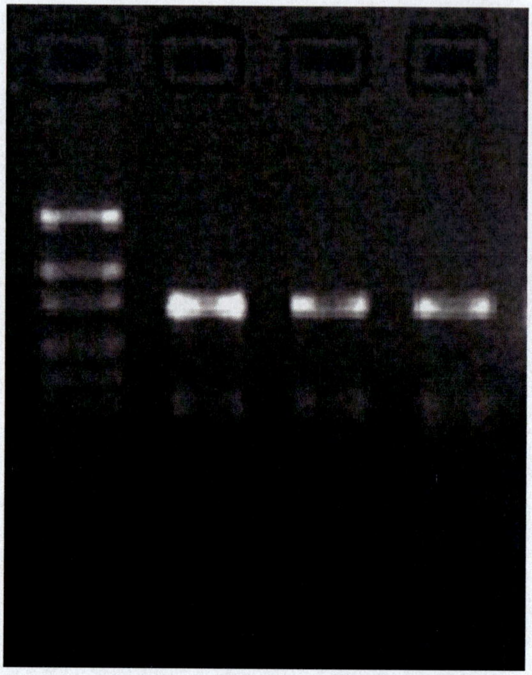

The analysis of the amplification product using the primer $(GTG)_5$ also gave identical results. The Dice Coefficient value of the CWF and SPWF was 10.7%. This primer revealed a further closeness (46.2%) of the SPWF with the MAU and CICR of Maharashtra (Fig. 7.3e).

A unique set of amplification products was obtained using the two single primers. Analysis using the Dice Coefficient values based on the amplification products obtained from the primer $(TCC)_5$ showed that the CWF and the SPWF shared no similarity between them. The Bhubaneshwar and Dholi populations shared 77.3% similarity and the Bangalore and Bhubaneshwar populations showed 67.3% similarity. The Bangalore and Dholi population shared 57.8% similarity. The percentage similarity of the collections from Bangalore, Bhubaneshwar and Dholi with the SPWF was 27.8, 20.2 and 27.2 respectively (Fig. 7.3f). The 28S r RNA PCR products of *B. tabaci* ranged from 600 to 800bp in length (Fig. 7.4).

7.2.5 Endosymbionts

Although symbionts of aphids and psyllids are rather widely studied, little is known about the composition and diversity of endosymbionts in whiteflies, one of their closest relatives. *B. tabaci* collections from cassava and sweet potato biotype were analyzed to investigate the diversity of associated prokaryotic endosymbionts.

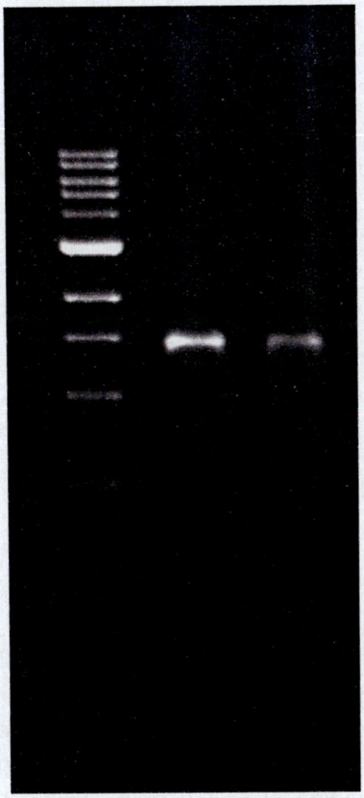

Fig. 7.5 Amplification 16S rRNA region of primary endosymbiont of *B. tabaci* (PCR product ~ 1,500 bp). Lane 1 – 1 Kb marker, Lane 2 – Cassava whitefly, Lane 3 – Sweet potato whitefly

Bacteria cultured from surface sterilized adult whitefly from cassava and sweet potato could be grouped into two. Two colonies were opaque and white in colour and the other two colonies were yellow in appearance. Two types of bacterial colonies were obtained from both the cassava and sweet potato whitefly. The bacterial colonies were of the bacilli and cocci forms. Further motility test showed that the colonies were non motile. As the next step the bacteria were tested with Gram's stain and were found to be all gram negative.

The primary endosymbionts were detected in both biotypes of *B. tabaci* using the primary endosymbiont specific 16S rDNA primers. The amplified product was obtained at the approx. 1,500 bp region. PCR amplification with secondary symbionts-specific primers revealed that symbionts were not associated with the two biotypes (Fig. 7.5). The primary endosymbiont was successfully amplified with the primers, but the secondary symbionts were not detected using the specific primers in both the biotypes. In India no work of this nature has so far been reported. Therefore as a preliminary step towards the study of endosymbionts we have succeeded in locating the mycetocytes (reported to be the house for endosymbionts) inside the ovum and also succeeded in detecting the presence of primary endosymbionts in both biotypes using molecular techniques.

7.2.6 *Parasitoids*

Among aphelinid parasitoids recorded for *B. tabaci, Eretmocerus mundus* (Mercet) a potential parasitoid could be multiplied only on SPWF but not on CWF (Antony and Palaniswami 2002; Antony et al. 2001, 2003, 2004).

7.3 Indian Cassava Mosaic Virus

Whitefly-transmitted viruses, belonging to a family of ssDNA viruses, the *Geminiviridae*, and within the genus *Begomovirus*, cause Cassava mosaic disease (CMD) and to date, it is the most significant constraint to cassava production in India and Africa (Palaniswami et al. 2005; Antony et al. 2009).

7.3.1 *ICMV in Cassava Plants*

ICMV culture was maintained in cassava plants (cv. H226 and M4) in a greenhouse (Temperature 30 ± 3°C and RH 70 ± 5%). These plants were used as source of inocula. Since ICMV is not transmitted through true cassava seed cassava seedlings raised in an insect proof cage were taken as non-viruliferous cassava plants. ICMV was purified from infected cassava and *Nicotiana benthamiana* L. leaves. When centrifuged in 10–40% sucrose density gradients for 3 h at 35,000 rpm, the virus was concentrated as a sharp light scattering band at the lower portion of the tube. Dialyzed supernatant was used for further detection by ELISA and ISEM. The presence of ICMV was verified in the purified samples by TAS-ELISA. ELISA plates were coated with polyclonal antibody of ACMV at 10^{-3} dilution in coating buffer and incubated at 37°C for 1 h. Purified extracts from infected cassava and tobacco leaves were tested for the presence of geminate particles by negative staining. The grids were then examined in HITACHI Electron Microscope (model no H-600) and photographs were taken at an accelerated voltage of 75 KV.

Virus particles isolated from cassava and tobacco showed positive in ELISA tests. Viral concentration was higher in diseased cassava plants and tobacco (Fig. 7.6). SDS-PAGE of purified ICMV showed a single band of estimated molecular weight 34 kDa (Fig. 7.7).When selected fractions of the sucrose gradient were analyzed, bipartite gemini particles of 18–20 nm almost exclusively spherical in size, were detected. Gemini particles of this size were found bound to the ICMV antibody, which in turn was bounded to the gold labelled anti-rabbit IgG. Gold particles were clearly visible and bound to the geminivirus particles, thus specific for ICMV. In the study ICMV was purified from whitefly transmitted infected cassava plants and mechanically inoculated tobacco seedlings. ELISA and ISEM further confirmed this.

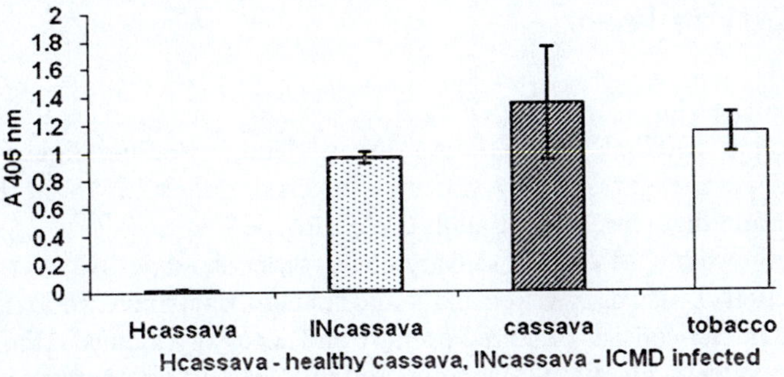

Fig. 7.6 Absorbance values (A405 nm) for ICMV using ICMV MAbs

Fig. 7.7 SDS-PAGE Electropherogram (10%) of purified ICMV from cassava showing single band at 34 kDa

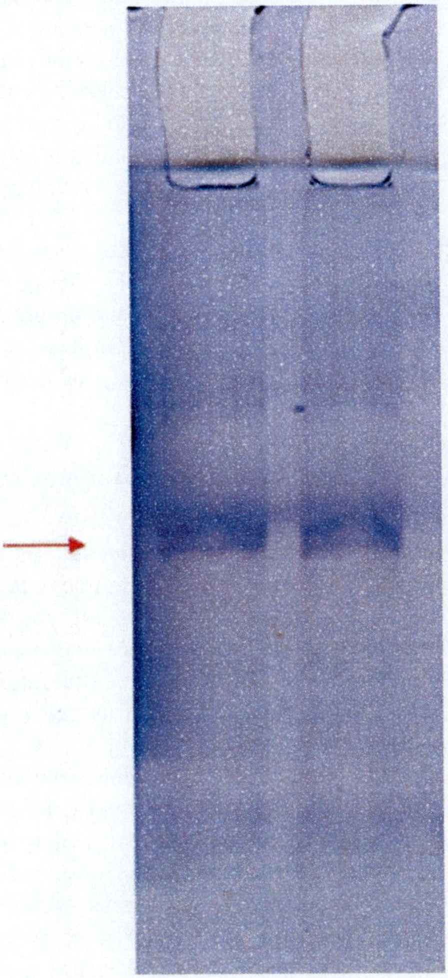

7.3.2 ICMV in Vector

ICMV a bipartite virus is transmitted by *B. tabaci* CWF in a persistent manner. Data on the detection of ICMV in *B. tabaci* salivary gland, stylet and digestive tract by PCR studies are meager. A study was undertaken to detect ICMV in the salivary gland, stylet and digestive tract of adult CWF biotype.

In order to locate ICMV in the salivary gland, stylet and digestive tracts a preliminary step section (1 μ), was prepared and studied under a light microscope. The salivary system of *B. tabaci* consists of paired primary and accessory glands in the prothorax. Each primary salivary gland (PSG) is composed of at least 13 cells varying in morphology and staining differentially, while the accessory glands are composed of four morphologically similar cells. The mouth leads to the pharynx, oesophagus, and filter chamber at the base of the abdomen. The mid gut has the filter chamber with descending and ascending mid gut. Ascending and descending mid gut connect at the filter chamber region and precedes the hindgut. Anterior portion of the hindgut broadens to form the ileum and the posterior portion narrows down to form the colon before joining the rectal sac. Longitudinal section through the abdominal region revealed ovum, ovariole, mycetosomes, mycetocyte inside the ovum, fat bodies and spermatheca.

ICMV DNA detection by PCR in organs dissected from a single viruliferous whitefly after 48 h acquisition access feeding period (AAFP) was determined. Nucleic acid extracted from viruliferous CWF after 48 h AAFP, when used with CTCRI-USIF primer S1 and A1 in PCR; produced an amplification band of 390 bp from the salivary gland, stylet and digestive tract. The product was not observed when template was omitted (N-negative control) and no product was observed when nucleic acid from uninfected CWF was used as template, indicating that the amplified product was specific to ICMV. The present study indicated detection of ICMV DNA in *B. tabaci*.

7.3.3 *Bemisia tabaci* Biotypes in ICMV Transmission

Biological studies were conducted to investigate the role of biotypes of *B. tabaci* as a vector of ICMV from cassava to cassava and tobacco. Whitefly colonies breeding only on cassava plants (cassava biotype, CWF) and egg plants (CWF) and colonies breeding only on sweet potato plants (sweet potato biotype, SPWF) were maintained separately. SPWF were cultured on sweet potato in confined cages for many generations and considered virus free whiteflies, and were used as negative control in all the PCR studies.

ICMV culture was maintained in cassava plants (cv. H226 and M4) in a greenhouse (Temperature 30±3°C and RH 70±5%). These plants were used as a source of inocula throughout the experiments. Since ICMV is not transmitted through true cassava seed, seedlings raised in an insect proof cage were taken as ICMV-free cassava plants. Non-viruliferous whitefly colony was established by collecting *B. tabaci* from egg plant stock colony (cv. Hema) (since CWF can also breed and multiply on egg plants) and released onto cassava seedlings raised in insect proof cages. Egg plant is not a host for cassava mosaic virus (Fauquet and Fargette 1990), and this ensured that CWF were free from ICMV. When 'red eyed' pupae emerged, they were taken out by using fine needle and placed into petri dishes, and when adult whiteflies emerged, these were used as non-viruliferous whiteflies throughout the experiments in ICMV transmission studies (Antony et al. 2006, 2009).

7.3.3.1 In Field

Cassava and tobacco seedlings were raised from true-seed grown in large pots (30 cm diameter) in an insect proof screenhouse. Seedlings were transplanted individually to pots at the three to five leaf growth stage. Plants [cassava (36) and tobacco (26) (*N. benthamiana*)] were distributed (75–90 cm) within a cassava field heavily infected with ICMV and infested with *B. tabaci*. Plants were observed for the expression of ICMD symptoms for as long as 3 months. TAS-ELISA was used to verify the presence of ICMV in leaf samples taken from plants expressing ICMD symptoms (Mean temperature and RH in the field during the experiments were 29±3°C and 70±5%, respectively). In the field experiments ICMV transmission from cassava to cassava plants on average was 58.33%. Out of the 36 cassava seedlings exposed in the field 21 expressed ICMD symptoms within 3 months. All the disease free healthy tobacco seedlings exposed in the field showed ICMD symptoms indicating 100% transmission.

7.3.3.2 Green House

Cassava seedlings raised and planted at two plants per pot were placed (44 nos.) into three rows in the greenhouse with high populations of adult *B. tabaci* and ICMD infected plants. ICMD symptom expression in the seedlings was recorded up to 3 months. (Mean temperature and RH in the greenhouse were 30±3°C and 70±5%, respectively). Of the 44 cassava seedlings in the greenhouse 28 expressed ICMD symptoms showing 63.6% transmission.

7.3.3.3 Leaf-Clip Cage Confined CWF- and SPWF-ICMV Transmission Studies

Virus free colonies of CWF and SPWF were maintained in the insect proof cages on cassava and sweet potato plants, respectively. For virus transmission, cassava plants were grown from seeds in 21 cm diameter earthen pots in insect proof cages. When 3–4 young leaves were formed, seedlings were carefully removed and placed into glass tubes (3.5 cm in diameter) containing soil. Nutrient (Hoagland's) solution was added to the soil at 3-day intervals. Ten adults, each of CWF and SPWF were collected from their respective colonies. After 3 h of starvation, they were released into leaf clip cages (3 cm in diameter) placed on the leaves of young cassava mosaic disease plants. After 48 h AAFP, whiteflies were aspirated from each leaf clip cage and starved again for 3 h. Thereafter CWF and SPWF were released onto non-diseased cassava seedlings raised in glass tubes (4 cm diameter). After 48 h inoculation access feeding period (IAFP) the insects were aspired out from the leaf clip cages and killed. Seedlings were carefully removed from the glass tubes and planted in plastic pots (14 cm in diameter) with soil and kept in the insect proof cages for the development of symptoms for 3 months in the screen house (25–32°C and 65–70% RH).

Results showed more than 85% adult mortality during 48 h AAFP in the leaf clip cage experiment using SPWF in ICMV transmission from cassava to cassava. Surviving adults of SPWF when released onto young cassava seedlings for IAFP (48 h) did not produce ICMD symptoms even after 3 months indicating non-transmission of ICMV by SPWF. In the experiment using CWF, there was 71.43% transmission of ICMV from cassava to cassava during the 3 month observation period.

7.3.3.4 Influence of Different Number of CWF on ICMV Transmission

In order to determine the optimum number of whiteflies required for the successful transmission of ICMV, different numbers of whiteflies ranging from 1 to 8 were given an AAFP and IAFP ranging from 10 to 72 h. After a fixed AAFP and IAFP, the insects were removed and plants were observed for the expression of ICMD symptoms and the percentage transmission was recorded.

The results indicated that even a single whitefly with 10 h AAFP/IAFP could transmit the virus causing 50% ICMD. It was observed that increasing the number of whitefly (1–8 nos.) did not influence percentage transmission. Percentage of transmission was directly related to AAFP and IAFP. Eight whiteflies with 42/48 h AAFP/IAFP could effect 100% transmission, whereas five whiteflies with 48/48 h AAFP/IAFP also could effect 100% transmission. Similarly four whiteflies with 64/72 h AAFP/IAFP could effect 100% ICMD spread. Transmission studies using 2, 3, 4, 5, 6 whiteflies under different AAFP/IAFP showed 33.30–100, 50–66.70, 50–100, 50–100 and 50% spread of ICMV in test plants respectively (Table 7.6).

7.3.3.5 Effect of Duration of AAFP and IAFP on ICMV Transmission

Importance of duration of AAFP and IAFP on ICMV transmission was determined by confining six adult CWF, after 3 h starvation period, in 4 cm diameter leaf clip cages each for 1, 2, 3, 5, 6, 10, 24, 42, 48, 62, 64 or 72 h AAFP on young ICMV-infected cassava leaves. CWF adults were taken each time fresh from the stock colony. After each AAFP, CWF adults were aspirated out from the leaf cages and starved for 3 h. Thereafter, CWF were released into glass tubes containing 3–5 leaf stage cassava seedlings for different IAFP (12–120 h). After each IAFP, the insects were aspirated out from the glass tubes and killed. Seedlings were removed from glass tubes and planted in pots and kept in insect proof cages for 90 days for development of ICMD symptoms under above mentioned greenhouse conditions.

Table 7.6 Influence of number of CWF under different AAFP and IAFP on ICMV transmission

No. of CWF released per plant	Hours: AAFP	IAFP	Percentage of ICMV transmission
1	10	10	50.00
2	42	48	66.70
	64	72	100.00
	62	48	33.30
3	42	48	66.70
	48	48	50.00
4	64	72	100.00
5	10	10	50.00
	24	40	50.00
	48	48	100.00
	72	72	100.00
6	24	40	50.00
	48	48	50.00
7	64	72	56.12
8	42	48	100.00

AAFP Acquisition access feeding period, *IAFP* Inoculation accessing feeding period, *ICMV* Indian cassava mosaic virus

When adults of CWF were given 1–6 h AAFP with 12 h IAFP; ICMV transmission ranged from 16.66% to 33.33%. At 10–72 h AAFP and 48–120 h IAFP the ICMV transmission ranged from 50% to 100%. Higher rate of transmission (71–100%) was observed with 42–48 h AAFP and 48–120 h IAFP . ICMD symptoms occurred within 9 days when AAFP and IAFP were 48/48 h, respectively. In contrast, it took 3 months for expression of ICMD when AAFP was 1–6 h with 12 h IAFP. Increasing the duration of AAFP and IAFP influenced ICMV transmission to a certain extent, as the experimental results indicated the longer the duration of AAFP and IAFP the earlier symptoms occurred and *vice versa* (Table 7.7).

7.3.4 *Activity of Cyanogenic Glycoside Detoxifying Enzymes in SPWF and CWF*

7.3.4.1 Rhodanase

Activity of rhodanase was determined according to Sorbo (1955). CWF and SPWF were collected separately. The immobilized whiteflies (2 mg each of CWF and SPWF) were placed separately in a microfuge tube and homogenized with a glass rod in 200 mM ice-cold sodium acetate buffer (pH 7.5). 2 mg of CWF and SPWF were homogenized in phosphate buffer and centrifuged at 1,000 rpm for 15 min.

Table 7.7 Effect of different hours of acquisition access and virus inoculation access periods on Indian cassava mosaic virus transmission by cassava biotype

Hours: AAFP	IAFP	Replications	Transmission of ICMV (%)
01	12	14	21.42
02	12	6	16.66
03	12	6	33.33
06	12	6	16.66
10	10	5	50.00
24	40	8	50.00
42	48	9	77.78
46	120	6	100.00
48	48	7	71.43
62	48	4	50.00
64	72	5	60.00
72	72	3	66.66

AAFP Acquisition access feeding period, *IAFP* Inoculation accessing feeding period, *ICMV* Indian cassava mosaic virus

The supernatant was collected and used to determine enzyme activity. The incubation mixture contained: 250 µl enzyme extract, 250 µl of buffered 125 mM potassium cyanide and 250 µl of 125 mM sodium thiosulphate. After 30 min of incubation at 30°C, the reaction was stopped by addition of 0.25 ml of 37% formaldehyde followed by 2.5 ml ferric nitrate reagent. Next, the reaction mixture was centrifuged at 12,000 rpm for 15 min and then absorbance at 460 nm was measured against a control containing formaldehyde. Sodium thiocyanate was used as a standard and activity of the rhodanase was expressed in µg of thiocyanate/min^{-1}/2 mg whitefly sample.

7.3.4.2 ß-Cyanoalanine Synthase (Beta-CAS)

The beta-CAS activity was determined using a slightly modified method of Bluementhal et al. (1968). The SPWF and CWF were homogenized in ice-cold 50 mM Tris-HCL buffer (pH 8.5). The crude homogenate was centrifuged at 10,000 rpm for 15 min and the supernatant was used as the enzyme source. The reaction mixture contained 0.25 ml enzyme extract, 125 µl of 50 mM NaCN and 125 µl of 10 mM L-Cysteine, both buffered in 100 mM Tris-HCL buffer (pH 8.5). This mixture was incubated in a closed vessel for 30 min at 30°C. The reaction was stopped by the addition of 0.5 ml of 20 mM N, N-dimethyl-*p*-phenylenediamine sulphate and 0.5 ml of 30 mM ferric chloride. The mixture centrifuged at 13,000 rpm for 15 min and the absorbance at 650 nm was measured. Sodium sulfide was used as a standard and activity of the beta-CAS was expressed in µg of H_2S/min^{-1}/2 mg whitefly sample (Binu and Palaniswami 2006) .

The results indicated activity of rhodanase and *β*–cyanoalanine synthase in SPWF and CWF. Rhodanase activity in CWF was 128.85 ± 5.15 µg thiocyanate/min^{-1}/2 mg

Table 7.8 Activity of cyanide metabolizing enzymes in the cassava reared (CWF) and sweet potato reared (SPWF) whitefly (2 mg) sample

Whitefly	Rhodanese [μg thiocyanate/ min^{-1}/2 mg tissue] (Mean ± SE)	*β*–cyanoalanine synthase [μg H_2S/ min^{-1}/2 mg tissue] (Mean ± SE)
CWF	128.85 ± 5.15*	1.70 ± 0.11*
SPWF	108.23 ± 1.37	1.28 ± 0.16

Significance determined by student's *t* test
*Significant at 0.05 levels

tissue compared to 108.23 ± 1.37 μg thiocyanate/min^{-1}/2 mg tissue in SPWF. *β*–cyanoalanine synthase activity in CWF was 1.70 ± 0.11 μg H_2S/min^{-1}/2 mg tissue when compared to 1.28 ± 0.16 μg H_2S/min^{-1}/2 mg tissue in SPWF (Table 7.8).

7.4 ICMV Detection

7.4.1 In Plants

7.4.1.1 ICMV Serological Assays – Dot-Blot Immuno Assay/TAS-ELISA

Presence of ICMV was verified in ICMD symptom expression plants using TAS-ELISA technique. ELISA plates were coated with polyclonal antibody of ACMV at 10^{-3} dilution in coating buffer and incubated at 37°C for 1 h. ICMD leaf samples were ground with extraction buffer (0.05 M Tris-HCl, pH 8.0, 0.005 M EDTA, 2% Polyvinyl pyrrolidone (PVP) and 0.05% Tween-20). Plates were washed with Phosphate buffer saline with 0.05% Tween-20 (PBS-T). Immediately after washing, 100 μl of test sample per well was added. There were three replicates for each dilution that were incubated overnight at 4°C, then blot dried followed by three rapid washes in PBS-T. The plate wells were blocked by adding 5% dried skim milk powder in PBS-T-PVP and incubated for 1 h at room temperature. The blocking solution was drained and 100 μl/well ICMV-monoclonal antibodies (MAbs) (SCR-60) at 10^{-3} dilution in PBS-T-PVP was added and the plate incubated for 3 h at 37°C. After washing in PBS-T, goat antimouse-alkaline phosphatase conjugate diluted to 1:15,000 in conjugate buffer (PBS-T-PVP and 0.2% ovalbumin) was added at 100 μl/well and incubated for 3 h at 37°C. Later three rapid washes were given with PBS-T, after blot drying, substrate buffer (pH 9.8) having PNPP as substrate at 100 μl/well was added. Reaction was read at 405 nm in ELISA reader after 1 h and overnight incubation at 4°C. Results of absorbance values of TAS-ELISA were analyzed by one-way analysis of variance (ANOVA), conducted using the *SYSTAT* program. Significant mean differences were accepted at 0.05% probability level.

Plant leaf samples from cassava and tobacco with ICMD symptoms collected from the fields and from the experimental plants reacted positively in ELISA tests. Viral concentration was higher in cassava leaf samples from the fields than that of tobacco and ICMV infected cassava seedlings (Fig. 7.6). In the different duration of AAFP/IAFP transmission experiments, the ICMD symptomatic plants when tested showed positive for the presence of ICMV. The absorbance value for 48/48 h AAFP/IAFP was highest compared to other AAFP/IAFP indicating higher concentration of ICMV (LSD at 0.05%=0.0203) (F=74.353; df=6, 35, P=0.000).

7.4.1.2 Polymerase Chain Reaction

Leaves from cassava seedlings infected with ICMV through CWF and non-infected cassava seedlings were ground (500 mg plant tissue) in liquid nitrogen to a fine powder using mortar and pestle. The tissue samples were transferred to a 15 ml centrifuge tube so as to allow the liquid nitrogen to evaporate and followed the protocol for isolation of DNA from plant tissue with the DNeasy Plant Maxi Kit (QIAGEN Inc., USA). Extracted total DNA was dissolved in buffer and kept in deep freeze (–20°C) for further studies. Leaves of cassava seedlings raised from true-seed in whitefly proof cages were used as healthy control. Based on the work of Deng et al. (1994) the two degenerate oligonucleotide primers (primer A and B) (SIGMA GENOSYS, UK) used were:

Primer A : 5′ – TAA TAT TAC CKG WKG VCC SC – 3′
Primer B : 5′ – TGG ACY TTR CAW GGB CCT TCA CA – 3′

Where K=G or T, R=A or G, S=C or G, W=A or T, Y=C or T, B=C, G or T &V=A, C or G

Standard 50 µl polymerase chain reaction (PCR) was then carried out in the Gen Amp 9,600 thermal cycler containing, 16 µl of Taq polymerase buffer (PCR buffer 5 µl, Q solution 10 µl and Taq 1 µl, QIAGEN, Inc. USA), 100 pmol each of primer A and B, 200 µM of each dNTP (Amersham Pharmacia Biotech, USA) and 10 µl of template DNA and final volume was made up to 50 µl with nuclease free water. They were overlaid with 20 µl mineral oil. PCR conditions were 35 cycles. The first reaction cycle comprised of 2 min at 94°C, 1 min at 52°C and 2 min at 72°C, and was followed by 35 cycles of 45 s at 94°C, 1 min at 55°C, and 2 min at 72°C, and final cycle of 45 s at 95°C, 1 min at 55°C, and 5 min at 72°C.

In the analysis, 6–8 µl samples of PCR products were analyzed by electrophoresis in 1.8% agarose gel in Tris-acetate-EDTA buffer stained with ethidium bromide and photographed on a UV transilluminator (Bio-Rad, Hercules, CA). Markers used were 100 bp DNA ladder (Amersham Pharmacia Biotech, USA).

In the experiment an amplification of 530 bp size fragment was observed when primer A and primer B were used in PCR with nucleic acid extracted from ICMV-infected cassava leaves . This product was not observed when template was omitted (N-negative control) and no products were observed when nucleic acid from uninfected leaves (H- healthy plants) was used as template. Results indicated that the amplified product obtained with nucleic acid extracted from ICMD leaves was specific to ICMV.

7.4.2 In B. tabaci – CWF

7.4.2.1 Dot-Blot Immunoassay

Effect of AAFP on ICMV transmission was determined by confining 50 CWF adults, after 3 h starvation period, in 4 cm diameter leaf clip cages for 30 min., 45 min., 1, 2, 3, 10, 24, 48, 72 h AAFP on young ICMV-infected cassava leaves. CWF adults (non-viruliferous) were taken each time from the stock colony. After each AAFP, using an aspirator, 25 CWF adults were collected and placed in a freezer for about 1 h. The immobilized whiteflies (25 nos.) of different AAFP were placed separately in a microfuge tube and homogenized with a glass rod in extraction buffer (TBS pH 7.5). Then homogenates were centrifuged at 10,000 rpm and 5 μl of supernatant was placed in a 3×4 cm nitrocellulose membrane (NCM). The NCM was allowed to air dry and immersed in blocking solution (spray dried milk (SDM) in TBS pH 7.5) by gentle shaking for 1 h at room temperature. The blocking solution was rinsed in TBS for 10 min and the NCM was incubated with primary Ab diluted in TBS-SDM. The NCM was washed three times with TBS and then transferred into enzyme labelled anti-rabbit IgG. Again the NCM was washed three times with TBS. The NCM was incubated in substrate solution (75 g NBT and 50 g BCIP in dimethyl formamide), respectively in substrate buffer (pH 9.5, 0.1 M Tris, 0.1 M NaCl and 5 mM $MgCl_2$). Finally NCM developed colour, was air dried and photographed. *B. tabaci* adults reared on sweet potato were taken as negative control. Similarly extracts from healthy and ICMV- infected plants were taken as negative and positive controls respectively.

In ICMV serological assays – Dot-blot immuno assay/TAS-ELISA, adult CWF, each with 30 min., 45 min., 1, 2, 3, 10, 24, 48, 72 h AAFPs reacted positive indicating the presence of ICMV in the CWF biotype in all cases. ICMV was detected by spotting 5 μl aliquot of a homogenate prepared by grinding 25 whiteflies of different AAFPs on ICMV-infected plants. No background reactions were observed in control whiteflies, the sweet potato biotype (non-viruliferous) and from the extracts of healthy plants.

7.4.2.2 Serological Assay for ICMV in *B. tabaci* – CWF

Adults CWF (50 nos.) were collected from the virus free stock culture of cassava plants. Following 3 h of starvation they were released into leaf clip cages on young ICMV infected cassava leaves for 24 and 48 h AAFP. Using an aspirator, 25 CWF adults were collected and placed in a freezer for about 1 h. The immobilized whiteflies (25 nos.) each of 24 h and 48 h AAFP were placed separately in a microfuge tube and homogenized with a glass rod. Homogenates were treated with extraction buffer and TAS-ELISA was conducted as described above. Virus free CWF adults were used as healthy control.

B. tabaci adults of CWF biotype each with AAFP of 24 and 48 h reacted positive with ICMV-ELISA indicating the presence of ICMV in CWF biotype in both cases. Maximum absorbance value was in 48 h AAFP as compared to 24 h AAFP suggesting higher concentration of ICMV at 48 h AAFP.

7.4.2.3 Polymerase Chain Reaction

Twenty adult CWF- (both male and female) collected from virus free culture, after given 30 min., 45 min., 1,2, 3, 5, 10 24, 48 and 72 h AAFP, were homogenized in a microfuge tube with a minidril in liquid nitrogen and transferred to a 2 ml microfuge tube so as to allow the liquid nitrogen to evaporate and followed the protocol for isolation of DNA from insect tissue with the DNeasy Tissue Kit (QIAGEN Inc., USA). Extracted total DNA were dissolved in buffer and kept in deep freeze (–20°C) for further studies. *B. tabaci* - SPWF reared on sweet potato were used as healthy control.

Nucleic acid extracted from CWF biotype (with varying AAFPs of 30 min., 45 min., 1,2, 3, 5, 10, 24, 48 and 72 h) on ICMV-infected plants, when used with primer A and B in PCR, produced an amplification band of 530 bp. The product was not observed when template was omitted (N-negative control). No products were observed when nucleic acid from uninfected SPWF was used as template, indicating that the amplified product was specific to ICMV.

7.4.3 DNA Sequencing

Degenerate primers A and B were used for the amplification of 530 bp DNA fragments. A total of 25 μl reaction mixture consisted of 8 μl Taq Polymerase buffer (PCR buffer 2.5 μl, Q solution 5 μl and Taq 0.5 μl, QIAGEN, Inc. USA), 50 pmol of each primer, 10 μM of each dNTP's and 5 μl of template and final volume was made up to 25 μl with nuclease free water. Template consisted of plant extracted DNA and insect extracted DNA. Samples of 6 μl PCR products were analyzed by electrophoresis in 1.8% agarose gel in Tris-acetate-EDTA buffer stained with ethidium bromide. Markers used were 100 bp DNA ladder (Amersham Pharmacia Biotech, USA). Bands were cut and Montage Gel Extraction Kit (MILLIPORE, USA) was used to clean all PCR products, which were then sequenced in both directions with the primers A and B using Big Dye terminators at half recommended volumes on an ABI Prism (Model 310) automated sequencer. Sequencing PCR was carried out, using 1.14 μl 10 mM $MgCl_2$, 2 μl sequencing mix (Perkin Elmer, Germany) 1 μl primer A and 3 μl purified template made up to 20 μl with sterile nuclease free water. PCR conditions were 35 cycles of 95°C denaturation (30 s), 52°C annealing (30 s) and 72°C extension (1 min) with initial denaturation for 2 min and final extension for 4 min. Oligonucleotide sequences obtained from the sequencing reactions were blasted (BLAST algorithm) in the NCBI home page (http://www.ncbi.nic.org) in order to determine the nucleotide similarity with other published sequences.

By using the degenerate primer pair A and B ~ 510 bp were sequenced. The comparison of nucleotide sequence showed no similarity with analyzed DNA and published sequences of CMV isolates originating from India, Africa and Sri Lanka. Even though we could observe the primer A sequence in the complete sequence, it didn't show any sequence similarity with published CMV sequences.

7.4.4 ICMV (Trivandrum) Nested Primer

Development of a nested PCR primer that facilitates the amplification of the coat protein (*Cp*) for field isolates of ICMV to achieve virus detection in extracts of infected plants and in viruliferous adult *B. tabaci* demonstrates that together with the sequence of the amplicon obtained using virus-specific sequencing primers, could allow sensitive detection and identification of ICMV isolates. The sequence for the ICMV *Cp* amplicon was thus determined to ascertain provisional virus identification, and the *Cp* sequence was used to investigate the phylogenetic relationships between ICMV and all known cassava-infecting begomoviruses for which sequences are available in the GenBank database.

7.4.4.1 Design of Oligonucleotide Primers

ICMV sequences obtained from GenBank and NCBI/GenBank databases were examined for the design of the oligonucleotide primers. NCBI/GenBank accession numbers AF423180 and AJ314739 were used. The *Cp* sequence for each virus was extracted from the complete DNA-A component sequence and sequences were aligned using CLUSTAL W (Bioedit, North Carolina State University, USA). Sequencing primers and a nested primer specific for ICMV were designed using the analysis program PRIMER PREMIER (PREMIER Bio-soft international, CA, USA). Sequencing primers were synthesized based on the ICMV-Maha sequence and nested primer was based on the ICMV-Tri isolate. Sigma Genosis, USA, synthesized the oligonucleotides.

Design of oligonucleotide primers: ~1,080 bp fragments 2,739 bp of ICMV-Maharashtra (NCBI/GenBank accession no. AJ314739) and ~771 bp of ICMV- Tri (NCBI/GenBank accession no. AF423180) were used for the primer design. GEM-CO S1, GEM-CO A2, GEM-CO A1 and GEM-CO A2 derived from ICMV-Maharashtra; whereas CTCRI-USIF nested S1, CTCRI-USIF nested A2, CTCRI-USIF nested A1 and CTCRI-USIF nested A2 were derived from ICMV-Tri 1 . The primer synthesizing regions, amplicon products and sequence no. are given (Tables 7.9 and 7.10). The newly nested primer is designated as CTCRI-USIF nested primer S1 and A1.

Table 7.9 Primers used for polymerase chain reaction to amplify the begomoviral coat protein gene and those used for nucleotide sequencing of amplicons, respectively

No	Oligo name	Primer sequence (5′–3′)	Strand	Positions	Amplicon length (bp)
1.	CTCRI-USIF Nested (*S1*)	CCTgggTAAgATATggATggA	Sense	348-368 (ICMV-Tri)	390
	CTCRI-USIF Nested (*A1*)	CTAATCTTCAgCgTAgCgTAT	Anti-Sense	736-716 (ICMV-Tri)	
2.	CTCRI-USIF Nested (*A2*)	TCCTTgTAAgggATCgTAggC	Sense	413-433 (ICMV-Tri)	310
	CTCRI-USIF Nested (A2)	ACAgggTTAgAggCATgAgTA	Anti-sense	715-695 (ICMV-Tri)	
3.	GEM-CO *S1*	CgAAgCGACCAgCAgATA	Sense	302-319 (ICMV-Maha)	605
	GEM-CO *A1*	CCCTAAAgAAACgCCTAACT	Anti-sense	905-886 (ICMV-Maha)	
4.	GEM-CO *A2*	CAgTATgCgAgCAAggAgC	Sense	865-883 (ICMV-Maha)	180
	GEM-CO A2	gATTCTAATCTTCAACgTAgCg	Anti-sense	1038-1017 (ICMV-Maha)	

Table 7.10 Percentage nucleotide identity and amino acid similarity (parenthetically) for cassava-infecting begomoviruses for which sequences were determined herein and/or are available in the GenBank database

	CP gene (~771 bp)			NCBI/GenBank
Virus isolates	1–258 bp	258–510 bp	510–777 bp	accession no.
ICMV-Tri (771 bp)	100(100)	99(98)	100(100)	AF423180
ICMV-Adivaram (771 bp)	97(98)	96(98)	98(100)	AJ575819
ICMV-Maha (771 bp)	92(93)	93(97)	93(100)	AJ314739
ICMV-CGCP (533 bp)	88(84)	93(97)	92(98)	AF075593
ICMV-CGA	91(91)	93(95)	91(98)	Z24758
SLCMV (771 bp)	90(90)	92(97)	93(100)	AJ314737
SACMV (777 bp)	77(79)	65(77)	73(79)	AF155807
ACMV-UG/Svr (777 bp)	81(77)	71(73)	68(76)	AF126802
ACMV-(Nam) (777 bp)	81(77)	71(73)	69(76)	AF423177
ACMV-CM/BB (777 bp)	81(77)	72(75)	69(76)	AF211464
ACMV-AK (777 bp)	81(75)	71(73)	70(76)	AY211461
ACMV-CM/YA (777 bp)	81(76)	72(75)	69(76)	AY211463
ACMV-CM/AK2 (777 bp)	81(75)	72(75)	70(76)	AY211460
ACMV-CM/39 (777 bp)	81(77)	72(76)	70(76)	AY211462
EACMV-UGV (777 bp)	77(75)	73(78)	68(75)	Z83257
EACMV-UG (Nam) (774 bp)	77(75)	73(76)	68(75)	AF423178
EACMV-[K2B] (774 bp)	75(74)	75(80)	72(79)	Z83258
EACMV-[Mtw] (774 bp)	75(74)	75(80)	72(79)	AF423179
EACMV-Ko (774 bp)	75(74)	71(77)	71(76)	AY211467
EACMV-BB (774 bp)	76(73)	71(75)	69(76)	AY211468
TLCV-Ban II (771 bp)	87(87)	79(96)	81(95)	U38239
TLCBV (771 bp)	82(86)	85(95)	80(91)	AF188481
MYVV [Y47] (771 bp)	81(84)	76(89)	84(89)	AJ457824
PLCBV (771 bp)	85(88)	83(95)	81(93)	AF314531

Figures in parentheses are percentage amino acid sequence identity

7.4.4.2 Coat Protein Gene Sequencing

Primers used for the amplification of Cp (~771 bp) were GEM-CO S1, A2, A1 and A2 and CTCRI nested S1 and A1 (Table 7.9). A total of 25 μl reaction mixture consisted of 8 μl Taq Polymerase buffer (PCR buffer 2.5 μl, Q solution 5 μl and Taq 0.5 μl, QIAGEN, Inc. USA), 50 pmol of each primer, 10 μm of each dNTP and 5 μl of template and final volume was made up to 25 μl with nuclease free water. Template consisted of plant extracted DNA and insect extracted DNA. PCR conditions for the primer GEM-CO S1 and A1 comprised of 2 min at 94°C, 1 min at 52°C and 2 min at

72°C, and was followed by 35 cycles of 45 s at 94°C, 1 min at 52°C, and 1 min at 72°C, and final cycle of 45 s at 95°C, 1 min at 55°C, and 5 min at 72°C. PCR conditions using the primer GEM-CO A2 and A2 comprised of 2 min at 94°C, 1 min at 52°C and 2 min at 72°C, and was followed by 35 cycles of 45 s at 94°C, 1 min at 50°C and 1 min at 72°C, and final cycle of 45 s at 95°C, 1 min at 55°C, and 4 min at 72°C. PCR conditions using CTCRI- USIF Nested primer S1 and A1 were same as described above. Samples of PCR products (6 μl) were analyzed by electrophoresis in 1.8% agarose gel in Tris-acetate-EDTA buffer stained with ethidium bromide. Markers used were 100 bp DNA ladder (Amersham Pharmacia Biotech). Bands were cut and Montage Gel Extraction Kit (Millipore, USA) was used to clean all PCR products, which were then sequenced in both directions with the primers (CTCRI-USIF Nested S1 and A1; GEM-CO S1, A2, A1 and A2) using Big Dye terminators at half recommended volumes on an ABI Prism (Model 310) automated sequencer. Sequencing PCR was carried out, using 1.14 μl 10 mM $MgCl_2$, 2 μl sequencing mix (Perkin Elmer, Germany), 1 μl primer (CTCRI-USIF Nested S1 and A1; GEM-CO S1, A2, A1 and A2) (5 pmol) and 3 μl purified template made up to 20 μl with sterile nuclease free water. PCR conditions were 35 cycles of 95°C denaturation (30 s), 50°C annealing (30 s) and 72°C extension (1 min) with initial denaturation for 2 min and final extension for 4 min. Oligonucleotide sequences obtained from the sequencing reactions were blasted (BLAST algorithm) in the NCBI home page (http://www.ncbi.nic.org) in order to determine the nucleotide similarity with primer designed ICMV strains. Nucleotide sequence was submitted to GenBank under accession no. AY312989.

PCR Amplification of ICMV in Different Cassava Varieties Using Nested Primer

When CTCRI-USIF nested primer S1 and A1 were used in PCR with nucleic acid extracted from ICMV-infected cassava leaves of six released varieties, an amplification of 390 bp size fragment was observed. This product was not observed when template was omitted (N-negative control) and no products were observed when nucleic acid from uninfected leaves (H- healthy plants) was used as template. Results indicated that the amplified product obtained with nucleic acid extracted from ICMV-infected leaves was specific to ICMV.

Detection of ICMV in Whiteflies Under Different AAFP Using Nested Primer

Nucleic acid extracted from CWF of varying AAFP on ICMV-infected leaves, when used with nested primer *S1* and *A1* in PCR, an amplification band of 390 bp was obtained. The product was not observed when template was omitted (N-negative control) and no product was observed when nucleic acid from uninfected SPWF was used as template, indicating that amplified product was specific to ICMV. On the other hand, using the nested primer pair *A2* and A2, a PCR amplified fragment of 310 bp was obtained from viruliferous whiteflies of varying AAFP (30 min., 45 min., 1, 2, 3, 5, 10, 24 and 48 h). No PCR amplification occurred, when using either of the primer pairs, with extracts from healthy SPWF not allowed access to ICMV.

7.5 *B. tabaci* Feeding Induced Pathogenesis Related Proteins in Cassava

Little work has been undertaken in cassava and pathogenesis related (PR) proteins in response to insect attack. Therefore an attempt was made to study the PR proteins in cassava plants due to *B. tabaci* attack. Experiments were conducted on CMD-free healthy cassava (raised from true cassava seed) as well as on CMD infected cassava plants. Cassava seedling raised plants (CSRP) in an insect proof cage were taken as CMD free healthy cassava plants. Adult *B. tabaci* (~300 nos.) were released into four cages (Test plants with whiteflies T1–T4) for 16 days while two cages were kept as controls (Control plants without whiteflies C1–C2). Number of nymphs, pupae and adults on the third and fourth leaves from the top were counted in all the experimental plants in all six cages (Tables 7.11 and 7.12) and leaf samples were analysed by SDS-PAGE for PR proteins, PAGE for the peroxidase, and colorimetric assay for the determination of total protein, peroxidase, β-1,3-glucanase and chitinase activity (Antony and Palaniswami 2006).

7.5.1 *Total Protein Determination and Electrophoresis*

Analysis of total protein in the leaves showed that leaf protein content in both CSRP (CMD free healthy) and CMD infected plants decreased significantly in whitefly infested cassava plants (T1–T4), when compared to the non-infested control plants (C1 and C2). The total protein content in the whitefly infested leaves decreased from 23.7 to 21.94 mg/g in CMD free healthy plants while in CMD plants the total protein decreased from 20.59 to 16.23 mg/g (Tables 7.13 and 7.14).

Appearance of a specific band of MW 48.00–52.48 kDa was noted in all CSRP infested with whiteflies. Whereas appearance of two specific bands of Rf values ranging from 0.46 to 0.476 and 0.545 to 0.563, respectively were noted in all whitefly infested CMD plants. These bands were absent in the control C1 and C2 plants. New cassava proteins, induced by whitefly feeding were obviously visible on the gel.

In the electropherogram, well-resolved peroxidase bands were observed in the infested and uninfested cassava plants. In CSRP non-infested plants (C1 and C2) three bands were obtained whereas in the test plants (T1–T4) an extra band was

Table 7.11 Nymphal, pupal and adult population of *B. tabaci* infested (T1–T4) and non-infested (C1–C2) cassava seedlings (CSRP)

	Population per three leaves			
Host	Nymph	Pupae	Adults	Total
T1	372	149	158	679
T2	272	169	141	582
T3	267	247	75	589
T4	321	166	113	600
C1	–	–	–	–
C2	–	–	–	–

Table 7.12 Nymphal, pupal and adult population of *B. tabaci* infested (T1–T4) and non-infested (C1–C2) cassava leaves with CMD infection

	Population per three leaves			
Host	Nymph	Pupae	Adults	Total population
T1	40	102	67	234
T2	50	90	36	131
T3	85	91	43	219
T4	19	25	25	69
C1	–	–	–	–
C2	–	–	–	–

observed. This extra band was expressed in all four groups of whitefly infested (T1–T4) CMD free cassava plants. However, no new bands were observed in the peroxidase gels of whitefly infested and non-infested CMD plants (Figs. 7.8–7.11).

7.5.2 *Enzyme Activity*

Enzymes activities expressed in specific activity in *B. tabaci* infested leaves of CSRP increased significantly with whitefly feeding. *β*-1, 3-glucanase increased seven-fold, chitinase activities increased almost threefold and peroxidase activity increased twofold. In the total activity all three enzymes also showed significant increased activities. *β*-1, 3-glucanase increased seven-fold, chitinase activity increased 3.5-fold, and peroxidase increased 2.5-fold. *β*-1, 3-glucanase, chitinase and peroxidase activities were significantly higher in whitefly infested leaves compared to non-infested plants (Table 7.13).

β-1, 3-glucanase, chitinase and peroxidase activity measurement for whitefly infested and uninfested leaf samples of CMD infected plants (diseased) are presented (Table 7.14). Enzymes activities expressed in specific activity in *B. tabaci*

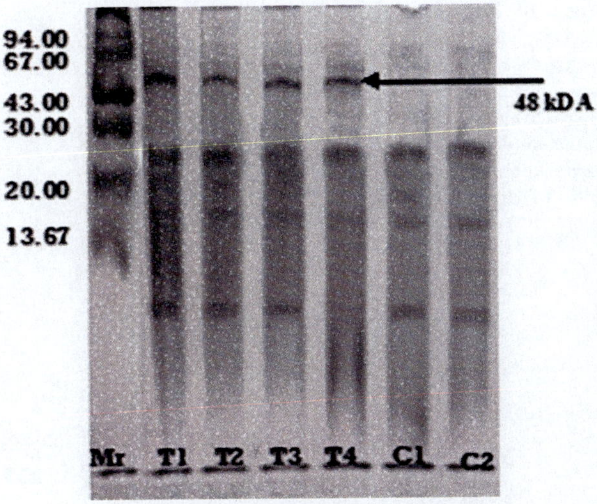

Fig. 7.8 SDS-PAGE Electropherogram of cassava seedlings raised (non-diseased) plant and CMD infected (diseased) from *Bemisia tabaci* infested (T1–T4) and non-infested plants (C1–C2)

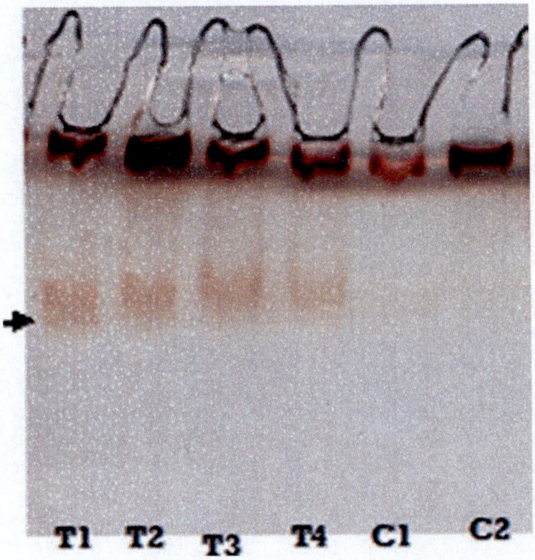

Fig. 7.9 SDS-PAGE Electropherogram of cassava seedlings raised (non-diseased) plant and CMD infected (diseased) from *Bemisia tabaci* infested (T1–T4) and non-infested plants (C1–C2)

infested leaves increased significantly with whitefly feeding. β-1, 3-glucanase approximately 4.5-fold, chitinase activities increased almost three-fold and peroxidase activity increased 2.5-fold. Based on total activity, β-1, 3-glucanase increased 3.5-fold, chitinase activity increased 1.75-fold, and peroxidase increased 2.5-fold. Total activities and specific activities of β-1, 3-glucanase, chitinase and peroxidase were significantly higher in whitefly infested leaves of CMD plants than the non-infested plants (Table 7.14).

Our studies involving three enzymes: β-1, 3-glucanase, chitinase and peroxidase indicated that there was an increase in the enzyme activities induced in cassava

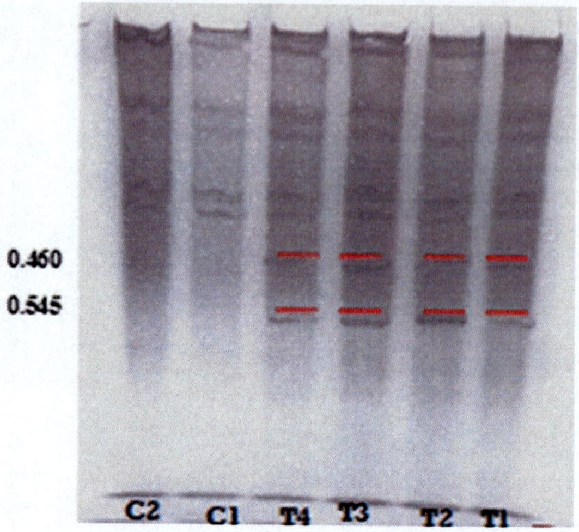

Fig. 7.10 PAGE Electropherogram of peroxidase profile of cassava seedlings raised (non-diseased) plant and CMD infected (diseased) from *Bemisia tabaci* infested (T1–T4) and non-infested plants (C1–C2)

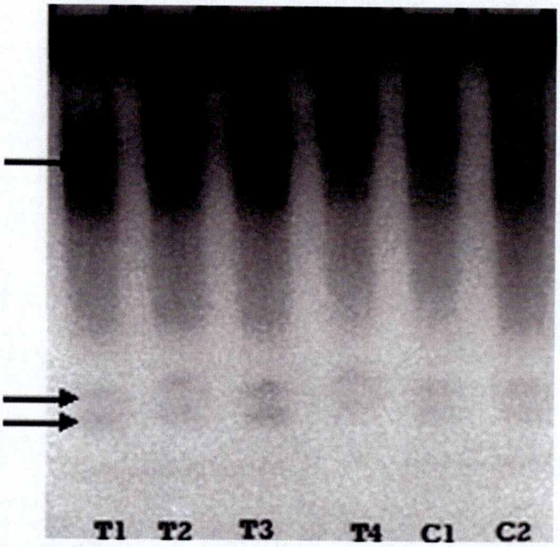

Fig. 7.11 PAGE Electropherogram of peroxidase profile of cassava seedlings raised (non-diseased) plant and CMD infected (diseased) from *Bemisia tabaci* infested (T1–T4) and non-infested plants (C1–C2)

leaves due to *B. tabaci* feeding. The result of leaf protein analysis indicates that total protein content is decreased in infested cassava plants suggesting that there may be suppression of some plant protein occurring simultaneously with the apparent induction of PR proteins. The decrease in leaf protein content might explain an increase in PR enzyme activities. The total activities of chitinase, β-1, 3-glucanase and peroxidase in infested plants were significantly higher than control plants ($P<0.001$). Hence it appears that whitefly feeding induces PR enzymes and reduces general protein levels in infested plant tissues (Binu and Palaniswami 2006).

Table 7.13 Total protein measurements and enzyme activity in cassava seedlings raised (non-diseased) plants from *Bemisia tabaci* infested and non-infested plants

		β-1, 3-glucanase			Chitinase			Peroxidase		
Sample	Total protein (mg/g tissue)	mg Glucan/ min/g fresh wt.	Total activity (mg Glucan/ min/g fresh wt.)	Specific activity	OD/h/g fresh wt.	Total activity (OD/h/g dry wt.)	Specific activity	OD/h/g fresh wt.	Total activity (OD/h/g dry wt.)	Specific activity
Infested	21.94 ± 0.68	47.00 ± 9.82	239.80 ± 50.12	2.87 ± 0.53	43.81 ± 13.01	222.97 ± 66.23	2.68 ± 0.72	36.22 ± 7.35	167.70 ± 34.03	1.73 ± 0.34
Uninfested	23.70 ± 0.10	8.76 ± 0.135	36.05 ± 0.54	0.43 ± 0.01	16.23 ± 2.92	66.79 ± 12.04	0.80 ± 0.76	14.42 ± 5.33	58.16 ± 21.52	0.82 ± 0.23
t	6.252	9.535	9.958	4.415	5.067	5.683	4.906	5.878	6.664	6.175
df	10	10	10	10	10	10	10	10	10	10
P value	P<0.001	P<0.001	P<0.001	P<0.001	P<0.001	P<0.001	P<0.001	P<0.001	P<0.001	P<0.001

Specific activity expressed in OD/h/g fresh wt./mg protein g^{-1} tissue
Significance determined by student's *t* test
Values represented as Mean ± SD

Table 7.14 Total protein measurements and enzyme activity in CMD infected (diseased) cassava from *Bemisia tabaci* infested and non-infested plants

		β-1, 3-glucanase			Chitinase			Peroxidase		
Sample	Total protein (mg/g tissue)	mg Glucan/ min/g fresh wt.	Total activity (mg Glucan/ min/g fresh wt.)	Specific activity	OD/h/g fresh wt.	Total activity (OD/h/g dry wt.)	Specific activity	OD/h/g fresh wt.	Total activity (OD/h/g dry wt.)	Specific activity
Infested	16.23 ± 3.15	38.68 ± 5.91	120.13 ± 18.34	3.59 ± 0.88	31.33 ± 7.53	115.01 ± 27.64	2.64 ± 0.73	40.82 ± 7.01	126.05 ± 19.21	2.81 ± 0.41
Uninfested	20.59 ± 0.93	15.68 ± 0.75	34.31 ± 1.63	0.76 ± 0.20	17.28 ± 0.53	71.40 ± 2.17	0.85 ± 0.73	22.62 ± 0.92	49.52 ± 2.01	1.18 ± 0.31
T	3.250	9.464	11.42	4.80	4.562	3.860	3.283	6.301	9.706	7.231
Df	10	10	10	10	10	10	10	10	10	10
P value	P < 0.05	P < 0.001	P < 0.001	P < 0.001	P < 0.001	P < 0.001	P < 0.001	P < 0.001	P < 0.001	P < 0.001

Specific activity expressed in OD/h/g fresh wt./mg protein g^{-1} tissue
Significance determined by student's *t* test
Values represented as Mean ± SD

7.6 Effect of ICMV on Vector

The population build up and development of *B. tabaci* on ICMV infected as well as on healthy cassava leaves revealed clearly the effect of ICMV on the insect biology (Tables 7.11 and 7.12). The population was generally higher on the CMD free healthy plants (CSRP) (ranging from 582 to 679 per three leaves) than CMD plants (ranging from 69 to 234 per three leaves). Present studies showed that total enzymatic activities increased in CMD infected plants and CSRP (healthy plants) in response to *B. tabaci* feeding. This may be explained on the basis that CMD infected plants already contained ICMV as pathogen and induced PR protein formation. Hence whitefly feeding induced less PR protein in the CMD infected plants.

7.7 Conclusion

In India, biotypes of *B. tabaci* (cassava biotype and sweet potato biotypes) have been established by Lisha et al. (2003). Present studies using TAS-ELISA and PCR confirmed that CWF is responsible for the spread of ICMV in India and SPWF do not transmit ICMV. Dot blot and TAS-ELISA tests corroborated the ICMV in experimental plants indicating vectoring efficiency of CWF. During the survey in Northern India the incidence of ICMD and associated *B. tabaci* population were found to be negligible on cassava. The absence of CWF and ICMD in these areas might have contributed to the non-spread of ICMD in these areas.

Higher percentage transmission of ICMV was due to the availability of the cassava whitefly biotype, with optimum AAFP/IAFP (48/48 h). ICMV transmission improved with increase in IAFP and AAFP to a certain extent. Low percentage of ICMV transmission at shorter AAFP and IAFP schedules might probably be the result of less than threshold acquisition of ICMV titers and less than required circulation time of ICMV in the CWF. Perhaps the time taken for circulation of ICMV in the vector before successful transmission, requires more than 12 h.

Nucleic acid extracts from viruliferous CWF adults and ICMV infected cassava leaves, when used with two degenerate primers in PCR gave a specific amplification of 530 bp size fragment. Experimental results confirmed that only *B. tabaci* – CWF biotype transmits ICMV from cassava to cassava and from cassava to tobacco.

The serological assay indicated that the whiteflies acquired a higher virus titer at 48 h AAFP, which resulted in higher percentage of ICMV transmission at 48/48 h AAFP/IAFP. TAS-ELISA assays also confirmed that the CWF acquired higher virus titer after 48 h AAFP compared to 24 h AAFP. This is the first report of ICMV transmission by the cassava biotype in India.

Present study showed rhodanase and *ß*–cyanoalanine synthase activities were higher ($P<0.05$) in CWF than in SPWF. Cassava leaf tissue contains high concentration of the cyanogenic glycoside, linamarin, localized in the symplast (Mkpong et al. 1990). It is possible that penetration of the phloem tissue by SPWF stylet and salivary secretions might lead to the limited hydrolysis of cyanogenic glycosides and consequent release of toxic cyanide. Survival on cassava would depend on mechanisms either to prevent the release of free cyanide or to effect its detoxification. Hence, this could cause the death of insects during acquisition feeding for a long duration on cassava. This may be the plausible explanation for the non-survival of SPWF feeding on cassava, and failure to transmit ICMV.

Even though several primers are available for PCR amplification of geminiviruses, very few are specific to ICMV. Furthermore some of these are degenerate and do not work effectively with ICMV in *B. tabaci*. In addition, there might be a possibility of emerging new strains either through mutational events or by recombination. Hence newly designed nested primer (CTCRI-USIF) and PCR protocols using these primer pairs, can be considered as a good tool for diagnosing ICMV in diseased plants as well as in the vector. Using the newly designed nested primer (CTCRI-USIF), nested S1 and A1, and GEM-CO S1, A1, S2 and A2, ICMV *CP* AV1 gene (771 bp) was sequenced and named as ICMV- Tri II. Oligonucleotide sequences obtained from the sequencing reactions were blasted (BLAST algorithm) in the NCBI. This nested primer when used in the detection of ICMV in different cassava varieties was found to be a good diagnostic tool for rapid and sensitive screening of ICMV. Using this nested primer the role of *B. tabaci* in ICMV transmission under different AAFP and IAFP was demonstrated. Nested primer gave an amplification of 390 bp and 310 bp fragments. ICMV *CP* AV1 gene (771 bp) was sequenced and is available in the GenBank database under accession no. AY323973. *CP* gene of ICMV showed 99% nucleotide identity to ICMV-Tri and 97% to ICMV-Adivaram, and 92% similarity with SLCMV. A comparison of deduced coat protein amino acid (*Cp*) sequences revealed that ICMV-Tri II is distantly related to EACMV isolates and ACMV. In contrast, the *Cp* for ICMV and SLCMV were most closely related to the begomovirus cluster containing Tomato leaf curl virus-Bangalore II, *Tomato leaf curl Bangladesh virus*, *Pepper leaf curl Bangladesh virus*, and *Malvastrum yellow vein virus* (MYVV). *Cp* amino acid comparisons also suggest a possible common ancestor for ICMV, SLCMV and TLCV-Bang II; which may have subsequently evolved independently possibly in relation to plant host and/or vector selection, together with geographical isolation. ICMV DNA in whitefly stylet, salivary gland and digestive tract was amplified and later sequenced using primer CTCRI-USIF nested S1 and A1, thereby proving ICMV transmission as circulative and persistent.

Whitefly feeding induced (PR) proteins in both ICMV-free and ICMV-infected cassava plants. Heavy infestation of *B. tabaci* showed increased levels of PR proteins. Quantitative measurement and electrophoretic detection of total protein, peroxidase, chitinase and β 1,3 glucanase in both ICMV-infected and ICMV-free cassava plants under whitefly infested and non-infested conditions revealed that cassava plants with whitefly infestation had an increased level of peroxidase, chitinase and β 1,3 glucanase, and decreased level of total proteins. The enzyme increased the specific activity two to fourfold in whitefly-infested plants compared to non-infested plants. Thus *B. tabaci* feeding induces PR proteins in cassava. Decreased protein content in leaf extracts from whitefly-infested plants was observed by comparison with those from uninfested plants. Among the three PR proteins studied, *B. tabaci* feeding induces β-1, 3-glucanase activities significantly higher when compared to the other two PR proteins. Hence it appears that it may be possible to utilize PR proteins in whitefly control programs through elevation of plant defensive proteins at appropriate times by application of elicitors in conjunction with biological control strategies using entomopathogens to increase the rate of infection. In addition, it is known that since *B. tabaci* possess peritrophic membranes, other possibilities such as the effective use of chitinases in control strategies exist.

There is however need for additional research along the following lines:

Molecular characterization of different collections of *B. tabaci;* Sequencing of the amplified product of the 28S rRNA region using the primers sequenced; Sequencing the 16S rRNA the region at the 5′ end of the mitochondrial 16S rRNA using the primers, LR-N-13398 and LR-J-12887; Sequencing the ITS region specific to *B. tabaci* designed from the 18S and 28S regions, and Mitochondrial cytochrome oxidase I gene; Studies on transfer of endosymbiotic organisms from adult *B. tabaci* to oocytes; Effect of antibacterial material on the biology of whitefly; Molecular characterization of endosymbionts of *B. tabaci*- sequencing of 16S rRNA of primary endosymbionts; Bio-efficiency of males and females in transmission of ICMV; Biological impact of ICMD leaves on longevity and fecundity of *B. tabaci* and Molecular approaches for the identification of aphelinid parasitoids.

Entrez Nucleotide Sequences at NCBI/EMBL/DDBJ Under the Project Accession nos. AY321152, AY517720, AY528819, AY356747, AY323973, AY518548, AY359237, AY360217, AY518547 and AY517721, AY 312989 and AY321252.

Acknowledgements The authors are grateful to USIF/USDA and ICAR, New Delhi for the grant and facilities in the conduct of the approved project.

References

Antony B, Palaniswami MS (2002) Influence of relative humidity and host biotypes on *Eretmocerus mundus* (Mercet) and aphelinid parasitoid of *Bemisia tabaci* (Genn.). Insect Environ 8:4–5

Antony B, Palaniswami MS (2006) *Bemisia tabaci* feeding induces pathogenesis-related proteins in cassava (Manihot esculenta Crantz). Indian J Biochem Bio 43:182–185

Antony B, Palaniswami MS, Henneberry TJ (2001) Hyperparasitism by an autoparasitoid *Encarsia transvena* (Timberlake) and their implication for the biological control of the whitefly, *Bemisia tabaci*. Entomon 26:11–16

Antony B, Palaniswami MS, Henneberry TJ (2003) *Encarsia transvena* (Hymenoptera: Aphelinidae) development on different *Bemisia tabaci* Gennadius (Homoptera: Aleyrodidae) instars. Environ Entomol 32:584–591

Antony B, Palaniswami MS, Kirk AA, Henneberry TJ (2004) Development of *Encarsia bimaculata* (Heraty & Polaszek) (Hymenoptera: Aphelinidae) in *Bemisia tabaci* Gennadius (Homoptera: Aleyrodidae) nymphs. Biol Control 30:546–55

Antony B, Lisha VS, Palaniswami MS, Sugunan VS, Makeshkumar T, Henneberry TJ (2006) *Bemisia tabaci* and Indian cassava mosaic virus transmission. Int J Trop Insect Sci 26:172–186

Antony B, Lisha VS, Palaniswami MS (2009) Evidences for transmission of Indian cassava mosaic virus through *Bemisia tabaci* – cassava biotype. Arch Phytopathol Plant Protect (UK) 42:922–929

Bartlett AC, Gawel NJ (1993) Determining whitefly species. Science 261:1333

Bluementhal SG, Hendrickson YPA, Conn EE (1968) Cynanide metabolism in higher plants III. The biosynthesis of beta-cyanoalanine. J Biol Chem 243:5302–5307

Brown JK (1994) A global position paper on the status of *Bemisia tabaci* Genn. as a pest and vector in world agroecosystems. FAO Plant Protect B 42:3–33

Brown JK, Coats SA, Bedford ID, Markham PG, Bird J, Frohlich DR (1995a) Characterization and distribution of esterase electromorphs in the whitefly, *Bemisia tabaci* (Genn.) (Homoptera: Aleyrodidae). Biochem Genet 33:205–214

Brown JK, Frolich DR, Rossell RC (1995b) The sweetpotato or silverleaf whiteflies. Biotypes of *Bemisia tabaci* or a species complex. Annu Rev Entomol 40:511–534

Burban C, Fishpool LDC, Fauquet C, Fargette D, Thouvenel JC (1992) Host-associated biotypes within West African populations of the whitefly *Bemisia tabaci* (Genn.) (Hom.; Aleyrodidae). J Appl Entomol 113:416–423

Byrne FJ, Denholm I, Birnie LC, Devonshire AL, Rowland MW (1992) Analysis of insecticide resistance in the whitefly, *Bemisia tabaci* (Genn.). In: Denholm L, Devonshire AL, Hollomon D (eds.) Achievements and developments in combating pesticide resistance. Elsevier, London

Costa HS, Brown JK (1991) Variation in biological characteristics and esterase patterns among populations of *Bemisia tabaci*, the association of one population with silverleaf symptom induction. Entomol Exp Appl 61:211–219

Costa HS, Russel M (1975) Failure of *Bemisia tabaci* to breed on cassava plants in Brazil (Homoptera, Aleyrodidae). Cienia Cult 27:388–390

Deng D, Mcgrath DF, Robinson DJ, Harrison BD (1994) Detection and differentiation of whitefly-transmissible geminiviruses in plants and vector insects by the polymerase chain reaction and degenerate primers. Ann Appl Biol 125:327–336

Fauquet CM, Fargette DD (1990) African cassava mosaic virus: etiology and control. Plant Dis 74:404–411

Faust RH (1992) Conference report and 5–year national research and action plan for development of management and control methodology for sweetpotato whitefly. ARS 107, pp 165

Legg JP (1996) Host-associated strains within Ugandan populations of the whitefly *Bemisia tabaci* (Genn.), (Hom., Aleyrodidae). J Appl Entomol 120:523–527

Lisha VS, Antony B, Palaniswami MS, Henneberry TJ (2003) *Bemisia tabaci* (Gennadius) biotypes (Homoptera: Aleyrodidae) in India. J Econ Entomol 96:322–327

Liu HY, Cohen S, Duffus JE (1992) The use of isozyme patterns to distinguish sweet potato whitefly (*Bemisia tabaci*) biotypes. Phytoparasitica 20:187–194

Mkpong OE, Yan H, Chism G, Sayre RT (1990) Purification, characterization and localization of linamarase in cassava. Plant Physiol 93:176–181

Palaniswami MS (2004) Natural enemies of *Bemisia tabaci* and their potential as biocontrol components of IPM. Final research project report. USIF/USDA funded IN-ARS-858/FG-IN-785, pp 407

Palaniswami MS, Nair RR (1995) Identification of vectors of CMD transmission in cassava and significance of biotypes of *Bemisia tabaci* Genn. In: Annual report (1994–95). CTCRI, Trivandrum, pp 27–28

Palaniswami MS, Nair RR, Pillai KS, Thankappan M (1996) Whiteflies on cassava and its role as vector of cassava mosaic disease in India. J Root Crops 22:1–8

Palaniswami MS, Antony B, Vijayan L, Henneberry TJ (2001) Sweet potato whitefly ecobiology, host interaction and its natural enemies. Entomon 26:256–262

Palaniswami MS, Makeshkumar T, Hegde V (2005) Management of vector borne diseases in horticultural crops. In: Chadha KL et al (eds) Crop improvement and production technology of horticultural crops, vol 1. Horticultural Society of India, New Delhi

Perring TM (2001) The *Bemisia tabaci* complex. Crop Prot 20:725–737

Perring TM, Cooper A, Kazmer DJ, Shields C, Sheilds J (1991) New strain of sweet potato whitefly invades California vegetables. Calif Agr 45:10–12

Perring TM, Cooper AD, Rodriguez RJ, Farrar CA, Bellows TS (1993) Identification of a whitefly species by genomic and behavioural studies. Science 259:74–77

Prabhaker N, Coudriet DL, Meyerdirk DE (1987) Discrimination of three whitefly species (Homoptera:Aleyrodidae) by electrophoresis of non-specific esterases. J Appl Entomol 103:447–451

Sorbo BJ (1955) Rhodanese. Method Enzymol 2:334–337

van Lenteren JC, Noldus LPJ (1990) Behavioural and ecological aspects of whitefly plant relationships. In: Gerling D (ed.) Whiteflies: their bionomics, pest status and management. Intercept Publications, Dorset

Chapter 8
The Performance of Viruliferous and Non-Viruliferous Cassava Biotype *Bemisia tabaci* on Amino Acid Diets

Winston M.O. Thompson

Abstract The role of amino acids is examined through a determination of their influence on *Bemisia tabaci* survival, honeydew secretion and oviposition. *B. tabaci* showed better performance when feeding on a diet of 20 amino acids more mimicking cassava host plants (*Manihot esculenta* Crantz.) as opposed to elevated levels of the amino acids. On a concentration range, 80% survival was observed at 94 h on all concentrations of serine, alanine, proline and phenylalanine. At the higher concentrations, asparagine, glutamic acid, aspartic acid, arginine and tryptophan caused poor survival of whiteflies. Greater oviposition occurred on the lower concentrations of aspartic acid, glutamic acid, arginine and histidine but the reverse was true for asparagine. On the basis of *B. tabaci* survival, honeydew secretion and oviposition, the better amino acids assorted as follows: asparagine, serine, proline, glutamic acid, glutamine, threonine, alanine, aspartic acid/cysteine and glycine; and these were predominantly non-essential amino acids. Performance of viruliferous and non-viruliferous *B. tabaci* on four amino acid diets *viz*., asparagine, serine, proline and alanine showed that *B. tabaci* survival was influenced by an interaction effect between whitefly viruliferous status and diet type. This accounted for the significantly higher survival of viruliferous whiteflies on the asparagine diet, and non-viruliferous whiteflies on the serine diet as compared to other treatment combinations. Honeydew secretion was influenced by a main diet-type effect. Significantly more honeydew droplets were secreted when feeding occurred on the asparagine diet as compared to the proline and serine diets. Oviposition was not influenced by either an interactive or main effect. Based on analysis, oviposition was not significantly different between viruliferous and non-viruliferous whiteflies. The physiological basis of amino acids is discussed and an insight is presented on the possible role of other nutrients, plant semiochemicals and endosymbionts in the interaction of *B. tabaci* with its host and begomoviruses.

W.M.O. Thompson (✉)
Plant Protection, National Agricultural Research Institute, Mon Repos, East Coast Demerara, Guyana, South America
15216 NE 8th St., H3 98007 Bellevue, Washington, USA
e-mail: winstont335@gmail.com

W.M.O. Thompson (ed.), *The Whitefly, Bemisia tabaci (Homoptera: Aleyrodidae) Interaction with Geminivirus-Infected Host Plants*,
DOI 10.1007/978-94-007-1524-0_8,

Keywords Amino acids • *Bemisia tabaci* • Honeydew secretion • Oviposition • Survival

8.1 Introduction

The interaction of *Bemisia tabaci* with its host plant and the begomoviruses transmitted result in consequences with either positive, negative or neutral effects to the vector. Such effects are related to the direct influences of the virus on the vector, the indirect induced responses of the host plant to infection that affect the vector, or a combination of these factors. In Chapter 3–7 this type of interaction is addressed as it relates to *B. tabaci* and *Tomato yellow leaf curl virus* (TYLCV); *B. tabaci* and *Cotton leaf curl virus* (CLCuV) and *B. tabaci* and some cassava mosaic viruses.

Direct effects on the vector could produce physiological changes that affect fecundity and thus reproductive capability and population levels of the vector. Neutral- type effects have also been observed where *B. tabaci* population levels have not been significantly different between uninfected and infected plants (Costa et al. 1991; Thompson 2002).

In other insect-plant-virus-interactions, virus infection causes an increase in the insect populations feeding on infected plants (Ajayi and Dewar 1983; Colvin et al. 1999). The influence on the insects appears to be related to changes in plant biochemistry resulting from infection (Ajayi and Dewar 1983; Colvin et al. 1999). Some observed chemical changes have in particular, involved changes in the amino acid composition of the diseased plants. Plants infected with maize rough dwarf virus showed an accumulation of asparagine (Harpaz and Applebaum 1961). An increase in the amino acid content in the phloem of plants in response to virus infection has been reported (Matthews 1981). Furthermore high population levels of *B. tabaci* has been associated with cassava plants infected with the Uganda variant of EACMV, which showed elevated levels of amino acids (Colvin et al. 1999).

In order to determine the possible influence of plant compounds, and in this case amino acids on the performance of whiteflies, it is necessary to investigate such influences *in vitro*, so as to isolated such effects outside of the plant environment. To this end, artificial feeding systems can be utilized.

8.2 Artificial Feeding Systems

Development of artificial feeding systems for phloem sap sucking insects date back to as early as 1927 when an artificial feeding system for the sugar beet leafhopper *Euttetix tenellus* was successfully developed (Carter 1927). Liquid diets in membrane feeding systems have been used successfully for rearing and studying the nutritional requirements of leafhoppers and planthoppers (Koyama 1984), as well as for rearing insects for disease transmission studies (Mitsuhashi 1979).

On the basis of research work conducted in the period 1963–1975, seven aphid species were successfully reared on artificial diets. These were *Aphis craccivora* Koch, *Aphis frangulae gossypii* Glover, *Cavariella aegopodii* (Scop), *Myzus ornatus* Laing, *Aulacorthum circumflexum* (Buckton) (Krieger 1971); *Myzus persicae* (Sulzer) (Mittler and Koski 1976); and *Acyrthosiphum pisum* (Harris) (Srivastava and Auclair 1971; Akey and Beck 1972).

The successful diets for these aphid species were based on modifications of three standard diets developed separately by Mittler and Koski (1976), Ehrhardt (1968) and Akey and Beck (1971, 1972). The three standard diets developed by these researchers included at least 18 protein amino acids, and sucrose at concentrations of 15%, 20% and 35% respectively. Other components included vitamins, and other organic and inorganic compounds (Kunkel 1976). Apart from such holidic diets, simple sucrose diets have been used successfully in studies involving planthoppers (Sogowa 1974; Kim et al. 1985) and aphids (Srivastava and Auclair 1971).

The development of a membrane feeding system for *B. tabaci* has been used to evaluate insecticides (Elbert and Nauen 1996) and natural toxins (Davidson et al. 1996) as well as to monitor the development of the juvenile stages of *B. tabaci* B-biotype (Jancovich et al. 1997). The extensive experiments using membrane feeding systems to determine the influence of amino acids on the performance of *B. tabaci* were executed to elucidate the interaction of *B. tabaci* with it host and vectored begomovirus, at the plant tissue and population levels with the possible involvement of amino acids (Thompson 2006; Thompson unpublished data). This investigation of the amino acids was conducted since previous studies reflected the influence of amino acids on phloem sap sucking insects (Douglas and Prosser 1992; Dorschner 1993; Simpson et al. 1995; Fujita and Mitsuhashi 1995).

8.3 Influence of Amino Acids on *B. tabaci*

8.3.1 Influence on Different Aged B. tabaci

In a case study to examine the effects of amino acids on a cassava biotype of *B. tabaci* from Namulonge Uganda, amino acid diets produced no differential effects between different aged whiteflies. Evaluating performance on a sucrose diet containing 17 amino acids, reflected no significant difference in survival between 2 and 4 day old whiteflies after 24, 48, 72 and 95 h of feeding. Also there were no significant differences in the amount of honeydew droplets secreted after a 95 h feeding period (Thompson unpublished data).

8.3.2 Influence of a Complete Range of Amino Acids

In follow-up studies to determine the influence of a complete range of 20 amino acids in sucrose solution 20% w/v, three treatments were examined: amino acids at the concentration levels found in the natural host based on previous data (Stevenson et al. unpublished data), amino acids at $\times10^{-1}$ the natural conc., and at $\times10^{1}$ the natural conc. Whitefly survival was significantly greater on the diluted ($\times10^{-1}$) and natural concentrations of amino acids compared to the 10^{1} concentration at 24, 48, 72 and 93 h This also coincided with significantly larger amounts of honeydew droplets secreted when feeding on the diluted and natural concentration of amino acid diets. These results indicate that *B. tabaci* show better performance in terms of survival and honeydew secretion when feeding on amino acid diets more mimicking the host plant as opposed to elevated levels of amino acids, in an *in vitro* environment and to the exclusion of other plant nutrients.

8.3.3 *Influence of Single Amino Acids at Varying Concentrations*

Subsequent studies concentrated on investigating each of the 20 amino acids separately, at four different concentration levels. The concentration range of each amino acid tested was based on the mean values of that amino acid found in cassava plants infected with mosaic virus, uninfected cassava plants cv. Ebwanateraka and other cassava leaf samples (Stevenson et al. unpublished data) (i.e. three different mean values). Four experimental concentrations were chosen for testing in 20% sucrose solution. One concentration was below the lowest mean value, one concentration was above the highest mean value and in most cases, two concentrations were between these two mean values since feeding diets were established within a 10× concentration series.

Results showed at least 80% survival after 94 h on all concentrations of the following amino acids: serine, alanine, proline and phenylalanine. The following amino acids: asparagine, glutamic acid, aspartic acid, arginine and tryptophan, caused a poor survival rate (40–79%) after 94 h at the higher concentration levels (Thompson 2006). It can be pointed out that on glutamic acid, survival was no more than 40% after the first 24 h on the two higher concentrations and this was also the case for cysteine at the highest concentration level (Thompson unpublished data). The leucine diet produced a trend of better survivorship at the higher concentration levels (Thompson 2006).

In terms of honeydew rating, more honeydew droplets were produced at the lower concentration range for glutamic acid, aspartic acid, histidine, cysteine, tyrosine and isoleucine. The reverse was true for leucine and valine. The other amino acids did not show a consistent trend (Thompson unpublished data).

B. tabaci oviposited larger number of eggs at the lower concentration range for aspartic acid, glutamic acid, arginine and histidine, while the reverse was true for asparagine. The other amino acids did not show a particular trend between oviposition and amino acid concentration (Thompson 2006). Data on preferred concentration for survival, honeydew secretion and oviposition for the various amino acids, are presented in Table 8.1.

Table 8.1 Amino acid concentrations (mg/ml) for highest survival, honeydew rating and oviposition by *B. tabaci* after 94 h (Thompson unpublished data)

Amino acid	Conc. for highest survival	Conc. for highest honeydew rating	Conc. for highest oviposition
Asparagine	0.07	0.07	70.0
Glutamine	0.02	0.02	0.02
Serine	2.0/20.0	20.0	2.0
Glutamic acid	0.3	0.3	0.03
Alanine	3.0	0.03	0.03
Aspartic acid	0.03	0.03	0.003
Proline	0.02	0.2	2.0
Arginine	0.01	0.01	0.001
Methionine	0.04	0.004	0.04
Histidine	0.05	0.005/0.05	0.05
Cysteine	0.04	0.04	0.4
Glycine	3.0	0.3	0.003
Tyrosine	30.0	0.03	0.3
Threonine	2.0	0.02	–
Lysine	0.05	0.05/5.0	–
Phenylalanine	0.003	0.03	0.0003
Isoleucine	0.004	0.004	0.4
Leucine	0.3	3.0	–
Valine	0.02	0.2/2.0	–
Tryptophan	0.03	0.03	3.0

– Eggs were not produced

8.3.4 Influence of Essential and Non-Essential Amino Acids

The data were collated to present a broader perspective on the role of amino acids. From the analysis, 14 of the amino acids showed (≥ 90%) survival of *B. tabaci* and these included five essential amino acids and nine non-essential amino acids. *B. tabaci* produced a honeydew rating of (≥ 5.0) when feeding on 14 of the amino acids. Four of these were essential while ten were non-essential. *B. tabaci* oviposited at least ten eggs on ten different amino acid diets, and only one of these amino acids was essential (Table 8.2) (Thompson unpublished data). These results show the predominance of influence of the non-essential amino acids.

The amino acids on which *B. tabaci* produced a honeydew rating of (≥ 5.0) and showed (≥ 90%) survival included: asparagine, glutamine, threonine, lysine, glutamic acid, glycine, serine, cysteine, proline and aspartic acid. Feeding on the asparagine diet in particular resulted in large amounts of honeydew droplets. The presence of large amounts of asparagine in honeydew has been observed in other studies (Sasaki et al. 1991; Douglas 1993).

The amino acids on which *B. tabaci* produced a honeydew rating of (< 5.0) but showed (≥ 90%) survival included: methionine, alanine, isoleucine and histidine

Table 8.2 Ranking of amino acids[a] based on percentage *B. tabaci* survival, honeydew droplet rating and total number of eggs produced

Highest mean % survival after 94 h		Highest mean honeydew rating after 94 h		Total eggs produced after 94 h		Overall ranking[b] in descending order
Threonine	96.7	Asparagine	8.0	Proline	64	Asparagine
						Serine
		Glutamine	7.8	Aspartic acid	48	Proline
Serine, Cysteine	95.0	Threonine	6.8	Alanine	44	Glutamic acid
						Glutamine
				Serine	30	Threonine
Asparagine, Alanine, Proline	93.3	Lysine, Glutamic acid	6.2	Asparagine	25	Alanine
		Glycine	6.0	Glutamic acid	21	Aspartic acid, Cysteine, Glycine
Histidine, Glutamic acid, Aspartic acid	91.7	Serine	5.8	Glycine	17	
		Tyrosine	5.7	Glutamine	16	Lysine
		Leucine	5.5	Methionine	15	Methionine
Glutamine, Glycine, Isoleucine, Lysine, Methionine	90.0	Cysteine, Phenylalanine, Arginine, Proline	5.3	Arginine	13	Arginine
				Cysteine	6	Tyrosine, Histidine
				Histidine	5	Phenylalanine, Isoleucine
						Leucine
Arginine	88.3	Aspartic acid	5.0	Isoleucine, Tyrosine	4	Valine
						Tryptophan
Phenylalanine, Leucine	86.7	Valine, Methionine, Alanine	4.5	Tryptophan, Phenylalanine	1	
Tyrosine	85.0	Isoleucine	4.3	Leucine, Valine, Lysine, Threonine	0	
Tryptophan	81.7	Tryptophan	4.2			
Valine	76.7	Histidine	3.7			

[a]The amino acids that are grouped together show similar results

[b]Overall ranking of the amino acids based on parameters of survival, honeydew rating and oviposition

(Table 8.2) (Thompson unpublished data). This difference in honeydew rating may be related to differences in assimilation and excretion of dietary compounds (Bragdon and Mittler 1963; Douglas 1993). Tryptophan and valine were noted at the lower end of the amino acid ranking both in terms of highest mean percentage survival of *B. tabaci* at 94 h (81.67% and 76.67% respectively), and honeydew droplet rating (< 5.0) (Thompson unpublished data).

Except for arginine, on which 13 eggs were oviposited with whiteflies showing a highest mean percentage survival of 88.3% at 94 h., all the amino acids resulting in whiteflies producing at least 10 eggs, were responsible for (≥90%) survival of the whiteflies. These included proline, aspartic acid, alanine, serine, asparagine, glutamine, glutamic acid, glycine and methionine. The amino acids permitting (≥ 90%) survival, and oviposition of (<10) eggs by *B. tabaci* included: cysteine, histidine and isoleucine. *B. tabaci* feeding on lysine diet oviposited no eggs, although (≥ 90%) survival was attained. Threonine was the highest ranking of the amino acids in terms of highest mean percentage survival of *B. tabaci* at 94 h, but the whiteflies oviposited no eggs on a diet containing this amino acid. Phenylalanine, leucine and tyrosine permitted intermediate survival of *B. tabaci* (85.0–86.7%), and a lower number of eggs (<10) (Thompson unpublished data). Valine was at the lowest ranking both in terms of mean *B. tabaci* survival at 94 h, and number of eggs oviposited (Table 8.2) (Thompson unpublished data).

When ranking the amino acids on the basis of highest mean survival at 94 h, highest mean honeydew rating and total number of eggs oviposited by *B. tabaci,* they assort in descending order: asparagine, serine, proline, glutamic acid, glutamine, threonine, alanine, aspartic acid/cysteine and glycine (Table 8.2). These amino acids are nearly similar to the amino acids showing high survival and oviposition by *B. tabaci*. The exceptions are related to the presence of threonine and cysteine, and the absence of methionine. Furthermore, this group of amino acids is dominated by the amino acids considered non-essential for higher animals. Like some aphid species performing well on diets containing non-essential amino acids (Douglas 1993), these results show that *B. tabaci* perform well on diets containing non-essential amino acids. The importance of these amino acids and methionine on the performance (feeding and growth) of aphids has been reported for decades by several researchers (Dadd and Krieger 1968; Van Emden and Bashford 1971; Douglas 1993).

8.3.5 Performance of Viruliferous and Non-Viruliferous *B. tabaci* on Amino Acid Diets

In a series of experiments to investigate viruliferous and non-viruliferous whitefly survival, honeydew secretion and oviposition on four amino acid diets: asparagine, serine, alanine and proline, results showed that on the diet containing asparagine, there were no significant differences in survival at 24, and 48 h but viruliferous whiteflies showed significantly better survival at 72 and 94 h. There were no significant differences in honeydew secretion and oviposition at 96 h between the two whitefly groups (Thompson unpublished data).

On the alanine diet there were no significant differences between viruliferous and non-viruliferous whiteflies in terms of honeydew secretion and oviposition at 96 h, and survival at 24, 48, 72 and 96 h. Similar observations were made where feeding occurred on the proline diet except that only non-virulifeous whiteflies oviposited eggs (Thompson unpublished data)

When the whiteflies were feeding on a diet containing serine, there were no significant differences in survival between viruliferous and non viruliferous whiteflies at 24 h and 96 h, though non-viruliferous whiteflies showed significantly higher survival at 48 and 72 h. No significant difference was observed in honeydew secretion. As on the diet containing proline, the eggs produced when feeding on the serine diet were oviposited only by non-viruliferous whiteflies (Thompson unpublished data).

The study was further advanced to test for interaction effects. Data for survival, honeydew rating and oviposition were analysed using a factorial design of 4 diet type (the four diets) × 2 whitefly groups (viruliferous and non-viruliferous whiteflies). When survival data (collectively at all time intervals) were arc-sine transformed and examined, the main effect of diet type was found to be highly significant, albeit the main effect of whitefly group was not significant. There was a very high significant interaction between diet type and whitefly group (Thompson unpublished data). Based on the Tukey HSD test for mean separation, whiteflies on asparagine showed significantly better survival than those on proline. Also whiteflies on serine showed significantly higher survival than those on proline.

When examining treatment combinations, significantly higher survival occurred where non-viruliferous whiteflies were feeding on a diet containing serine as opposed to any other diet, or viruliferous whiteflies feeding on a diet containing proline. Survival was also significantly higher when viruliferous whiteflies were feeding on asparagine as compared to non-viruliferous whiteflies feeding on alanine or both whitefly groups feeding on proline. In general, highest survival was seen when non-viruliferous whiteflies were feeding on serine and viruliferous whiteflies were feeding on asparagine (Thompson unpublished data).

After honeydew rating was square root transformed, only a diet type main effect was significant. Based on the Tukey HSD test, significantly more honeydew droplets were produced where feeding occurred on the asparagine diet as compared to the proline and serine diets. Based on the amount of honeydew droplets produced by the whiteflies, the diets appeared in the following descending order: asparagine, alanine, proline and serine (Thompson unpublished data).

Oviposition data for the asparagine and alanine diets were pooled since eggs were oviposited by both groups of whiteflies only where feeding occurred on these diets. After the data were transformed by the square root transformation, no significant interaction existed between whitefly group and diet type. Also there were no significant differences between the two diets, or between viruliferous and non-viruliferous whiteflies in terms of the number of eggs oviposited (Thompson unpublished data).

These studies show that diet type could provide a survival advantage of viruliferous whiteflies over non-viruliferous whiteflies or *vice versa,* but ovipositional differences were not significant and this may in part explain insignificant differences in population levels of viruliferous and non-viruliferous whiteflies as seen in some pathosystems.

8.4 Physiological Basis of Amino Acids Relative to Other Phloem Sap Sucking Insects

The ten amino acids uppermost on the ranking scale (Table 8.2) played a pivotal role in *B. tabaci* survival and/or oviposition, and this emphasizes their physiological importance. In studies of several aphid species, the influence of these amino acids has been observed.

The role of asparagine, aspartic acid, glutamine, glutamic acid, serine, alanine, methionine and cysteine in the growth of the aphid, *M. persicae* has been reported (Dadd and Krieger 1968). Secondly, threonine and a combination of glutamine and asparagine were positively correlated to the relative growth rate of *Brevicoryne brassicae*, while a combination of glutamine/asparagine coupled with methionine were positively correlated to the relative growth rate of *M. persicae* (Van Emden and Bashford 1971).

Furthermore, elevated levels of amino acids have been associated with higher population densities of phloem sap sucking insects. The younger leaves of *Prunus domestica* that contained higher concentrations of glutamine, asparagine and methionine were exclusively colonized by the aphid *Hyalopterus pruni.* (Douglas 1993).

Alanine and glutamine were found in consistently higher levels in wheat plants infected with *Barley yellow dwarf virus* as compared to healthy wheat plants (Ajayi 1986). Based on the findings of Ajayi and Dewar (1983) these observations may partly explain the development of larger populations of *Sitobion avenae* and *Metopalophium dirhodum* on infected wheat plants, and the nutritional importance of these amino acids.

Glycine and threonine also show positive effects on the aphid, *Drepanosiphum platanoides,* since these dominated the amino acid composition in the phloem of the host plant, *Acer pseudoplanatus* during the September to October period, a period when the nutritional quality of the host plant was superior to that during the summer months of July to August (Douglas 1993).

For the other amino acids appearing lower down the scale on the overall ranking (Table 8.2), lysine, histidine and isoleucine were important for (≥ 90%) survival although *B. tabaci* feeding on these did not produce many eggs. These amino acids may have an effect on the growth of whiteflies. Isoleucine and histidine were reported as essential for the growth of *M. persicae*, while a moderate reduction in weight occurred in the absence of lysine (Dadd and Krieger 1968). In other work, isoleucine was important in enhancing the acceptability of sucrose by *M. persicae* (Mittler 1967).

The performance of *B. tabaci* on tyrosine, phenylalanine, leucine and arginine was intermediate in terms of survival (85.0–88.3%). Whiteflies on these amino acids produced high amounts of honeydew (5.33–5.67). *B. tabaci* on arginine oviposited many eggs (>10), but few on tyrosine and phenylalanine (<10). Eggs were not produced on leucine (Thompson unpublished data). These amino acids may be required in lower concentrations if at all in the diet of *B. tabaci.* Dadd and Krieger (1968) for example indicated that tyrosine, phenylalanine, leucine, arginine, valine and tryptophan were not essential for the growth of *M. persicae*. Secondly, arginine was shown to have a slight deterrent effect on diet acceptability by *M. persicae* (Mittler 1967). In the study results (Thompson unpublished data) the performance of *B. tabaci* on valine and tryptophan on an overall basis was particularly poor.

The direct influence of tryptophan and its derivatives on the reproduction of *B. tabaci* has been demonstrated in other studies. The conversion of tryptophan to the indole-alkaloid, tryptamine by the enzyme Tryptophan decarboxylase (TDC) from *Catharanthus roseus* significantly decreased reproduction of *B. tabaci* by up to 70% on transgenic tobacco as compared to the controls (Thomas et al. 1995).

8.5 Distribution and Varying Influence of the Amino Acids

Within the host plant *per se*, the free amino acids cannot serve as a cue to the distribution of amino acids within *B. tabaci*. In a study to determine the influence of host plant nitrogen status on the content and distribution of free amino acids in whiteflies and their honeydew, results showed that low nitrogen status of the host resulted in decreased free amino acids within the insect. Large amounts of asparagine were excreted where feeding occurred on high nitrogen status plants, and amino acid excretion was minimal to cessation on low nitrogen status plants. It was observed that the amino acid distribution in whiteflies and their honeydew was not closely associated to that of the host plant (Crafts-Brandner 2002). In this study glutamine, alanine and proline accounted for the major pool of amino acids within whiteflies regardless of nitrogen status of the plants. This underscores the importance of these amino acids and in the study (Thompson unpublished data), where alanine, glutamine and proline appeared uppermost on the amino acid ranking scale with regards to relative performance to *B. tabaci* (Table 8.2).

Thus *B. tabaci* nutritive requirements of glutamine, alanine and proline account for their dominance within the free pool of amino acids and is also linked to the relative assimilation and excretion of the various amino acids. It should be noted that asparagine was excreted in larger amounts as nitrogen status of the host plant increased (Crafts-Brandner 2002) and as it increased in concentration levels in artificial diets (Thompson unpublished data).

One significant observation is the varying influences of the same amino acid composition within the plant as opposed to an artificial diet. The performance of phloem sap sucking insects on the natural host could be different from that on an artificial diet mimicking the amino acid composition within the host (Sandstrom 1994).

In a study, on the natural host, *Cucumis melo* L. essential amino acids reflected a positive influence on *B. tabaci.* The predominantly essential amino acids were a good predictor of adult body weight. As the relative concentration of these decreased, body weight also decreased. Furthermore these amino acids were implicated in developmental time and percentage emergence. As the relative concentration decreased, developmental time increased and as relative concentration increased, percentage emergence increased. The phloem sap concentration of amino acids predominated by aspartic acid however, portrayed a negative association with percentage emergence (Blackmer and Byrne 1999).

These results show that within the natural host the essential amino acids were the dominant ones influencing growth and development. In the study (Thompson unpublished data), *B. tabaci* in general showed better oviposition and survival when feedings on diets containing non-essential amino acids.

The resultant effects of amino acids within the natural host as opposed to *in vitro*, are more than likely linked to the presence of other plant compounds within the host, albeit absent from the artificial diets.

8.6 Concluding Remarks

8.6.1 The Amino Acids

The various amino acids produce varying effects on *B. tabaci* survival and oviposition. On some amino acid diets whitefly performance was better at lower concentration range and on other diets performance was better at higher concentrations. In general the non-essential amino acids accounted for better survival and oviposition of *B. tabaci.*

Data on whitefly preferred concentrations of the various amino acids for parameters of survival and oviposition (Table 8.1) is an important guide for the development of holidic diets for *B. tabaci* and possible other phloem sap sucking insects. Such diets could be utilized in bioassay studies evaluating whitefly performance on different compounds, and in *in vitro* screening trials of biorational pesticides.

Interaction of *B. tabaci* with infected host plants and the possible involvement of amino acids has produced different results. Colvin et al. (1999) observed higher populations of *B. tabaci* on EACMV-UG infected cassava plants and associated higher levels of amino acids. Costa et al. (1991), observed varying results when examining *B. tabaci* performance on host plants separately infected with six different viruses versus respective healthy host plants. The researchers observed no correlation between levels of amino acids and oviposition or survival rates of *B. tabaci.* Thus from this observation along with the study results (Thompson 2006; Thompson unpublished data) where particular amino acids elicited improved performance at high concentrations and other amino acids were required at lower concentrations it would appear that the component amino acids at specific concentrations is key to the interaction of insect vector with its host. Influence of the amino acids is complex and this could be expected since they constitute a large group. Although the interaction of *B. tabaci* with its host plant and the begomoviruses transmitted involve amino acids at least in some pathosystems, the involvement of other factors should be considered.

8.6.2 Other Nutrient Factors

Other important factors may include micro nutrients, carbohydrates and the ratio of carbohydrates to amino acids. The nutritional importance of these is recognized, as they form the essential components of successful diets that have been used for phloem sap feeders (Mittler and Koski 1976; Kunkel 1976).

Carbohydrates occur in relatively larger amounts within the phloem sap (Law et al. 1992; Kehr 2006) and as such may not be a limiting factor with regards to performance of phloem sap feeders. In fact the sugars are often present in larger amounts than is required. For example, a study showed that plants subjected to varying levels of water stress, exhibited different concentration levels of sugars. Highly water stressed plants showed higher concentration levels of sugars and resulted in lower feeding by whiteflies and higher honeydew production of predominantly the oligosaccharide trehalulose. Whitefly performance did not vary in response to plants with different water stress levels, and it was concluded that performance was not affected because through osmoregulation, the insects were able to maintain their required water potential (Isaacs et al. 1998). The role of trehalulose in osmoregulation has been indicated in other studies (Byrne et al. 2003).

Thus carbohydrates usually occur in sufficient quantities and at high concentration levels whiteflies are not negatively affected since they are able to physiologically process the excessive amounts.

Consideration should be given to the compatibility of nutritive components, where these exist at optimum concentration levels and at desired proportionality. A typical example is proportion of carbohydrates to amino acids. Blackmer and Lee (2008) pointed out that the carbohydrate to amino acid ratio could be an influencing factor on *B. tabaci* feeding and oviposition.

8.6.3 Role of Plant Semiochemicals

Another area of importance is the characteristic plant compounds of specific plant species. It is evident that the selection of particular crop varieties by specific herbivores is in part related to plant chemicals. Examples include the glucosinolates in Brassica genotypes (Hopkins et al. 1997; Hopkins et al. 2009); and the terpenoids in tomatoes (Bleeker et al. 2009). Since these chemical compounds could influence the population level of a pest it can be argued that specific plant compounds can serve as population triggers or inhibitors. A beneficial interaction of an insect pest with its host in the presence of plant semiochemicals, will rely on its ability to effectively detoxify a chemical inhibitor, e.g. the cyanide detoxifying enzymes found in the cassava biotype (Antony et al. 2006), or to strive in the presence of beneficial compounds e.g. the epicuticular lipid extracts and lipid components on plant surfaces that could improve insect feeding and oviposition (Eigenbrode and Espelie 1995). This is an added dimension to the role of not only the amino acids, but of plant chemicals in general and the underlying tenets as related to the complex interactions involving *B. tabaci*, transmitted begomoviruses and host plant response mechanisms.

8.6.4 The Role of Endosymbionts

Quite a number of phloem sap feeders contain endosymbionts that play an important role in the nutritional requirements of the insect (Spaulding and von Dohlen 1998). The prokaryotic bacteria, *Buchnera aphidicola* is known to synthesize essential amino acids for the aphid, *Acyrthosiphon pisum* (Jimenez et al. 2000). In whiteflies the primary endosymbiont, is different to that of aphids and mealybugs, in that it lacks the external membrane of the gram negative cell wall (von Dohlen et al. 2001; Thao and Baumann 2004). The primary endosymbiont in whiteflies is carried in intact form to the oocytes and is transmitted transovarially to the progeny (Costa et al. 1997). There are also secondary endosymbionts such as *Chlamydia* spp. and *Wolbachia* spp.(Thao et al. 2000; Thao and Baumann 2004). The role of these is not fully understood, but if whitefly endosymbionts play a pivotal role in the nutritional requirements of *B tabaci*, then it is expected that factors that affect these will ultimately influence whitefly survival and reproduction.

In a study investigating the effects of antibiotics against endosymbionts, it was found that antibiotics negatively affected the internal microbes and thus the survival of whiteflies (Costa et al. 1997). It is important to investigate the effects of begomoviruses on endosymbionts since internal microbes represent a significant source of essential nutrients for *B. tabaci* survival and reproduction. Such research ventures will facilitate an improved understanding of the interaction at different trophic levels. Additionally, the application of antibiotic based pesticides will create avenues for alternative approaches to management, through novel measures or complementary/integrative methods.

References

Ajayi O (1986) The effect of barley yellow dwarf virus on the amino acid composition of spring wheat. Ann Appl Biol 108:145–149

Ajayi O, Dewar AM (1983) The effect of barley yellow dwarf virus on field populations of the cereal aphids *Sitobion avenae* and *Metopolophium dirhodum*. Ann Appl Biol 103:1–11

Akey DH, Beck SD (1971) Continuous rearing of the pea aphid, *Acyrthosiphon pisum*, on a holidic diet. Ann Entomol Soc Am 64:353–356

Akey DH, Beck SD (1972) Nutrition of the pea aphid *Acyrthosiphon pisum*: requirements for trace metals, sulphur and cholesterol. J Insect Physiol 8:1901–1914

Antony B, Lisha VS, Palaniswami MS, Sugunan VS, Makeshkumar T, Henneberry TJ (2006) *Bemisia tabaci* (Homoptera: Aleyrodidae) and Indian cassava mosaic virus transmission. Int J Trop Insect Sci 26:176–182

Blackmer JL, Byrne DN (1999) The effect of *Bemisia tabaci* on amino acid balance in *Cucumis melo*. Entomol Exp Appl 93:315–319

Blackmer JL, Lee LL (2008) Nutritional factors influencing whitefly development and flight behaviour: the search for a suitable bioassay to test hypothesis. Bemisia international workshop proceedings. J Insect Sci 8:6–7

Bleeker PM, Diergaarde PJ, Ament K, Guerra J, Weidner M, Schutz S, de Both MTJ, Haring MA, Schuurink RC (2009) The role of specific tomato volatiles in tomato whitefly interaction. Plant Physiol 151:925–935

Bragdon JC, Mittler TE (1963) Differential utilisation of amino acids by *Myzus persicae* (Sulzer) fed on artificial diets. Nature 198:209–210

Byrne DN, Hendrix DL, William LH III (2003) Presence of trehalulose and other oligosaccharides in hemipteran honeydew, particular Aleyrodidae. Physiol Entomol 28:144–149

Carter W (1927) A technique for use with Homopteran vectors of plant disease with special reference to the sugar beet leafhopper, *Euttetix tenellus* Baker. J Agric Res 34:449–451

Colvin J, Otim-Nape GW, Holt J, Omongo C, Seal S, Stevenson P, Gibson G, Cooter RJ, Thresh JM (1999) Factors driving the current epidemic of severe cassava mosaic disease in East Africa. In: VIIth international plant virus epidemiology symposium. Plant virus epidemiology: current status and future prospects, Aguadulce (Almeria)

Costa HS, Brown JK, Byrne DN (1991) Life history traits of the whitefly, *Bemisia tabaci* (Homoptera: Aleyrodidae), on six virus-infected or healthy plant species. Environ Entomol 20:1102–1107

Costa HS, Henneberry TJ, Toscano NC (1997) Effects of antibacterial materials on *Bemisia argentifolii* (Homoptera: Aleyrodidae) oviposition, growth, survival and sex ratio. Ann Entomol Soc Am 90:333–339

Crafts-Brandner SJ (2002) Plant nitrogen status rapidly alters amino acid metabolism and excretion in *Bemisia tabaci*. J Insect Physiol 48:33–41

Dadd RH, Krieger DL (1968) Dietary amino acid requirements of the aphid *Myzus persicae*. J Insect Physiol 14:741–764

Davidson EW, Patron RB, Lacey LA, Frutos R, Vey A, Hendrix DL (1996) Activity of natural toxins against the silverleaf whitefly, *Bemisia argentifolii* using a novel feeding bioassay system. Entomol Exp Appl 79:25–32

Dorschner KW (1993) Survival, growth and reproduction of two aphid species on sucrose solutions containing host and non host honeydews. Entomol Exp Appl 68:31–41

Douglas AE (1993) The nutritional quality of phloem sap utilised by natural aphid populations. Ecol Entomol 18:31–38

Douglas AE, Prosser WA (1992) Synthesis of the essential amino acid tryptophan in the pea aphid (*Acyrthosiphon pisum*) symbiosis. J Insect Physiol 38:565–568

Ehrhardt P (1968) Die Wirkung Verschiedener Spurenelemente auf Wachstum, Reproduktion und Symbionten von *Neomyzus circumflexus* Buckt (Aphidae, Homoptera, Insecta) bei Kunstlicher Ernahrung. Z Vergl Physiol 58:47–75

Eigenbrode SD, Espelie KE (1995) Effects of plant epicuticular lipids on insect herbivores. Annu Rev Entomol 40:171–194

Elbert A, Nauen R (1996) Bioassays for imidacloprid for resistance monitoring against the whitefly Bemisia tabaci. In: Brighton crop protection conference: pests and diseases, vol 2, Proceedings of an international conference. Brighton, 18–21 Nov, pp 731–738

Fujita A, Mitsuhashi J (1995) Effects of dietary amino acids on the production of sexual morphs by the green peach aphid *Myzus persicae*. Arch Insect Biochem Physiol 29:259–268

Harpaz I, Applebaum SW (1961) Accumulation of asparagine in maize plants infected by maize rough dwarf virus and its significance in plant virology. Nature 192:780–782

Hopkins RJ, Birch ANE, Griffiths DW, Baur R, Stadler E, Mc Kinlay RG (1997) Leaf surface compounds and oviposition preference of turnip root fly *Delia floralis*. The role of glucosinolate and nonglucosinolate compounds. J Chem Ecol 23:629–643

Hopkins RJ, van Dam NM, van Loon JJA (2009) Role of glucosinolates in insect-plant relationships and multitrophic interactions. Annu Rev Entomol 54:57–83

Isaacs R, Byrne DN, Hendrix DL (1998) Feeding rates and carbohydrate metabolism by *Bemisia tabaci* (Homoptera: Aleyrodidae) on different quality phloem sap. Physiol Entomol 23:241–248

Jancovich JK, Davidson EW, Lavine M, Hendrix DL (1997) Feeding chamber and diet for culture of nymphal *Bemisia argentifolii* (Homoptera: Aleyrodidae). J Econ Entomol 90:628–633

Jimenez N, Gonzalez-Candelas F, Silva FJ (2000) Prephenate dehydratase from the aphid endosymbiont (*Buchnera*) displays changes in the regulatory domain that suggests its desensitization to inhibition by phenylalanine. J Bacteriol 182:2967–2969

Kehr J (2006) Phloem sap proteins, their identities and potential roles in the interaction between plants and phloem-feeding insects. J Exp Bot 57:767–774

Kim M, Koh H-S, Fukami H (1985) Isolation of C-glycosyl flavones as probing stimulants of plant hoppers in rice plant. J Chem Ecol 11:441–453

Koyama K (1984) Nutritional Physiology of the brown plant hopper, *Nilaparvata lugens* Stal. (Hemiptera: Delphacidae). Chin J Entomol 4:93–107

Krieger DL (1971) Rearing several aphid species on synthetic diet. Ann Entomol Soc Am 64:1176–1177

Kunkel H (1976) Membrane feeding systems in aphid research. In: Harris KF, Maramorosch K (eds) Aphids as virus vectors. Academic, New York

Law JH, Ribeiro JMC, Wells MA (1992) Biochemical insights derived from insect diversity. Annu Rev Biochem 61:87–111

Matthews REF (1981) Plant virology, 2nd edn. Academic, New York

Mitsuhashi J (1979) Artificial rearing and aseptic rearing of leafhopper vectors: applications in virus and MLO research. In: Maramorosch K, Harris KF (eds) Leafhopper vectors and plant disease agents. Academic, New York

Mittler TE (1967) Gustation of dietary amino acids by the aphid *Myzus persicae*. Entomol Exp Appl 10:87–96

Mittler TE, Koski P (1976) Development of meridic and oligidic diets for rearing the aphid *Myzus persicae*. J Insect Physiol 22:1135–1141

Sandstrom J (1994) Performance of pea aphid (*Acyrthosiphon pisum*) clones on host plants and synthetic diets mimicking the same plants phloem amino acid composition. J Insect Physiol 40:1051–1057

Sasaki T, Hayashi H, Ishikawa H (1991) Growth and reproduction of the symbiotic and asymbiotic pea aphids, *Acyrthosiphon pisum* maintained on artificial diets. J Insect Physiol 37:749–756

Simpson SJ, Abisgold JD, Douglas AE (1995) Response of the pea aphid *Acyrthosiphon pisum* to variation in dietary levels of sugar and amino acids- the significance of amino acid quality. J Insect Physiol 41:71–75

Sogowa K (1974) Studies on the feeding habits of the brown planthopper, *Nilaparvata lugens* (Stal) (Hemiptera: Delphacidae). IV. probing stimulant. Appl Entomol Zool 9:204

Spaulding AW, von Dohlen CD (1998) Phylogenetic characterization and molecular evolution of bacterial endosymbionts in Psyllids (Hemiptera: Sternorrhyncha). Mol Biol Evol 15:1506–1513

Srivastava PN, Auclair JL (1971) Influence of sucrose concentration on diet uptake and performance by the pea aphid *Acyrthosiphon pisum*. Ann Entomol Soc Am 64:739–743

Thao ML, Baumann P (2004) Evolutionary relationships of primary prokaryotic endosymbionts of whiteflies and their hosts. Appl Environ Microbiol 70:3401–3406

Thao ML, Moran NA, Abbot P, Brennan EB, Burckhardt DH, Baumann P (2000) Cospeciation of Psyllids and their prokaryotic endosymbionts. Appl Environ Microbiol 66:2898–2905

Thomas JC, Adams DG, Nessler CL, Brown JK, Bohnert HJ (1995) Tryptophan decarboxylase, tryptamine, and reproduction of the whitefly. Plant Physiol 109:717–720

Thompson WMO (2002) Comparison of *Bemisia tabaci (*Homoptera: Aleyrodidae) development on uninfected cassava plants and cassava plants infected with *East African cassava mosaic virus*. Ann Entomol Soc Am 95:387–394

Thompson WMO (2006) Influence of amino acids on cassava biotype *Bemisia tabaci* (Gennadius) (Homoptera: Aleyrodidae) when feeding on an artificial system. J Entomol 3:198–203

Van Emden HF, Bashford MA (1971) The performance of *Brevicoryne brassicae* and *Myzus persicae* in relation to plant age and leaf amino acids. Entomol Exp Appl 14:349–360

von Dohlen CD, Kohler S, Alsop ST, McManus WR (2001) Mealybug β proteobacterial endosymbionts contain γ- proteobacterial symbionts. Nature 412:433–436

Chapter 9
Bemisia tabaci, the Capacity to Invade

P.J. De Barro

Abstract *Bemisia tabaci* (Gennadius) (Hemiptera: Aleyrodidae) is a cryptic species complex composed of numerous morphologically indistinguishable species, a number of which have been shown to be either completely or partially reproductively isolated. Several members of the complex have invaded beyond their home ranges and two, Middle East-Asia Minor 1 (commonly known as the B biotype) and Mediterranean (commonly known as the Q biotype), have invaded globally through the international trade in ornamental plants. Over the past decade our knowledge of the factors influencing the capacity of different members of the complex to invade has increased substantially. This review discusses the roles traits associated such as asymmetrical mating interference, competitive male behavior, host range, insecticide resistance and interactions with Begomoviruses play in the capacity of different members of the complex to invade. In addition, the use of different methods to identify and distinguish between different members of the complex is discussed and recommendations as to what approaches should be adopted to address different aspects of *B. tabaci* invasion are made. Finally, the invasions associated with the spread of *pepper yellow leaf curl Indonesia virus* and cassava mosaic disease are described and discussed.

9.1 Introduction

Bemisia tabaci (Gennadius) (Hemiptera: Aleyrodidae) is a cryptic species complex (Dinsdale et al. 2010) composed of at least 24 morphologically indistinguishable species, a number of which have been shown to be either completely or partially reproductively isolated (Xu et al. 2010). It feeds on the phloem of primarily

P.J. De Barro (✉)
CSIRO Ecosystem Sciences, 120 Meiers Road, Indooroopilly 4068, Australia
e-mail: Paul.debarro@csiro.au

W.M.O. Thompson (ed.), *The Whitefly, Bemisia tabaci (Homoptera: Aleyrodidae) Interaction with Geminivirus-Infected Host Plants*,
DOI 10.1007/978-94-007-1524-0_9,

herbaceous species and some members of the complex are pests of a number of vegetable, grain legume and fiber crops either through feeding or the transmission of plant pathogenic begomoviruses (Oliveira et al. 2001; Jones 2003). The complex has a global distribution with the genetic structure of the complex showing a marked geographic structure with primarily East and West Africa, Asia, Mediterranean Basin/Middle East/Asia Minor and the New World being the major species groupings within the complex (Boykin et al. 2007; Dinsdale et al. 2010). As a pest, *B. tabaci* first became internationally apparent in the late 1970s in Sudan, but widespread global impact began in the late 1980s in the Southwestern USA. Since then it has earned a well deserved reputation for being invasive. The rise in awareness of *B. tabaci* as a global pest also saw the spawning of the idea that *B. tabaci* was composed of numerous biotypes. While there is no doubt biological differences exist between different members of the complex, there is a remarkable lack of quantified measures based on biological data that allow different members of the complex to be distinguished (De Barro et al. 2011). Instead, the vast majority of biotypes are based primarily on genetic markers be they protein or DNA and not biological data and so the term biotype is really a surrogate term for genetic group. Furthermore, quite a number of the biotypes are in fact the same, they have just been identified using either different methods or are the result of the failure to fully explore the available literature (De Barro et al. 2011). So, while the usage of biotype terminology has become entrenched in the vernacular of the *B. tabaci* lexicon, its validity is doubtful. Moreover, biotype and species are not interchangeable. For example B and B2 fall within the same species group as do Q, J and L. Similarly, K, P, ZHJ2 and PCG-2 are in most cases identical. Rather than expound on the biotype issue in detail here, the reader is instead referred to De Barro et al. (2011) for a detailed review of evidence as to why the use of biotype is now inappropriate and that it is far more correct to view *B. tabaci* as a cryptic species complex than a complex species made up of different biotypes. Throughout this review we will use the putative species structure proposed by Dinsdale et al. (2010) although we will draw the connection between the putative species and the more widely known and understood biotype name if there is one available.

The international rise in awareness of *B. tabaci* is firmly linked to the global spread over the past 20 or so years of the Middle East-Asia Minor 1 putative species (known commonly as the B biotype, but is also identical to the B2 biotype) principally via the trade in ornamentals (Cheek and Macdonald 1994). This species originated in the Middle East – Asia Minor region and has since spread to at least 50 countries in Africa, the Americas, Asia, Australia and Europe (De Barro et al. 2011). More recently, a second member of the cryptic species complex, the Mediterranean putative species (commonly known as the Q biotype, but is also identical to the L biotype) has also begun to invade and has now been detected in at least ten countries in the Americas, Asia and New Zealand (De Barro et al. 2011) again via the pathway of trade in ornamental species (Dalton 2006). This species was initially thought to have originated in the countries bordering the Mediterranean Basin, but increasingly there is evidence to suggest that it originated in East and West Africa before spreading northwards. Rather than trying to laboriously record all the examples and consequences of invasions by *B. tabaci*, this review will focus on the underlying process of invasion with a view to identifying the key traits that enable invasions to succeed. The review will also consider the recent outbreak of "Penyakit Kuning" in Indonesia and the debate as to whether pandemic cassava mosaic disease in Uganda is due to an invasion process.

9.2 Molecular Techniques Utilized in Whitefly Identification and Classification

The history of the use of molecular markers in regards to *B. tabaci* in a sense charts the history of our understanding of the genetic complexity of *B. tabaci*. The early markers, both allozyme and RAPD-PCR contributed to our recognition of the genetic complexity of the species complex, but also added layers of confusion about what differences in the profiles produced by these early markers really meant. In effect the train of thinking progressed from the use of markers to show there were differences between entities with different phenotypes e.g. A verses B, to a situation where a simple difference in banding profile in the absence of any biological data was sufficient grounds for identifying yet another biotype. This is covered extensively in De Barro et al (2011), but in summary the failures of application of these and other markers were,

1. The failure to consider genetic limits. The key issue here is the question of how different two profiles or sequences needed to be before they were considered sufficiently different to be considered different biotypes, races or species? This lack of consistency in process lead Dinsdale et al. (2010) to for the first time, develop a consistent, quantified measure of group separation.
2. The failure to consider what went before. Here we saw new biotypes being raised simply because there was no type standard against which to compare. This meant that a biotype named on the basis of one diagnostic method e.g. esterase profile may then have been given a different named based on the use of a different method e.g. RAPD PCR only to be named yet again using a third method e.g. sequencing of mitochondrial cytochrome oxidase one (mtCO1); a process that yields three biotypes instead of one.
3. The failure to consider the globally available data in full. Here, a biotype named by one research group was given a different name by another even when the same marker was being used and in the case of mtCO1, identical haplotypes were being published and placed into GenBank. This has further contributed to the proliferation of biotype names.

These lapses continued on into the use of phylogenetic tools. Again, De Barro et al. (2011) detailed the errors of application. Two gene regions, mtCO1 and nuclear ribosomal internally transcribed space 1 (rITS1), have been primarily used to construct evolutionary relationships. Both give similar evolutionary reconstructions (De Barro et al. 2005), but their analysis in numerous papers has been flawed by a poor adherence to fairly basic phylogenetic rules. Here are some of the major rule violations that have been committed and are still being committed,

1. Poor outgroup selection. *Trialeurodes vaporariorum* is not a good outgroup for *B. tabaci*, it is too different and does not act as a good anchor. Outgroup selection is an essential part of phylogenetic analysis, poor selection has the potential to affect tree topology i.e. evolutionary relationships.
2. Multiple inclusions of the same sequence. A phylogenetic reconstruction should be composed of unique sequences; duplicates can distort the relationships between groups.
3. Sequences for coding genes such as mtCO1 should have no gaps and no stop codons, if they do then they are wrong and must not be used; either, they are due to sequence read errors or they are pseudogenes; mitochondrial DNA that has been incorporated into the nuclear DNA. Three key steps here are to make sure that (a) sequence quality is checked, poor quality sequences will lead to incorrect assignment of bases, (b) check the trace file against the sequence read as this will help eliminate misreads and (c) sequence in both directions as the terminal ends of reads will often have read errors.
4. Arbitrary assignment of group structure. Here we see groups being assigned on the basis of the quite different levels of sequence divergence; this means that we see potential groups being ignored and other groups being identified for no apparent reason. These inconsistencies have created confusion in terms of the relationships between groups and in our overall understanding of evolutionary relationships.
5. Limited consideration of genetic diversity. The use of small numbers of different haplotypes out of the available pool of haplotypes, limited taxon sampling, has been shown to cause significant increases in phylogenetic estimation error (Pollock and Bruno 2000; Pollock et al. 2002; Zwickl and Hillis 2002). Furthermore, Pollock et al. (2002) showed that the benefits of using more different sequences were similar to adding an equivalent amount of sequence length for the same taxa. In other words, assigning unknown sequences to particular groups is prone to error if the full range of genetic diversity has not been considered

Dinsdale et al. (2010) addressed all these and also offered a good solution in terms of assigning unknown sequences to a species group. Firstly, Dinsdale et al (2010) eliminated all duplicate sequences and those that were problematic in terms of gaps, ambiguous bases and stop codons. Secondly, while they selected appropriate outgroups to anchor the tree they also recognized that more work needed to be done in the area of outgroup choice in order to achieve a reliable tree topology. Thirdly, the analysis encompassed a very wide portion of the available genetic variability. Fourthly, it created a rule set upon which to base sequence assignment, in other words, for the first time the genetic bounds of each group were explicitly identified. This offered a considerable advantage to those wanting to identify unknown sequences. Rather than undertaking a phylogenetic analysis, a sequence can be compared against the consensus sequences. A positive match is the closest match where sequence divergence is <3.5%. If divergence exceeds 3.5% then it becomes a new species group. Table 9.1 details the DNA markers used to study *B. tabaci* and provides a brief outline of uses and issues to consider.

Table 9.1 Summary of the different molecular methods used to study the *Bemisia tabaci* species complex together with comments describing each method and its appropriateness

Marker		Comments
RAPDS	Random amplification of polymorphic DNA	A rapid, low cost diagnostic. To be avoided where possible due to problems with reproducibility except in select cases where reproducibility has been confirmed through extensive cross checking with another method. They still have a role in diagnostics provided banding patterns are stable, but SCAR and CAPS remove many of the problems associated with reproducibility and it is strongly suggested that these be instead used.
SCAR	Sequence characterized amplified regions	Used primarily as a rapid, low cost alternatives to sequencing of mtCO1. SCARs are DNA fragments amplified by the PCR using specific 15–30 bp primers, designed from nucleotide sequences established in cloned RAPD fragments linked to a trait of interest. By using longer PCR primers, SCARs do not face the problem of low reproducibility generally encountered with RAPDs. To develop a SCAR marker, diagnostically useful bands from a RAPD profile are cut from the gel, cloned and then sequenced. Long primers are then developed on the basis of the sequence. Like any diagnostic they need to be screened against a broad range of the available genetic diversity to ensure reliable amplification and capacity for unambiguous separation. Ideally they are used after mtCO1 has been used to determine species identity thoroughly. (Chu et al. 2004; Khasdan et al. 2005; Zang et al 2006; Boukhatem et al. 2007; Ko et al. 2007; Shatters et al. 2009).
CAPS or RFLP	Cleaved amplified polymorphic sequences or restriction fragment length polymorphism	Used primarily as a rapid, low cost diagnostic alternatives to sequencing of mtCO1. Involves amplifying mtCO1 and then using restriction enzymes to cleave the product. Restriction enzymes bind to DNA at specific locations; if the site is absent no binding and therefore no cleavage takes place. Digestion with different restriction enzymes, or combinations of enzymes enables the mtCO1 to be cleaved into different length fragments; these can then be separated using electrophoresis. The different banding patterns can then be used to separate the different species. Care needs to be taken to ensure that all mtCO1 haplotypes for a particular species are cleaved. As the number of different species increases within-species haplotype variability can lead to increasing difficulty in getting reliable separation as there is insufficient range of variability in cleavage sites. It is therefore essential that one has a good understanding of the underlying variability within a particular species. As a result this approach is most reliable when used to separate more distantly related species as closely related species share many of their restriction sites thereby preventing reliable separation. This approach is best used to separate invader from indigenous species as the invader population is usual composed on a very limited number of haplotypes and these tend to be quite different from the indigenous haplotypes.

(continued)

Table 9.1 (continued)

Marker		Comments
		Ideally they are used after mtCO1 has been used to determine species identity thoroughly. (Khasdan et al. 2005; Ma et al. 2009)
AFLP	Amplified fragment length polymorphism	AFLP-PCR was originally described by Zabeau and Vos in 1993. The procedure involves first digesting of total DNA with one or more restriction enzymes and ligation of restriction half-site specific adaptors to all restriction fragments. This is followed by the selective amplification of some of these fragments with two PCR primers that have the corresponding adaptor and restriction site specific sequences. The amplified products are then separated using electrophoresis followed by visualisation of the band pattern. It has been rarely used in *B. tabaci* research and there seems no need to consider it. (Cervera et al. 2000)
mt16S	Mitochondrial 16s ribosomal RNA	The mitochondrial 16S gene has been used in early phylogenetics studies. It mutates at a slower rate than mtCO1 and is less useful at resolving species level differences. It is likely to be a more useful tool with which to examine higher level evolution. (Fohlich et al. 1999)
mtCO1	Mitochondrial cytochrome oxidase 1	This gene is widely used in phylogenetic analysis of species. It has been used as a diagnostic (sequence comparison and CAPS) and to explore evolutionary relationships. MtCO1 has several advantages over other approaches, • It is readily amplified and easy to direct sequence. • There is a very large body of literature that explores the use, analysis and interpretation of data. • There is a large set of mtCO1 sequences accessible through GenBank. However, there are some critical issues that those using this gene need to bear in mind and these are addressed in detail in the text, but briefly, • mtCO1 sequence when aligned should have no gaps in the alignment as it is a coding gene; sequences with gaps should either be discarded or checked against the trace file in case there has been a misread, • The number of ambiguous bases should be $<1\%$, • Sequence quality needs to be high to avoid miscalling bases; it is critical that peak quality is considered and that the trace fill be compared against the automated sequence output, • Automated sequencing quality is often poor for the first 15–40 bases and >700 bases; it is therefore critical that sequences are obtained for both DNA strands. One should aim for at least 500 contiguous unambiguous bases.

		• There should be no stop codons, the presence of these should be checked for and any sequences with stop codons deleted from the analysis. • Sequences uploaded into GenBank should have the following information, country of origin, location of collection, host plant and date of collection should be clearly indicated. Uploading the trace file would also be advantageous. • The most widely used primer set is C1-J-2195 [5TGATTTTTTGGTCAYCCWGAAGT3] in combination with the reverse primer TL2-N-3014 [5 TCCAATGCACTAATCTGCCATATTA3] (Simon et al. 1994) • Duplicate haplotypes should not be used in phylogenetic analyses. • Outgroup selection is critical; *Trialeurodes* spp. is not a good outgroup.
rITS1	Nuclear ribosomal ITS1	Internally transcribed spacer 1 of the nuclear ribosomal RNA gene lies between the 18s and 5.8s subunits. It is used to analyse evolutionary (phylogenetic) relationships. It shows a very similar power of species level resolution as mtCO1. However, as there are multiple copies of the ribosomal gene within a single cell, direct sequencing can be difficult due to within nucleus variation. Overcoming this requires cloning of the PCR amplicon prior to sequencing which adds to the cost although sequence quality is usually high. Whitefly ITS1 is also GC rich which necessitates additional steps in the PCR process to ensure amplification. The other drawback is that there are fewer than 260 sequences in GenBank which limits the coverage of genetic diversity. (De Barro et al. 2000; Wu et al. 2003; Abdullahi et al. 2003; De Barro et al. 2005; Li et al. 2007)
Microsatellites	Also known as simple sequence repeats or tandem repeats – usually repeating sequences of <7 base pairs	See en.wikipedia.org/wiki/Microsatellite Used primarily in *B. tabaci* research to consider population structure. (De Barro et al. 2003; Tsagkarakou and Roditakis 2003; De Barro 2005; De Barro et al. 2008; Delatte et al. 2005, 2006; Dalmon et al. 2008, Tsagkarakou et al. 2007; Gauthier et al 2008)

9.3 Reproductive Isolation

Since 1993 a series of studies (summarised in Xu et al. 2010) have shown that members of the different genetic groups are for the most part either completely or partially reproductively incompatible. Studies such as Costa et al. (1993) which indicate some small capacity to produce F1 females in crosses between B and A showed that the F1 females always had the same profiles as the female parent and never any of the intermediate or segregating patterns that one would expect if hybridization had occurred. This suggests contamination rather than a true capacity to hybridize. Further, Perring and Symmes (2006) never observed copulation between A and B. De Barro and Hart (2000) showed that F1 females could occasionally be produced, but were always sterile. The only studies to show some capacity to copulate and produce viable F1 progeny come from experiments involving the individuals from the Mediterranean/Asia Minor/Africa (B), Mediterranean (Q and L) and Indian Ocean (MS) (see Xu et al. 2010 for summary, Elbaz et al. 2010). Delatte et al. (2006) has also observed this partial compatibility in the field. Similarly, Maruthi et al. (2001, 2004) observed partial mating incompatibility between different putative species from Sub-Saharan Africa. Partial mating compatibility between the more closely related genetic groups is unsurprising and appears to occur when groups have divergences below 10% and under such circumstance incomplete reproductive isolation can be expected (Mallet 2005).

9.4 Factors Influencing Global Geographic Structure – Clues to an Invasion Process

In terms of capacity to invade, the allopatric phylogeographical structure raises two key questions. The first is how has this pattern persisted? In cases such as Australia, Indian Ocean and New World geographic barriers would appear to offer an explanation, yet geographical barriers to spread are unlikely explanations when it comes to considering the failure of groups such as the Asia I, Asia II, China, Mediterranean, Middle East-Asia Minor, Sub-Saharan Africa and Uganda to invade each other's space. Further, while mixed infestations occur (Delatte et al. 2006), they usually fail to persist (Brown et al. 1995; Perring 1996; Simon et al. 1999; Legg et al. 2002; Lima et al. 2002; Pascual and Callejas 2004; Rehka et al. 2005; Liu et al. 2007), members of one genetic group eventually displace the other. This also raises the interesting question of scale. In several cases putative species have been considered sympatric because they occur in the same country, but subsequent examination shows that they occupy different geographic locations. The key issue is whether two putative species can persist over time sharing the same space, which may well be at the scale of the leaf.

Given the global structure displayed by *B. tabaci,* this does not correspond to any obvious global climatic regions; it would appear that abiotic factors are unlikely to be significant contributors to the observed patterns. Furthermore, given the apparent incapacity for neighboring genetic groups to invade each others' space effectively, why have at least two members of the *B. tabaci* complex, Middle East-Asia Minor 1 (B) and Mediterranean (Q), both closely related sibling groups (Boykin et al. 2007; Dinsdale et al. 2010), shown such a remarkable capacity to spread well beyond their respective home ranges, but not so successfully into the space occupied by their immediate neighbours (De Barro et al. 2011)?

The failure to invade and establish persistently in neighbouring regions and the related strong allopatric phylogeographical structure suggest that for *B. tabaci* as a whole, the capacity to invade is uncommon. Further, the fact that the best known examples of invasion are associated with genetic groups that utilize globally traded ornamental plants suggests that the capacity to invade may be linked to their ability to access a pathway that allows them to be transplanted into non-neighboring geographic regions. The phylogeographical structure and the pattern of global colonization of these invaders tends to suggest that if a *B. tabaci* is transplanted into a neighbouring geographic region then it may well fail to establish with any long term success (Simon et al. 1999), but if transplanted into a non-neighbouring geographic region then successful establishment is more likely. This set of observations suggests that the underlying mechanism for invasion may be connected to biological interactions and draws upon the ideas of local species adaptation. Here, species which co-evolve in the same space tend to better adapted to each other and so are better able to compete against each other (see Kniskern and Rausher 2001 for review). In contrast, species transplanted into a geographic space unrelated to their evolutionary history or that of the indigenous occupants, may bring with them a set of traits which are unfamiliar to the indigenous population, this lack of familiarity may confer a selective advantage over the traits of the invader (Cox 2004). Further, as Holway and Suarez (1999) have shown in their review, invasive animals often thrive at the expense of indigenous, closely related organisms, and insight into the causes of animal invasions often hinges on detailed assessments of behavioral mechanisms. Similarly, Hardin (1960) in his discussion of the competitive exclusion principal concluded that complete competitors were incapable of sympatric co-existence. Likewise, Odum (1971) observed that closely related organisms with similar ecological requirements were unable to occur in the same place.

9.5 The Capacity to Invade

9.5.1 Early Evidence of Biological Interactions

The first detailed observation on mating behavior of *B. tabaci* was made by Li et al. (1989) who showed that single males were capable of interrupting courtship, an observation that has also been made for *Trialeurodes vaporariorum* (Las 1979; Ahman and Ekbomm 1981). This was followed by Perring et al. (1994) and Perring and Symmes (2006) both of whom reported that (B) males actively interfered with the courting and mating of New World (A) females. This is not unexpected as closely related species often have incompletely isolated mate recognition systems and so are susceptible to reproductive interactions and interference (Butlin 1995). This lack of capacity to discriminate during courting and mating was identified by Reitz and Trumble (2002) as one factor contributing to competitive displacement through reproductive interference. Another study of mating interactions between the invading Middle East – Asia Minor 1 (B) and the indigenous Australian (AN) putative species (De Barro and Hart 2000) observed that the interaction reduced overall population increase through a marked increase in the proportion of male progeny and the production of no fertile hybrid females. Further, De Barro et al. (2006) suggested that a threshold existed that required a minimum number of the invader adults relative to indigenous individuals before establishment occurred. If the invaders are considered propagules, then De Barro et al. (2006) showed that propagule pressure could contribute to establishment success, i.e. if the number of invaders relative to the number of indigenous competitors was too low then establishment would fail. This agrees with Lockwood et al. (2005) who concluded that one of the key factors influencing successful establishment was propagule pressure, either large numbers of individuals on infrequent occasions or small numbers of individuals frequently.

The interaction between Middle East – Asia Minor 1 (B) and Mediterranean (Q) is a useful one to explore in greater detail. The two are very closely related (Dinsdale et al. 2010), occupy neighbouring geographic ranges and are reproductively isolated (Elbaz et al. 2010). Pascual and Callejas (2004) studied the interaction between the two and showed that Middle East – Asia Minor 1 (B) was able to displace Mediterranean (Q). Further, they showed that on tomato Middle East – Asia Minor 1 (B) had higher fecundity, lower juvenile mortality and a higher reproductive potential. Differential female fecundity conforms with one of the mechanisms identified by Reitz and Trumble (2002) as contributing to competitive displacement and is linked to differential resource acquisition where one species makes better use of a resource than another leading to higher growth rates and greater survivorship. Several studies Bethke et al. (1991, cotton and poinsettia), Costa and Brown (1991, cotton, poinsettia and pumpkin), Bedford et al. (1994 bean, tomato, tobacco, thornapple, cotton, squash, capsicum and Chinese violet), Nombela et al. (2001, tomato), Pascual and Callejas (2004, tomato), Zang et al. (2006, cabbage, cotton, kidney bean, squash, tobacco) have shown differences in fecundity and survivorship of Middle East – Asia Minor 1 (B) and a number of other different genetic groups across a range of host plants. The results from Pascual and Callejas (2004) suggest that Middle East – Asia Minor 1 (B) has the capacity to replace Mediterranean (Q) as do two later studies, Crowder et al (2009) and Crowder et al. (2010a). However, the situation in the field is variable. Guirao et al. (1997) showed that between 1993 and 1995 there was no evidence for displacement of Mediterranean (Q) by Middle East – Asia Minor 1 (B). Further, Simón et al. (1999) showed that by 1999 Mediterranean (Q) was the dominant genetic group in southern Spain. In China, where Middle East – Asia Minor 1 (B) established several years earlier, Mediterranean (Q) is now displacing Middle East – Asia Minor 1 (B), yet in the USA Mediterranean (Q) has been unable to do so.

So, there appear to be several processes going on which influence the outcome between Middle East – Asia Minor 1 (B) and Mediterranean (Q) and these will be explored along with other mechanisms that contribute to invasion success.

9.5.2 Asymmetrical Mating Interactions

Evidence for asymmetrical mating interactions first came to notice through the work of Mabbett (2004) and Pascual (2006) both of whom noted that not only did individual Middle East – Asia Minor 1 (B) males and females spend more time courting than Mediterranean (Q) males and females, but Middle East – Asia Minor 1 (B) males spent longer courting Mediterranean (Q) females than Mediterranean (Q) males spent courting Middle East – Asia Minor 1 (B) females, suggesting a degree of asymmetry in the mating interactions. Perring et al. (1994) and Perring and Symmes (2006) both showed Middle East-Asia Minor 1 (B) males actively interfering with the courtship of pairs of New World (A) individuals. Asymmetrical mating interference was identified by Reitz and Trumble (2002) as a mechanism for competitive displacement. Zang and Liu (2007) and Ruan et al. (2007) also observed the capacity for males to interfere in courtship; when two males from different genetic groups and a female of a given genetic group were placed together, the female was frequently courted by both males and courting and copulation could be interrupted by the second male. Similarly, Crowder et al. (2010a) observed asymmetrical interference in the interactions between Middle East-Asia Minor 1 (B) and Mediterranean (Q). Liu et al. (2007) also showed that this type of reproductive interference was able to influence the relative availability of males and females in whiteflies and could have profound ecological impacts on whitefly abundance and the capacity for one species to displace another. The ability of Middle East-Asia Minor 1 (B) to utilize mating interference to disrupt with the courtship of pairs of New World (A) individuals (Perring et al. 1994; Perring and Symes 2006), Mediterranean (Q) (Mabbett 2004; Pascual 2006; Pascual and Callejas 2004; Crowder et al. 2010a), Australia (AN) (De Barro and Hart 2000; Liu et al. 2007) and Asia II 3 (ZHJ1) (Liu et al. 2007; Ruan et al. 2007; Zang and Liu 2007; Luan et al. 2008) indicates that this is a key mechanism contributing to the invasion success of Middle East-Asia Minor 1 (B). As whiteflies are haplo-diploid, a key feature of the interaction involved the initial shift in progeny sex ratio from a female to male bias (De Barro and Hart 2000; Liu et al. 2007; Crowder et al. 2010a). However, unilaterally increasing copulations by Middle East – Asia Minor 1 (B) females, but not Australia 1 (AN), Asia II 3 (ZHJ1) or Mediterranean (Q) females, elevated the proportion of female progeny in Middle East – Asia Minor 1 (B) to levels above that seen in pre-competition populations. Moreover, copulation by indigenous individuals was more frequently interrupted by invader males than that of the invader individuals interrupted by the indigenous males (Liu et al. 2007; Crowder et al. 2010a). This is further evidence of species level divergence and gives an insight into the mechanism that could be driving both species dominance and speciation.

The consequences of these asymmetric mating interactions have obvious impact on the numerical changes of the invader and indigenous populations because the increase in the proportion of invader females and the concomitant decrease in the proportion of indigenous females results in an immediate higher population growth rate for Middle East – Asia Minor 1 (B) and a lower growth rate for the indigenous population. With this mechanism operating, even a small number of Middle East – Asia Minor 1 (B) in a new location may succeed by rapidly producing female progeny and thus achieve rapid population growth. As the abundance of B increases relative to the indigenous individuals, the increased allocation of eggs to female progeny and the active interference of mating of indigenous individuals by Middle East – Asia Minor 1 (B) males combine to drive the indigenous population to local extinction. These results help to reveal behavioral mechanisms underlying the Middle East – Asia Minor 1's (B) capacity to invade and displace indigenous populations. The strong competitive ability of Middle East – Asia Minor 1 (B) results partly from its capacity to adjust sex ratio in favor of its population increase, and partly from its capacity to interfere with the mating of indigenous individuals.

9.5.3 Host Plant Mix Mediates Establishment

Zang et al. (2005) showed that host plant contributed to the rate at which the invasive Middle East – Asia Minor 1 (B) displaced its indigenous competitor, Asia II 3 (ZHJ1). In cage studies using equal numbers of individuals from the two genetic groups on cotton, displacement took six generations whereas it required only two generations on squash. Similarly, while De Barro and Hart (2000) showed that a minimum number of invaders relative to indigenous individuals were required before establishment could take place; De Barro et al. (2006) showed that this minimum propagule number could be mitigated by resource availability. In a study of the influence of one bottom factor contributing to establishment, they combined the use of two host plants, *Euphorbia cyathophora* – a host well suited to both the invading Middle East – Asia Minor 1 (B) and indigenous Australian (AN) individuals and the other cotton, a host more suited to the invader. They showed that the presence of cotton enabled (B) to establish at much lower numbers than would occur using *E. cyathophora* alone.

Middle East – Asia Minor 1 (B) is frequently stated as having a much wider host range than individuals from other genetic groups (see Brown et al. 1995; Perring 2001 for reviews) and while these claims are based more on assumptions than any actual data (De Barro et al. 2011), in those studies that have been published, as a general rule Middle East – Asia Minor 1 (B) tends to outperform its competitors on the same host. As previously mentioned several studies have shown differences in fecundity and survivorship (see section 8.5.1). Furthermore, host plant variety may also be important as Bethke et al. (1991) showed Middle East – Asia Minor 1 (B) outperforming the New World A biotype on cotton whereas Costa and Brown (1991) showed the reverse. This suggests that differential host acceptability may influence the capacity for the invader to either interfere or escape interference. If one considers the process of invasion from a landscape perspective, then landscape can be divided into patches that are suitable and unsuitable. If, as in the case of between genetic group differential host acceptability, the suitable and unsuitable patches do not overlap entirely then there will be spaces in the landscape which are either unavailable to or poorly utilized by either the invader or the indigenous *B. tabaci*. If these patches are more acceptable to the invader then it in effect escapes interference by entering interference free space and it will be more able to establish; the converse may also be the case.

Evidence of this has recently been published. De Barro and Bourne (2010) studied the ability of Middle East – Asia Minor 1 (B) to displace the indigenous Australian (AN) competitor and the role different host plant mixes played in the displacement process. They found that when only a mutually acceptable host was available, invader and indigenous adults were equally spread over the available plants and females oviposited equally on all plants. However, when given the choice of a host that was equally acceptable and one that was acceptable to the invader only, adult invaders still distributed themselves evenly across both hosts, but shifted their oviposition away from the mutually acceptable host and instead laid mostly on the host poorly utilized by the indigenous competitor. As De Barro and Hart (2000), De Barro et al. (2006) and Liu et al. (2007) had already shown, there is a complex set of competitive interactions between Middle East – Asia Minor 1 (B) and Australian (AN) that have an initial negative effect on the rate of population increase of both the invader and indigenous species. What De Barro and Bourne (2010) found was that the invader's broader host range conferred an advantage through plasticity in choice of oviposition host. By choosing the host that was not utilized by the indigenous competitor, the invader was able to avoid the negative consequences of the competitive interactions. The results showed that the avoidance of competition, by escaping into competition free space, during the early stages of establishment through the agency of a broader host range increased the invader's capacity to establish and more rapidly displace its competitor. These results provide a good example of why the consideration of scale is a key factor in more fully understanding ecological processes such as biological invasions. At the scale of the plant, a minimum number of invaders were required to overcome the numerical advantage of the indigenous competitor (De Barro et al. 2006). However, at wider scales the invader's broader host range enables numerical advantage at the plant scale to be overcome and shifts the advantage towards the invader.

9.5.4 *Insecticide Resistance*

The role of insecticides in influencing the capacity for invasion and persistence of different genetic groups of *B. tabaci* has been considered largely in regards to the interaction between the Middle East-Asia Minor 1 (B) and Mediterranean (Q) and to a lesser extent Middle East-Asia Minor 1 (B) and New World (A). Costa et al. (1993)

and Coats et al. (1994) both suggested that the capacity of Middle East-Asia Minor 1 (B) to invade the USA was linked to Middle East-Asia Minor 1's (B) higher level of resistance relative to the indigenous genetic group. Horowitz et al. (2003, 2005) found that higher levels of resistance to pyriproxyfen and neonicotinoids in Mediterranean (Q) were associated with the greater abundance of Mediterranean (Q) relative to Middle East – Asia Minor 1 (B) in Israel. Similarly, the abundance of Mediterranean (Q) biotype in southern Spain has been linked to high resistance and cross-resistance to pyriproxyfen and neonicotinoids in southern Spain (Nauen et al. 2002; Rauch and Nauen 2003; Pascual 2006). Abundance of Middle East-Asia Minor 1 (B) and Mediterranean (Q) in Spain has been monitored since 1993 (Guirao et al. 1997; Simon et al. 1999) and Mediterranean (Q) has consistently been the more abundant. Neonicotinoids have been in use in Spain throughout that period (Elbert and Nauen 2000) and it is possible that they have contributed to the dominance of Mediterranean (Q). However, whether insecticides do influence the capacity to invade on a regional scale is difficult to determine. A recent study, Crowder et al (2010b) provides some insight into the role of insecticide resistance in the invasion potential of Middle East-Asia Minor 1 (B) and Mediterranean (Q). Here, the study found that while Middle East-Asia Minor 1 (B) would ordinarily be expected to displace Mediterranean (Q), but in situations where insecticides were used such as in Israel, the higher levels of resistance in Mediterranean (Q) enabled it to displace Middle East-Asia Minor 1 (B). Similarly, in the USA where Middle East-Asia Minor 1 (B) has evolved high levels of resistance, Mediterranean (Q) has been unable to displace Middle East-Asia Minor 1 (B) (Crowder et al. 2010b).

9.5.5 Whitefly-Begomovirus Interactions

An additional factor that may aid in invasion and displacement is the acceleration of population increase gained by the invader through its mutualism with the begomoviruses it transmits (Costa et al. 1991; McKenzie 2002; Zhang et al. 2000; Jiu et al. 2007). Costa et al. (1991) observed that Middle East – Asia Minor 1 (B) showed greater egg to adult survival on pumpkin infected with watermelon curly mottle strain of *Squash leaf curl virus* compared with uninfected plants. Zhang et al. (2000) demonstrated both increased vector fecundity and vector density for *B. tabaci* feeding on cassava infected with cassava mosaic virus and speculated that the interaction between virus and vector accelerated the spread of the vector and therefore the virus. Similarly, Middle East – Asia Minor 1 (B) laid more eggs on tomato plants infected with *Tomato mottle virus* than they did on uninfected plants (McKenzie 2002).

The connection between changes in whitefly performance and virus infection in the host plant was made by Jiu et al. (2007) and Liu et al. (2009). *Tobacco curly shoot virus* (TbCSV), *Tomato yellow leaf curl virus* (TYLCV) and *Tomato yellow leaf curl China virus* (TYLCCNV) are all whitefly-transmitted begomoviruses that have become widespread in south China following the invasion and establishment of the B biotype. Jiu et al. (2007) and Liu et al. (2009) compared the performance of the invasive Middle East-Asia Minor 1 (B) and indigenous Asia II 3 (ZHJ1) putative species on healthy and infected plants. The results vary in regards to Middle East-Asia Minor 1 (B) in that it showed either increased (Jiu et al. 2007) or no significant reduction in performance, but were consistent for Asia II 3 (ZHJ1) which showed marked declines in performance (Jiu et al. 2007; Liu et al. 2009) as a result of feeding on infected plants. These observations indicate that invasions of Middle East-Asia Minor 1 (B) may be facilitated by some begomoviruses which may act to increase the fitness of the invader while at the same time decreasing the fitness of indigenous competitors. Furthermore, in a positive feedback, increasing abundance of Middle East-Asia Minor 1 (B) will in turn lead to the increased spread of begomoviruses and so further compound the negative influence on the fitness of the indigenous while at the same time compounding the negative consequences of asymmetrical mating interference. In a similar situation to Middle East-Asia Minor 1 (B), Mediterranean (Q) has also been shown to derive a greater fitness benefit from plants infected by begomoviruses than another indigenous competitor Asia II 1 (ZHJ2) (Liu et al. 2010).

9.6 Pepper Yellow Leaf Curl Indonesia Virus– A Different Type of Invasion

Not all invasions by *B. tabaci* result in displacement. De Barro et al. (2008) consider the recent outbreak of "Penyakit Kuning" in Indonesia. The disease caused by the begomovirus *Pepper yellow leaf curl Indonesia virus* (PYLCIV) first appeared in 1999 in the important cash crop of hot chilli (*Capsicum annum*, *C. frutescens*) and has since led to the banning of the planting of the crop in some regions of Sumatra. Hot chillies are an important source of cash income for small crop vegetable farmers in Indonesia with production occupying at least 155,000 ha and involving >500,000 farmers (Vos and Duriat 1995; Mustafa et al. 2006). Prior to 1995 symptoms related to PYLCV had never been observed in chilli from Indonesia (Vos 1994). Between 1999 and 2000 symptoms were uncommon, but post 2000 symptoms became more widespread and census data for the period from 2000 to 2006 showed that the disease had undergone a 5.4 fold increase between 2000 and 2001, 4.6 fold between 2001 and 2002, 2.5 fold between 2002 and 2003 and an average 1.4 fold increase between 2003 and 2006 (unpublished data, Indonesian Ministry of Agriculture). The disease, based on visual symptoms, is now widespread over western Indonesia where much of the chilli production takes place, but less common in Sulawesi and Irian Jaya and has not been observed in Maluku and East and West Nusa Tenggara.

Unlike the previous examples of *B. tabaci* invasions which involved members of two different allopatric genetic groups, this one involved individuals from the same genetic group (Asia I), but with parapatric origins – central Thailand and Indonesia. The study compared 15 microsatellite from individual *B. tabaci* collected from

Indonesia (Bali, Java, Sumatra), Malaysia, Thailand and Vietnam. The results showed that in 1999 alleles shared by individuals from parts of Thailand around Bangkok were present in Sumatra, but not Bali or Java; 4 years later these alleles were widespread across Java; at the same time symptoms of PYLCV spread across Sumatra, Bali and Java. The novel feature of this invasion was that there was no sign of local population extinction, but instead there was evidence that genetic material from the Thai invader becoming incorporated into the indigenous population suggesting an invasion of the genome had occurred. In other words, instead of displacing the indigenous population as usually occurs with the invasions by the *B. tabaci* that we have previously considered, it appears that genes from the Thai *B. tabaci* have introgressed, i.e., invaded the genome of the indigenous population in a situation not dissimilar to the invasion of Africanized honeybee genes into the genomes of honeybees of Eurasian ancestry (Whitfield et al. 2006). An additional feature of the study was that the closest genetic relative to PYLCV also came from the same region in Thailand as the Thai alleles detected in the Indonesian *B. tabaci*. The fact that PYLCV has moved so rapidly through Indonesia raises the question as to why; one possibility is that PYLCV confers a selective advantage on the vectoring individuals and this has promoted spread.

9.7 Severe *Cassava Mosaic Virus* Epidemic in Uganda – Is This the Consequence of an Invasion?

In the late 1980s cassava mosaic disease (CMD) began to outbreak in northern Uganda and over the ensuing years spread southward through the central, southern and eastern parts of the country (Legg and Ogwal 1998). Two features of the outbreak, apart from apparent changes in the virus responsible for the disease, were a more rapid rate of spread and increased *B. tabaci* population densities (Legg and Ogwal 1998; Colvin et al. 2004). Further, the individuals involved in the outbreak were more fecund than those prior to the outbreak (Otim-Nape et al. 1998). By sampling *B. tabaci* along a transect that passed through the invasion front and then using mtCO1 Legg et al. (2002) demonstrated that at or before the invasion front, the *B. tabaci* (Uganda 1, Ug1) present were genetically distinct from those at or behind the front (Uganda 2, Ug2) and showed approximately 8% divergence. Maruthi et al. (2001) has cast doubt on the new invader hypothesis after using RAPD-PCR to compare individuals collected in outbreak and non-outbreak areas and showed no differences between the distributions of banding profiles. However, the non-specific nature of the RAPD-PCR primers, the unknown origin of the bands, difficulties with reproducibility and the lack of robustness of the clustering analysis used require a degree of caution when interpreting the results. A further critical issue is that none of the RAPD-PCR profiles have been matched to mtCO1 sequences so it is not clear whether the assumption that RAPD profiles for putative invaders and non-invaders will be different is valid.

Following Maruthi et al. (2001), (2004) compared the capacity for different Ugandan *B. tabaci* to interbreed and used UgCas-Nam which was identical (based on mtCO1) to one of the Legg et al. (2002) Ug2 invader samples, 5BNama. They also chose two other populations TzCas-Mtw and GhCas-Acc both of which fall into the same Ug1 cluster and show between 7% and 8% sequence divergence from Ug2. They found that while the control crosses produced a roughly 1:1 male/female sex ratio, the mixed crosses produced a 3:1 male/female sex ratio indicating a marked shift to male production in the progeny with only around half the replicates producing any female F1 progeny at all. The results were similar for the F2 generation and agreed with those of Maruthi et al. (2001) which also showed an overall trend towards the reduced proportion of female progeny and a higher level of reproductive incompatibility than observed in the single line crosses. Maruthi et al. (2001, 2004) concluded that as viable F1 and F2 females were produced the putative invaders and non-invaders were not different.

However, their conclusion seems to ignore their observations of the two generation shift in sex ratio to a strong male bias and the failure to produce female progeny in many of the replicates, both of which suggest a degree of incompatibility. The production of F1 females between the closely related genetic groups Mediterranean/Asia Minor/Africa (B), Mediterranean (Q and L) and Indian Ocean (MS), have been reported previously (Byrne et al. 1995; Ronda et al. 1999; Moya et al. 2001; Delatte et al. 2006) and the levels of divergence between the putative invader and non-invaders are in line with those found between these groups. We know from several studies between B and Q which show 5% divergence, that female F1 progeny are readily produced in the laboratory. Delatte et al. (2006) has shown evidence from her microsatellite study that unidirectional introgression between MS and B with 8% divergence, occurs in field populations. As Mallett (2005) discussed in his review of hybridization, closely related species often have incomplete reproductive isolation and so we can reinterpret the results in this light. One further deficiency in the conclusions made is that neither study considered the fitness of the progeny nor determined whether introgression was occurring in the field.

Whether the outbreak of pandemic CMD in Uganda is due to an invader *B. tabaci* is still open to debate. However, the data from Legg et al. (2002) and that from Maruthi et al. (2004) indicate that Ug1 and Ug2 both are sufficiently distinct genetically and are sufficiently incompatible in reproduction to conclude that they are distinct entities. The doubt that remains is whether Ug2 is really invading or whether other factors (e.g. Colvin et al. 2004; Sseruwagi et al. 2006) have simply influenced the relative abundance of the different *B. tabaci* that were already there.

9.8 Conclusions

Table 9.2 summarizes the mechanisms identified as contributing to the capacity to invade. The obvious conclusion from this is that invasion success is multifactorial. Asymmetrical mating interference and aggressive male behaviour, couple with

Table 9.2 Summary of mechanisms identified as contributing to the capacity to invade

Factors influencing competitive displacement	Genetic groups	References
Differential female fecundity	Middle East-Asia Minor 1 (B) vs Mediterranean (Q)	Pascual and Callejas (2004), Crowder et al. (2010b)
Differential insecticide resistance	Middle East-Asia Minor 1 (B) vs Mediterranean (Q)	Horowitz et al. (2003, 2005), Pascual and Callejas (2004), Pascual (2006), Crowder et al. (2010b)
Differential resource utilization	Middle East-Asia Minor 1 (B) vs Mediterranean (Q), Middle East-Asia Minor 1 (B) vs Australia (AN)	Pascual and Callejas (2004), De Barro et al. (2006), De Barro and Bourne (2010)
Courting and mating interference	Middle East-Asia Minor 1 (B) vs Mediterranean (Q), Middle East-Asia Minor 1 (B) vs New World (A), Middle East-Asia Minor 1 (B) vs Australian (AN), Middle East-Asia Minor 1 (B) vs Asia II 3 (ZHJ1), Middle East-Asia Minor 1 (B) vs Mediterranean (Q)	Pascual and Callejas (2004), Perring et al. (1994), Perring and Symes (2006), De Barro and Hart (2000), Liu et al. (2007), Crowder et al. (2010a)
Increased frequency of copulation	Middle East-Asia Minor 1 (B) vs Asia II 3 (ZHJ1), Middle East-Asia Minor 1 (B) vs Australia (AN)	Liu et al. (2007), Crowder et al. (2010a)
Indirect effects of begomovirus infection in the host plant	Middle East-Asia Minor 1 (B) vs Asia II 3 (ZHJ1), Mediterranean (Q) vs Asia II 1 (ZHJ2)	Jiu et al. (2007), Liu et al. (2009, 2010)

differential host plant utilization and a linked capacity to avoid interference through escape to competition free space to directly improve the fitness of the invader while at the same time decreasing the fitness of native competitors. These traits are further modified by differences in resistance to insecticides and the asymmetrical fitness benefits/deficits associated with the transmission of begomoviruses. Understanding these traits and how they interact will continue to increase our understanding of the invasion process. The future though no doubt lies in understanding the underpinning genomic contribution to invasion success. Early work such as Mahadav et al. (2009) while quite preliminary, is likely to contribute even more to our understanding of the capacity to invade.

References

Abdullahi I, Winter S, Atirim GI, Thottappilly G (2003) Molecular characterization of whitefly, *Bemisia tabaci* Hemiptera, Aleyrodidae populations infesting cassava. Bull Entomol Res 93:97–106

Ahman I, Ekbomm BS (1981) Sexual behaviour of the greenhouse whitefly *Trialeurodes vaporariorum*, orientation and courtship. Entomol Exp Appl 29:330–338

Bedford ID, Briddon RW, Brown JK, Rosell RC, Markham PG (1994) Geminivirus transmission and biological characterisation of *Bemisia tabaci* Gennadius from different geographic regions. Ann Appl Biol 125:311–325

Bethke JA, Paine TD, Nuessly GS (1991) Comparative biology, morphometrics, and development of two populations of *Bemisia tabaci* Homoptera, Aleyrodidae on cotton and poinsettia. Ann Entomol Soc Am 84:407–411

Boukhatem N, Jdaini S, Mukovski Y, Jacquemin JM, Bouali A (2007) Identification of *Bemisia tabaci* (Gennadius) (Homoptera: Aleyrodidae) based on RAPD and design of two SCAR markers. J Biol Res (Thessalon) 8:167–176

Boykin LM, Shatters RG, Rosell RC, McKenzie CL, Bagnall RA, De Barro PJ, Frohlich DR (2007) Global relationships of *Bemisia tabaci* Hemiptera, Aleyrodidae revealed using Bayesian analysis of mitochondrial CO1 DNA sequence. Mol Phylogenet Evol 44:1306–1319

Brown JK, Frohlich DR, Rosell RC (1995) The sweetpotato or silverleaf whiteflies, Biotypes of *Bemisia tabaci* or a new species complex. Annu Rev Entomol 40:511–534

Butlin R (1995) Genetic variation in mating signals and responses. In: Lambert DM, Spencer HG (eds) Speciation and the recognition concept, theory and application. The Johns Hopkins University Press, Baltimore, pp 327–366

Byrne FJ, Cahill M, Denholm I, Devonshire AL (1995) Biochemical identification of interbreeding between B-type and non B-type strains of the tobacco whitefly *Bemisia tabaci*. Biochem Genet 33:13–23

Cervera MT, Cabezas JA, Simon B, Martinez-Zapater JM, Beitia F, Cenis JL (2000) Genetic relationships among biotypes of *Bemisia tabaci* Hemiptera, Aleyrodidae based on AFLP analysis. Bull Entomol Res 90:391–396

Cheek S, Macdonald O (1994) Extended summaries SCI pesticides group symposium management of *Bemisia tabaci*. Pestic Sci 42:135–142

Chu D, Zhang YJ, Cong B, Xu BY, Wu QJ (2004) Developing sequence characterized amplified regions (SCARs) to identify *Bemisia tabaci* and *Trialeurodes vaporariorum*. Plant Prot 30:27–30

Coats SA, Brown JK, Hendrix DL (1994) Biochemical characterisation of biotype-specific esterases in the whitefly, *Bemisia tabaci* Genn Homoptera, Aleyrodidae. Insect Biochem Mol 24:723–728

Colvin J, Omongo CA, Maruthi MN, Otim-Nape GW, Thresh JM (2004) Dual begomovirus infections and high *Bemisia tabaci* populations, two factors driving the spread of a cassava mosaic disease pandemic. Plant Pathol 53:577–584

Costa HS, Brown JK (1991) Variation in biological characteristics and esterase patterns among populations of *Bemisia tabaci*, and the association of one population with silverleaf symptom induction. Entomol Exp Appl 61:211–219

Costa HS, Brown JK, Byrne DN (1991) Life history traits of the whitefly, *Bemisia tabaci* (Homoptera, Aleyrodidae) on six virus-infected or healthy plant species. Environ Entomol 20:1102–1107

Costa HS, Brown JK, Sivasupramaniam S, Bird J (1993) Regional distribution, insecticide resistance and reciprocal crosses between the 'A' and 'B' biotypes of *Bemisia tabaci*. Insect Sci Appl 14:127–138

Cox GW (2004) Alien species and evolution. The evolutionary ecology of exotic plants, animals, microbes, and interacting native species. Island Press, Washington, DC

Crowder DW, Horowitz AR, Tabashnik BE, Dennehy TJ, Denholm I, Gorman K, Carrie're Y (2009) Analyzing haplodiploid inheritance of insecticide resistance in whitefly biotypes. Bull Entomol Res 99:307–315

Crowder DW, Horowitz AR, Showalter AM, Kontsedalov S, De Barro PJ, Liu SS, Liu J, Carrière Y (2010a) Behavior and life-history predict competitive displacement by an invasive whitefly. J Anim Ecol 79:563–570. doi:10.1111/j.1365-2656.2010.01666.x

Crowder DW, Sitvarin MI, Carrie're Y (2010b) Plasticity in mating behaviour drives asymmetric reproductive interference in whiteflies. Anim Behav 79:579–587. doi: 10.1016/j.anbehav.2009.11.025

Dalmon A, Halkett F, Granier M, Delatte H, Peterschmitt M (2008) Genetic structure of the invasive pest Bemisia tabaci: evidence of limited but persistent genetic differentiation in glasshouse populations. Heredity 100:316–325

Dalton R (2006) The christmas invasion. Nature 443:898–900

De Barro PJ (2005) Genetic structure of the whitefly Bemisia tabaci in the Asia-Pacific region revealed using microsatellite markers. Mol Ecol 14:3695–3718

De Barro P, Bourne A (2010) Ovipositional host choice by an invader accelerates displacement of its indigenous competitor. Biol Invasions 12:3013–3023

De Barro PJ, Hart PJ (2000) Mating interactions between two biotypes of the whitefly, *Bemisia tabaci* Hemiptera, Aleyrodidae in Australia. Bull Entomol Res 90:103–112

De Barro PJ, Driver F, Trueman JWH, Curran J (2000) Phylogenetic relationship of world populations of *Bemisia tabaci* (Gennadius) using ribosomal ITS1. Mol Phylogenet Evol 16:29–36

De Barro PJ, Scott KD, Graham GC, Lange CL, Schutze MK (2003) Isolation and characterisation of microsatellite loci in Bemisia tabaci. Mol Ecol Notes 3:40–43

De Barro PJ, Trueman JWH, Frohlich DR (2005) *Bemisia argentifolii* is a population of *B tabaci*, the molecular genetic differentiation of *B tabaci* populations around the world. Bull Entomol Res 95:193–203

De Barro PJ, Bourne A, Khan SA, Brancatini VAL (2006) Host plant and biotype density interactions – their role in the establishment of the invasive B biotype of *Bemisia tabaci*. Biol Invasions 8:287–294

De Barro PJ, Hidayat SH, Frohlich D, Subandiyah S, Ueda S (2008) A virus and its vector, pepper yellow leaf curl virus and *Bemisia tabaci*, two new invaders of Indonesia. Biol Invasions 10:411–433

De Barro P, Liu SS, Bourne A (2010) Age-based differential host acceptability and human mediated disturbance prevent establishment of an invasive species and displacement of a native competitor. Biol Invasions. doi:10.1007/s10530-010-9741-8

De Barro PJ, Liu SS, Boykin L, Dinsdale A (2011) Bemisia tabaci: a statement of species status. Annu Rev Entomol 56:1–19

Delatte H, Reynaud B, Granier M, Thornary L, Lett JM et al (2005) A new silverleaf-inducing biotype Ms of Bemisia tabaci (Hemiptera: Aleyrodidae) indigenous to the islands of the south-west Indian Ocean. Bull Entomol Res 95:29–35

Delatte H, David P, Granier M, Lett JM, Goldbach R, Peterschmitt M, Reynaud B (2006) Microsatellites reveal extensive geographical, ecological and genetic contacts between invasive and indigenous whitefly biotypes in an insular environment. Genet Res 87:109–124

Dinsdale A, Cook L, Riginos C, Buckley YM, De Barro P (2010) Refined global analysis of *Bemisia tabaci* (Hemiptera: Sternorrhyncha: Aleyrodoidea: Aleyrodidae) mitochondrial cytochrome oxidase 1 to identify species level genetic boundaries. Ann Entomol Soc Am 103:196–208. doi:10.1603/AN09061

Elbaz M, Lahav N, Morin S (2010) Evidence for pre-zygotic reproductive barrier between the B and Q biotypes of *Bemisia tabaci*. Bull Entomol Res. doi:10.1017/S0007485309990630

Elbert A, Nauen R (2000) Resistance of *Bemisia tabaci* Homoptera, Aleyrodidae to insecticides in southern Spain with special reference to neonicotinoids. Pest Manag Sci 56:60–64

Frohlich DR, Torres-Jerez I, Bedford ID, Markham PG, Brown JK (1999) A phylogeographical analysis of the *Bemisia tabaci* species complex based on mitochondrial DNA markers. Mol Ecol 8:1683–1691

Gauthier N, Dalleau-Clouet C, Bouvret M-E (2008) Twelve new polymorphic microsatellite loci and PCR multiplexing in the whitefly, *Bemisia tabaci*. Mol Ecol Resour 8:1004–1007

Guirao P, Beitia F, Cenis JL (1997) Biotype determination of Spanish populations of *Bemisia tabaci* Hemiptera, Aleyrodidae. Bull Entomol Res 87:587–593

Hardin G (1960) The competitive exclusion principle. Science 131:1291–1297

Holway DA, Suarez AV (1999) Animal behavior, an essential component of invasion biology. Trends Ecol Evol 14:328–330

Horowitz AR, Gorman K, Ross G, Denholm I (2003) Inheritance of pyriproxyfen resistance in the whitefly, *Bemisia tabaci* (Q biotype). Arch Insect Biochem Physiol 54:177–186

Horowitz AR, Kontsedalov S, Khasdan V, Ishaaya I (2005) Biotypes B and Q of *Bemisia tabaci* and their relevance to neonicotinoid and pyriproxyfen resistance. Arch Insect Biochem Physiol 58:216–225

Jiu M, Zhou XP, Tong L, Yang X, Wan FH, Liu SS (2007) Vector-virus mutualism accelerates population increase of an invasive whitefly. PLoS ONE 2:e182. doi:10.1371/journal.pone.0000182

Jones DR (2003) Plant viruses transmitted by whiteflies. Eur J Plant Pathol 109:195–219

Khasdan V, Levin I, Rosner A, Morin S, Kontsedalov S et al (2005) DNA markers for identifying biotypes B and Q of *Bemisia tabaci* (Hemiptera: Aleyrodidae) and studying population dynamics. Bull Entomol Res 95:605–613

Kniskern J, Rausher MD (2001) Two modes of host-enemy coevolution. Popul Ecol 43:3–14

Ko CC, Hung YC, Wang CH (2007) Sequence characterized amplified region markers for identifying biotypes of *Bemisia tabaci* (Hem., Aleyrodidae). J Appl Entomol 131:542–547

Las A (1979) Male courtship persistence in the greenhouse whiteβy, *Trialeurodes vaporariorum* Westwood (Homoptera, Aleyrodidae). Behaviour 72:107–125

Legg JP, Ogwal S (1998) Changes in the incidence of African cassava mosaic virus disease and the abundance of its whitefly vector along south-north transects in Uganda. J Appl Entomol 122:169–178

Legg JP, French R, Rogan D, Okao-Okuja G, Brown JK (2002) A distinct *Bemisia tabaci* Hemiptera, Sternnorryncha, Aleyrodidae genotype cluster is associated with the epidemic of severe cassava mosaic virus disease in Uganda. Mol Ecol 11:1219–1229

Li TY, Vinson SB, Gerling D (1989) Courtship and mating behaviour of *Bemisia tabaci* Homoptera, Aleyrodidae. Environ Entomol 18:800–806

Li ZX, Lin HZ, Guo XP (2007) Prevalence of Wolbachia infection in *Bemisia tabaci*. Curr Microbiol 54:467–471

Lima LHC, Campos L, Moretzsohn MC, Návia D, de Oliveira MRV (2002) Genetic diversity of *Bemisia tabaci* Genn. populations in Brazil revealed by RAPD markers. Genet Mol Biol 25:217–223

Liu SS, De Barro PJ, Xu J, Luan JB, Zang LS, Ruan YM, Wan FH (2007) Asymmetric mating interactions drive widespread invasion and displacement in a whitefly. Science 318:1769–1772

Liu J, Zhao H, Jiang K, Zhou XP, Liu SS (2009) Differential indirect effects of two plant viruses on an invasive and an indigenous whitefly vector: implications for competitive displacement. Ann Appl Biol 155:439–448

Liu J, Li M, Li JM, Huang CJ, Zhou XP, Xu FX, Liu SS (2010) Viral infection of tobacco plants improves performance of *Bemisia tabaci* but more so for an invasive than for an indigenous biotype of the whitefly. J Zhejiang Univ Sci B 11:30–40

Lockwood JL, Cassey P, Blackburn T (2005) The role of propagule pressure in explaining species invasions. Trends Ecol Evol 20:223–228

Luan JB, Ruan YM, Zang L, Liu SS (2008) Precopulation intervals, copulation frequencies, and initial progeny sex ratios in two biotypes of whitefly, Bemisia tabaci. Entomol Exp Appl 129:316–324

Ma DY, Li XC, Dennehy TJ, Lei CL, Wang M et al (2009) Utility of mtCO1 polymerase chain reaction-restriction fragment length polymorphism in differentiating between Q and B whitefly *Bemisia tabaci* biotypes. Insect Sci Appl 16:107–114

Mabbett T (2004) Mating interactions of *Bemisia tabaci* biotypes in Cyprus. Resistant Pest Manage Newsl 13:3–4

Mahadav A, Kontsedalov S, Czosnek H, Ghanim M (2009) Thermotolerance and gene expression following heat stress in the whitefly Bemisia tabaci B and Q biotypes. Insect Biochem Mol Biol 39:668–676

Mallet J (2005) Hybridization as an invasion of the genome. Trends Ecol Evol 20:229–237

Maruthi MN, Colvin J, Seal S (2001) Mating incompatibility, life-history traits, and RAPD–PCR variation in *Bemisia tabaci* associated with the cassava mosaic disease pandemic in east Africa. Entomol Exper Appl 99:13–23

Maruthi MN, Colvin J, Thwaites RM, Banks GK, Gibson G, Seal SE (2004) Reproductive incompatibility and cytochrome oxidase I gene sequence variability amongst host-adapted and geographically separate *Bemisia tabaci* populations Hemiptera, Aleyrodidae. Syst Entomol 29:560–568

McKenzie CL (2002) Effect of tomato mottle virus (Tomov) on *Bemisia tabaci* biotype B (Homoptera: Aleyrodidae) oviposition and adult survivorship on healthy tomato. Fla Entomol 85:367–368

Moya A, Guirao P, Cifuentes D, Beitia F, Cenis JL (2001) Genetic diversity of Iberian populations of *Bemisia tabaci* Hemiptera, Aleyrodidae based on random amplified polymorphic DNA–polymerase chain reaction. Mol Ecol 10:891–897

Mustafa U, Ali M, Kuswanti H (2006) Indonesia. In Ali M (ed.) Chilli (*Capsicum* spp) food chain analysis: setting research priorities in Asia. Shanhua, Taiwan: AVRDC – The World Vegetable Center, Technical Bulletin No. 38, AVRDC Publication 06-678, p 253

Nauen R, Stumpf N, Elbert A (2002) Toxicological and mechanistic studies on neonicotinoid cross resistance in Q type *Bemisia tabaci*. Pest Manag Sci 58:868–875

Nombela G, Beitia F, Muñiz M (2001) A differential interaction study of *Bemisia tabaci* Q-biotype on commercial tomato varieties with and without the *Mi* resistance gene, and comparative host responses with the B-biotye. Entomol Exp Appl 98:339–344

Odum HT (1971) Environment power and society. Wiley, New York

Oliveira MRV, Henneberry TJ, Anderson P (2001) History, current status, and collaborative research projects for *Bemisia tabaci*. Crop Prot 20:709–723

Otim-Nape GW, Thresh JM, Bua A, Baguma Y, Shaw MW (1998) Temporal spread of cassava mosaic virus disease in a range of cassava cultivars in different agro-ecological regions of Uganda. Ann Appl Biol 133:415–430

Pascual S (2006) Mechanisms in competition, under laboratory conditions, between Spanish biotypes B and Q of *Bemisia tabaci* Gennadius. Span J Agric Res 44:351–354

Pascual S, Callejas C (2004) Intra- and interspecific competition between biotypes B and Q of *Bemisia tabaci* Hemiptera, Aleyrodidae from Spain. Bull Entomol Res 94:369–375

Perring TM (1996) Biological differences of two species of Bemisia that contribute to adaptive advantage. In: Gerling D, Mayer D (eds.) Bemisia, 1995. Taxonomy, biology, damage, control and management. Intercept Ltd, Andover

Perring TM (2001) The *Bemisia tabaci* species complex. Crop Prot 20:725–737

Perring TM, Symmes EJ (2006) Courtship behavior of *Bemisia argentifolii* Hemiptera, Aleyrodidae and whitefly mate recognition. Ann Entomol Soc Am 99:598–606

Perring TM, Farrar CA, Cooper AD (1994) Mating behaviour and competitive displacement in whiteflies. In: Henneberry TJ, Toscano NC, Faust RM, Coppedge JR (eds) Supplement to the five-year national research and action plan. USDA-ARS, ARS-125, Washington, DC

Pollock DD, Bruno WJ (2000) Assessing an unknown evolutionary process: effect of increasing site-specific knowledge through taxon addition. Mol Biol Evol 17:1854–1858

Pollock DD, Zwickl DJ, McGuire JA, Hillis DM (2002) Increased taxon sampling is advantageous for phylogenetic inference. Syst Biol 51:664–671

Rauch N, Nauen R (2003) Identification of biochemical markers linked to neonicotinoid cross resistance in *Bemisia tabaci* Hemiptera, Aleyrodidae National Meeting of the Entomological Society of America, Symposium – Biorational Insecticides – mechanism and application, Fort Lauderdale, Florida, USA, November 2002. Arch Insect Biochem 54:165–176

Reitz SR, Trumble JT (2002) Competitive displacement among insects and Arachnids. Annu Rev Entomol 47:435–465

Rekha AR, Maruthi MN, Muniyappa V, Colvin J (2005) Occurrence of three genotypic clusters of *Bemisia tabaci* and the rapid spread of the B biotype in south India. Entomol Exp Appl 117:221–233

Ronda MA, Adan A, Cifuentes D, Cenis JL, Beitia F (1999) Laboratory evidence of interbreeding between biotypes of *Bemisia tabaci* Homoptera, Aleyrodidae present in Spain. In: V11th international plant virus epidemiology symposium – plant virus epidemiology, current status and future prospects 1999, Aguadulce, pp 83–84

Ruan YM, Luan JB, Zang LS, Liu SS (2007) Observing and recording copulation events of whiteflies on plants using a video camera. Entomol Exp Appl 124:229–233

Shatters RG Jr, Powell CA, Boykin LM, He LS, McKenzie CL (2009) Improved DNA barcoding method for *Bemisia tabaci* and related Aleyrodidae: development of universal and *Bemsia tabaci* biotype-specific mitochondrial cytochrome c oxidase chain reaction primers. J Econ Entomol 102:750–758

Simon C, Frati F, Becknenbach A, Crespi B, Liu H, Flook P (1994) Evolution, weighting and phylogenetic utility of mitochondrial gene sequences and a compilation of conserved polymerase chain reaction primers. Ann Entomol Soc Am 87:651–701

Simon B, Moriones E, Soria C, Beitia F, Bosco D, Cenis JL (1999) Variación genética de poblaciones de *Bemisia tabaci* Gennadius en la cuenca del Mediterráneo occidental. In: Resúmenes del Congreso Nacional de Entomología Aplicada VII Jornadas Científicas de la Sociedad Española de Entomología Aplicada, Junta de Andalucía, Consejería de Agricultura y Pesca, Aguadulce, 8–12 Nov 1999, p 20

Sseruwagi P, Maruthi MN, Colvin J, Rey MEC, Brown JK, Legg JP (2006) Colonization of non-cassava plant species by cassava whiteflies *Bemisia tabaci* in Uganda. Entomol Exp Appl 119:145–153

Tsagkarakou A, Roditakis N (2003) Isolation and characterization of microsatellite loci in Bemisia tabaci (Hemiptera: Aleyrodidae). Mol Ecol Notes 3:196–198

Tsagkarakou A, Tsigenopoulous CS, Gorman K, Lagnel J, Bedford ID (2007) Biotype status and genetic polymorphism of the whitefly Bemisia tabaci (Hemiptera: Aleyrodidae) in Greece: mitochondrial DNA and microsatellites. Bull Entomol Res 97:29–40

Vos JGM (1994) Integrated crop management of hot pepper *Capsicum* spp in tropical lowlands. PhD dissertation, Wageningen Agricultural University

Vos JGM, Duriat AS (1995) Hot pepper *Capsicum* spp production on Java, Indonesia, toward integrated crop management. Crop Prot 14:205–213

Whitfield CW, Behura SK, Berlocher SH, Clark AG, Johnston JS, Sheppard WS, Smith DR, Suarez AV, Weaver D, Tsutsui ND (2006) Thrice out of Africa, ancient and recent expansions of the honey bee, *Apis mellifera*. Science 314:642–645

Wu X, Li Z, Hu D, Shen Z (2003) Identification of Chinese populations of *Bemisia tabaci* Gennadius by analyzing ITS1 sequence. Prog Nat Sci 13:276–281

Xu J, De Barro PJ, Liu SS (2010) Reproductive incompatibility among genetic groups of *Bemisia tabaci* supports the proposition that the whitefly is a cryptic species complex. Bull Entomol Res 100:359–366. doi:10.1017/S0007485310000015

Zang LS, Liu SS (2007) A comparative study on mating behaviour between the B biotype and a non-B biotype of *Bemisia tabaci* (Hemiptera, Aleyrodidae) from Zhejiang, China. J Insect Behav 20:157–171

Zang LS, Liu SS, Liu YQ, Ruan YM, Wan FH (2005) Competition between the B biotype and a non-B biotype of the whitefly, *Bemisia tabaci*, (Homopterea, Aleyrodidae) in Zhejiang, China. Biodivers Sci 13:181–187

Zang LS, Chen WQ, Liu SS (2006) Comparison of performance on different host plants between the B biotype and a non-B biotype of *Bemisia tabaci* from Zhejiang, China. Entomol Exp Appl 121:221–227

Zhang XS, Holt J, Colvin J (2000) A general model of plant-virus disease infection incorporating vector aggregation. Plant Pathol 49:435–444

Zwickl DJ, Hillis DM (2002) Increased taxon sampling greatly reduces phylogenetic error. Syst Biol 51:588–598

Chapter 10
Global Emergence and Spread of Whitefly (*Bemisia tabaci*) Transmitted Geminiviruses

Anupam Varma, Bikash Mandal, and Manoj K. Singh

Abstract Whitefly-transmitted geminiviruses (WTGs) have emerged as the most destructive pathogens, particularly in the tropics and subtropics. The epidemics caused by the emerging WTGs have spread even in the regions that were free from these viruses earlier. The most seriously affected crops include cassava, cotton, grain legumes, and cucurbitaceous, malvaceous and solanaceous vegetables. Cassava mosaic disease pandemic in East Africa in the early 1990s was caused by the emergence of a highly virulent recombinant WTG, having sequences from African and East African cassava mosaic viruses. Since then, the recombinant WTGs, aided by sharp increase in whitefly population, have spread over large cassava growing areas of Africa, leading to acute food shortages in the affected region. Cotton leaf curl disease has been endemic in the Sudan for a long time. In the 1990s it caused severe epidemics in the north-western region of the Indian subcontinent resulting in enormous economic losses. The expanding host range of WTGs infecting legumes, along with their spread to new geographical regions, limits production of this important group of crops. The most dramatic emergence of WTGs affecting tomato has been in Asia and the Americas during the last two decades.

The major contributory factors for the emergence and spread of new WTGs are (a) evolution of variants of WTGs through mutations, recombination and pseudo-recombination, (b) acquisition of satellite-like DNA molecules, (c) appearance of aggressive biotypes and increase in populations of *Bemisia tabaci*, (d) changes in the cropping systems, (e) introduction of new crops, (f) introduction of host susceptibility genes, and (g) the movement of infected planting materials. In addition, favourable climatic changes and human activity have also played an important role in the emergence of serious WTG associated diseases across the globe.

A. Varma (✉) • B. Mandal • M.K. Singh
Advanced Centre for Plant Virology, Indian Agricultural Research Institute,
New Delhi 110 012, India
e-mail: av.acpv@gmail.com; anupamvarma@vsnl.net; leafcurl@rediffmail.com

W.M.O. Thompson (ed.), *The Whitefly, Bemisia tabaci (Homoptera: Aleyrodidae) Interaction with Geminivirus-Infected Host Plants*,
DOI 10.1007/978-94-007-1524-0_10, © Springer Science+Business Media B.V. 2011

This chapter highlights the emergence and spread of WTGs in selected crops threatened by this large group of plant viruses.

Keywords Whitefly-transmitted viruses • WTGs • Begomoviruses • *Bemisia tabaci* • Whitefly • Emerging viruses • Leaf curl • Golden mosaic • Yellow mosaic • Yellow vein mosaic • Plant diseases

10.1 Introduction

Whitefly (*Bemisia tabaci*) transmitted geminiviruses (WTGs), belonging to genus *Begomovirus,* Family *Geminiviridae,* have emerged as the most threatening group of plant viruses globally during the post-Green Revolution period. The diseases caused by WTGs, have been known for a long time, but during the last three to four decades these diseases have assumed devastating proportions due to the emergence and spread of new WTGs and their variants, causing severe disease epidemics, particularly in the tropics and semi-tropics. In recent years, some WTGs have also spread to temperate regions causing serious diseases of horticultural crops grown under protected conditions. The frequency with which new WTGs are appearing shows that these viruses are still evolving and pose a serious threat to sustainable agriculture. During the last two decades severe disease epidemics, caused by newly emerged WTGs, have threatened crops like cassava, cotton, cucurbits, grain legumes, potato, malvaceous vegetables and tomato. Economic losses due to WTGs are estimated to be US $1,300–2,300 million in cassava in Africa (Thresh et al. 1998), US $5 billion in cotton in Pakistan during 1992–1997 (Briddon and Markham 2000), US $300 million per annum in grain legumes in India (Varma et al. 1992) and US $140 million in tomato alone in Florida, USA (Moffat 1999). In Brazil, since 1972, bean production has been severely reduced by the golden mosaic disease caused by a WTG, which results in annual reduction in yields in the range of 90,000–280,000 tons (Aragão and Faria 2009). Despite the efforts to contain WTGs and *B. tabaci*, menacing disease epidemics caused by newly emerging or re-emerging WTGs are becoming frequent and spreading even in the regions, which were previously free from such diseases.

WTGs have unique twin particle morphology and their genome is either monopartite or bipartite ssDNA (van Regenmortel et al. 2000). A majority of the WTGs have bipartite genomes designated as DNA-A and DNA-B and infect dicotyledonous plants. Two virion-sense and four complementary- sense ORFs are located in DNA-A, whereas DNA-B has one virion-sense and one complementary-sense ORF. The two genomic components share a common region containing the origin of replication and regulatory regions for bi-directional transcription. So far, the ORF AV2 of DNA-A has been found only in the WTGs distributed in the Old World and not in the New World (Harrison and Robinson 1999). It is suggested that this ORF could have made DNA-B of some of the Old World WTGs redundant, resulting in the evolution of monopartite WTGs, which spread singly or in combination with satellite-like betasatellite and alphasatellite molecules (Briddon et al. 2003, 2004).

The diseases caused by WTGs are easily recognized by their distinctive symptoms. Broadly, the symptoms induced by WTGs are of three types: (a) vein yellowing, (b) yellow mosaic, and (c) leaf curl. In recent years, there has been a tremendous increase in the number of WTGs spreading globally. In the early 1960s about 27 WTGs were known (Varma 1963), the number increased to more than 100 by the end of the last century (van Regenmortel et al. 2000), by 2006 the number further increased to more than 300 (Fauquet et al. 2008), and since then many more distinct WTGs spreading in economically important crops have been detected. In this chapter, the major plant disease problems associated with WTGs and underlying factors leading to their emergence and spread are discussed.

10.2 WTGs Associated with Cassava Mosaic Disease

Cassava (*Manihot esculenta*), an important food crop for millions of people living in the tropics, is severely affected by cassava mosaic disease (CMD) induced by WTGs. The affected plants develop mosaic, mild to severe leaf distortion, short internodes and the yield of tubers is considerably reduced (Fig. 10.1). CMD, a major constraint in cassava production, is an important example of emergence of a disease, due to the introduction of a crop susceptible to the indigenous viruses in the new region. Cassava was introduced into Africa in the sixteenth century and South-East Asia in the eighteenth century from the Americas; later cassava was also brought to India from Africa in late nineteenth century. CMD was first reported in East Africa in 1894 and since then it has spread to all the cassava growing areas of the continent (Legg 1999). In India, CMD was recognized as a major threat to the cultivation of cassava in the early 1940s (Abraham 1956). The crop, however, is not affected by CMD in the Americas and South-East Asia (Fig. 10.2), showing its susceptibility to the WTGs occurring in Africa, India and Sri Lanka but not to those spreading in the other cassava growing areas.

Nine distinct bipartite WTGs and a large number of their strains and isolates are found associated with CMD in Africa and the Indian sub-continent (Table 10.1). A majority of these WTGs have come to light during the last decade. Seven WTGs, ACMV, EACMCV, EACMKV, EACMMV, EACMV, EACMZV and SACMV,[1] are spreading in Africa and two WTGs, ICMV and SLCMV occur in the Indian sub-continent. Recently, ICMV has also moved to West Africa (Adjata et al. 2008), apparently with the vegetative planting material. Phylogenetically, the WTGs associated with CMD form distinct clusters, irrespective of the geographic origin of their isolates (Ndunguru et al. 2005) suggesting independent origin of the WTGs associated with CMD. The primary spread of WTGs is by whitefly transmission, but their inter- and intra-continental dissemination could result from inadvertent exchange of infected planting materials, as symptomless infection of cassava is common (Malathi et al. 1989).

[1]Full names of the viruses are listed in the connected tables.

Fig. 10.1 Cassava mosaic disease (CMD) caused by *Indian cassava mosaic virus* (ICMV). (**a**) Severe CMD symptoms; (**b** and **c**) Severe leaf distortion; (**d**) A commercial cassava crop affected by CMD; (**e**) Differential response of susceptible and resistant cassava varieties to ICMV (Pictures courtesy Dr. T. Makesh kumar, CTCRI, Kerala, India)

Fig. 10.2 Distribution of WTGs associated with cassava mosaic disease (CMD) in different parts of the world. Cassava is also grown extensively in the encircled areas, but these areas are CMD-free

10.2.1 Emergence and Spread of WTGs in Cassava in Africa

The epidemics of CMD have been frequent in Africa. A severe epidemic devastated cassava crops in eastern Uganda in 1933–1944. The use of resistant varieties and roguing of infected plants reduced CMD incidence, but the disease re-emerged in severe epidemic form in the 1980s. In 1990–1992, the overall incidence of CMD was greater in the northern region (78%) than in the southern region (40%). The high incidence in the northern region was considered to be due to favourable environmental conditions supporting high populations of *B. tabaci*. Gradually, the WTGs associated with CMD have spread across Uganda and moved into western Kenya along a broad front, at the rate of 20–30 km/year, supported by the large populations of whiteflies and resulting in a high incidence of CMD in Kenya. The severely affected fields produced no or poor yields, causing huge economic losses, acute food shortages and famine in some areas (Otim-Nape et al. 1998; Legg 1999; Rybicki and Pietersen 1999). The cassava WTGs have steadily spread to all the major cassava growing regions of Africa (Fargette et al. 2006). Apart from Uganda and Kenya, cassava WTGs have also emerged as a constraint in cassava production in to Angola, Burkina Faso, Cameroon, Democratic Republic of Congo, Ghana, Guinea, Ivory Coast, Nigeria, Sudan, Tanzania, Togo, Madagaskar, Malawi, Rwanda, South Africa and Zimbabwe (Table 10.1; Fig. 10.2). Of the eight cassava WTGs reported to occur in Africa, ACMV and EACMV strains Ke and UG are most widespread (Table 10.1). The CMD pandemic in Uganda and other areas in the region have resulted from the emergence of highly virulent recombinants and pseudo-recombinants between EACMV and ACMV (Harrison et al. 1997b; Zhou et al. 1997). No sequence of ACMV has been found to have a recombinant fragment, but an isolate of ACMV was the donor of core coat protein (*CP*) gene sequences resulting in the evolution of EACMV-UG2, which was associated with the CMD epidemic in Uganda in the 1990s. Unlike ACMV, EACMV-like viruses show multiple putative recombinations resulting in the evolution of EACM strains and SACMV (Ndunguru et al. 2005; Fargette et al. 2006). The appearance of EACMV-UG in Rwanda appears to have resulted from the introduction of germplasm from Uganda (Legg 1999). Similar spread might also have occurred in other areas not connected with the pandemic. Mixed infection by ACMV and the EACMV variants results in severe disease symptoms. Major damage, however, appears to be caused by pseudo-recombinants between different isolates of EACMV (Pita et al. 2001). The other factors leading to the pandemic seem to be increase in whitefly populations and dual infections by ACMV and EACMV-UG. The plants showing severe symptoms not only had higher concentration of virus but also supported high populations of whiteflies. CMD affected plants seem to provide an advantage for the buildup of whitefly populations. The efficiency of transmission of ACMV by whitefly from plants with dual infections was much less than the transmission of EACMV-UG, which apparently resulted in significant increases in the percentages of plants infected with EACMV-UG (Colvin et al. 2004).

Table 10.1 WTGs naturally occurring in cassava (Source: GenBank, Fauquet et al. (2008) and other references given in the table)

S.N.	Acronym	Virus	Country	Host	Location	Isolates	Year
1	ACMV	*African cassava mosaic virus*	Cameroon	Cassava	–	5	1998
			Co^te d'Ivoire	Cassava	–	1	1990
			Kenya	Cassava	–	1	1992
			Nigeria	Cassava	–	2	1990
			Nigeria	Soybean	–	1	2008 (Mgbechi-Ezeri et al. 2008)
			Nigeria	*Senna occidentalis*	–	1	2006 (Ogbe 2006)
			Tanzania	Cassava	–	1	2001
			Uganda	Cassava	–	1(Severe)	1997
				Cassava	–	1(Mild)	1997
2	EACMCV	*East African cassava mosaic Cameroon virus*	Cameroon	Cassava	–	2	1998
			Co^te d'Ivoire	Cassava	–	1	1998
			Nigeria	Cassava	–	1	–
			Tanzania	Cassava	–	2	2001
3	EACMKV	*East African cassava mosaic Kenya virus*	Kenya	Cassava	Kathiana	2	2002
				Cassava	Kehancha	3	2002
				Cassava	Matuu	3	2002
				Cassava	Migori	2	2002
				Cassava	Mitaboni	1	2002
				Cassava	Tala	3	2002
4	EACMMV	*East African cassava mosaic Malawi virus*	Malawi	Cassava	–	2	1996

5	EACMV	*East African cassava mosaic virus*	Nigeria	*Combretum confertum*	–	1	2006 (Ogbe 2006)
5.1	EACMV-KE	East African cassava mosaic virus-Kenya	Angola	Cassava	–	1	–
			Cameroon	Cassava	–	1	–
			Ghana	Cassava	–	1	–
			Guinea	Cassava	–	1	–
			Kenya	Cassava	Boa	2	2001
				Cassava	Boundary	1	2001
				Cassava	–	1	1996
				Cassava	Katumani	1	2001
				Cassava	Kibaoni	1	2001
				Cassava	Kinyumbini	1	2001
				Cassava	Kitui	2	2002
				Cassava	Kwale	1	2002
				Cassava	Mamba	1	2002
				Cassava	Migori	1	2002
				Cassava	Migwani	3	2002
				Cassava	Misakwakwani	2	2001
				Cassava	Msabaha	1	2001
				Cassava	Msambweni	2	2002
				Cassava	Mwezangombe	1	2001
				Cassava	Perani	1	2001
				Cassava	Sharian	1	2001
				Cassava	Shirachi	1	2001
				Cassava	Shirachi	1	2001
			Malawi	Cassava	–	1	–
			Madagasker	Cassava	–	1	–
			Nigeria	Cassava	–	1	–

(continued)

Table 10.1 (continued)

S.N.	Acronym	Virus	Country	Host	Location	Isolates	Year
			Tanzania	Cassava	–	1	1996
			Togo	Cassava	–	1	–
			Uganda	Cassava	–	1	1997
			Zimbabwe	Cassava	–	1	–
5.2	EACMV-TZ	East African cassava mosaic virus-Tanzania	Tanzania	Cassava	–	1	–
5.3	EACMV-UG	East African cassava mosaic virus-Uganda	Angola	Cassava	–	1	2009 (Kumar et al. 2009a, b)
			Burkina Faso	Cassava	–	1	–
			Democratic republic of Congo	Cassava	–	1	–
			Kenya	Cassava	Bungoma	3	2002
				Cassava	Busia	4	2002
				Cassava	Funyula	1	2002
				Cassava	Katumani	1	2001
				Cassava	Malaba	1	2002
				Cassava	Migori	2	2002
				Cassava	Mumias	2	2002
				Cassava	Sega	2	2002
				Cassava	Ugunja	2	2002
				Cassava	Wote	2	2002
			Sudan	Cassava	–	1	–
			Tanzania	Cassava	–	1	–
			Uganda	Cassava	–	2 (mild)	1997
					–	2 (severe)	1997
					–	1	1996

6	EACMZV	*East African cassava mosaic Zanzibar virus*	Kenya	Cassava	Felunzi	1	2001
				Cassava	Kasumalini	1	2001
				Cassava	Kibwezi	3	2002
				Cassava	Kilifi	1	1999
				Cassava	Kwakadzengo	1	2001
				Cassava	Kwamugomba	1	2001
				Cassava	Machakos	1	2002
				Cassava	Malindi	1	2001
				Cassava	Msambweni	1	2002
				Cassava	Vipingo	1	2001
			Tanzania	Cassava	Uguja	1	1998
7	ICMV	*Indian cassava mosaic virus*	India	Bitter gourd	Tamilnadu	1	2007 (Rajinimala and Rabindran 2007)
				Cassava	Kerala	3	2002
				Cassava	Maharashstra	2	1998
				Cassava	Trivandrum	1	1986
				Jatropha	Dharwad	1	2010 (Gao et al. 2010)
				Jatropha	Lucknow	1	2008 (Raj et al. 2008a, b)
			Togo	Cassava	–	1	2008 (Adjata et al. 2008)
8	SACMV	*South African cassava mosaic virus*	Madagascar	Cassava	–	1	
			South Africa	Cassava	–	1	–
			Zimbabwe	Cassava	–	1	–

(continued)

Table 10.1 (continued)

S.N.	Acronym	Virus	Country	Host	Location	Isolates	Year
9	SLCMV	*Sri Lankan cassava mosaic virus*	India	Cassava	Adivaram	1	–
				Cassava	Kerala	3	–
				Cassava	Kattukuda	1	–
				Cassava	Muvattupucha	1	2004
				Cassava	Salem	1	–
				Cassava	Tamilnadu	3	–
			Sri Lanka	Cassava	Colombo	1	1998

In western Kenya also increased incidence and severity of CMD in the main intensive cassava growing area has been attributed to the spread of EACMV-UG and ACMV causing single or mixed infections. In this region EACMV-UG2 is the main driver of the epidemic as it was detected in 69% of the CMD affected plants. However, in the coastal area of Kenya, where cassava fields are scattered and isolated, EACMV is endemic but CMD is not a serious concern (Were et al. 2003). In Angola, mixed infection by ACMV, EACMV and EACMV-UG is common, but ACMV is found to be most predominant, being associated with nearly 85% of the CMD affected plants, followed by EACMV (63%) and EACMV-UG (18.5%). In the northern area, however, EACMV-UG was most common (Kumar et al. 2009a, b).

In contrast to the devastating pandemic of CMD in East Africa, the disease has not been threatening in West Africa due to limited opportunity for developing virulent recombinants of the causal viruses as only ACMV was associated with the disease in this region (Swanson and Harrison 1994). However, the emergence of EACMV (Legg 1999) and ICMV (Adjata et al. 2008) in West Africa, and their occurrence singly and in combination has increased the possibility of development of virulent recombinants in this region too. Indeed, the studies have shown that increase in CMD severity in Togo is due to the emergence of recombinants rather than synergy between different viruses (Adjata et al. 2009).

10.2.2 Emergence and Spread of CMD Associated WTGs in Asia

In South-East Asia cassava is grown in China, India, Indonesia, Malaysia, the Philippines, Sri Lanka, Thailand and Vietnam in 3.5 million ha, but CMD occurs only in Southern India and Sri Lanka, which account for <10% of the area under cassava in Asia (Howeler 2005). In India, ICMV has been spreading in the southern region, the main cassava growing area of the country, for a long time. The disease has also been introduced into geographically distant north-eastern region of India through infected planting material, but it has not spread further as *B. tabaci* is not very active in this region (Varma and Malathi 2003). Recently, another CMD-WTG SLCMV has emerged in India. It is more closely related to ICMV than ACMV. Interestingly SLCMV DNA A alone could cause upward leaf roll symptoms in *N. benthamiana*, behaving like a monopartite WTG. It could also produce a viable pseudo-recombinant with ACMV DNA-B but not with ICMV DNA-B. SLCMV DNA-A co-inoculated with ageratum yellow vein disease betasatellite produced typical yellow vein symptoms in *Ageratum conyzoides* (Saunders et al. 2001a, b). This led to the speculation that betasatellites could get associated with the cassava WTGs in the Indian Sub-continent (Varma and Malathi 2003), but so far no betasatellite has been detected with CMD in India (Makesh Kumar, personal communication), although CMD associated WTGs can trans-replicate DNA satellite molecules (Patil and Fauquet 2010). The interaction between CMD-WTGs and different DNA satellite molecules is shown to result in variation in symptoms in the test plants (Patil and Fauquet 2010).

Until recently, natural occurrence of CMD-WTGs was found only in cassava. However, the picture is changing as ACMV has been detected in some leguminous species (Ogbe 2006; Mgbechi-Ezeri et al. 2008), and ICMV seems to be widespread in Jatropha in India (Raj et al. 2008a, b; Gao et al. 2010) and a cucurbit (Rajinimala and Rabindran 2007) (Table 10.1). Natural hosts, other than cassava, of CMD-WTGs or their ancestors are expected both in Africa and Asia as these viruses must be indigenous in these regions before they moved to cassava. So far, South-East Asia, where cassava is being grown in large areas for food, starch and animal feed, is free from CMD.

10.3 WTGs Spreading in Cucurbits

A variety of cucurbits are grown across the world. Cultivation of cucurbits is affected by a large number of viruses, including 21 WTGs (Table 10.2). In the Indian sub-continent, yellow vein mosaic of pumpkin (*Cucurbita pepo)*, caused by a WTG, has been known for over 60 years (Varma 1963). The disease was mainly confined to central-western India. In 1990, the WTGs causing diseases in cucurbits spread in epidemic proportions in northern India, causing diseases like leaf curl in muskmelon (*Cucumis melo*), yellow vein mosaic of cucumber (*Cucumis sativus*) and yellow vein mosaic in pumpkin (Varma 1990; Raj and Singh 1996). The spread of WTGs in 1990 in cucurbits coincided with sudden increase in whitefly populations early in the cucurbit growing season. Since then, the diseases (Fig. 10.3) caused by WTGs have emerged as a major constraint in the production of cucurbits in northern India (Varma and Giri 1998). In 2001 over 50% of the commercial crops of pumpkin, muskmelon, watermelon (*Citrullus vulgaris*) and bottlegourd (*Lagenaria siceraria*) were severely affected by WTGs in northern India (Varma and Malathi 2003). Another cucurbit, sponge gourd (*Luffa aegytiaca*), is grown throughout India as a common vegetable. Its cultivation is adversely affected by high incidence (>90%) of leaf distortion mosaic disease, caused by *Tomato leaf curl New Delhi virus* (ToLCNDV) (Sohrab et al. 2003). Since then, ToLCNDV has also spread to bitter gourd (*Momordica charantia*), bottle gourd, cucumber, ivy gourd (*Coccinia* sp.), long melon (*Cucumis melo* cv. Utilissimus), pumpkin, ridge gourd (*Luffa acutangula*) and watermelon in Northern India and chayote (*Sechium edule*) in North-Western India (Sohrab et al. 2006, 2010; Mandal et al. 2004; Tiwari et al. 2010a, b). Similar spread of ToLCNDV has also occurred in bottlegourd, cucumber and muskmelon in Thailand (Ito et al. 2008). A new ToLCNDV isolate has been found associated with a severe disease of oriental melon in Taiwan (Chang et al. 2010). Thus, ToLCNDV appears to emerge as the most serious constraint to cucurbit production in Asia. Resistance to ToLCNDV in sponge gourd is shown to be controlled by a single dominant gene (Islam et al. 2010).

Squash leaf curl China virus (SLCCNV), which emerged in China in the 1990s (Hong et al. 1995), is another WTG, which is fast spreading in the Indian

Table 10.2 WTGs associated with the diseases of cucurbits (Source: GenBank, Fauquet et al. (2008) and other references given in the table)

S.N.	Acronym	Virus	Country	Host	Location	Isolates	Year
1	BGYVV	Bitter gourd yellow vein virus[a]	Pakistan	Bitter gourd		1	2010 (Tahir et al. 2010a)
2	BGYBV	*Bitter gourd yellow blotch virus*[b]		–	–	1	–
3	ChaYMV	*Chayote yellow mosaic virus*	Nigeria	Chayote	Ibadan	1	–
4	CuLCrV	*Cucurbit leaf crumple virus*	USA	Squash/Melon	Arizona	1	1991
			USA	Squash/Melon	California	1	1998
			Mexico				
5	LYMV	*Luffa yellow mosaic virus*	Vietnam	Luffa	–	1	–
6	MCLCuV	*Melon chlorotic leaf curl virus*	Costa Rica	Melon	Guanacaste	1	1998
			Guatemala	Melon	–	1	2000
7	MCMV	*Melon chlorotic mosaic virus*[b]		–	–	1	–
8	MLCuV	*Melon leaf curl virus*[b]		–	–	1	–
9	PepLCBDV	*Pepper leaf curl Bangladesh virus*	India	Bitter gourd	–	1	2010 (Raj et al. 2010)
10	PYMV	*Pumpkin yellow mosaic virus*	India	Pumpkin			2003 (Maruthi et al. 2003)
11	SLCCNV	*Squash leaf curl China virus*	China	Squash	Guangxi	1	2002
			China	Squash	Guangxi	1	2005
			China	Squash	Hainan	1	2005
			India	Pumpkin	Coimbatore	1	–
			India	Pumpkin	Lucknow	1	2008 (Singh et al. 2008)
			India	Pumpkin	Varansi		2009 (Singh et al. 2009)
			Mexico	Watermelon			

(continued)

Table 10.2 (continued)

S.N.	Acronym	Virus	Country	Host	Location	Isolates	Year
			Pakistan	Squash	Lahore	1	2004
							2010 (Tahir et al. 2010b)
			Vietnam:B	Squash	–	1	–
			Vietnam:K	Squash	–	1	–
12	SLCChiV	*Squash leaf curl Chinese virus*[b]		–	–	1	–
13	SLCIV	*Squash leaf curl Israel virus*[b]		–	–	1	–
14	SLCPHV	*Squash leaf curl Philippines virus*	Philippines	Squash	–	1	–
			Taiwan	Pumpkin	–	1	2005
15	SLCV	*Squash leaf curl virus*	USA	Squash	Imperial Valley	1	1979
				Cucumber	–	1	–
				Melon	–	1	–
			Egypt		–	1	–
			Jordan		–	1	–
			Palestinian			1	– (Ali-Shtayeh et al. 2010)
16	SLCYNV	*Squash leaf curl Yunnan virus*	China	Squash	Yunnan	1	2000
17	SMLCV	*Squash mild leaf curl virus*	USA	Squash	Imperial Valley	1	1979
			Mexico				
18	SYMoV	*Squash yellow mottle virus*[b]			–	1	–
19	ToLCNDV	*Tomato leaf curl New Delhi virus*	India	Bitter gourd	–	1	2010 (Tiwari et al. 2010a, b)
				Bitter gourd	Pnipat,		2006 (Sohrab et al 2006)
				Bottle gourd	Panipat		2006 (Sohrab et al 2006)
				Chayote	Kalimpong		2004 (Mandal et al 2004)

				Cucumber	Panipat		2006 (Sohrab et al 2006)
				Ivy gourd	Panipat		2006 (Sohrab et al 2006)
				Luffa	New Delhi		2003
				Pumpkin	New Delhi		2006 (Sohrab et al 2006)
				Ridge gourd	Sriganganagar		2006 (Sohrab et al 2006)
				Watermelon	Sirsa		2006 (Sohrab et al 2006)
			Pakistan	Bitter gourd		1	2007 (Tahir and Haider 2007)
				Luffa		1	2004
			Taiwan	Oriental melon			2010 (Chang et al 2010)
			Thailand	Luffa	–	1	
				Bottle gourd	–	1	2008 (Ito et al. 2008)
				Cucumber	–	1	2008 (Ito et al. 2008)
				Muskmelon	–		2008 (Ito et al. 2008)
20	WmCSV	*Watermelon chlorotic stunt virus*	Iran	Watermelon	–	1	1997
			Sudan	–	–	1	–
			Yemen	–	–	1	–
21	WmCMV	*Watermelon curly mottle virus*[b]		–	–	1	–

[a]Proposed new species
[b]Unassigned isolates (Fauquet et al. 2008)

Fig. 10.3 WTG-induced diseases of cucurbits. (**a**) Yellow vein and (**b**) severe leaf curl disease of pumpkin; (**c**) Leaf distortion disease of Luffa caused by ToLCNDV

sub-continent and South-East Asia in cucurbits (Table 10.2; Fig. 10.4). In India, yellow vein mosaic disease of pumpkin (Singh et al. 2009), and in China, Pakistan and Vietnam yellow vein mosaic disease of squash is caused by SLCCNV (Tahir et al. 2010b). The disease caused by SLCCNV in pumpkin results in up to 90% loss in yield (Singh et al. 2009). Yellow vein disease of wild bittergourds (*Momordica charantia*) is common in West Africa (Varma 1984), but it does not seem to have spread in commercial cucurbits. A similar disease in Pakistan is caused by Bitter gourd yellow vein virus (BGYVV) (Tahir et al. 2010a) and Tomato leaf curl Palampur virus (ToLCPalV) (Ali et al. 2010). ToLCPaLV has spread fast in North-Eastern region of the Indian sub-continent. In 2007, it caused severe disease of bittergroud. The affected plants developed leaf chlorosis,

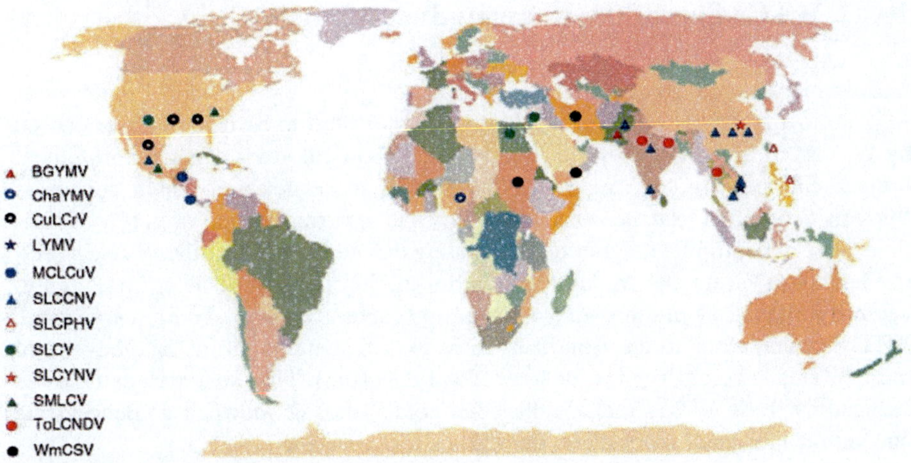

Fig. 10.4 Distribution of WTGs associated with the diseases of cucurbits

crumpling and vein thickening (Ali et al. 2010). In India, bitter gourd cultivation is affected by *Pepper leaf curl Bangladesh virus* (Raj et al. 2010) and ToLCNDV (Tiwari et al. 2010a, b).

The emergence of cucurbit diseases caused by WTGs in northern India in the 1990s coincided with the emergence of similar diseases in the middle east, although in Israel, the yellow vein mosaic disease of cucumber was reported 50 years ago (Cohen and Nitzany 1960). Three different cucurbit infecting WTGs were first observed in the Yemen in 1993 (Bedford et al. 1994). Since then severe outbreaks of the diseases caused by WTGs in melon and watermelon have occurred in central and eastern Sudan and southern Iran (Kheyr-Pour et al. 2000). In 2008 and 2009, high (24 to 100%) incidence of an isolate of *Squash leaf curl virus* (SLCV), closely related (sharing 96 to 98% sequence identity) to the isolates prevalent in Jordan, Egypt and the US, was observed affecting cucumber and melon in the Palestine (Ali-Shtayeh et al. 2010). Several WTGs, *Cucurbit leaf crumple virus* (CuLCrV), SLCV and *Squash mild leaf curl virus* (SMLCV) are known to cause serious diseases in squash and melon in the USA and Mexico for over 30 years (Idris et al. 2008). Although the B biotype of *B. tabaci* is a serious problem of cucurbits in the New World, it has not resulted in the emergence of new viruses or an increase in the severity of SLCV, indicating a lack of diversity in the WTGs infecting cucurbits in the New World, compared to those spreading in cucurbits in the Old World. However, the situation may change as in the beginning of this century a highly virulent WTG, *Melon chlorotic leaf curl virus* (MCLCuV), was found spreading in cantaloupe, melon and watermelon in Guatemala. MCLCuV is shown to have high potential for inter-specific re-assortment of its genomic components leading to expansion of its natural host range (Idris et al 2008). Worldwide distribution of WTGs affecting cucurbits is shown (Fig. 10.4).

10.4 WTGs Spreading in Leguminous Crops

Yellow mosaic and golden mosaic diseases of grain legumes and Leguminous vegetables are serious problems in the tropics and subtropics. The diseases caused by WTGs in grain legumes were first recorded about 50 years ago. The prominent bright yellow and golden mosaic symptoms of these diseases which appear as 'blooms', could not have been missed if they had occurred earlier, even if the infections were of negligible magnitude, suggesting the recent origin of the WTGs affecting legumes (Varma and Malathi 2003). So far, 30 distinct WTGs are reported to naturally infect Leguminous plants globally (Table 10.3; Fig. 10.5); 13 of these WTGs are spreading in the Americas, three in Africa and eight in Asia. Several of these WTGs have emerged in this decade (Table 10.3). The most serious diseases induced by these WTGs include the bean golden mosaic, cowpea golden mosaic and yellow mosaic of blackgram, mungbean, and soybean.

10.4.1 Bean Golden Mosaic

Bean golden mosaic disease (BGMD), a major constraint in bean production, in the Americas was first observed in the 1960s in South America (Costa 1965). The affected plants develop brilliant golden mosaic, remain stunted, have prolonged vegetative growth and produce fewer pods of poor quality. The area affected by the disease has continuously increased. Four WTGs, *Bean calico mosaic virus* (BCaMV), *Bean dwarf mosaic virus* (BDMV), *Bean golden mosaic virus* (BGMV) and *Bean golden yellow mosaic virus* (BGYMV), are found to be associated with the disease. BGYMV isolated from Lima bean in Puerto Rico is sap transmissible, and it is more virulent than the non-sap-transmissible BGMV isolate from Brazil. Considerable variability has been found in the isolates of BGYMV, but not in BGMV. The lack of variability in BGMV may be due to the absence of resistant varieties of beans that would create selection pressure for the development of more aggressive isolates (Faria and Maxwell 1999). In Brazil, bean production has been severely reduced due to the spread of BGMV attributed to increase in whitefly populations. The country has developed promising transgenic bean, which may help in recovery of about 180,000 ha rendered unfit for bean cultivation due to the occurrence of BGMV (Francisco et al. 2009). In North America, BGMD first emerged in Florida in 1993, leading to complete destruction of crops in some fields, and considerable reduction in bean production (Blair et al. 1995). The epidemic was apparently caused by the introduction of BGYMV into Florida in the late summer of 1992 by viruliferous whiteflies blown in by hurricane Andrew from the Caribbean basin, and successive plantings of beans provided favorable conditions for the establishment of the virus in the area (Blair et al. 1995). In North America, beans are also infected with a non-legume WTGs. In 2006, beans in a field in Alachua County, Florida, USA were found infected with *Sida golden mosaic virus* (SiGMV). The affected plants developed foliar mottling, puckering and curling symptoms (Durham et al. 2010).

Table 10.3 WTGs associated with the diseases of leguminous plants (Source: GenBank, Fauquet et al. (2008) and other references given in the table)

S.N.	Acronym	Virus	Country	Host	Location	Isolates	Year
1	BCaMV	*Bean calico mosaic virus*	Mexico	Bean	Sonora	1	1986
2	BDMV	*Bean dwarf mosaic virus*	Colombia	Bean	–	1	1987
3	BGMV	*Bean golden mosaic virus*	Brazil	Bean	Campinas	1	1978
				Soybean			2009 (Fernandes et al. 2009)
4	BGYMV	*Bean golden yellow mosaic virus*	Cuba	Bean	–	1	–
			Dominican Republic	Bean	–	2	1987
			Guatemala	Bean	–	1	1987
			Mexico	Bean	–	1	–
			Puerto Rico	Bean	–	1	–
			USA	Bean	–	1	2005
5	BMFlV	*Bean mosaic Florida virus*[a]		–	–	1	–
6	CPGMV	*Cowpea golden mosaic virus*	Nigeria	Cowpea	Nsukka	1	1990
7	DoYMV	Dolichos yellow mosaic virus	Bangladesh	Dolichos	Gazipur	1	
			India	Dolichos	Bangalore	2	2004
					Mysore	1	2004
					New Delhi	1	2000
						1	
8	HgYMV	*Horsegram yellow mosaic virus*	India	Horsegram	Coimbatore	1	2010 (Barnabas et al. 2010)
			Srilanka	Horsegram	–	1	2009 GU323321

(continued)

Table 10.3 (continued)

S.N.	Acronym	Virus	Country	Host	Location	Isolates	Year
9	MYMIV	*Mungbean yellow mosaic India virus*	India	Mungbean	Sriganganagar	1	1996
					Akola	1	–
					Jabalpur	1	–
					Punjab	1	2005
				Cowpea	Anand	1	2005
					Kanpur	1	2005
					New Delhi	3	98,04,05
					Varanasi	1	–
				Soybean	New Delhi	1	1999
				Blackgram	New Delhi	1	1991
				Dolichos	Varanasi	1	–
			Pakistan	Mungbean	–	4	–
				–	Islamabad	1	2000
				Cowpea	–	1	2000
				Moth bean	–		2006 (Qazi et al. 2006)
			Bangladesh	Mungbean	–	1	1998
			Nepal	Mungbean	Lalitpur	1	–
10	MYMV	*Mungbean yellow mosaic virus*	India	Mungbean	Haryana	1	2001
					Namakkal	2	2005
					Vamban	1	2005
				Soybean	Madurai	2	–
					Maharashstra	1	1999
				Vigna	Vamban	4	–
						1	–
			Pakistan	Soybean	Islamabad	2	2000
			Thailand	Mungbean	–	2	–

11	OMoV	Okra mottle virus[b]	Brazil	Soybean	–	1	2009 (Fernandes et al. 2009)
12	RhGMV	*Rhynchosia golden mosaic virus*	Honduras	Rhynchosia	Comayagua	1	1999
			Mexico	Soybean	–	1	2005
13	RhGMHaV	Rhynchosia golden mosaic Havana virus[b]	Cuba	Rhynchosia	–	1	2010 (Fiallo-Olive et al. 2010)
14	RhGMSV	*Rhynchosia golden mosaic Sinaloa virus*	Mexico	Rhynchosia	Sinaloa	1	2005
15	RhMTV	*Rhynchosia minima Trinidad virus*[a]			–	1	–
16	RhMV	*Rhynchosia mosaic virus*[a]			–	1	–
17	RhRGMV	Rhynchosia rugose golden mosaic virus[b]	Cuba	Rhynchosia	–	1	–
18	RYMV	Rhynchosia yellow mosaic virus[b]	Pakistan	Rhynchosia	–	1	2009 (Ilyas et al. 2009)
19	RhYMIV	Rhynchosia yellow mosaic India virus[b]	India	Rhynchosia	Trivendrapurum	1	2010 HM 777508, HM777509
20	RhYMYuV	Rhynchosia yellow mosaic Yucatan virus[b]	Mexico	Rhynchosia	–	1	2010 (Hernández-Zepeda et al. 2010b)
21	SbBMV	*Soybean blistering mosaic virus*	Argentina		–	1	2005
22	SbCLV	*Soybean crinkle leaf virus*	Japan		–	1	

(continued)

Table 10.3 (continued)

S.N.	Acronym	Virus	Country	Host	Location	Isolates	Year
23	SbCBV	Soybean chlorotic blotch virus[b]	Nigeria	Soybean	–	1	
24	SiGMV	*Sida golden mosaic virus*	USA	Bean	–	1	
				Snap bean			2010 (Durham et al.2010)
25	SiMMV	Sida Micrantha mosiac virus[b]	Brazil	Soybean	–	1	2009 (Fernandes et al. 2009)
26	SbMV	*Soybean mosaic virus*	Argentina	–	–	1	
27	SbMMV	Soybean mild mottle virus[b]	Nigeria	Soybean	–	1	2010 (Alabi et al. 2010)
28	TYLCV	*Tomato yellow leaf curl virus*	Spain	Bean	–	1	2005 (Monci et al 2005)
29	TYLCMalV	*Tomato yellow leaf curl Malaga virus*	Portugal	Bean	–	1	2002 (Louro et al. 2002)
30	VBSMV	Velvet bean severe mosaic virus[b]	India	Velvet bean	–	1	2010 (NC_013414)

[a]Unassigned isolates (Fauquet et al. 2008)
[b]Proposed new species

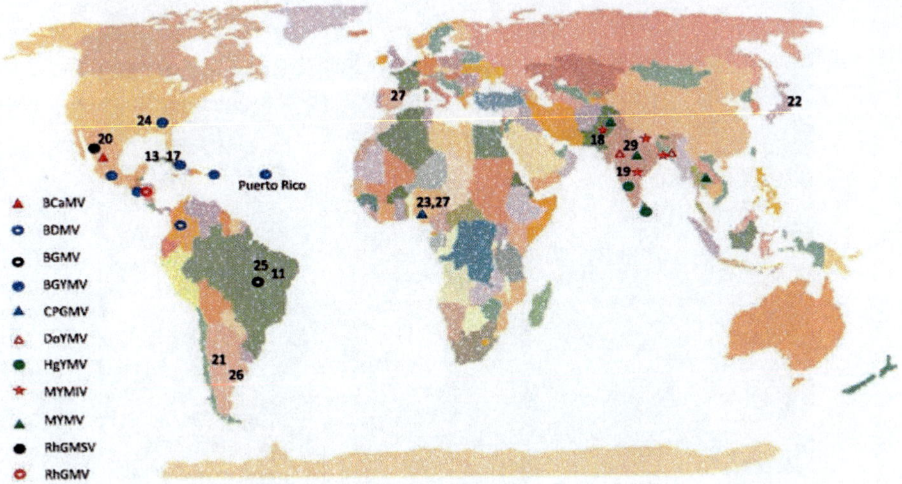

Fig. 10.5 Distribution of WTGs associated with the diseases of Leguminaceous crops. Numbers correspond to the virus names in Table 10.3

10.4.2 Cowpea Golden Mosaic

Cowpea golden mosaic disease (CPGMD), caused by WTGs occurs in Africa, Asia and South America (Hampton and Thottapilly 2003). In Africa CPGMD is caused by *Cowpea golden mosaic virus* CPGMV, which is endemic in the region (Singh and Allen 1979; Winter et al. 1999), and in some regions the incidence of the disease reaches epidemic proportions as was observed in 1984 in the Sudan and the savannah region of northern Nigeria (Rossel and Thottapilly 1985). In India, only a mild yellow fleck disease of cowpea transmitted by whiteflies (Sharma and Varma 1976) was known up to 1978, when CPGMD (Fig. 10.6) was first detected in cowpea germplasm introduced from West Africa. By 1984, CPGMD emerged as a major problem of cowpea cultivation in northern India (Varma and Reddy 1984), but southern India remained relatively free from CPGMD (Swanson et al. 1992). In North-Western India and Bangladesh the disease is caused by a minor variant of *Mungbean yellow mosaic Indian virus* (MYMIV) (Roy and Malathi 2001; Maruthi et al. 2007; John et al. 2008) showing the association of different WTGs with CPGMD in Africa and Asia.

10.4.3 Yellow Mosaic Diseases of Grain Legumes

Yellow mosaic disease (YMD) apparently caused by a WTG was first observed in Lima bean (*Phaselus lunatus*) in the late 1940s in western India (Capoor and Varma 1948) and in mungbean (*Vigna radiata*) in the 1950s in northern India (Nariani 1960).

Fig. 10.6 WTG-induced diseases of legumes. Mungbean yellow mosaic disease (**a**) and Cowpea golden mosaic disease (**b**) caused by MYMIV

Since then YMD has emerged as a major threat to the production of a variety of grain legumes, like Frenchbean (*P. vulgaris*), blackgram (*Vigna mungo*), cluster bean (*Cyamopsis tetragonoloba*), groundnut (*Arachis hypogea*), horsegram (*Macrotyloma uniflorum*), hyacinth bean (*Lablab purpurea*), moth bean (*Vigna aconitifolia),* mungbean, Lima bean, pigeonpea *(Cajanus cajan*) and soybean *(Glycine max)* (Nene 1973; Varma et al. 1992; Varma and Malathi 2003). YMD is also a major constraint in the production of mungbean in Bangladesh, Pakistan and Sri Lanka (Sivanathan 1977; Jalaluddin and Shaikh 1981; Malik 1992). A serious outbreak of YMD in mungbean occurred in northern Thailand in 1977, resulting in a considerable reduction in crop area and a shift in the cropping pattern. Subsequently, the disease incidence became sporadic and did not recur between 1981 and 1991, although mungbean continued to be grown in northern Thailand (Chiemsombat 1992). It is unusual for a disease caused by a WTG to completely disappear from an area in spite of the presence of susceptible plants and the vector. Factors leading to this situation could provide valuable clues for the management of WTGs.

Unlike the situation in Thailand, in India YMD not only persisted but also spread to new crops and regions. In northern India, YMD was a major constraint to the production of grain legumes grown during the wet season (monsoon period, July-October), but not in the preceding dry and hot summer season (April-June). However, in the early 1980s, YMD (Fig. 10.6) emerged in epidemic proportions even in the summer mungbean crop, due to the gradual buildup of *B. tabaci* population and virus inoculum in the weed hosts growing along the irrigation channels, resulting in a decline in summer mungbean production (Varma et al. 1992). Soybean could not be grown in northern India due to severe incidence of YMD infection, but the crop was successful in central India. However, the emergence of YMD in the late 1980s virtually destroyed soybean production in some parts of central India too. The yield losses due to YMD in blackgram, mungbean and soybean were estimated to be about $300 million (Varma et al. 1992). The overall losses due to YMD must be greater as the disease also affects several other grain legumes.

In the Indian sub-continent, YMD of grain legumes is caused by four distinct bipartite WTGs (Table 10.3), two of these WTGs, MYMIV and *Mungbean yellow mosaic virus* (MYMV) (Honda and Ikegami 1986; Mandal et al. 1997) are most destructive and widespread. MYMIV is prevalent in northern India and Pakistan (Hameed and Robinson 2004; Varma and Biswas 2009), whereas MYMV mainly occurs in southern India. MYMV was earlier reported from Thailand, but has ceased to be a problem. Several variants of WTGs associated with YMD of grain legumes in different geographical regions of India have been detected (Varma et al. 1998). Antigenically, these variants fall into two major groups (Swanson et al. 1992). The variants of MYMIV from blackgram, mung bean and soybean from northern India, representing one major antigenic group, have >95% sequence identity. Their sequences mainly differ in ORF AV2 (Usharani et al. 2001). The MYMIV variants share 81% sequence identity with DNA A and 72–89% identity with DNA B of the MYMV variants reported from Thailand and southern India (Usharani et al. 2001). In recent years, MYMV has assumed serious proportion causing up to 80% losses in the production of Frenchbean in southern India. Molecular marker for resistance to MYMV in Frenchbean excession IC 525260 has been identified (Ravishankar et al. 2009), which will be useful in molecular breeding for resistance. Resistance in mungbean to MYMIV is found to be of recessive nature (Kundagrami et al. 2009).

MYMIV, MYMV, *Soybean blistering mosaic virus* (SbBMV), Soybean chlorotic blotch virus (SbCBV), *Soybean crinkle leaf virus* (SbCLV) and Soybean mild mottle virus (SbMMV) are known to infect soybean globally (Table 10.3). MYMIV is a major constraint in soybean cultivation in north-western parts of the Indian sub-continent, and SbBMV is a constraint in Argentina (Rodríguez-Pardina et al. 2011). Recently, two new WTGs, SbMMV and SbCBV have emerged in newly introduced soybean crops in Nigeria (Alabi et al. 2010). These viruses appear to infect wild leguminous hosts and moved to soybean crops. SbCBV has been shown to infect

Centrosema pubescens under natural conditions (Alabi et al. 2010). Semi-perennial leguminous crops like pigeon pea may also play an important role in the spread WTGs like MYMIV to annual crops like soybean and mungbean (Biswas et al. 2008). Sources of resistance to WTGs in soybean have been identified. The soybean varieties resistant to MYMIV are found to degrade viral transcripts earlier compared to the susceptible varieties (Yadav et al. 2009).

10.4.4 Non-legumo WTGs Spreading in Grain Legumes

In Spain and Portugal, monopartite WTGs, *Tomato yellow leaf curl virus* (TYLCV) and *Tomato yellow leaf curl Malaga virus* (TYLCMalV), infect *P. vulgaris* causing bean leaf crumple disease (BLCD). The affected plants show stunted growth, flower abortion and pod deformation. The disease results in nearly 100% loss in production. Resistance to TYLCV in beans appears to be conferred by a single dominant gene (Monci et al. 2005). In Portugal also TYLCV causes sporadically a novel disease of common bean (Louro et al. 2002). Sida micrantha mosaic virus and Okra mottle virus infect soybean sporadically in Brazil (Fernandes et al. 2009). These are just a few indicators that non-legumo WTGs, particularly TYLCV, may cause serious disease problems in grain legumes as it is the most widely distributed WTG.

10.5 WTGs Causing Diseases in Malvaceous Crops

Twenty-six WTGs have been identified to cause diseases in Malvaceous crops, like cotton and okra, in different parts of the world (Table 10.4). WTGs occurring in cotton and okra (commonly known as 'Bhindi' in the Indian sub-continent) cause leaf curl, leaf crumple, leaf blistering, vein yellowing, yellow crinkle, yellow mosaic and yellow mottle diseases. Of these, cotton leaf curl disease (CLCuD) is most serious.

10.5.1 WTGs Causing Disease in Cotton

CLCuD-affected plants develop curling of leaves, thickening of veins, enations and stunting (Fig. 10.7). The most diagnostic symptom is the dark green colour of the veins which is easily observed in transmitted light. Severely affected plants look bushy, due to shortening of internodes and twisting of petiole, peduncle and young stems, and suppression of bud formation and fruiting. Almost no yield is obtained from severely affected plants, resulting in enormous losses to the growers. CLCuD has been known to occur in Sudan and Nigeria and some pockets in Pakistan for many years, but it emerged in a severe form in the cotton belt of North-Western parts of the Indian sub-continent in the late 1980s and early 1990s. Several factors

Table 10.4 WTGs causing diseases in Malvaceous crops (cotton, jute, mesta and okra) (Source: GenBank, Fauquet et al. (2008) and other references given in the table)

S.N.	Acronym	Virus	Country	Host	Location	Isolates	Year
1	BYVMV[c]	*Bhindi yellow vein mosaic virus*	India	Okra	Madurai	1	–
			Pakistan	Okra	–	1	1996
			Thailand		–	1	–
2	CoGMV	*Corchorus golden mosaic virus*	Vietnam				2005
			India	Corchorus	Kolkata		2008 (Ghosh et al. 2008)
3	CoYSV	*Corchorus yellow spot virus*	Mexico				2005
4	CoYVV	*Corchorus yellow vein virus*	Vietnam				2000
5	CLCrV	*Cotton leaf crumple virus*			–	1	–
5.1	CLCrV-AZ	Cotton leaf crumple virus-Arizona	Mexico	Cotton	Sonora	1	1991
			USA	Cotton	Arizona	1	1991
			USA	Cotton	California	1	1991
5.2	CLCrV-TX	Cotton leaf crumple virus-Texas	USA	Cotton	Texas	1	1991
6	CLCuV	*Cotton leaf curl virus*	Pakistan	Radish		1	2000 (Mansoor et al. 2000)
			Pakistan	Chili peppers		1	2003 (Hussain et al. 2003)
7	CLCuAV	*Cotton leaf curl Alabad virus*	Pakistan	Cotton	–	2	1996
8	CLCuBV	*Cotton leaf curl Bangalore virus*	India	Cotton	Bangalore	1	2004

(continued)

Table 10.4 (continued)

S.N.	Acronym	Virus	Country	Host	Location	Isolates	Year
9	CLCuBuV	Cotton leaf curl Burewala virus[a]	Pakistan	Cotton	Faisalabad	1	2010 (Amrao et al. 2010a)
10	CLCuGV	*Cotton leaf curl Gezira virus*	Burkina Faso	Okra	–	1	2010 (Tiendrebeogo et al. 2010)
			Egypt	Okra	Aswan	2	–
			Egypt	Hollyhock	Cairo	1	–
			Mali		–	1	–
			Niger		–	1	2009 (Shih et al. 2009)
			Sudan	Okra	Gezira	2	
			Sudan	Sida	Gezira	1	–
			Sudan	Cotton	Gezira	1	–
11	CLCuKV	*Cotton leaf curl Kokhran virus*	Pakistan	Cotton	Faisalabad	1	–
			Pakistan	Cotton	Kokhran	1	1995
			Pakistan	Cotton	Manisal	1	1996
			India	Cotton	Dabawali	1	–
			India	Soybean	–	1	2006 ([Raj et al. 2006b)
12	CLCuMV	*Cotton leaf curl Multan virus*					–
			China	Cotton		1	2010 (Cai et al. 2010)
12.1		CLCuMV -Bhatinda	India	Cotton	Bhatinda	1	–
12.2		CLCuMV -Faisalabad	Pakistan	Cotton	Dera Ghazi Khan	1	1995
			Pakistan	Cotton	Faisalabad	2	–
			Pakistan	Cotton	Yazman	1	1995

12.3		CLCuMV -Hisar	India	Cotton	Hisar	1	1999
			India	Cotton	Ludhiana	1	1999
			India	Cotton	New Delhi	1	1999
			Pakistan	Cotton	Faisalabad	1	
			Pakistan	Okra	Multan	1	1996
			Pakistan	Cotton	Multan	1	
12.4		CLCuMV -Rajasthan	India	Cotton	Abohar	1	2003
			India	Cotton	Hisar	1	2003
			India	Cotton	New Delhi	1	2003
			India	Cotton	Sirsa	1	1999
			India	Cotton	Sriganganagar	1	1994
			Pakistan	Tomato	Faisalabad	1	2009 (Shahid et al. 2009)
13	CLCuShV	Cotton leaf curl Shadadpur virus[a]	Pakistan	Cotton	–	1	2010 (Amrao et al. 2010b)
14	JLMV	Jute leaf mosaic virus[a]	Bangladesh				2008 (Haque et al. 2008)
15	KeLCuV	Kenaf leaf curl virus[a]	India	Kenaf	–	1	2009 (Paul et al. 2009b)
16	MeYVMV	Mesta yellow vein mosaic virus[a]	India	Mesta	–	1	2008 (Das et al. 2008)
17	OkLCuIV	*Okra leaf curl India virus*[b]			–	1	–
18	OkLCuV	*Okra leaf curl virus*[b]			–	1	–
19	OkMMV	*Okra mosaic Mexico virus*[b]			–	1	–
20	OkYMoIV	*Okra yellow mottle Iguala virus*[b]			–	1	–
21	OkYMoV	*Okra yellow mottle virus*[b]			–	1	–

(continued)

Table 10.4 (continued)

S.N.	Acronym	Virus	Country	Host	Location	Isolates	Year
22	OYCrV	*Okra yellow crinkle virus*	Mali	Okra	–	2	2005
23	OYMMV	*Okra yellow mosaic Mexico virus*	Mexico	Okra	Mazatepec	1	2004
			USA		–	1	2010 (Hernandez-Zepeda et al. 2010a)
24	OYMoIV	*Okra yellow mottle Iguala virus*	Mexico	Okra	Iguala	1	–
25	OYVMV	*Okra yellow vein mosaic virus*	Pakistan	Okra	Faisalabad	1	1995
26	ToLCJV	*Tomato leaf curl Joydebpur virus*	India	Kenaf	–	1	2009 (Paul et al. 2009a)

[a]Proposed new species
[b]Unassigned isolates (Fauquet et al. 2008)
[c]Bhindi has been misspelt in ICTV list as 'Bhendi'

Fig. 10.7 WTG-induced diseases of Malvaceous crops. (**a**) A cotton plant affected by leaf curl disease; (**b**) A leaf of cotton from a plant severely affected by leaf curl disease; (**c**) An okra plant affected by BYVMV; (**d**) An okra plant affected by BYVMV and OLCuV; (**e**) An okra plant affected by OLCuV

like the emergence of new viruses/strains, increase in *B. tabaci* population, changes in cropping systems and introduction of susceptible genotypes, singly or in combination, seem to have led to the emergence of CLCuD in this region (Varma and Malathi 2003).

CLCuD was confined to the Sudan and Nigeria (Varma 1963), up to early 1960s, when it was first observed in small pockets in the Multan area of Pakistan. It remained confined to the area until 1988 when about 60 ha of a newly released cotton variety, S12, was affected by a very high incidence of the disease and within 5 years the causal WTGs spread to >0.2 million ha, resulting in losses of US$5 billion between 1992 and 1997 (Briddon and Markham 2000).

Unlike the situation in the Multan region, CLCuD was not found in India, until 1989, when it was first noticed in some exotic accessions of *Gossypium barbadense* grown for the first time in the experimental fields of Indian Agricultural Research Institute (IARI), New Delhi (Varma 1990, 1993; Varma et al. 1993). A few plants were also observed to develop yellow vein and blister mosaic diseases with which

WTGs were associated; these diseases were, however, not observed in subsequent years although the incidence of CLCuD increased. Regular removal of symptomatic plants and avoidance of cultivation of *G. barbadense* restricted the incidence of CLCuD in the fields of IARI. However, in 1994, CLCuD appeared in an alarming proportion in the cotton belt along the border with Pakistan, mainly affecting the fields of *G. hirsutum.* A distinct gradient in the incidence of CLCuD was observed in 1994 and 1995. The incidence and severity of the disease was highest in areas close to the border with Pakistan and it gradually declined in areas further away from the border. The appearance of CLCuD coincided with the first North-Easterly dust storms about 2 weeks earlier, which apparently blew in viruliferous whiteflies from the neighbouring fields across the border where CLCuD has become endemic. In both the years the incidence was greater in fields adjoining citrus and mango orchards, which provided wind breaks for the landing of wind-blown whiteflies. A high incidence of a leaf curl disease was also observed in okra (*Abelmoschus esculentus),* which is commonly grown alongside cotton in that area (Varma et al. 1995; Varma and Malathi 2003).

A strategy comprising cultivation of *G. arboreum* and resistant varieties of *G. hirsutum* in areas most prone to CLCuD, avoidance of cultivation of *G. barbadense* and their hybrids in North-West India, avoidance of cultivation of okra in disease prone and adjoining areas, and integrated management of weeds and the whitefly *B. tabaci,* helped in slowing disease spread, although by 1997 over 0.2 million ha were affected (Narula et al. 1999). Since then the varietal spectrum has changed in the region due to the introduction of Bt-hybrids, but the gradient and area affected by the disease has remained unchanged. In a survey in October 2010, incidence of CLCuD in the cotton belt of North-West India, most severely affected in 1994–1995, was found to vary from 50% to 80% in different Bt-hybrids, except the hybrid JK 1947, which showed high degree of resistance with <1% infection (SS Sohrab, personal communication). Considering the high incidence of CLCuD in the region, promotion of a wider use of resistant hybrids will be useful for containing the disease. Fortunately, CLCuD has not spread to commercial crops in South India, although the disease is prevalent in the backyard gardens in Karnataka in southern India (Nateshan et al. 1996).

The WTGs associated with CLCuD are highly divergent and form two distinct phylogenetic clusters (Kirthi et al. 2004; Varma and Praveen 2010). Most of the cotton-WTGs have a restricted distribution, except *Cotton leaf curl Gezira virus* (CLCuGV), *Cotton leaf curl Kokhran virus* (CLCuKV) and *Cotton leaf curl Multan virus* (CLCuMV) (Fig. 10.8). CLCuGV is widespread in Africa, CLCuKV and CLCuMV are prevalent in north-western Indian sub-continent, and the latter has also spread to China (Cai et al. 2010). Seven WTGs have been found associated with CLCuD in the Indian Sub-continent (Table 10.4).Three of these WTGs: *Cotton leaf curl Alabad virus* (CLCuAV), CLCuKV and CLCuMV, and their variants were associated with CLCuD epidemics of the 1990s in the North-Western cotton growing region of the Indian sub-continent (Malathi et al. 2003).

Recently, Cotton leaf curl Burewala virus (CLCuBuV), a recombinant WTG, has emerged as a major threat to cotton in Pakistan (Akhtar et al. 2010), and it is

Fig. 10.8 Distribution of WTGs associated with the diseases of Malvaceous crops. Numbers correspond to the virus names in Table 10.4

spreading fast in the region. The CLCuV isolate from Sudan, where CLCuD has been endemic for over 70 years, is even more distinct (Idris and Brown 2002). Variability in the WTGs associated with CLCuD may have resulted from the development of natural recombinants (Varma and Malathi 2003). One of the isolates of WTGs associated with CLCuD in Pakistan shares over 50% of its DNA A, covering ORFs AV1, AV2, AC2 and AC3, with that of *Bhindi yellow vein mosaic virus* (BYVMV) (Zhou et al. 1998), which is endemic in the region (Varma 1963). The recombinant WTGs could have developed in okra plants infected with both the viruses. Recently, occurrence of recombinants of different isolates originating in north-western parts of the Indian sub-continent have also been observed (Kumar et al. 2010). Another possibility is that BYVMV might have recombined with another endemic begomovirus resulting in the development of a variant that could infect cotton (Zhou et al. 1998). The recombinant WTGs, like CLCuBuV, causing CLCuD could be common as the WTGs associated with CLCuD have a wide host range including bean, tobacco, tomato, okra and hollyhock (Harrison et al. 1997a; Radhakrishnan et al. 2001a), and co-infection with different WTGs is common in areas affected by CLCuD (Sanz et al. 2000).

Various isolates of WTGs associated with CLCuD are known to have only DNA A (Nadeem et al. 1997; Zhou et al. 1998; Briddon et al. 2000; Radhakrishnan et al. 2001a). The inability of DNA A of CLCuD associated WTGs to induce symptoms, led to the detection of satellite-like ssDNA molecules (Briddon et al. 2000), now referred to as alphasatellite and betasatellite. The alphasatellites can replicate autonomously in plant cells, but requires the viral DNA A for movement in plants and encapsidation in the viral CP to facilitate whitefly transmission. However, the plants co-infected with WTG DNA A and the associated alphasatellite do not develop typical CLCuD symptoms (Mansoor et al. 1999), whereas co-infection with the betasatellite, which, requires the viral DNA A both for replication and whitefly transmission, induced typical CLCuD symptoms (Briddon et al. 2001; Radhakrishnan et al. 2001a). Thus, CLCuD is caused by a monopartite WTG in association with betasatellite molecules. The CLCuD complex may be still evolving. Some of the molecules associated with the disease are chimeras of sequences derived from the CLCuD associated WTG, betasatellite and alphasatellite (Briddon et al. 2001). The high levels of recombination between various molecules associated with CLCuD may be an important factor in the evolution of CLCuD associated WTGs.

Cotton leaf crumple virus (CLCrV) causing leaf crumple disease of cotton (CLCrD), also known as 'Acromania', or 'crazy top', has been known to occur in south-western US since the early 1920s (Cook 1924). Host range of the virus is restricted to plant families, Malvaceae and Fabaceae. Ratooned cotton and malvaceous weeds are important sources of virus inoculum, which is efficiently spread by B-biotype of whitefly identified as *B. argentifolii.* Restriction on ratooning of cotton crop helped in restricting the spread of CLCrV in the US. Sources of resistance to the virus have been identified in *G. barbadense* and *G. arboretum* (Malathi et al. 2003). Phylogenetic analyses indicate CLCrV might have evolved through recombination and reassortment events involving ancestors of WTGs belonging to SLCV and *Abutilon mosaic virus* clades for DNA A, and SLCV and some unknown clades for DNA B (Idris and Brown 2004).

10.5.2 WTGs Causing Diseases in Okra

Okra leaf curl disease (OLCuD) is a serious disease of okra (*Abelmoschus esculentus*) in West and North-East Africa. In Egypt, Niger and Sudan, CLCuGV and betasatellite CLCuGB are found to be associated with the disease (Shih et al. 2009), indicating that CLCuGV and its betasatellite are spreading in West and North-East Africa. In Mali, OLCuD is shown to be caused by *Okra yellow crinkle virus* (OYCrV) and a recombinant isolate of CLCuGV along with a variant of CLCuGB and an alphasatellite of distinct West African lineage (Kon et al. 2009). CLCuGV and several betasatellites and alphasatellites are associated with OLCuD in Burkina Faso, highlighting the complexity of WTGs causing OLCuD (Tiendrebeogo et al. 2010). In 2009, up to 75% of okra cv Green Emerald plants in Texas, USA were found infected with *Okra yellow mosaic Mexico virus* (OYMMV). The leaves of infected plants developed irregular yellow patches, yellow margins and chlorosis. The DNA A component was 99% identical to OYMMV, whereas the closest relative of DNA B was SiGMV indicating that a pseudo-recombinant is spreading in the US (Hernandez-Zepeda et al. 2010).

Bhindi yellow vein mosaic disease (BYVMD), also known as yellow vein mosaic disease of Okra (OYVMD) is a serious constraint of okra production in South-East Asia. BYVMD has been known to occur in India for nearly 90 years (Kulkarni 1924). Two distinct WTGs, BYVMV and *Okra yellow vein mosaic virus* (OYVMV) have been found associated with OYVMD (Table 10.4). BYVMV is the most common WTG spreading in okra crops in the Indian Sub-continent. BYVMV also requires a betasatellite molecule for symptom development (Jose and Usha 2003). Diversity in BYVMV has attracted limited attention, but frequent breakdown of resistance of okra varieties indicates occurrence of BYVMV variants in the Indian Sub-continent. The Indian Sub-continent was free of OLCuD until the late 1980s, when the disease was first noticed in okra genotypes introduced from West Africa. The disease must have been caused by indigenous WTGs. Soon the varieties commonly grown in the region were also affected by OLCuD (Varma and Malathi 2003). In Pakistan, OLCuD is shown to be caused by CLCuMV and in India by *Okra leaf curl India virus* (OKLCuIV) (Varma and Mandal 2003). Mixed infection with WTGs causing OYVMD and OLCuD result in severe disease and complete loss in yield (Fig. 10.7). In Thailand resistance to BYVMV in okra has been developed by mutation breeding (Phadvibulya et al. 2009). Durability of resistance in these lines remains to be seen.

10.5.3 WTGs Spreading in Kenaf

Kenaf (*Hibiscus cannabinus*), an important 'non-tree' source of newsprint, is affected by monopartite WTGs causing severe leaf curl and yellow vein mosaic diseases (Fig. 10.9). Three different viruses are shown to cause yellow vein and leaf curl diseases of Kenaf in India. In eastern and southern India, Mesta yellow vein mosaic virus (MeYVMV) and *Tomato leaf curl Joydebpur virus* (ToLCJV) are spreading in Kenaf crops, whereas in northern India, Kenaf leaf curl virus (KeLCuV) is spreading (Chatterjee and Ghosh 2007; Paul et al. 2009a, b).

Fig. 10.9 Yellow vein (**a**) and leaf curl (**c**) diseases of mesta caused by MeYVMV and KeLCuV respectively; (**b**) initial symptoms on a mesta plant experimentally inoculated with MeYVMV (Pictures courtesy Dr. Anirban Roy, NBPGR, New Delhi, India)

10.6 WTGs Spreading in Solanaceous Crops

Solanaceous plants, particularly tomato, appear to be the most favoured host of WTGs, as these crops are affected by 117 WTGs and a large number of their strains/isolates. The number of WTGs infecting tomato is increasing at a very fast rate. At the beginning this century, 39 WTGs were reported to naturally infect tomato (Varma and Malathi 2003), and within a decade the number increased to 112 (Table 10.5). A large number of the new WTGs affecting tomato have been detected in Asia and the Americas (Fig. 10.10; Table 10.5). It is likely that many more await detection considering the limited number of isolates characterized so far. Most of the tomato infecting WTGs induce characteristic leaf curl symptoms, including severe reduction in leaf size, downward curling, crinkling of inter-veinal areas, inter-veinal and marginal chlorosis, occasional development of enations, purple discolouration of the abaxial surface of leaves, shortening of internodes, development of small branches and reduced fruiting (Fig. 10.11). Some of these viruses also cause bright yellow spots on leaves. A majority of tomato diseases caused by WTGs are collectively described as either 'leaf curl' or 'yellow leaf curl' based on the subtle differences in symptoms. It is not possible to distinguish the viruses causing these diseases in tomato by symptoms or other biological properties such as host range and transmission. Reactions with panels of monoclonal antibodies (MAbs) distinguished these viruses to a limited extent (McGrath and Harrison 1995), but they are best distinguished by DNA hybridization, PCR and nucleotide sequence analysis (Torres-Pacheco et al. 1996; Rojas et al. 2000). The most devastating WTGs affecting tomato are those with generic names 'tomato leaf curl virus' and 'tomato yellow leaf curl virus' (Varma and Malathi 2003). These WTGs, commonly referred to as ToLCVs and TYLCVs, are widely distributed in Africa, the Americas, Asia, Australia and parts of Europe (Fig. 10.10). TYLCV seems to have evolved in the Middle East between 1930 and 1950 (Lefeuvre et al. 2010), and has spread to almost all the tomato producing parts of the world.

10.6.1 WTGs Spreading in Tomato in Central and North America

More than 20 tomato infecting WTGs are spreading in the Americas (Table 10.5). WTGs emerged as a problem in tomato cultivation in the US in 1989, when *Tomato mottle virus* (ToMoV) was first observed in southern Florida. By 1991, annual economic losses due to ToMoV were estimated to be around US $125 million as a result of reduced fruit yields and the cost of pesticide application to control whiteflies. The emergence of ToMoV was attributed to the appearance of the B-biotype of *B. tabaci* on greenhouse ornamentals and tomato in 1986–1987. The main spread of ToMoV was by whiteflies migrating from the abandoned fields, particularly in the spring crops (Polston et al. 1996), as no significant virus reservoir in weed plants has been found.

Table 10.5 WTGs associated with the diseases of Solanaceous crops (Source: GenBank, Fauquet et al. (2008) and other references given in the table)

S.N.	Acronym	Virus	Country	Host	Location	Isolate	Year
1	ChiLCV	*Chilli leaf curl virus*	India	Chilli	–	1	2005
				Chilli	Varanasi	1	2006
				Papaya	–	1	2005
				Tomato	–	1	2005
			Pakistan	Chilli	Khanewal	1	2004
				Chilli	Multan	1	1998
				Potato	–		2009 (Mubin et al. 2009)
2	ChiLCMV	*Chilli leaf curl Multan virus*	Pakistan	Chilli	–	1	2009 (Akhter et al. 2009)
3	CdTV	*Chino del tomate virus*	Mexico	Soybean	Sinaloa	1	2005
				Tomato	Cinvestav	1	
				Tomato	–	1	2005
				Tomato	Sinaloa	5	1983
4	PepGMV	*Pepper golden mosaic virus*	Costa Rica	Pepper		1	
			Mexico	Pepper	Tamaulipas	1	
			USA	Pepper	Serano	1	1989
			USA	Pepper	–	1 Distortion 2 Mosaic	1987
5	PepHYVV	*Pepper huasteco yellow vein virus*	Mexico	Pepper	Sinaloa	1	1988
					Tamaulipas		
6	PepLCBDV	*Pepper leaf curl Bangladesh virus*	Bangladesh	Pepper	Bogra	1	1999
			India	Bitter gourd		1	2010 (Raj et al. 2010)
			Pakistan	Pepper	Khanewal	1	2004

(continued)

Table 10.5 (continued)

S.N.	Acronym	Virus	Country	Host	Location	Isolate	Year
7	PepLCLV	*Pepper leaf curl Lahore virus*	Pakistan	Pepper	Lahore	1	2004
8	PepLCV	*Pepper leaf curl virus*	Malaysia	Pepper	–	1	1997
			Thailand	Pepper	–	1	–
9	PepMTV	*Pepper mild tigre' virus*[a]		Pepper	–	1	–
10	PepRAV	*Pepper rizado amarillo virus*[a]		Pepper	–	1	–
11	PepYLCIV	*Pepper yellow leaf curl Indonesia virus*	Indonesia	Pepper	–	1	–
		PepYLCIV-A		Ageratum	–	1	2005
		PepYLCIV-A		Tomato	–	1	2005
		PepYLCIV-B		Pepper	–	1	2005
12	PepYVV	*Pepper yellow vein virus*[a]		Pepper	–	1	
13	PepYVMV	*Pepper yellow vein Mali virus*	Mali	Pepper	–	1	
			Burkina Faso			1	2008 (Tiendrebeogo et al. 2008)
14	PYMPV	*Potato yellow mosaic Panama virus*	Panama	Potato	Divisa	1	
15	PYMV	*Potato yellow mosaic virus*	Guadeloupe	Tomato	–	1	
			Puerto Rico	Tomato	–	1	2004
			Trinidad	Tomato	–	1	
			Venezuela	Potato	–	1	

16	SGMV	*Serrano golden mosaic virus*	USA	Tomato	–	1	1990 (Brown 1990)
			Mexico	Tomato	–	1	
17	TbASV	*Tobacco apical stunt virus*[a]	Mexico	Tobacco	Chiapas	1	1991
18	TbCSV	*Tobacco curly shoot virus*	China	Ageratum	Yunnan	1	2003
				Pepper		1	2010 (Qing et al. 2010)
				Mirabilis jalapa			2010 (Xiong et al. 2010)
				Tobacco	Yunnan	2	1999,2001
				Tomato	Yunnan	1	2001
			India	Wild Sunflower		1	2010 HQ407395
19	TbLCJV	*Tobacco leaf curl Japan virus*	Japan	Honeysuckle	–		2007 (Ogawa et al. 2007)
				Tomato	–	1	
				Tobacco	–	2	
20	TbLCuCUV	*Tobacco leaf curl Cuba virus*	Cuba	Tobacco	Taguasco	1	2005
			Jamaica	*Malachra alceifolia*	–	1	2006 (Hall et al. 2008)
21	TbLCIV	*Tobacco leaf curl India virus*[a]	India	Tobacco	–	1	
22	TbLCPuV	Tobacco leaf curl Pusa virus[b]	India	Tobacco	–	1	2010 HQ180391
23	TbLCThV	Tobacco leaf curl Thailand virus[b]	Thailand	Tobacco	–	1	2007 (Knierim and Maiss 2007)

(continued)

Table 10.5 (continued)

S.N.	Acronym	Virus	Country	Host	Location	Isolate	Year
24	TbLCYnV	*Tobacco leaf curl Yunnan virus*	China	Ageratum	Yunnan	1	2004
				Tobacco	Yunnan	3	1999,2002
				Tomato	Yunnan	1	2003
25	TbLCZV	*Tobacco leaf curl Zimbabwe virus*	Zimbabwe	Tobacco	–	2	2002
				Tobacco	Comoros Archipelago		2006 (Lett et al. 2006)
26	TbLRV	Tobacco leaf rugose virus[b]	Cuba	Tobacco	–	1	2002 (Domínguez et al. 2002)
27	TbMLCV	Tobacco mottle leaf curl virus[b]	Cuba	Tobacco	–	1	2009 (Dominguez et al. 2009)
28	TbYCrV	Tobacco yellow crinkle virus[b]	Cuba	Tobacco	–	1	2009 (Fiallo-Olive et al. 2009b)
29	ToChLPV	*Tomato chino La Paz virus*	Mexico	Tomato	Baja La Paz	2	2002
30	ToClVV	*Tomato chlorotic vein virus*[a]	Brazil	Tomato	Brasilia	1	1994
31	ToCMoV	*Tomato chlorotic mottle virus*	Brazil	Tomato	Seabra	1	1996
				Tomato	Betim	1	1996
				Tomato	Igarape	1	1996
32	ToCrV	*Tomato crinkle virus*[a]	Brazil	Tomato	Pesqueira	1	1998
33	ToCrLYV	*Tomato crinkle leaf yellows virus*	Brazil	Tomato	–		2003 (Rafaelo et al. 2003)

34	ToCYLV	*Tomato crinkle yellow leaf virus*[a]	Brazil	Tomato	Vicosa	1	1999
35	ToCSV	*Tomato curly stunt virus*	South Africa	Tomato	Onderberg	1	1998
36	TGMV	*Tomato golden mosaic virus*	Brazil	Tomato	–	1 Common	1984
				Tomato	–	1 Yellow vein	
37	ToGMoV	*Tomato golden mottle virus*	Guatemala	Tomato	–	1	1994
			Mexico	Tomato	–	1	2005
38	ToGVV	*Tomato golden vein virus*[a]	Brazil	Tomato	–	1	2003
39	ToIYV	*Tomato infectious yellows virus*[a]	Brazil	Tomato	–	1	1999
40	ToLCAnV	Tomato leaf curl Antsiranana virus[b]	Ghana	Tomato	–	1	2008 (Osei et al. 2008)
41	ToLCArV	*Tomato leaf curl Arusha virus*	Tanzania	Tomato	Tengelu	1	2005
42	ToLCBV	*Tomato leaf curl Bangalore virus*	India			1	
		ToLCBV-A		Tomato	Bangalore	1	
		ToLCBV-A		Tomato	Kerala	1	2005
		ToLCBV-A		Tomato	Kolar	1	
		ToLCBV-B		Tomato	Bangalore	1	
		ToLCBV-B		Cotton	Fatehabad	1	
		ToLCBV-C		Tomato	Bangalore	1	1997
					Bangalore	1	

(continued)

Table 10.5 (continued)

S.N.	Acronym	Virus	Country	Host	Location	Isolate	Year
43	ToLCBDV	*Tomato leaf curl Bangladesh virus*	Bangladesh	Tomato	–	1	
44	ToLCCNV	*Tomato leaf curl China virus*	China	Tomato	–	3	2002
				Tomato	–	1	2003
45	ToLCKMV	*Tomato leaf curl Comoros virus*	Mayotte	Tomato	–	1	2003
46	ToLGhV	Tomato leaf curl Ghana virus[b]	Ghana		–	1	2008 (Osei et al. 2008)
47	ToLCGuV	*Tomato leaf curl Guangdong virus*	China	Tomato	Guangzhou	1	2003
48	ToLCGxV	*Tomato leaf curl Guangxi virus*	China	Tomato	Guangxi	3	2003
49	ToLCGV	*Tomato leaf curl Gujarat virus*	India	Tomato	Mirzapur	1	1999
				Tomato	Vadodara	1	1999
				Tomato	Varanasi	1	2000
				Datura metel	New Delhi		2007 (Sivalingam and Varma 2007b)
			Nepal	Tomato	Panchkhal	1	2000
50	ToLCHaV	Tomato leaf curl Hainan virus[b]	China		–	1	2010 (Zhang et al. 2010)
51	ToLCHsV	*Tomato leaf curl Hsinchu virus*	China	Tomato	Fujian	1	2005
			Taiwan	Tomato	Hsinchu	1	2005

52	ToLCIV	*Tomato leaf curl India virus*[a]	India	Tomato	–	1	–
53	ToLCIDV	*Tomato leaf curl Indonesia virus*[a]	Indonesia	Tomato	–	1	–
54	ToLCJV	*Tomato leaf curl Java virus*	Indonesia	Tomato	–	1	–
		Tomato leaf curl Java virus-A		Tomato	–	1	–
		Tomato leaf curl Java virus-B		Ageratum	–	1	–
55	ToLCJoV	*Tomato leaf curl Joydebpur virus*	Bangladesh	Tomato	–	1	–
			India	Chilli	Punjab	1	2007 (Shih et al. 2007)
				Kenaf	–	1	2009 (Paul et al. 2009a)
				Tomato	Kalyani	1	2006
56	ToLCKV	*Tomato leaf curl Karnataka virus*	India	Tomato	Bangalore	1	1993
				Soybean	–	1	2006 (Raj et al. 2006a)
			India	Tomato	Janti	1	2005
			Iran	Tomato	–	1	–
57	ToLCKeV	*Tomato leaf curl Kerala virus*	India	Tomato	–	1	2005
58	ToLCKuV	Tomato leaf curl Kumasi virus[b]	Ghana		–	1	2008 (Osei et al. 2008)
59	ToLCLV	*Tomato leaf curl Laos virus*	Laos	Tomato	–	1	–
60	ToLCMGV	*Tomato leaf curl Madagascar virus*	Madagascar	Tomato	–	2	2001

(continued)

Table 10.5 (continued)

S.N.	Acronym	Virus	Country	Host	Location	Isolate	Year
61	ToLCMYV	*Tomato leaf curl Malaysia virus*	Malaysia	Tomato	–	1	1997
62	ToLCMaV	*Tomato leaf curl Mayotte virus*	Mayotte	Tomato	–	1	2003
63	ToLCNV	*Tomato leaf curl Nicaragua virus*[a]	Nicaragua	Tomato	–	1	–
64	ToLCNDV	*Tomato leaf curl New Delhi virus*	Bangladesh	Tomato	Jessore	1 severe	2005
			India	Cotton	Hissar	1	2005
				Bitter gourd		1	2010 (Tiwari et al. 2010a, b)
				Bottle gourd			2006 (Sohrab et al 2006)
				Bimili jute	Lucknow	1	2007 (Raj et al. 2007)
				Chilli		1	2006 (Khan et al. 2006)
				Cucumber			
				Chayote	Kalimpong	1	2000 (Mandal et al. 2004)
				Ivy gourd	Panipat	1	2006 (Sohrab et al. 2006)
				Luffa	New Delhi		2005 (Sohrab et al. 2003)
				Luffa	Sonepat	1	2005
				Pigeonpea	Lucknow		2005 (Raj et al. 2005)
				Potato	Meerut	1	2002
				Potato	Happur	1	2005
				Potato	Meerut	1	2005

		Papaya	New Delhi	1	2005 (Raj et al. 2008a, b)
		Pumpkin	New Delhi		2006 (Sohrab et al. 2006)
		Ridge gourd			2006 (Sohrab et al. 2006)
		Solanum nigrum		1	200 (Sivalingam and Varma 2007b)
		Tomato	New Delhi	1	2005
		Tomato	Lucknow	1	
		Tomato	New Delhi	1	
		Tomato	New Delhi	1 Mild	1992
		Tomato	New Delhi	1 Severe	1992
		Watermelon			
	Pakistan	Bitter gourd		1	2005 (Tahir and Haider 2007)
		Bell pepper		1	2006 (Tahir and Haider 2006)
		Chilli	Khalawal	1	2004
		Eclipta prostrata		2	2005 (Haider et al. 2005)
		Luffa	Multan	1	2004
		Solanum nigrum		1	1997,2004
		Tomato	Islamabad	1	2000
		Tomato	Dargai	1	2001
		Tomato	Lahore	1	2004
	Taiwan	Oriental melon			2010 (Chang et al 2010)
	Thailand	Bottle gourd		1	2008 (Ito et al. 2008)

(continued)

Table 10.5 (continued)

S.N.	Acronym	Virus	Country	Host	Location	Isolate	Year
				Cucumber		1	2008 (Ito et al. 2008)
				Luffa		1	
				Muskmelon		1	2008 (Ito et al. 2008)
65	ToLCPKV-	*Tomato leaf curl Pakistan virus*	Pakistan	Tomato	Rahim Yar Khan	1	2004
			India	Mentha		1	2009 (Samad et al. 2009)
66	ToLCPalV	Tomato leaf curl Palampur virus[b]	India	Tomato		1	2008 (Kumar et al. 2008)
			Pakistan	Bitter Gourd		1	2010 (Ali et al. 2010)
			Iran	Cucumber,Melon		1	2009 (Heydarnejad et al. 2009)
67	ToLCPaV	Tomato leaf curl Patna virus[b]	India			1	2009, 2010 (Kumari et al. 2009, 2010)
68	ToLCPV	*Tomato leaf curl Philippines virus*	Philippines			1	
		ToLCPV-A		Tomato	LosBanos	1	1995
		ToLCPV-B		Tomato	LosBanos	1	
		ToLCPV-C		Tomato	San Leonardo	1	2005
69	ToLCPuV	*Tomato leaf curl Pune virus*	India	Tomato	Pune	1	2005
70	ToLCPlaV	*Tomato leaf curl Playifas virus*				1	
71	ToLCrV	Tomato leaf crumple virus	Nicaragua			1	2000 (Rojas et al. 2000)
72	ToLCRaV	*Tomato leaf curl Rajasthan virus*	India	Tomato	Rajasthan	1	2005

73	ToLCSV	*Tomato leaf curl Senegal virus*[a]				1	–
74	ToLCSinV	*Tomato leaf curl Sinaloa virus*	Nicaragua	Tomato	–	2	–
75	ToLCSLV	*Tomato leaf curl Sri Lanka virus*	Sri Lanka	Tomato	–	1	1997
76	ToLCSDV	*Tomato leaf curl Sudan virus*	Sudan	Tomato	Gezira	1	1996
				Tomato	Shambat	1	1996
			Yemen	Tomato	Tihamah	1	2006
77	ToLCSuV	Tomato leaf curl Sulawesi virus[b]	Indonesia	Tomato	–	1	2009 (Tsai et al. 2009)
78	ToLCTWV	*Tomato leaf curl Taiwan virus*	Taiwan			1	
		ToLCTWV-A		Tomato	Changua	1	2005
				Tomato	Guangdong	1	2005
				Tomato	Hsinchu	1	2005
				Tomato	Hualian	1	2005
				Tomato	Tainan	1	2005
				Tomato	Taitung	1	2005
				Tomato	Taoyuan	1	2005
				Tomato		1	2005
		ToLCTWV-B		Tomato	Hualian	1	2005
		ToLCTWV-C		Tomato	Chiayi	1	2005
79	ToLCTZV	*Tomato leaf curl Tanzania virus*[a]			–	1	–

(continued)

Table 10.5 (continued)

S.N.	Acronym	Virus	Country	Host	Location	Isolate	Year
80	ToLCUV	*Tomato leaf curl Uganda virus*	Uganda	Tomato	Iganga	1	2005
81	ToLCVV	*Tomato leaf curl Vietnam virus*	Vietnam	Tomato	Hanoi	1	1998
				Tomato	Hanoi	1	2005
82	ToLCV	*Tomato leaf curl virus*	Australia	*Solanum* spp.	–	2	
				Tomato	–	1	
83	ToMYLCAV	*Tomato mild yellow leaf curl Aragua virus*	Venezuela	Tomato	–	1	
84	ToMYMoV	*Tomato mild yellow mottle virus*[a]	Honduras		–	1	1996
					–		
85	ToMHV	*Tomato mosaic Havana virus*	Cuba	Tomato	–	1	
			Nicaragua		–	1	2007 (Monger et al. 2007)
86	ToMBV	*Tomato mosaic Barbados virus*[a]			–	1	
87	ToMoTV	*Tomato mottle Taino virus*	Cuba	Tomato	–	1	
88	ToMoLCV	*Tomato mottle leaf curl virus*[a]	Brazil	–	Mossoró	1	1999
89	ToMoV	*Tomato mottle virus*	Puerto Rico	Tomato	–	1	2004
			USA	Tomato	–	1	1989
90	ToRMV	*Tomato rugose mosaic virus*	Brazil	Tomato	–	1	1996

91	ToRMV	Tomato rugose mosaic virus[b]			–	1	2006 (Fernandes et al. 2006)
92	ToSMV	*Tomato severe mosaic virus*[a]	Brazil		–	1	1999
93	ToSLCV	*Tomato severe leaf curl virus*	Guatemala		Sansirisay	1	1996
			Mexico		Rioverde	2	2005
			Nicaragua		Condega	1	
					Santa Lucia	1	
94	ToSRV	*Tomato severe rugose virus*	Brazil	Pepper	Petrolina de Go	1	2003
				Tomato	Uberlandia	1	2000
95	ToYLDV	Tomato yellow leaf distortion virus[b]	Cuba		–	1	2009 (Fiallo-Olive et al. 2009a)
96	ToYMV	*Tomato yellow mosaic virus*[a]	Brazil		–	1	
97	ToYMoV	*Tomato yellow mottle virus*[a]	Brazil		–	1	
98	ToYSV	*Tomato yellow spot virus*	Brazil	Tomato	Bicas	1	1999
99	ToYVSV	*Tomato yellow vein streak virus*[a]	Brazil			1	1995
100	TYLCAxV	*Tomato yellow leaf curl Axarquia virus*	Spain	Tomato	Algarrobo	1	2000
101	TYLCCNV	*Tomato yellow leaf curl China virus*	China				
				Bean	Yunnan	1	2004

(continued)

Table 10.5 (continued)

S.N.	Acronym	Virus	Country	Host	Location	Isolate	Year
				Datura spp.	Yunnan	1	2005
				Kidney bean			2007 (Dong et al. 2007)
				Solanum spp.	Yunnan	1	2005
				Sieges beckia	Yunnan	1	2001
				Tomato	Yunnan	1	2000
				Tobacco	Yunnan	1	2005
				Tobacco	Yunnan	1	2001
				Tobacco	Yunnan	2	1999
				Tomato	Guangxi	1	2004
				Tomato	Guangxi	1	
				Tobacco	Yunnan	2	2005
				Tobacco	Yunnan	1	2000
				Tobacco	Yunnan	1	2001
					Yunnan	1	2001
				Tobacco	Yunnan	1	2000
102	TYLCGuV	*Tomato yellow leaf curl Guangdong virus*	China	Tomato	Guangzhou	1	2003
103	TYLCIDV	*Tomato yellow leaf curl Indonesia virus*	Indonesia	Tomato	Lembang	1	2005
104	TYLCKaV	*Tomato yellow leaf curl Kanchanaburi virus*	Thailand	Tomato	Kanchanaburi	1	2001
			Thailand	Eggplant	Kanchanaburi	1	2001
			Vietnam	Tomato	–	1	2005
			Vietnam	Eggplant	Binhduong	1	2005
105	TYLCKWV	*Tomato yellow leaf curl Kuwait virus*[a]			–	1	

106	TYLCMalV	*Tomato yellow leaf curl Malaga virus*	Spain	Tomato		1	1999
107	TYLCMLV	*Tomato yellow leaf curl Mali virus*	Ethiopia	Tomato	Melkassa	1	2005
			Mali	Tomato	–	1	2003
108	TYLCNV	*Tomato yellow leaf curl Nigeria virus*[a]			–	1	
109	TYLCSV	*Tomato yellow leaf curl Sardinia virus*	Greece				2007 (Papayiannis et al. 2007a)
			Israel	Tomato	Rehovot	1	2005
			Italy	Tomato	Sicily	1	
			Italy	Tomato	Sardinia	1	1988
			Morocco	Tomato	Agadir	1	2002
			Spain	Tomato	Almeria	1	1992
			Spain	Tomato	Canary	1	
			Spain	Tomato	Murcia	1	1992
			Tunisia	Tomato	Bkalta	1	2002
110	TYLCSAV	*Tomato yellow leaf curl Saudi Arabia virus*[a]			–	1	
111	TYLCTZV	*Tomato yellow leaf curl Tanzania virus*[a]			–	1	
112	TYLCTHV	*Tomato yellow leaf curl Thailand virus*			–	1	
		Strains TYLCTHV-A	Thailand	Tomato		2	
		TYLCTHV-B	China	Tomato	Yunnan	1	2002
		TYLCTHV-B	Myanmar	Tomato	Yangon	1	1999
		TYLCTHV-B	Thailand	Tomato	Chiang Mai	2	
		TYLCTHV-B		Tomato	NongKhai	1	

(continued)

Table 10.5 (continued)

S.N.	Acronym	Virus	Country	Host	Location	Isolate	Year
		TYLCTHV-C	Thailand	Tomato	Sakon Nakhon		
113	TYLCVNV	*Tomato yellow leaf curl Vietnam virus*	Vietnam	Tomato	Hanoi	1	2005
114	TYLCV	*Tomato yellow leaf curl virus*			–		
			Ethopia		–	1	2006 (Shih et al. 2006)
			Greece	Common Bean	–	1	2007 (Papayiannis et al. 2007b)
			Guatemala		–	1	2010 (Salati et al. 2010)
			Guadeloupe		–	1	2007 (Akhtar et al. 2007)
			Hawaii		–	1	2010 (Melzer et al. 2010)
			Kuwait		–	1	2007 (Akhtar et al. 2007)
			Mauritius		–	1	2010 (Lobin et al. 2010)
			Mexico		Sinaloa	1	2009 (Gamez-Jimenez et al. 2009)
			Netherlands		–	1	2009 (Botermans et al. 2009)
			Portugal		–	1	2007 (Akhtar et al. 2007)
			Reunion		–	1	2007 (Akhtar et al. 2007)
			Spain	Tobacco	–	1	2005 (Font et al. 2005)
			Saudi Arabia		–		2007 (Ajlan et al. 2007)
			USA		South Carolina	1	2006 (Ling et al. 2006)

	USA		California	1	2007 (Rojas et al. 2007)
	USA		Kentucky	1	2008 (de Sá et al. 2008)
	USA		North Carolina	1	2002 (Polston et al. 2002)
	Venezuela		–	1	2007 (Zambrano et al. 2007)
Strains					
TYLCV – Gezira	Sudan	Tomato	–	1	1996
TYLCV – Iran	Iran	Tomato	Iranshahr	1	1998
TYLCV – Israel	China	Tomato	Shanghai	1	2005
	Cuba	Tomato	–	1	
	Dominican Republic	Tomato	–	1	
	Egypt	Tomato	Ismaelia	1	1991
	Israel	Tomato	Rehovot	1	1986
	Italy	Tomato	Sicily	1	2004
	Japan	Tomato	Haruno	1	2005
	Japan	Tomato	Misumi	1	
	Japan	Tomato	Miyazaki	1	
	Japan	Tomato	Omura	2	
	Japan	Tomato	Tosa	1	2005
	Jordan	Tomato	–	1	2005
	Lebanon	Tomato	–	1	2005
	Mexico	Tomato	Culiacan	1	2005
	Morocco	Tomato	Berkane	1	2005
	Puerto Rico	Tomato	–	1	2001
	Spain	Pepper	Almeria	1	1999
	Tunisia	Tomato	–	1	2005

(continued)

Table 10.5 (continued)

S.N.	Acronym	Virus	Country	Host	Location	Isolate	Year
			Turkey	Tomato	Mersin	1	2005
			USA	Tomato	Florida	1	1997
		TYLCV-Mild	Israel	Tomato	–	1	1993
			Japan	Tomato	Aichi	1	
				Tomato	Aichi2	1	2003
				Tomato	Atumi	1	
				Tomato	Daito	1	
				Tomato	Kisozaki	1	
				Tomato	Osuka	1	
				Tomato	Shimizu	1	
				Tomato	Shizuoka	1	
				Tomato	Yaizu	1	
			Jordan	Cucumber	–	1	2005
			Jordan	Tomato	Homra	1	2003
			Jordan	Tomato	–	1	2005
			Lebanon	Tomato	–	1	2005
			Portugal	Tomato	–	1	1995
			Reunion	Tomato	–	1	2002
			Spain	Tomato	–	1	1997
			Spain	Tomato	Almeria	1	1999
		TYLCV-Oman	Oman	Tomato	Al-Batinah	1	2005
115	TYLCYV	*Tomato yellow leaf curl Yemen virus*[a]			–	1	–
116	TYMLCV	*Tomato yellow margin leaf curl virus*	Venezuela	Tomato	Merida	1	–
117	TYVSV	*Tomato yellow vein streak virus*[b]	Brazil				2010 (Albuquerque et al. 2010)

[a]Unassigned isolates (Fauquet et al. 2008)
[b]Proposed new species

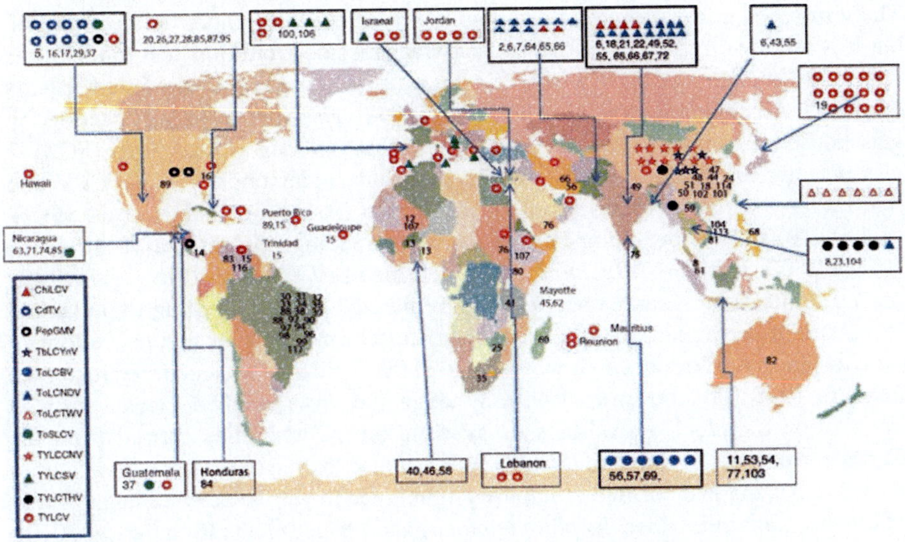

Fig. 10.10 Distribution of WTGs and their strains/isolates associated with the diseases of Solanaceous crops. Numbers correspond to the virus names in Table 10.5

Fig. 10.11 WTG-induced diseases of Solanaceous crops. (**a**) Tomato leaf curl disease; (**b**) Tobacco leaf curl disease; (**c**) Chilli plants affected by leaf curl disease; (**d**) Apical leaf curl disease of potato

The virus does infect tropical soda apple (*Solanum viarum)* under field conditions, but it is an unlikely source of the virus as transmission from this host to tomato is difficult (McGovern et al. 1994).WTGs have also emerged in tomato in other parts of the US. An epidemic of *Pepper huasteco yellow vein virus* (PHYVV) occurred in glasshouse tomatoes in 1997 in Texas, and *Serrano golden mosaic virus* (SGMV) was detected in tomato in Arizona (Polston and Anderson 1997). PHYVV and SGMV have also emerged in tomato in Mexico (Ascencio-Ibanez et al. 1999). In 2009, TYLCV appeared in Hawaii and it is spreading (Melzer et al. 2010).

The WTG, causing TYLCD in Israel, identified as TYLCV-Is (Pico et al. 1996), moved to the Dominican Republic through infected tomato seedling imports, and from there it seems to have spread to other Caribbean countries and in the tomato growing areas of Florida, USA, where up to 100% incidence was recorded in tomato fields in 1998 (Polston et al. 1999). Whether the virus reached Florida through importation of infected seedlings or by wind-blown whiteflies carried from the Caribbean basin is not known. In 2006, TYLCV also moved to Guatemala in Central America causing leaf curling, crumpling, light green to yellow mosaics and stunting of tomato, tomatillo (*Physalis philadelphica*) and pepper plants (Salati et al. 2010). TYLCV is also spreading in tomatillo in Mexico indicating that TYLCV is getting established in different crops in the Americas (Gamez-Jimenez et al. 2009). In central and eastern Cuba also high (80–100%) incidence of TYLCV in tomato was recorded (Martinez et al. 2003), which later displaced the bipartite WTGs (Martinez et al. 2006).

10.6.2 WTGs Spreading in Tomato in Latin America

In 2006, TYLCV infection of tomato was found in Venezuela. This seems to be the first report of the occurrence of TYLCV in South America (Zambrano et al. 2007). *Tomato yellow vein streak virus* (ToYVSV) has also spread in tomato and potato crops in recent years in Brazil causing serious reduction in the production of these crops (Albuquerque et al. 2010). In Nicaragua, the increase in whitefly populations in the 1980s resulted in an epidemic caused by WTGs in 1988, affecting nearly 100% of the tomato plants in various regions of the country. The epidemic was caused by *Tomato leaf curl virus* (ToLCV), *Tomato mild mottle virus* (ToMMoV), *Tomato leaf crumple virus* (ToLCrV) and several *Sida* infecting WTGs (Rojas et al. 2000). Later, *Tomato severe leaf curl virus* (ToSLCV), *Tomato leaf curl Sinaloa virus* (ToLCSinV), *Pepper golden mosaic virus* (PepGMV) and a new (provisional) species *Tomato leaf curl Playifas virus* (ToLCPlaV), were found spreading in tomato in Nicaragua (Ala-Poikela et al. 2005). The genome organization of ToLCSinV and ToSLCV is similar to the other bipartite WTGs occurring in the Americas. These viruses, however, belong to different clades; ToLCSinV groups with the Abutilon mosaic virus clade and ToSLCV with the Squash leaf curl virus clade. Recombination in the AV1 gene of ToSLCV has resulted in the evolution of the two strains of the virus (Rojas et al. 2005). WTGs are also spreading in tomato crops in Venezuela causing severe diseases (Faria and Nava 2009).

In Brazil, 15 tomato infecting WTGs are spreading (Table 10.5). Some of these, occurring in the North-Eastern and South-Eastern regions, have arisen as a result of recombinant events. Their emergence has been attributed to the introduction of a polyphagous biotype of *B. tabaci* (Ribeiro et al. 2003). *Tomato chlorotic mottle virus* (ToCMoV), a bipartite WTG, has spread widely and emerged in new geographical areas, which were free from infection before 2004. This emergence has not been attributed to any recent recombinant events (Ribeiro et al. 2007). *Tomato rugose mosaic virus* (ToRMV) causes golden mosaic and leaf distortion in tomato, and also infects beans, which may provide alternate source of infection for neighbouring tomato crops (Fernandes et al. 2006).

Potato yellow mosaic virus (PYMV) causes bright yellow mosaic, leaf distortion, and a chlorotic mottle, leaf distortion, leaf rolling and stunting in tomato. In the 1990s, it emerged as a serious problem in tomato cultivation in parts of South America and the Caribbean basin (Polston et al. 1998). The virus may not be a problem in potato as the crop is generally grown in areas and seasons which do not favour whitefly activity. The incidence of *Tomato golden mosaic virus* (TGMV), which was first reported in Brazil in 1975, increased sharply in the early 1990s, coinciding with the appearance of the B-biotype of *B. tabaci* (Ribeiro et al. 1998).

10.6.3 WTGs Spreading in Tomato in Europe, Mediterranean and Middle East

TYLCD, considered to be one of the most devastating plant diseases in the world (Glick et al. 2009), is reported to cause serious losses in tomato production in Cyprus, Egypt, Israel, Italy, Jordan, Lebanon, Saudi Arabia, Spain and Tunisia (Noris et al. 1994; Pico et al. 1996). The losses are greatest when infection occurs prior to flowering. TYLCD was first observed in Israel in 1939–1940 coinciding with an increase in the whitefly population. About 20 years later the disease destroyed tomato crops in the Jordan Valley. In Morocco, tomato is the second most important crop for export. The occurrence of TYLCD was first reported from the area south of Casablanca in 1996–1997, and by 1998 it spread to all the tomato growing areas (Jebbour 2001). The tomato producing areas in central, southern and south-eastern Iran are threatened by the high incidence of TYLCD, caused by TYLCV-Gezira, which is prevalent in the Mediterranean region and ToLCNDV, prevalent in the Indian sub-continent (Fazeli et al. 2009). TYLCV and ToMoV are spreading in protected crops of tomato in Kuwait (Al-Ali et al. 2009).

In Europe, TYLCV, was first found in Sardinia in 1988 and subsequently in Sicily and southern Italy. In Spain, emergence of TYLCV coincided with the increase in the population of *B. tabaci* (Pico et al. 1996). It emerged in southern

Portugal in 1995. Since then it has caused severe epidemics annually (Louro et al. 2002), especially in green house crops (Ramos et al. 2002). TYLCD was also introduced into commercial plantations in France through infected seedlings, however, the disease was soon eradicated (Dalmon and Cailly 2001), and there is no report of its re-emergence. In 2007, for the first time, TYLCV induced characteristic symptoms were observed in tomatoes growing in greenhouses located in the western parts of the Netherlands. The virus was found in 19 of the 27 glasshouses growing tomato in the region. Measures like the destruction of infected plants and eradication of *B. tabaci* helped in eliminating TYLCV infection, and no infection was found in 2008 (Botermans et al. 2009).

Several WTGs, driven by recombinant events, have emerged in tomato around the Nile and Mediterranean during the last 30 years. These viruses fall into two phylogenetic clades represented by TYLCV and *Tomato yellow leaf curl Sardinia virus* (TYLCSV). Six recombination hot spots have been identified in these viruses indicating that the recombinant events are not random (Fauquet et al. 2005). In Sicily, mixed infection by TYLCSV and TYLCV in 'protected' crops of tomato was found common in 2004. In addition to these parent species, two recombinants were also detected. One recombinant produced symptoms typical of the parent viruses, but the second recombinant was more virulent as it induced severe downward curling and rugosity of the leaves (Davino et al. 2009). In 2003–2004, TYLCSV also appeared in Castrovillari area in the Calabria region of southern Italy, in hydroponically grown tomatoes in glasshouses. Maximum incidence was in plants grafted on Beanfort DRS tomato rootstock obtained from Sicily. The virus might have been introduced with the rootstocks and *B. tabaci* must have spread it further (Crescenzi et al. 2004). A year later, up to 100% infection by TYLCSV was observed in protected tomato crops in the Ionic Coast of Basilicata region of Southern Italy. ToLCSV seems to have been carried by viruliferous whiteflies from Sicily to Basilicata via Calabria region (Comes et al. 2009).

10.6.4 WTGs Spreading in Tomato in Africa

In Africa TYLCD was first observed in the Sudan (Yassin and Nour 1965). In the late 1970s, a high incidence of the disease was observed throughout Nigeria. The incidence was greater in southern Nigeria than in the north. Association of a geminivirus with the disease was shown by electron microscopy; the virus formed typical crystalline inclusions in the nuclei of infected cells of tomato (Varma 1984). Later the virus was identified as TYLCV (Czosnek and Laterrot 1997). TYLCD has also been reported from Burkina Faso (Konate et al. 1995), Cape Verde, the Ivory Coast, Mali (d'Hondt and Russo 1985), Egypt (Czosnek et al. 1990), and Cameroon (Czosnek and Laterrot 1997). A recombinant WTG, *Tomato yellow leaf curl Mali virus* (TYLCMLV), evolved due to the recombination of sequences from TYLCV-mild and *Hollyhock leaf crumple virus* (HoLCrV), has spread throughout West Africa causing serious diseases in association with Cotton leaf curl Gezira betasatellite (CLCuGB) (Chen et al. 2009). TYLCV is also reported to have caused severe disease in 2009 in the Southern parts of Mauritius (Lobin et al. 2010).

10.6.5 Tomato WTGs Spreading in Tomato in Asia

In Asia, more than 30 WTGs are associated with ToLCD and TYLCD. Many of these WTGs have emerged recently (Table 10.5). In the Indian Sub-continent, the occurrence of ToLCD, has been known for over 60 years (Vasudeva and Samraj 1948). The disease is widespread, causing enormous losses to growers. It has also emerged in other parts of the sub-continent like Bangladesh (Begum and Khan 1996), Bhutan (Thinlay-Penjore 1996), Nepal (Joshi et al. 1996), Pakistan (Hameed 1996) and Sri Lanka (Zoysa 1996). So far, 16 WTGs associated with ToLCD are known to spread in the Indian Sub-continent (Table 10.5). Eleven WTGs cause ToLCD in India, six in Pakistan, four in Bangladesh and one each in Nepal and Sri Lanka.

ToLCNDV is the most widely distributed WTG infecting tomato crops in Bangladesh, India and Pakistan (Table 10.5). ToLCPalV also appears to be widely distributed as it also occurs in Iran (Kumar et al. 2008; Heydarnejad et al. 2009; Ali et al. 2010). Considerable variability has been found in the WTGs associated with ToLCD in India. For example *Chilli leaf curl virus* (ChiLCV), *Tomato leaf curl Gujarat virus* (ToLCGV) and ToLCNDV are bipartite WTGs, whereas *Tomato leaf curl Bangalore virus* (ToLCBV), *Tomato leaf curl Joydebpur virus* (ToLCJoV) and *Tomato leaf curl Karnataka virus* (ToLCKV) are monopartite WTGs. ToLCV, originating from Australia, and *Tomato leaf curl Taiwan virus* (ToLCTWV) are also monopartite WTGs. Some uncharacterized WTGs are also affecting tomato crops in India (Sivalingam and Varma 2007a). The question whether the WTGs associated with ToLCD in India originated from a common ancestor is difficult to address, but the differences between the tomato isolates from different parts of the country suggest their diverse origin. The divergence between tomato leaf curl viruses in India cannot be explained on the basis of geographical separation (Varma and Malathi 2003). Earlier reports suggested that the bipartite WTGs affect tomato crops in Northern India, whereas monopartite WTGs occur in Southern India (Sivalingam and Varma 2007a), but recently monopartite ToLCJoV and ToLCKV have been found spreading in Northern India (Tiwari et al. 2010a, b). Sequencing of a larger number of isolates may throw more light to determine the basis of their emergence in the Indian sub-continent.

An epidemic of severe ToLCD in southern India in 1999 coincided with nearly 1000-fold increase in the whitefly populations and appearance of the B-biotype of *B. tabaci* (Banks et al. 2001). Tomato infecting WTGs have also been detected in weeds like *Datura metel* and *Solanum miasum*, which may play an important role in the spread of these viruses (Sivalingam and Varma 2007b). In northern India, ToLCNDV is the most serious constraint in tomato production. To minimize the losses caused by this virus, sources of resistance to ToLCNDV in tomato have been identified (Tripathi and Varma 2003) and transgenic lines resistant to ToLCNDV have been developed (Varma and Shelly 2006; Shelly et al. 2005).

ToLCD is also widespread in South-East Asia. In Indonesia, 100% infection of tomato by ToLCSV has been observed (Tsai et al. 2009). TYLCV appeared in tomato fields of Anhui and Shandong Provinces of China in 2008 (Yu et al. 2009). It has also spread to Nanjing, Suzhon, Changzhou, Xuzhon, Taizhon and Yancheng in 2008 (Sun et al. 2009). In China, the virus is reported to be spreading towards northern China from southern regions (Yu et al. 2009). *Tomato leaf curl Guangxi virus* (ToLCGxV), a recombinant monopartite WTG containing sequences from *Ageratum yellow vein virus, Euphorbia leaf curl virus* and *Tomato leaf curl China virus* (ToLCCNV), emerged in Guangxi province of China as a serious problem (Xu and Zhou 2007). *Malvastrum coromandelianum*, an alternative host of *Tomato yellow leaf curl China virus* (TYLCCNV) may be playing an important role in the perpetuation and spread of TYLCCNV in China (Liu et al. 2009a, b). The TYLCV infection in tomato in Japan was also identified for the first time in 2007 in Okirawa Prefecture. TYLCV seems to have spread further through domestic movement of transplants and transmission by B-biotype of *B. tabaci* (Kato et al. 2009).

10.6.6 WTGs Spreading in Other Solanaceous Crops

WTGs also cause serious diseases in other solanaceous crops, like potato, chilli and tobacco (Fig. 10.11). In 1999, foliar chlorosis, curling, smalling of leaves, fruit distortion and reduction in yield of chillies caused by *Pepper yellow leaf curl Indonesia virus* (PepYLCIV) was observed for the first time in West Java, Indonesia. Since then, the disease incidence has grown 5.4-folds. Emergence of the disease in Indonesia has been assigned to 'invasion' of *B. tabaci* and the virus from central Thailand between 1994 and 1999 (De Barro et al. 2008). WTG infection of chilli pepper (*Capsicum annuum*) in Taiwan is spreading fast. In 2007, less than 10% incidence of *Tomato yellow leaf curl Thailand virus* (TYLCThV) in Tainan county was observed, but by 2009, the incidence increased to 70% in some fields spreading over four different counties (Shih et al. 2010). WTG infection in *Capsicum frutescens* in Sichuan Province in south-western China also emerged in 2009. This Province was apparently free from WTGs earlier. Infection in *C. frutescens* was caused by the spread of *Tobacco curly shoot virus* (TbCSV), resulting in leaf yellowing, mild curling and stunting of the affected plants (Qing et al. 2010).

In 2007 and 2008, high incidence of leaf curl disease of chillies resulted in enormous yield losses in the North-Western parts of the Indian sub-continent. The epidemic was caused by the spread of ToLCNDV and *Chilli leaf curl Multan virus* (ChiLCMV) in Pakistan (Akhter et al 2009) and ToLCNDV in India (Senanayake et al. 2007). The WTGs associated with ToLCD and TYLCD have a wide natural host range (Diaz-Pendon et al. 2010), which favours their perpetuation and also helps in the development of natural variants.

Potato is grown in regions and seasons, which do not favour active whitefly populations, but some WTGs are shown to cause bright yellow mosaic, leaf distortion, leaf curling and stunting of potato plants in the Americas and Indian sub-continent. PYMV is associated with the disease in Venezuela, *Potato yellow mosaic Panama virus* (PYMPV) in Panama and *Potato yellow mosaic Trinidad virus* (PYMTV) in Trinidad and Tobago (Varma and Malathi 2003). These WTGs do not seem to be serious problems in potato production. In India, however, slight advancement in the date of planting of potato crops in Northern plains resulted in the emergence of a serious apical leaf curl disease in the late 1990s (Garg et al. 2001), and it continues to be a serious problem. The disease is shown to be caused by an isolate of ToLCNDV (Usharani et al. 2004).

The earliest known written record of a plant viral disease is of the yellowing disease of *Eupatorium lindleyanum* (Osaki et al. 1985); the disease is caused by a WTG, which is also associated with the leaf curl disease of tobacco (TbLCD). TbLCD (Fig. 10.11) has been affecting tobacco cultivation for over a century (Storey 1931; Pal and Tandon 1937). TbLCD is reported to occur in Africa, the Americas, South-East Asia and some parts of Europe. Nearly 20 WTGs have been found associated with TbLCD in various parts of the world (Table 10.5).

10.7 Diseases Caused by WTGs in Other Selected Crops

WTGs have been spreading in a variety of crops (Table 10.6). Emergence of WTG induced diseases in crops, like cabbage and radish, which are grown in seasons not favourable for whiteflies, is alarming as the WTGs are likely to spread in a new range of crops. Climate change resulting in warmer winters during the current decade may have been an important factor leading to expansion in the period of activity and increase in populations of whiteflies, and emergence of WTG-induced diseases in temperate crops. In recent years, WTG-induced diseases have also emerged in some new warm weather crops.

10.7.1 Cabbage

Cole-crops, which are generally grown in temperate climates, were found affected by a leaf curl disease in the US in 1990s. The leaves of affected plants develop vein clearing, mosaic, curling and puckering up to 100% plants were found affected by the disease in some fields in Georgia. The disease is caused by *Cabbage leaf curl virus* (CabLCuV) (Mandal et al. 2001).

10.7.2 Jute

Jute, an important fibre crop in South-East Asia, is affected by WTG-induced diseases in Bangladesh, India and Vietnam (Haque et al. 2008; Ghosh et al. 2008; Ha et al. 2008).

Table 10.6 WTGs associated with the diseases of cabbage, papaya, radish and sweet potato (Source: GenBank, Fauquet et al. (2008) and other references given in the table)

S.N.	Acronym	Virus	Country	Host	Location	Isolate	Year
1	CabLCuJV	*Cabbage leaf curl Jamaica virus*	Jamaica	Cabbage		3	2005
2	CabLCuV	*Cabbage leaf curl virus*	Jamaica	Cabbage	Douglas Castle	1	2005
			USA	Cabbage	Florida	1	1996
				Cabbage	Georgia		2001 (Mandal et al. 2001)
3	CrYVMV	*Croton yellow vein mosaic virus*					
4	ICrLCuV	*Ipomoea crinkle leaf curl virus*					1997 (Cohen et al. 1997)
5	ILCuV	*Ipomoea leaf curl virus*					2004 (Rodrigo et al. 2004)
6	IYVV	*Ipomoea yellow vein virus*	Spain			1	1998
7	PaLCuCNV	*Papaya leaf curl China virus*	China	*Ageratum* spp.	Guangxi	3	2002
				Papaya	Guangxi	2	2002
				Tomato	Guangxi	3	2002
				Tomato		1	2003
			Vietnam	Tobacco		1	2005
8	PaLCuCoV	*Papaya leaf curl Coimbtore virus*					
9	PaLCuGuV	*Papaya leaf curl Guangdong virus*	China	Papaya	Guangdong	1	2002
			China	Papaya	Guangzhou	1	2004

10	PaLCuV	*Papaya leaf curl virus*	India	Papaya	Lucknow	1	
				Tobacco	Bihar		2010 (Kumar et al. 2009a, b)
			Pakistan	Cotton		1	2002
			Taiwan				2003
11	RaLCV	Radish leaf curl virus*	India	Radish	Varanasi	1	2007 (Singh et al. 2007)
12	SPLCGV	*Sweet potato leaf curl Georgia virus*	USA			1	
13	SPLCV	*Sweet potato leaf curl virus*	China				2009 (Yang et al. 2009)
			Japan				1991 (Osaki and Inouye 1991)
			India				2007 (Makeshkumar et al. 2007)
			Italy			1	2002
					Sicily		2006 (Briddon et al. 2006)
			Kenya				2006 (Miano et al. 2006)
			Peru				2003 (Fuentes and Salazar 2003)
			Republic of Korea				2007 (Kwak et al. 2007)
			Taiwan				1985 (Chung et al. 1985)
			USA			1	1994

*Proposed new species

The WTGs associated with yellow mosaic and yellow vein diseases in India and Vietnam have been identified as *Corchorus yellow vein virus* (CoYVV) and *Corchorus golden mosaic virus* (CoGMV) (Ha et al. 2008; Ghosh et al. 2008). Interestingly both these viruses resemble the New World WTGs as they lack ORF AV2.

10.7.3 Papaya

The leaf curl disease of papaya is emerging as a serious problem. The affected plants develop severe leaf curling and remain dwarf with no fruiting. The disease, first reported in 1939 in India, is caused by six different WTGs (Table 10.6). ToLCNDV and *Papaya leaf curl virus* (PaLCuV) are associated with the disease in Northern India and *Croton yellow vein mosaic virus* (CrYVMV) and *Papaya leaf curl Coimbtore virus* (PaLCuCoV) are associated with the disease in Southern India (Reddy et al. 2010; Srivastava et al. 2010). In Taiwan and Pakistan, PaLCuV and in China, *Papaya leaf curl China virus* (PaLCuCNV) and *Papaya leaf curl Guangdong virus* (PaLCuGuV) are associated with the disease (Table 10.6).

10.7.4 Radish

Radish is cultivated in winter months in India. Generally crops grown in the winter season are not affected by WTGs in the sub-continent, but recent emergence of Leaf curl disease of radish caused by a new WTG, Radish leaf curl virus (RaLCV) in North-Western region of the sub-continent is alarming. The affected plants develop typical leaf curling, leaf distortion and enations (Singh et al. 2007; Mansoor et al. 2000).

10.7.5 Sweet Potato

Sweet potato (*Ipomoea batatas*), the seventh most important food crop in the world, is affected by three WTGs. In North America, *Sweet potato leaf curl virus* (SPLCV) and *Sweet potato leaf curl Georgia virus* (SPLCGV) have been found associated with leaf curl disease of sweet potato (Lotrakul et al. 1998). Similar diseases also occur in Brazil (Paprotka et al. 2010), India (Makesh kumar 2007). Taiwan (Chung et al. 1985), Japan (Osaki and Inouye 1991) Israel (Cohen et al. 1997), China (Luan et al. 2006), Korea (Miano et al. 2006), Sicily (Briddon et al. 2006) and Spain (Lozano et al. 2009) The occurrence of WTG-induced diseases of sweet potato in geographically widely separated regions suggests that such diseases may be even more widely distributed in sweet potato growing areas. In the USA, the SPLCV was initially isolated from two breeding lines that showed typical leaf curl symptoms.

It is possible that the virus might have been overlooked earlier as it remains symptomless in some accessions of sweet potato (Lotrakul et al. 1998). The genome organization of the isolate from the USA is similar to monopartite begomoviruses of the Old World, including the presence of ORF V2 (Lotrakul and Valverde 1999), suggesting the possibility of its introduction with infected planting material from the old world (Varma and Malathi 2003). However, recent phylogenetic studies suggest that *Ipomoea*-infecting WTGs might have developed before the old and New World WTGs separated or might have evolved from an ancestral Old World WTG (Lozano et al. 2009).

The incidence of SPLCV in South California has increased rapidly in recent years. The increase is attributed to the use of infected planting material and increase in whitefly populations (Ling et al. 2010). *Ipomoea leaf curl virus* (ILCV) has also been found associated with sweet potato leaf curl disease in the USA. The genome organization of ILCV is similar to SPLCV (Lotrakul et al. 2001). Recently, three new WTGs and a strain of SPLCV have emerged in Spain, apparently driven by extensive recombinant events (Lozano et al. 2009). Considerable diversity has been found in the WTGs associated with sweet potato germplasm in Brazil (Paprotka et al. 2010). SPLCV has also been found to infect tall morning glory (*Ipomoea purpurea*) plants in Fujian, China (Yang et al. 2009). Wide global distribution of SPLCV and its variants and increase in whitefly populations could lead to the emergence of even more serious diseases of this important crop.

10.7.6 *Mentha*

Mentha spicata (family Lamiaceae), an important medicinal plant,is widely cultivated in the foot-hills of Himalayas in northern India. In 2007, the crop was severely damaged by a WTG, tentatively identified as *Tomato leaf curl Pakistan virus* (ToLCPKV) (Samad et al. 2009).

10.8 Underlying Factors Leading to the Emergence and Spread of WTGs

In recent years, WTGs have emerged in threatening proportions in various agricultural systems around the world. Several factors have been identified to drive the evolution of new WTGs (Varma and Malathi 2003; Fargette et al. 2006; Varma 2010). These include evolution of new variants, evolution of satellite-like DNA molecules, appearance of whitefly biotypes, weather factors, changing cropping systems, movement of infected planting material and introduction of susceptible genotypes of cultivated plants. New WTGs and variants of the known WTGs are evolving at a very fast rate, as has been observed in the tomato infecting WTGs. At the beginning of the current century about 35 WTGs (Varma and Malathi 2003) were known to cause disease in tomato, but within a few years, the number has increased to nearly 120 (Table 10.5), and the number is growing. Similar diversity is also observed in WTGs spreading in other crops.

10.8.1 Evolution of Variants of WTGs

Enormous variability occurs in WTGs causing diseases in different groups of crop plants. Various isolates of WTGs differ in symptom development, disease severity, host range and transmissibility by whitefly. Variants of WTG evolve as a result of mutation, pseudo recombination and/or genomic recombination. The variability is greater amongst the inter-regional isolates compared to the intra-regional isolates, as the genotypes of some crops resisting WTG infection in one geographical region may succumb to infection in another region. For example, variability in MYMIV in northern India was apparent when different varieties of *V. mungo* and *V. radiata* were found to be resistant at one location and highly susceptible at another (Varma et al. 1992), although limited differences in their genomic sequences were detected (Usharani et al. 2001; Hameed and Robinson 2004).

Deployment of virus-resistant genotypes is an ideal approach to manage plant viral epidemics. However, it exerts selection pressure leading to the emergence of resistant breaking strains, like the development of recombinant variants of ToYLCD associated viruses in Ty-1 resistant plants (Garcia-Andres et al. 2009), and evolution of MYMIV isolates overcoming resistant genotypes within 5–6 years of their deployment (Varma et al. 1992).

The rate of intra-isolate mutations in WTGs is very high. Considerable intra-species variability is also observed in WTGs due to fast evolution resulting from high rates of nucleotide substitutions. It is estimated that the rate of substitution in DNA A and DNA B of EACMVs is 1.60×10^{-3} and 1.33×10^{-4} substitutions per site per year (Duffy and Holmes 2009). Similar rates of substitution were estimated earlier for TYLCV emerging in North America. An isolate of TYLC, maintained by insect transmission in a greenhouse in Israel for over 30 years, showed loss of virulence due to experimental passage in susceptible hosts. The loss in virulence over the years indicates natural mutation of the virus genome. Similar, changes in TYLCV have also been found in agroinoculated plants (Czosnek and Laterrot 1997). Naturally occurring variants also arise as a result of 'silent mutation' which are phenotypically indistinguishable (Harrison and Robinson 1999).

The ease with which genomic recombination in WTGs occurs not only between variants of the same virus but also between virus species, has resulted in fast development of new forms, leading to the emergence of one of the largest families of plant viruses. Recombination events in WTGs are considered (Varma and Malathi 2003) analogous, in their consequences, to the 'free sex' in *Orchidaceae.* Indeed, phylogenetic analysis have shown that the two genomic components of bipartite WTGs have evolved separately under different evolutionary pressures, and genomic component exchanges have played an important role in the diversification of WTGs (Briddon et al. 2010).

New variants or even new WTGs emerge through natural recombination, as mixed begomoviruses infections are common (Varma and Malathi 2003). Many recombinant WTGs have been detected recently. The most devastating recombinant WTGs evolved in cassava by the recombination between EACMV and ACMV giving rise to the highly virulent EACMV-Uganda variant which lead to a pandemic in East Africa (Fargette et al. 2006). Other serious recombinants of WTGs causing diseases in Malvaceous, Solanaceous and Leguminaceous crops have also appeared. A variant of CLCuV in Pakistan has been shown to have more than 50% of its genomic DNA A identical with that of BYVMV (Zhou et al. 1998). This variant could have primed the cotton leaf curl epidemic in the North-West of the Indian sub-continent. Subsequently, several different DNA A recombinants were found to be associated with CLCuD in the sub-continent having fragments from CLCuV and BYVMV (Harrison and Robinson 1999) and CLCuV and *Ageratum yellow vein virus* (AYVV) (Briddon et al. 2001; Saunders et al. 2001a, b). CLCrV in the US also seems to have evolved through recombination events (Idris and Brown 2004). TYLCV – Israel, which is the most widely distributed strain of TYLCV (Table 10.5), is a natural recombinant between the TYLCV – mild and a WTG related to ToLCKV (Navas-Castillo et al. 2000). ToLCGxV, which recently emerged as a serious problem in China is also a recombinant, containing sequences of two non-tomato- and one tomato-WTG (Xu et al. 2007). Recombination studies have shown that the region around Iran is the site of intensive TYLCV evolution leading to the emergence of TYLCV strains through inter- and intra-species recombination. However, the variants that evolved in this region are not of much epidemiological significance globally, as the TYLCV variants are mainly spreading from the Mediterranean basin (Lefeuvre et al. 2010). These are just a few examples of known natural recombinants. As more information on WTG sequences unfolds, it would not be surprising if many serious disease problems caused by WTG are found related to recombination events.

As predicted (Varma and Malathi 2003), serious plant diseases caused by recombinant WTGs have emerged in different parts of the world. An on-going epidemic of ToLCD in Iran is reported to be caused by a variant of ToLCPalV (Heydarnejad et al. 2009). This virus seems to have recombinant DNA-A originating from ToLCNDV and another tomato infecting WTG occurring in the Indian sub-continent. ToLCPalV was earlier found only in India, but how it moved to Iran, where it is devastating not only tomato but also cucumber and melon farms in Southern Iran (Heydarnejad et al. 2009), is not known.

10.8.2 Acquisition of Betasatellites and Alphasatellites by WTGs

The Old World begomoviruses seem to have the ability to dispense with the DNA-B component of their genome as the ORF AV2 in the DNA A codes for a protein that participates in virus movement (Padidam et al. 1996). This enables DNA A alone to cause a systemic infection, as has been shown for a Thailand isolate of TYLCV (TYLCV-Th) (Rochester et al. 1994). Some WTGs might have had similar genomic organization but gradually dispensed with DNA B to become monopartite viruses, and acquired satellite DNA molecules as in the WTGs associated with CLCuD in the Indian sub-continent (Mansoor et al. 1999; Briddon et al. 2001; Radhakrishnan et al. 2001b), to cause infection. The associated satellite molecules have been identified as betasatellite, nanovirus-like alphasatellite and defective interfering (DI) DNAs. DNA A of the monopartite WTGs can infect their natural or experimental hosts, but require the presence of betasatellites, which seem to have co-evolved with cognate viral DNA-A, to induce disease (Briddon et al. 2001; Saunders et al. 2000; Zhou et al. 2003; Sharma et al. 2010). Betasatellites are also associated with the disease like ToLCD caused by bipartite WTGs (Sivalingam et al. 2004). The plants co-infected with the two genomic components of ToLCNDV and the associated betasatellite not only increased symptom severity but also increased viral replication and efficiency of transmission of the virus by *B. tabaci* (Sivalingam et al. 2010). Considerable molecular diversity occurs in the betasatellites associated with ToLCD and other WTG-induced diseases in India (Sivalingam et al. 2010). Phylogenetically betasatellite form two distinct clusters. One cluster is represented by the betasatellite associated with WTGs infecting watermelon hosts and the second group is associated with WTGs infecting solanaceous and compositae. The betasatellites appear to have co-evolved with the helper WTGs (Briddon et al. 2003; Zhou et al. 2003). Recombination between the viral DNA A, betasatellite and sequences of unknown origin results in the development of DI molecules of size similar to betasatellite (Stanley et al. 1997). The DI molecules are shown to replicate in *trans* by DNA A, and the co-inoculated plants develop milder symptoms than produced when DNA A and betasatellite are co-inoculated. In the disease complex, betasatellites and its recombinants facilitate amplification of DNA A to wild-type levels which may be required for symptom development. In bipartite begomoviruses DNA B performs this function. It is suggested that betasatellites might have evolved through a satellite-like DNA progenitor, although DNA B and betasatellite may not be related evolutionarily (Saunders et al. 2001a, b). The alphasatellites (also termed DNA1 components), are like the betasatellites in size and replicate autonomously. Their size apparently helps in encapsidation by the associated WTG particles, thereby facilitating transmission by whiteflies. In China, alphasatellites have been found associated with WTG-infected plants, which also contained betasatellites, but only some plants that contained betasatellites contained alphasatellite (Xie et al. 2010). However, the role of alphasatellites in disease complex is not understood.

10.8.3 Viruses Infecting New Hosts and Emergence of New Virus Disease Problems

PYMV caused serious problems in parts of the Americas when it moved to tomato from potato (Polston et al. 1998), the WTGs associated with CLCuD cause severe leaf curl in okra and soybean (Zhou et al. 1998; Varma et al. 1995; Raj et al. 2006b) and

ToLCNDV is causing serious disease problems in potato and cucurbits in the Indian sub-continent (Varma and Malathi 2003) and in cucurbits in Thailand (Ito et al. 2008). In the Americas WTGs were a problem in leguminous crops but not in tomatoes, which were only sporadically affected, but the emergence of TYLCV in the 1990s resulted in serious problems in tomato cultivation (Polston and Anderson 1997). A large number of new WTGs have evolved rapidly in areas conducive to year round production of crops susceptible to these viruses, particularly after the 'Green Revolution' period (Varma 2010). The most dramatic increase is in the number of WTGs infecting tomato globally (Table 10.5). Some WTGs, like those associated with CMD have also moved to new hosts. For example ACMV has emerged in several leguminous hosts in West Africa (Ogbe 2006; Mgbechi-Ezeri et al. 2008) and ICMV in Jatropha and a cucurbit (Raj et al. 2008a, b; Gao et al. 2010; Rajinimala and Rabindran 2007). It is difficult to determine whether these viruses moved to new hosts or they were the original hosts of the viruses from which the viruses moved to cassava. In India, potato cultivation in the northern plains was not affected by any WTG, but in the late 1990s ToLCNDV moved to potato causing severe apical leaf curl disease, ToLCNDV emerged in potato a new host, due to advancement in the sowing of the crop coinciding with increase in *B. tabaci* populations (Varma and Malathi 2003). Similarly, ToLCPKV has emerged in *Mentha* spp. crops in serious proportions (Samad et al. 2009). Emergence of new WTGs affecting temperate vegetables, cabbage and radish (Table 10.6) is threatening, but not surprising in regions like the Indian sub-continent where the vector whiteflies are now active throughout the year.

10.8.4 Movement of Infected Planting Material

Human activity has played a significant role in the global spread and emergence of serious pathogens like WTGs. TYLCV emerged as a serious pathogen in the Americas after it was introduced, through infected transplants from Israel (Polston et al. 1999). The emergence of TYLCV in Portugal, Spain, France China, Japan and recently the Netherland, and several other countries/regions (Table 10.5), may also have followed a similar path. Intra-country spread of TYLCVs in countries like China, Italy and Japan is also attributed to the movement of infected transplants or rootstocks (Crescenzi et al. 2004). The spread of WTGs through infected planting material could be even more serious in vegetatively propagated crops like cassava, potato and sweet potato. Recent emergence of ICMV in West Africa could have occurred due to inadvertent introduction of infected planting material from India, which might have been collected from symptomless plants. SPLCV, notwithstanding its evolutionary history, also seems to have spread across different continents through infected planting material of sweet potato. In India, a serious disease of potato caused by WTG, ToLCNDV, emerged in North-Western plains of India at the beginning of this century (Garg et al. 2001), but within a few years, the disease assumed serious proportions and spread to other parts of the country through the infected seed tubers.

10.8.5 Insect Vector and Weather Factors

B. tabaci populations vary in their efficiency to transmit WTGs. B-biotype of *B. tabaci* is an efficient transmitter of WTGs occurring in the Americas. In the 1980s, appearance of B-biotype of *B. tabaci,* which transmits WTGs efficiently and has a wide host range, lead to the emergence of previously unreported viruses such as ToMoV in Florida (Polston and Anderson 1997; Polston et al. 1996), and the re-emergence of TGMV in Brazil (Ribeiro et al. 1998). But non-B biotype populations from Pakistan and Spain have been shown to transmit the WTGs occurring in Pakistan more efficiently than the B-biotype population from USA, South Africa and Mexico, although fitness of the B-biotypes was better than the non-B biotypes (Haider et al. 2003). Random amplified polymorphic DNA (RAPD) analysis has indicated diversity in *B. tabaci* populations. Phylogenetic analysis of mitochondrial cytochrome oxidase I shows at least three genetic lineages corresponding to the Indian, South Asian and Mediterranean-African clades. All three clades occur in Pakistan, but the Indian clade was in the area where WTGs associated with CLCuD are spreading, suggesting correlation between these viruses and the vector *B. tabaci* haplotypes (Simon et al. 2003).

In China, six biotypes, B, Q, ZHJ1, ZHJ2, ZHJ3 and FJ1 of *B. tabaci* have been identified the most dominant B-biotype of *B. tabaci* appeared in China in the late 1990s and by 2005–2006 it was the only biotype of *B. tabaci* prevalent in many regions of the country (Wan et al. 2009). Emergence of B-biotype coincided with the WTG epidemics in many crops. The B-biotypes not only transmitted WTGs efficiently, but also showed better survival on infected plants compared to the indigenous biotype ZHJ1. The population of B–biotype grew at a similar rate on TYLCCNV infected and healthy tomato plants, whereas biotype ZHJ1 did not do well on infected plants (Liu et al. 2009a, b) and also transmitted TYLCV less efficiently (Li et al. 2010). However, no such preference was observed with B-biotype of whitefly (Matsuura and Hoshino 2009). The indigenous *B. tabaci* occurring in north-western parts of India also prefer healthy cotton plants compared to leaf curl affected plants. This preference may provide advantage for rapid spread of WTGs in cotton crops (Mann et al. 2009).

Unlike the situation in the Americas, where *B. tabaci* had a narrow host range before the B-biotype appeared, in India *B. tabaci* has been known to be polyphagous with a very wide host range. This may explain the occurrence of a large number of WTGs in the region (Varma and Malathi 2003). During the last three decades, however, there has been a perceptible change in the biology of *B. tabaci* in India. In northern India, until the late 1970s, whitefly population build-up on plants coincided with the onset of WTG infection in the freshly sown *kharif* (wet-season) crops. The ‘summer’ crops, that preceded wet-season crops, remained free from WTGs, as did the winter crops that followed wet-season crops. The crops that suffered most during the wet-season were the grain legumes, particularly mungbean. During the 1960s, mungbean varieties were developed that could withstand the high day time temperatures (>40°C) of the summer months so that it could be grown during the vector-free period to avoid damage by MYMIV. The area under summer mungbean cultivation increased until the mid 1980s when MYMIV began affecting summer crops too (Varma et al. 1992), as the whiteflies became active even in summer months and their population increased nearly 20-fold (Varma and Malathi 2003). Like the Indian sub-continent, *B. tabaci* in West Africa also have wide host range. The *B. tabaci* populations feeding on cassava in Southern and Western Africa are found to have five major haplotypes based on mitochondrial cytochrome oxidase I sequence identity (Berry et al. 2004). The emergence and prevalence of WTGs associated with tomato diseases in Israel, Spain and Nicaragua is also correlated with increase in *B. tabaci* populations (Varma and Malathi 2003).

Whitefly populations are greatly influenced by weather conditions. Their populations build up under conditions of high humidity and high temperatures. The pandemic of cassava mosaic disease in East Africa and the increase in begomovirus problems in the Americas, are also associated with unusually high populations of whiteflies (Fargette et al. 2006; Polston and Anderson 1997). Emergence of WTGs in cucurbits in 1990 in Northern India is attributed to unusual increase in whitefly populations early in the cucurbit growing season due to unexpected rise in day and night temperatures. Normally, whiteflies do not fly long distances but hover above the crop canopy during the day and settle on the abaxial leaf surfaces during the night (Varma and Malathi 2003). Whiteflies are, however, passively weather-driven over long distances during the day, resulting in the spread of WTGs. The emergence of BGMV in Florida in 1992 was attributed to viruliferous whiteflies blown in from the Caribbean basin by Hurricane Andrew (Blair et al. 1995). In the north-west of the Indian subcontinent dust storms are common phenomena during the summer months. The emergence of CLCuD and OLCuD in the mid 1990s in the cotton belt of India was attributed to viruliferous whiteflies being blown in from Pakistan, by the dust-storms in May and June (Varma et al. 1995; Varma and Malathi 2003). Thus the diseases caused by WTGs can occur in new geographical areas through the introduction of wind-blown viruliferous *B. tabaci*. In Indonesia, increase of WTG infections in chillies and tomato has been assigned to the 'invasion' of mainland Asian *B. tabaci* and Asian WTGs from Central Thailand. The studies have also shown integration of the genome of Asian *B. tabaci* with the Indonesia *B. tabaci* (De Barro et al. 2008).

10.8.6 Introduction of Crops or Genotypes Susceptible to Viruses Endemic in a Region

Cassava provides a good example of indigenous viruses infecting a new host. Cassava has been under cultivation in Latin America for over 4000 years and was taken to Africa, by the European traders in the sixteenth century, and it was introduced into Asia in the eighteenth century as a food and starch crop. By the nineteenth and twentieth centuries, it becomes an important crop in Southern China, Southern India, Indonesia, Malaysia, the Philippines and Thailand (Howeler 2005). Interestingly cassava plantations in the Americas and South-East Asia are free from any disease caused by WTG, but the crop is severely affected by CMD in Africa, Southern India and Sri Lanka (Fig. 10.2) where the disease is caused by indigenous viruses. This is a classic example of the emergence of threatening WTG problems by the introduction of new crops in agricultural systems. Cassava mosaic diseases continue to be serious problems in the new cropping systems due to the evolution of virulent variants, as discussed earlier in this chapter.

Exchange of germplasm, an essential component of crop improvement programmes for introducing desirable genes from diverse sources, may also lead to the introduction of undesirable genes including the genes for susceptibility to biotic stresses (Varma and Malathi 2003). This was amply demonstrated by emergence of three different diseases caused by WTGs in the Indian cropping systems. MYMIV, endemic in the Indian subcontinent, did not infect cowpea (*V. unguiculata*), under field or experimental conditions until the late 1970s. In 1978, many exotic accessions of cowpea were introduced from West Africa, and for the first time CPGMD was observed in a few of these plants. By 1984, CPGMD became the most severe constraint to cowpea production in India. The disease was caused by a variant of MYMIV that emerged after the introduction of susceptible cowpea germplasm (Varma et al. 1992; Roy and Malathi 2001).

Leaf curl diseases are the most serious diseases caused by WTGs. In India, cotton and okra were not affected by leaf curl until late 1980s. CLCuD was first observed in an experimental field at New Delhi in a genotype of *G. barbadense*. As WTGs are not seed transmissible, CLCuD must have been caused by an endemic WTG which could infect the susceptible genotype of *G. barbadense,* but not the genotypes of *G. hirsutum*. However, in subsequent years, *G. hirsutum* genotypes were also infected (Varma 1993). During the same period, leaf curl disease also emerged in some okra accessions introduced from West Africa, where it is the most severe disease of the crop (Varma 1984, 1990). Mixed infection by the WTGs causing yellow vein mosaic and leaf curl diseases results in very severe disease rendering the plants non-productive (Varma and Malathi 2003).

10.8.7 Emergence of Geminivirus Problems with Changes in the Cropping Systems

An increase in the area growing soybean, a favoured host of *B. tabaci*, resulted in an increase in vector populations and the emergence of BDMV and BGMV as serious problems in Argentina, Bolivia and Brazil (Morales and Anderson 2001). In India, introduction of summer crop of mungbean under irrigated conditions resulted in the unseasonal appearance of MYMIV (Varma et al. 1992), and advancement of potato planting time in the North Indian plains resulted in the emergence of a serious apical leaf curl disease (Varma and Malathi 2003). Changes in the farming systems in

Soroti district of Uganda lead to an outbreak of devastating cassava mosaic disease and subsequent decline in the production of cassava (Hall and Clark 2010).

In the wake of developing 'green' technologies to check global warming bio-fuel crops are being increasingly grown across the world. One of the important biofuel source, *Jatropha curcas*, which is being promoted in India, has been found infected with an isolate of ICMV (Gao et al. 2010). It may have a far reaching consequence of increase in cassava mosaic disease.

10.9 Concluding Remarks

Epidemics caused by the emergence and spread of WTGs are a serious concern to sustainable crop production, particularly in the tropics and sub-tropics. Some of the WTG associated disease problems have emerged due to (a) the development of virulent strains/isolates by recombinant events and mutations under selection pressure created by the deployment of resistant genotypes of the hosts, (b) acquisition of satellite DNA molecules with the ability to enhance virulence of the associated WTGs, (c) introduction of crops/genotypes susceptible to the indigenous WTGs, (d) changes in the cropping system, (e) appearance of *B. tabaci* biotypes with greater ability to survive and transmit WTGs, (f) weather factors, and (g) inadvertent movement of infected planting material across the world. The rate at which WTG associated plant disease problems are evolving is alarming. There is an urgent need to have a better understanding of the factors which lead to the increase in the WTG related problems in diverse cropping systems. The changing scenario of the emergence and spread of WTGs particularly in the regions favoring the spread of WTGs suggests greater increase in the occurrence of WTG associated epidemics in the coming years.

Deployment of plant varieties resistant to the WTGs and the insect vector *B. tabaci* is the most preferred approach to contain the WTG associated plant disease epidemics. However, the development of WTG resistant varieties requires a ceaseless effort as the fast evolution of resistance breaking strains/isolates, necessitate quick varietal replacement. Resistance to WTGs has been mainly developed by traditional breeding. Recently, transgenic varieties have also shown promise to resist WTG infection; some of these are expected to come into commercial use in the coming years. Notwithstanding the strength of using varietal resistance to contain WTGs, long term solutions require integrated approaches to manage the complexities of the emerging WTG associated disease problems, through concerted efforts of the virologists, biotechnologists, breeders, entomologists and agronomists.

Acknowledgements The authors thank Dr. Winston Thompson for his kind invitation to contribute this Chapter, and Mr. Zile Singh for his help in various ways in the preparation of the manuscript.

References

Abraham A (1956) Tapioca cultivation in India, Farm bulletin no.17. Indian Council of Agricultural Research, New Delhi, pp 20

Adjata KD, Muller E, Peterschmitt M, Aziadekey M, Gumedzoe YMD (2008) Incidence of cassava viral diseases and first identification of East African cassava mosaic virus and Indian cassava mosaic virus by PCR in cassava (Manihot esculenta Crantz) fields in Togo. Am J Plant Physiol 3:73–80

Adjata KD, Muller E, Peterschmitt M, Traore O, Gumedzoe YMD (2009) Molecular evidence for the association of a strain of Uganda variant of East African Cassava Mosaic Virus to symptom severity in cassava (Manihot esculenta Crantz) fields in Togo. Am J Biochem Biotechnol 5:196–201

Ajlan AM, Ghanem GAM, Abdulsalam KS (2007) Tomato yellow leaf curl virus (TYLCV) in Saudi Arabia: Identification, partial characterization and virus-vector Relationship. Arab J Biotechnol 10:179–192

Akhtar JK, Al-Saady NA, Al-Mahruki MS, Al-Qufi M, Al-Subhi AM (2007) Molecular characterization of *Begomovirus* infecting sweet pepper in Oman. Indian J Biotechnol 6:45–51

Akhtar KP, Haidar S, Khan MKR, Ahmad M, Sarwar N, Murtaza MA, Aslam M (2010) Evaluation of *Gossypium* species for resistance to cotton leaf curl Burewala virus. Ann Appl Biol 157:135–147

Akhter A, Qazi J, Saeed M, Mansoor S (2009) A severe leaf curl disease on Chilies in Pakistan is associated with multiple begomovirus components. Plant Dis 93:962

Alabi OJ, Kumar PL, Mgbechi-Ezeri JU, Naidu RA (2010) Two new 'legumoviruses' (genus Begomovirus) naturally infecting soybean in Nigeria. Arch Virol 155:643–656

Al-Ali EHM, Al-Hashash H, Hejji AB (2009) Epidemiology of TYLCV and ToMoV in greenhouse grown tomato in Kuwait. Aspects Appl Biol 94:55–62

Ala-Poikela M, Svensson E, Rojas A, Horko T, Paulin L, Valkonen JPT, Kvarnheden A (2005) Genetic diversity and mixed infections of begomoviruses infecting tomato, pepper and cucurbit crops in Nicaragua. Plant Pathol 54:448–459

Albuquerque LC, Martin DP, Avila AC, Inoue-Nagata AK (2010) Characterization of tomato yellow vein streak virus, a begomovirus from Brazil. Virus Genes 40:140–147

Ali I, Malik AH, Mansoor S (2010) First report of Tomato leaf curl Palampur virus on bitter gourd in Pakistan. Plant Dis 94:276

Ali-Shtayeh MS, Jamous RM, Husein EY, Alkhader MY (2010) First report of *Squash leaf curl virus* in squash (*Cucurbita pepo*), melon (*Cucumis melo*), and cucumber (*Cucumis sativa*) in the Northern West Bank of the Palestinian Authority. Plant Dis 94:640

Amrao L, Amin I, Shahid MS, Briddon RW, Mansoor S (2010a) Cotton leaf curl disease in resistant cotton is associated with a single begomovirus that lacks an intact transcriptional activator protein. Virus Res 152:153–163

Amrao L, Akhter S, Tahir MN, Amin I, Briddon RW, Mansoor S (2010b) Cotton leaf curl disease in Sindh province of Pakistan is associated with recombinant begomovirus components. Virus Res 153:161–165

Aragão FJL, Faria JC (2009) First transgenic geminivirus-resistant plant in the field. Nat Biotechnol 27:1086–1088

Ascencio-Ibanez JT, Diaz-Plaza R, Mendez-Lozano J, Monsalve-Fonnegra ZI, Arguello-Astorga GR, Rivera Bustamante RF (1999) First report of Tomato yellow leaf curl geminivirus in Yucatan, Mexico. Plant Dis 83:1178

Banks GK, Colvin J, Reddy RVC, Maruthi MN, Muniyappa V, Venkatesh HM, Kumar KM, Padmaja AS, Beitia FJ, Seal SE (2001) First report of the *Bemisia tabaci* B biotype in India and an associated *Tomato leaf curl virus* disease epidemic. Plant Dis 85:231

Barnabas AD, Radhakrishnan GK, Ramakrishnan U (2010) Characterization of a begomovirus causing horsegram yellow mosaic disease in India. Eur J Plant Pathol 127:41–51

Bedford ID, Briddon RW, Jones P, Alkaff N, Markham PG (1994) Differentiation of three whitefly-transmitted geminiviruses from Republic of Yemen. Eur J Plant Pathol 100:243–257

Begum SN, Khan MS (1996) Tomato leaf curl virus in Bangladesh. In: Proceedings of the phase I final workshop of the South Asian vegetable research network, Kathmandu, pp 210–215

Berry SD, Fondong VN, Rey C, Rogan D, Fauquet CM, Brown JK (2004) Molecular evidence for five distinct *Bemisia tabaci* (Homoptera: Aleyrodidae) geographic haplotypes associated with cassava plants in sub-Saharan Africa. Ann Entomol Soc Am 97:852–859

Biswas KK, Malathi VG, Varma A (2008) Diagnosis of symptomless Yellow mosaic begomovirus infection in pigeonpea by using cloned Mungbean yellow mosaic India virus as probe. J Plant Biochem Biot 17:9–14

Blair MW, Bassett MJ, Abouzid AM, Hiebert E, Polston JE, McMillan RT Jr, Graves W, Lamberts M (1995) Occurrence of bean golden mosaic virus in Florida. Plant Dis 79:529–533

Botermans M, Verhoeven JThJ, Jansen CCC, Roenhorst JW, Stijger CCMM, Pham KTK (2009) First report of Tomato yellow leaf curl virus in tomato in the Netherlands. Plant Dis 93:1073

Briddon RW, Markham PG (2000) Cotton leaf curl virus disease. Virus Res 71:151–159

Briddon RW, Mansoor S, Bedford ID, Pinner MS, Markham PG (2000) Clones of cotton leaf curl geminivirus induce symptoms atypical of cotton leaf curl disease. Virus Genes 20:19–26

Briddon RW, Mansoor S, Bedford ID, Pinner MS, Saunders K, Stanley J, Zafar Y, Malik KA, Markham PG (2001) Identification of DNA components required for induction of cotton leaf curl disease. Virology 285:234–243

Briddon RW, Bull SE, Amin I, Idris AM, Mansoor S, Bedford ID, Dhawan P, Rishi N, Siwatch SS, Abdel Salam AM, Brown JK, Zafar Y, Markham PG (2003) Divesity of DNA β, a satellite molecule associated with some monopartite begomoviruses. Virology 312:106–121

Briddon RW, Bull SE, Amin I, Mansoor S, Bedord ID, Rishi N, Siwatch SS, Zafar Y, Abdel Salam AM, Markham PG (2004) Divesity of DNA 1: a satellite-like molecule associated with nonopartite begomovirus-DNA β complexes. Virology 324:462–474

Briddon RW, Bull SE, Bedford ID (2006) Occurrence of *Sweet potato leaf curl virus* in Sicily. Plant Pathol 55:286

Briddon RW, Patil BL, Bagewadi B, Nawaz-ul-Rehman MS, Fauquet CM (2010) Distinct evolutionary histories of the DNA-A and DNA-B components of bipartite begomoviruses. BMC Evol Biol 10:97

Brown JK (1990) Serrano Golden mosaic virus: a newly identified whitefly-transmitted geminivirus of pepper and tomato in the United States and Mexico. Plant Dis 74:720

Cai JH, Xie K, Lin L, Qin BX, Chen BS, Meng JR, Liu YL (2010) *Cotton leaf curl Multan virus* newly reported to be associated with cotton leaf curl disease in China. Plant Pathol 59:794–795

Capoor SP, Varma PM (1948) Yellow mosaic of *Phaseolus lunatus*. L Curr Sci 17:152–153

Chang Ho-Hsiung Ku, Hsin-Mei Tsai Wen-Shi, Rui-Che C, Fuh-Jyh J (2010) Identification and characterization of a mechanical transmissible begomovirus causing leaf curl on oriental melon. Eur J Plant Pathol 127:219–228

Chatterjee A, Ghosh SK (2007) Association of a satellite DNA β molecule with Mesta yellow vein mosaic disease. Virus Genes 35:835–844

Chen LF, Rojas M, Kon T, Gamby K, Xoconostle-Cazares B, Gilbertson RL (2009) A severe symptom phenotype in tomato in Mali is caused by a reassortant between a novel recombinant begomovirus (Tomato yellow leaf curl Mali virus) and a betasatellite. Mol Plant Pathol 10:415–430

Chiemsombat P (1992) Mungbean yellow mosaic disease in Thailand: a review. In: Green SK, Deo-Hwan (eds.) Mungbean yellow mosaic disease: Proceedings of an international workshop, Asian Vegetable Research and Development Center, Taipei, pp 54–58

Chung ML, Liao CH, Chen MJ, Chiu RJ (1985) The isolation, transmission and host range of sweet potato leaf curl agent in Taiwan. Plant Prot Bull (Taiwan) 27:333–341

Cohen S, Nitzany FE (1960) A whitefly transmitted virus of cucurbits in Israel. Phytopathol Mediterr 1:44–46

Cohen J, Milgram M, Antignus Y, Pearlsman M, Lachman O, Loebenstein G (1997) Ipomoea crinkle leaf curl caused by a whitefly-transmitted gemini-like virus. Ann Appl Biol 131:273–282
Colvin J, Omongo CA, Maruthi MN, Otim-Nape GW, Thresh JM (2004) Dual begomovirus infections and high *Bemisia tabaci* populations: two factors driving the spread of a cassava mosaic disease pandemic. Plant Pathol 53:577–584
Comes S, Pacella R, Fanigliulo A, Crescenzi A (2009) Further spreading of Tomato yellow leaf curl Sardinia virus in Southern Italy. Acta Hortic 808:199–202
Cook OF (1924) Acromania or 'crazytop', a growth disorder of cotton. J Agric Res 28:803–828
Costa AS (1965) Three whitefly-transmitted virus diseases of beans in Sao Paulo, Brazil. FAO Plant Prot B 13:121–130
Crescenzi A, Comes S, Napoli C, Fanigliulo A, Pacella R, Accotto GP (2004) Severe outbreaks of tomato yellow leaf curl Sardinia virus in Calabria, Southern Italy. Commun Agric Appl Biol Sci 69:575–580
Czosnek H, Laterrot H (1997) A worldwide survey of tomato yellow leaf curl viruses. Arch Virol 142:1391–1400
Czosnek H, Navot N, Laterrot H (1990) Geographical distribution of tomato yellow leaf curl virus. A first survey using a specific DNA probe. Phytopathol Mediterr 24:1–6
D'Hondt MD, Russo M (1985) Tomato yellow leaf curl in Senegal. Phytopathol Z 112:153–160
Dalmon A, Cailly M (2001) Risk of introduction of TYLCV in France. In: Abstracts of the EPPO workshop on TYLCV, Faro, p 8
Das S, Ghosh R, Paul S, Anirban R, Ghosh SK (2008) Complete nucleotide sequence of a monopartite begomovirus associated with yellow vein mosaic disease of mesta from north India. Arch Virol 153:1791–1796
Davino S, Napoli C, Dellacroce C, Miozzi L, Noris E, Davino M, Accotto GP (2009) Two new natural begomovirus recombinants associated with the tomato yellow leaf curl disease co-exist with parental viruses in tomato epidemics in Italy. Virus Res 143:15–23
De Barro PJ, Hidayat SH, Frohlich D, Subandiyah S, Ueda S (2008) A virus and its vector, pepper yellow leaf curl virus and Bemisia tabaci, two new invaders of Indonesia. Biol Invasions 10:411–433
de Sá PB, Seebold KW, Vincelli P (2008) First report of tomato yellow leaf curl virus in greenhouse tomatoes in Kentucky. Plant Health Prog. doi:10.1094/PHP-2008-0819-01-RS
Diaz-Pendon JA, Canizares MC, Moriones E, Bejarano ER, Czosnek H, Navas-Castillo J (2010) Tomato yellow leaf curl viruses: menage a trois between the virus complex, the plant and the whitefly vector. Mol Plant Pathol 11:441–450
Dominguez M, Ramos PL, Sanchez Y, Crespo J, Andino V, Pujol M, Borroto C (2009) Tobacco mottle leaf curl virus, a new begomovirus infecting tobacco in Cuba. Plant Pathol 58:786
Domínguez M, Ramos PL, Echemendía AL, Peral R, Crespo J, Andino V, Pujol M, Borroto C (2002) Molecular characterization of tobacco leaf rugose virus, a new begomovirus infecting tobacco in Cuba. Plant Dis 86:1050–1050
Dong JH, Luo YQ, Ding M, Zhang ZK, Yang CK (2007) First report of *Tomato yellow leaf curl China virus* infecting kidney bean in China. Plant Pathol 56:342
Duffy S, Holmes EC (2009) Validation of high rates of nucleotide substitution in geminiviruses: phylogenetic evidence from East African cassava mosaic viruses. J Gen Virol 90:1539–1547
Durham TC, Baker C, Jones L, Snyde LU (2010) First report of *Sida golden mosaic virus* infecting snap bean (*Phaseolus vulgaris*) in Florida. Plant Dis 94:487
Fargette D, Konate G, Fauquet C, Muller E, Peterschmitt M, Thresh JM (2006) Molecular ecology and emergence of tropical plant viruses. Annu Rev Phytopathol 44:235–260
Faria JC, Maxwell DP (1999) Variability in geminivirus isolates associated with *Phaseolus* spp. in Brazil. Phytopathology 89:262–268
Faria RAE, Nava A (2009) PCR detection of begomoviruses in tomato crop producing areas of the Venezuelan Andes. Rev Fac Agron 26:179–195
Fauquet CM, Sawyer S, Idris AM, Brown JK (2005) Sequence analysis and classification of apparent recombinant begomoviruses infecting tomato in the Nile and Mediterranean Basins. Phytopathology 95:549–555

Fauquet CM, Briddon RW, Brown JK, Moriones E, Stanley J, Zerbini M, Zhou X (2008) Geminivirus strain demarcation and nomenclature. Arch Virol 153:783–821
Fazeli R, Heydarnejad J, Massumi H, Shaabanian M, Varsani A (2009) Genetic diversity and distribution of tomato-infecting begomoviruses in Iran. Virus Genes 38:311–319
Fernandes JJ, Carvalho MG, Andrade EC, Brommonschenkel SH, Fontes EPB, Zerbini FM (2006) Biological and molecular properties of *Tomato rugose mosaic virus* (ToRMV), a new tomato-infecting begomovirus from Brazil. Plant Pathol 55:513–522
Fernandes FR, Cruz ARR, Faria JC, Zerbini FM, Aragao FJL (2009) Three distinct begomoviruses associated with soybean in central Brazil. Arch Virol 154:1567–1570
Fiallo-Olive E, Martinez-Zubiaur Y, Rivera-Bustamante RF (2009a) Tomato yellow leaf distortion virus, a new bipartite begomovirus infecting tomato in Cuba. Plant Pathol 58:785
Fiallo-Olive E, Rivera-Bustamante RF, Martinez-Zubiaur Y (2009b) Tobacco yellow crinkle virus, a new bipartite begomovirus infecting tobacco and pepper in Cuba. Plant Pathol 58:785
Fiallo-Olive E, Navas-Castillo J, Moriones E, Martı'nez-Zubiaur Y (2010) Two novel begomoviruses belonging to different lineages infecting *Rhynchosia minima*. Arch Virol 155:2053–2058
Font MI, Córdoba C, García A, Santiago R, Jordá C (2005) First report of tobacco as a natural host of *tomato yellow leaf curl virus* in Spain. Plant Dis 89:910
Fuentes S, Salazar LF (2003) First report of sweet potato leaf curl virus in Peru. Plant Dis 87:98
Gamez-Jimenez C, Romero-Romero JL, Santos-Cervantes ME, Leyva-Lopez NE, Mendez-Lozano J (2009) Tomatillo (Physalis ixocarpa) as a natural new host for Tomato yellow leaf curl virus in Sinaloa. Mexico Plant Dis 93:545
Gao ShiQiang Qu, Jing Chua Nam-Hai, Jian Y (2010) A new strain of Indian cassava mosaic virus causes a mosaic disease in the biodiesel crop Jatropha curcas. Arch Virol 155:607–612
Garcia-Andres S, Tomas DM, Navas-Castillo J, Moriones E (2009) Resistance-driven selection of begomoviruses associated with the tomato yellow leaf curl disease. Virus Res 146:66–72
Garg ID, Paul-Khurana SM, Kumar S, Lakra BS (2001) Association of a geminivirus with potato apical leaf curl in India and its immuno-electron microscopic detection. J Indian Potato Assoc 28:227–232
Ghosh R, Paul S, Das S, PalitP AS, Das A, Mir JI, Ghosh SK, Roy A (2008) Molecular evidence for existence of a New World begomovirus associated with yellow mosaic disease of *Corchorus capsularis* in India. Australas Plant Dis Notes 3:59–62
Glick E, Levy Y, Gafni Y (2009) The viral etiology of tomato yellow leaf curl disease – a review. Plant Prot Sci 45(3):81–97
Ha C, Coombs S, Revill P, Harding R, Vu M, Dale J (2008) Molecular characterization of begomoviruses and DNA satellites from Vietnam: additional evidence that the New World geminiviruses were present in the Old World prior to continental separation. J Gen Virol 89:312–326
Haider MS, Bedford ID, Evans AAF, Markham PG (2003) Geminivirus transmission by different biotypes of the whitefly *Bemisia tabaci* (Gennadius). Pak J Zool 35:343–351
Haider MS, Tahir M, Latif S, Briddon RW (2005) First report of Tomato leaf curl New Delhi virus infecting Eclipta prostrata in Pakistan. New Dis Rep 11:39
Hall A, Clark N (2010) What do complex adaptive systems look like and what are the implications for innovation policy? J Int Dev 22:308–324
Hall GC, Graham AP, Roye ME (2008) Tobacco leaf curl Cuba virus infects the weed *Malachra alceifolia* in Jamaica. Plant Pathol 57:398
Hameed S (1996) Leaf curl virus of tomatoes and chillies in Pakistan. In: Proceedings of the phase I final workshop of the South Asian vegetable research network, Kathmandu, pp 242–259
Hameed S, Robinson DJ (2004) Begomoviruses from mungbeans in Pakistan: epitope profiles, DNA A sequences and phylogenetic relationships. Arch Virol 149:809–819
Hampton RO, Thottapilly G (2003) Cowpea. In: Thottappilly G, Loebenstein G (eds.) Virus and virus-like diseases of major crops in developing countries. Kluwer, Dordrecht
Haque AHMM, Saha R, Alam MM, Khalequzzaman KM, Yasmin L (2008) Study of insect transmission of Jute leaf mosaic virus and management through use of insecticide. Int J Sustain Crop Prod 3:64–67

Harrison BD, Robinson DJ (1999) Natural genomic and antigenic variation in whitefly-transmitted geminiviruses (Begomoviruses). Annu Rev Phytopathol 37:369–398

Harrison BD, Liu YL, Khalid S, Hameed S, Otim-Nape GW, Robinson DJ (1997a) Detection and relationships of cotton leaf curl virus and allied whitefly transmitted geminiviruses occurring in Pakistan. Ann Appl Biol 130:61–75

Harrison BD, Zhou X, Otim-Nape GW, Liu Y, Robinson DJ (1997b) Role of a novel type of double infection in the geminivirus-induced epidemic of severe cassava mosaic in Uganda. Ann Appl Biol 131:437–448

Hernandez-Zepeda C, Isakeit T, Scott A Jr, Brown JK (2010) First report of *Okra yellow mosaic Mexico virus* in okra in the United States. Plant Dis 94:924

Hernández-Zepedao C, Brown JK, Moreno-Valenzuela OA, Argüello-Astorga G, Idris AM, Carnevali G, Rivera-Bustamante RF (2010) Characterization of Rhynchosia yellow mosaic Yucatan virus, a new recombinant begomovirus associated with two fabaceous weeds in Yucatan, Mexico. Arch Virol 155:1571–1579

Heydarnejad J, Mozaffari A, Massumi H, Fazeli R, Gray AJA, Meredith S, Lakay F, Shepherd DN, Martin DP, Varsani A (2009) Complete sequences of tomato leaf curl Palampur virus isolates infecting cucurbits in Iran. Arch Virol 154:1015–1018

Honda Y, Ikegami M (1986) Mungbean yellow mosaic virus. AAB Description Plant Viruses 323:4

Hong Y, Wang X, Tian B, Cai J (1995) Chinese squash leaf curl virus: a new whitefly- transmitted geminivirus. Sci China B38:179–186

Howeler R (2005) Cassava in Asia: present situation and its future potential in agro-industry. In: Setiawan A, Fuglie K (eds.) Proceedings of the international symposium on Sweet potato, Agricultural University, Bogor, International Potato Center CIP 2005, pp 36. ciat-library.ciat.cgiar.org

Hussain M, Mansoor S, Amin I, Iram S, Zafar Y, Malik KA, Briddon RW (2003) First report of cotton leaf curl disease affecting chili peppers. New Dis Rep 7:31

Idris AM, Brown JK (2002) Molecular analysis of cotton leaf curl virus – Sudan reveals an evolutionary history of recombination. Virus Genes 24:249–256

Idris AM, Brown JK (2004) Cotton leaf crumple virus is a distint Western Hemisphere begomovirus species with complex evolutionary relationship indictive of recombination and reassortment. Phytopathology 94:1068–1074

Idris AM, Lujan KM, Martin K, Brown JK (2008) *Melon chlorotic leaf curl virus*: characterization and differential reassortment with closest relatives reveal adaptive virulence in the *squash leaf curl virus* clade and host shifting by the host-restricted *bean calico mosaic virus*. J Virol 82:1959–1967

Ilyas M, Qazi J, Mansoor S, Briddon RW (2009) Molecular characterisation and infectivity of a "Legumovirus" (genus Begomovirus: family Geminiviridae) infecting the leguminous weed Rhynchosia minima in Pakistan. Virus Res 145:279–284

Ito T, Sharma P, Kittipakorn K, Ikegami M (2008) Complete nucleotide sequence of a new isolate of tomato leaf curl New Delhi virus infecting cucumber, bottle gourd and muskmelon in Thailand. Arch Virol 153:611–613

Jalaluddin M, Shaikh MAQ (1981) Evaluation of mungbean *(Vigna radiata* (L.) Wilczek) germplasm for resistance to mungbean yellow mosaic virus. Sabrao J 13:61–68

Jebbour F (2001) Situation du TYLCV au Maroc et les methodes du diagnostic utilisees. In: Abstract of the EPPO workshop on TYLCV, Faro, p 3

John P, Sivalingam PN, Haq QM, Kumar N, Mishra A, Briddon RW, Malathi VG (2008) Cowpea golden mosaic disease in Gujarat is caused by a Mungbean yellow mosaic India virus isolate with a DNA B variant. Arch Virol 153:1359–1365

Jose J, Usha R (2003) Bhendi yellow vein mosaic disease in India is caused by association of a DNA β satellite with a begomovirus. Virology 305:310–317

Joshi S, Shrestha K, Timila RD, Shrestha SK (1996) Occurrence of tomato leaf curl virus in Nepal. In: Proceedings of the phase I final workshop of the South Asian vegetable research network, Kathmandu, pp 35–241

Kato S, Torisawa E, Yoshida K, Saito Y, Suganuma K, Iida F, Fuji S, Fukuta S, Fukaya M (2009) Infection cycle of tomato yellow leaf curl virus (TYLCV) and preventing methods of TYLCV on nursery period. Res Bull Aichi Agric Res Cent 40:69–76

Khan MS, Raj SK, Singh R (2006) First report of *Tomato leaf curl New Delhi virus* infecting chilli in India. Plant Pathol 55:289

Kheyr-Pour A, Bananej K, Dafalla GA, Caciagli P, Noris E, Ahoonmanesh A, Lecoq H, Gronenborn B (2000) *Watermelon chlorotic stunt virus* from the Sudan and Iran: sequence comparisons and identification of a whitefly-transmission determinant. Phytopathology 90:629–635

Kirthi N, Priyadarshini CGP, Sharma P, Maiya SP, Hemalatha V, Sivaraman P, Dhawan P, Rishi N, Savithri HS (2004) Genetic variability of begomoviruses associated with cotton leaf curl disease originating from India. Arch Virol 149:2047–2057

Knierim D, Maiss E (2007) Application of Phi29 DNA polymerase in identification and full-length clone inoculation of tomato yellow leaf curl Thailand virus and tobacco leaf curl Thailand virus. Arch Virol 152:941–954

Kon T, Rojas MR, Abdourhamane IK, Gilbertson RL (2009) Roles and interactions of begomoviruses and satellite DNAs associated with okra leaf curl disease in Mali, West Africa. J Gen Virol 90:1001–1013

Konate G, Barro N, Fargette D, Swanson MM, Harrison BD (1995) Occurrence of whitefly-transmitted geminiviruses in crops in Burkina Faso and their serological detection and differentiation. Ann Appl Biol 126:121–129

Kulkarni CS (1924) Mosaic and other related diseases of crops in the Bombay Presidency. Poona Agric Coll Mag, p 16

Kumar Y, Vipin H, Zaidi AA (2008) Molecular characterization of a distinct bipartite begomovirus species infecting tomato in India. Virus Genes 37:425–431

Kumar J, Kumar A, Khan JA, Aminuddin (2009a) First report of Papaya leaf curl virus naturally infecting tobacco in India. J Plant Pathol 91(S4):107

Kumar PL, Akinbade SA, Dixon AGO, Mahungu NM, Mutunda MP, Kiala D, Londa L, Legg JP (2009b) First report of the occurrence of *East African cassava mosaic virus*-Uganda (EACMV-UG) in Angola. Plant Pathol 58:402

Kumar A, Kumar J, Khan JA (2010) Sequence characterization of cotton leaf curl virus from Rajasthan: phylogenetic relationship with other members of geminiviruses and detection of recombination. Virus Genes 40:282–289

Kumari P, Chattopadhyay B, Singh AK, Chakraborty S (2009) A new begomovirus species causing tomato leaf curl disease in Patna. India Plant Dis 93:545

Kumari P, Singh AK, Chattopadhyay B, Chakraborty S (2010) Molecular characterization of a new species of Begomovirus and betasatellite causing leaf curl disease of tomato in India. Virus Res 152:19–29

Kundagrami S, Basak J, Maiti S, Kundu A, Das B, Ghose TK, Pal A (2009) Agronomic, genetic and molecular characterization of MYMIV-tolerant mutant lines of *Vigna mungo*. Int J Plant Breed Genet 3:1–10

Kwak HR, Jung MN, Kim MK, Lee M, Park JW, Lee SH, Choi HS (2007) Symptoms and complete nucleotide sequence analysis of sweet potato leaf curl virus transmitted by *Bemisia tabaci*. J Insect Sci 8(4). In: Stansly SA, McKenzie CL (eds.) Fourth international Bemisia workshop international whitefly genomics workshop. Available online: <insectscience.org/8.04>. Abstract, p 26

Lefeuvre P, Martin DP, Harkins G, Lemey P, Gray AJA (2010) The spread of Tomato yellow leaf curl virus from the Middle East to the World. PLoS Pathog 6(10):e1001164. doi:10.1371/journal.ppat.1001164

Legg JP (1999) Emergence, spread and strategies for controlling the pandemic of cassava mosaic virus disease in east and central Africa. Crop Prot 18:627–637

Lett JM, Lefeuvre F, Naze F, Delatte H, Mohamed-Ali Y, Reynaud B (2006) First report of *Tobacco leaf curl Zimbabwe virus* affecting tobacco in the Comoros Archipelago. Plant Pathol 55:567

Li M, Hu J, Xu FC, Liu SS (2010) Transmission of Tomato yellow leaf curl virus by two invasive biotypes and a Chinese indigenous biotype of the whitefly Bemisia tabaci. Int J Pest Manag 56:275–280

Ling KS, Simons AM, Hassell RL, Keinath AP, Polston JE (2006) First report of *Tomato yellow leaf curl virus* in South Carolina. Plant Dis 90:379

Ling KS, Jackson DM, Harrison H, Simmons AM, Pesic- VanEsbroeck Z (2010) Field evaluation of yield effects on the U.S.A. heirloom sweetpotato cultivars infected by Sweet potato leaf curl virus. Crop Prot 29:757–765

Liu P, Xie Y, Zhou XP (2009a) *Malvastrum coromandelianum* is an alternative host of *Tomato yellow leaf curl China virus*. Plant Pathol 58:403

Liu J, Zhao H, Jiang K, Zhou XP, Liu SS (2009b) Differential indirect effects of two plant viruses on an invasive and an indigenous whitefly vector: implications for competitive displacement. Ann Appl Biol 155:439–448

Lobin K, Druffel KL, Pappu HR, Benimadhu SP (2010) First report of *Tomato yellow leaf curl virus* in Tomato in Mauritius. Plant Dis 94:1261

Lotrakul P, Valverde RA (1999) Cloning of a DNA-A like genomic component of sweet potato leaf curl virus: nucleotide sequence and phylogenetic relationships. Mol Plant Pathol. http://www.bspp.org.uk/mppol/1999/0206LOKTRAKUL

Lotrakul P, Valverde RA, Clark CA, Sim J, De La Torre R (1998) Detection of a geminivirus infecting sweet potato in the United States. Plant Dis 82:1253–1257

Lotrakul P, Valverde RA, Clark CA, Fauquet CM (2001) Properties of a new geminivirus species from sweet potato infected with sweet potato leaf curl virus. In: Abstracts of the 3rd international geminivirus symposium, John Innes Centre, Norwich, p 110

Louro D, Fernandes JE, Quinot A (2002) Current situation of *Tomato yellow leaf curl virus* in Portugal. EPPO Bull 32:47

Lozano G, Trenado HP, Valverde RA, Navas-Castillo J (2009) Novel begomovirus species of recombinant nature in sweet potato (*Ipomoea batatas*) and *Ipomoea indica*: taxonomic and phylogenetic implications. J Gen Virol 90:2550–2562

Luan YS, Zhang J, An LJ (2006) First report of *Sweet potato leaf curl virus* in China. Plant Dis 90:1111

Makesh kumar T, Prakash Krishnan BS, Hegde V, Jeeva ML, Edison S (2007) Sweet potato leaf curl disease – a new emerging virus problem of sweet potato in India. In: Program of abstracts, 10th international plant virus epidemiology symposium, controlling epidemics of emerging and established plant virus diseases – the way forward ICRISAT, India, pp 1–33

Malathi VG, Varma A, Nambisan B (1989) Detection of Indian cassava mosaic virus by ELISA. Curr Sci 58:149–150

Malathi VG, Radhaakrishan G, Varma A (2003) Cotton. In: Thottapilly G, Lobenstein G (eds.) Virus and virus-like diseases of major crops in developing countries. Kluwer, Dordrecht

Malik IA (1992) Breeding for resistance to MYMV and its vector in Pakistan. In: Green SK, Doo-Hwan Kin (eds.) Mungbean yellow mosaic disease: Proceedings of an international workshop, Asian Vegetable Research and Development Center, Taipei, pp 41–53

Mandal B, Varma A, Malathi VG (1997) Systemic infection of *Vigna mungo* using the cloned DNAs of the blackgram isolate of mungbean yellow mosaic geminivirus through agroinoculation and transmission of the progeny virus by whiteflies. J Phytopathol 145:505–510

Mandal B, Langston DB Jr, Pappu HR (2001) First report of *Cabbage leaf curl virus* (Family *Geminiviridae*) in Georgia. Plant Dis 85:561

Mandal B, Mandal S, Sohrab SS, Pun KB, Varma A (2004) A new yellow mosaic disease of chayote in India. Plant Pathol 53:797

Mann RS, Sidhu JS, Butter NS (2009) Settling preference of the whitefly *Bemisia tabaci* (Hemiptera: Aleyrodidae) on healthy versus cotton leaf curl virus-infected cotton plants. Int J Trop Insect Sci 29:57–61

Mansoor S, Khan SH, Bashir A, Saeed M, Zafar Y, Malik KA, Briddon R, Stanley J, Markham PG (1999) Identification of a novel circular single-stranded DNA associated with cotton leaf curl disease in Pakistan. Virology 259:190–199

Mansoor S, Mukhtar S, Hussain M, Amin I, Zafar Y, Malik KA, Markham PG (2000) Widespread occurrence of *Cotton leaf curl virus* on Radish in Pakistan. Plant Dis 84:809–809

Martinez Y, Quinones M, Fonseca D (2003) National surveys of tomato begomovirus in Cuba. Rev Proteccion Vegetal 18:168–175

Martinez Y, Quinones M, Palenzuela I, Muniz Y (2006) Diversity of begomoviruses present in Cuba. Rev Proteccion Vegetal 21:149–154

Maruthi MN, Colvin J, Briddon RW, Bull SE, Muniyappa V (2003) Pumpkin yellow vein mosaic virus: a novel begomovirus infecting cucurbits. J Plant Pathol 85:64–65

Maruthi MN, Rekha AR, Mirza SH, Alam SN, Colvin J (2007) PCR-based detection and partial genome sequencing indicate high genetic diversity in Bangladeshi begomoviruses and their whitefly vector, *Bemisia tabaci*. Virus Genes 34:373–385

Matsuura S, Hoshino S (2009) Effect of tomato yellow leaf curl disease on reproduction of *Bemisia tabaci* Q biotype (Hemiptera: Aleyrodidae) on tomato plants. Appl Entomol Zool 44:143–148

McGovern RJ, Polston JE, Danyluk GM, Hiebert E, Abouzid AM, Stansly PA (1994) Identification of natural weed host of tomato mottle geminivirus in Florida. Plant Dis 78:1102–1106

McGrath PF, Harrison BD (1995) Transmission of tomato leaf curl geminiviruses by *Bemisia tabaci*: effects of virus isolate and vector biotype. Ann Appl Biol 126:307–316

Melzer MJ, Ogata DY, Fukuda SK, Shimabuku R, Borth WB, Sether DM, Hu JS (2010) First report of Tomato yellow leaf curl virus in Hawaii. Plant Dis 94:641

Mgbechi-Ezeri JU, Alabi OJ, Naidu RA, Kumar PL (2008) First report of the occurrence of African cassava mosaic virus in a mosaic disease of soybean in Nigeria. Plant Dis 92:1709

Miano DW, LaBonte DR, Clark CA, Valverde RA, Hoy MW, Hurtt S, Li R (2006) First report of a begomovirus infecting sweet potato in Kenya. Plant Dis 90:832

Moffat AS (1999) Geminiviruses emerge as serious crop threat. Science 286:1835

Monci F, Garcia-Andres S, Maldonado JA, Moriones E (2005) Resistance to monopartite begomoviruses associated with the bean leaf crumple disease in *Phaseolus vulgaris* controlled by a single dominant gene. Phytopathology 95:819–826

Monger WA, Mumford RA, Antonio García E, Boa E (2007) Occurrence of Tomato mosaic Havana virus in Nicaragua. New Dis Rep 16:10

Morales FJ, Anderson PK (2001) The emergence and dissemination of whitefly-transmitted geminiviruses in Lain America. Arch Virol 146:415–441

Mubin M, Briddon RW, Mansoor S (2009) Complete nucleotide sequence of chili leaf curl virus and its associated satellites naturally infecting potato in Pakistan. Arch Virol 154:365–368

Nadeem A, Weng Z, Nelson MR, Xiong Z (1997) Cotton leaf crumple virus and cotton leaf curl virus are two distantly related geminiviruses. Mol Plant Pathol. http//www.bspp.org.uk/mppol.1997/0612nadeem

Nariani TK (1960) Yellow mosaic of mung (*Phaseolus aureus* L.). Ind Phytopathol 13:24–29

Narula AM, Monga D, Chauhan MS, Raj S (1999) Cotton leaf curl virus disease in India – the challenge ahead. J Cotton Res Dev 13:129–138

Nateshan HM, Muniyappa V, Swanson MM, Harrison BD (1996) Host range, vector relations and serological relationships of cotton leaf curl virus from southern India. Ann Appl Biol 128:233–244

Navas-Castillo J, Sanchez-Campos S, Noris E, Louro D, Accotto GP, Moriones E (2000) Natural recombination between *Tomato yellow leaf curl virus*-Is and *Tomato leaf curl virus*. J Gen Virol 81:2797–2801

Ndunguru J, Legg JP, Aveling TAS, Thompson G, Fauquet CM (2005) Molecular biodiversity of cassava begomoviruses in Tanzania: evolution of cassava geminiviruses in Africa and evidence for East Africa being a center of diversity of cassava geminiviruses. Virol J 2:21

Nene YL (1973) Viral diseases of some warm weather pulse crops in India. Plant Dis Rep 57:463–467

Noris E, Hidalgo E, Accotto GP, Moriones E (1994) High similarity among the tomato yellow leaf curl virus isolates from west Mediterranean Basin: the nucleotide sequence of an infectious clone from Spain. Arch Virol 135:165–170

Ogawa T, Sharma P, Ikegami M (2007) First report of a strain of *Tobacco leaf curl Japan virus* associated with a satellite DNA in honeysuckle in Japan. New Dis Rep 15:8

Ogbe FO (2006) Status of cassava begomoviruses and their new natural hosts in Nigeria. Plant Dis 90:548–553

Osaki T, Inouye T (1991) Transmission characteristics and cytopathology of a whitefly-transmitted virus isolated from the sweet potato leaf curl disease. Bulletin, University of Osaka Prefecture, series B. Agric Biol 43:11–19

Osaki T, Yamada M, Inouye T (1985) Whitefly transmitted viruses isolated from sweet potato, Abutilon pictum and Eupatorium japonica (preliminary report). Ann Phytopathol SOC Jpn 51:82 (Abstr. in Japanese)

Osei MK, Akromah R, Shih SL, Lee LM, Green SK (2008) First report and molecular characterization of DNA A of three distinct begomoviruses associated with Tomato leaf curl disease in Ghana. Plant Dis 92:1585

Otim-Nape GW, Thresh JM, Shaw MW (1998) The incidence and severity of cassava mosaic virus disease in Uganda: 1990–1992. Trop Sci 38:25–337

Padidam M, Beachy RN, Fauquet CM (1996) The role of AV2 ("precoat") and coat protein in viral replication and movement in tomato leaf curl geminivirus. Virology 224:390–404

Pal BP, Tandon RK (1937) Types of tobacco leaf curl in Northern India. Indian J Agr Sci 7:363–393

Papayiannis LC, Avgelis AD, Ioannou N, Katis NI (2007a) First report of Tomato yellow leaf curl Sardinia virus (TYLCSV) infecting tomato crops in Greece. Plant Pathol 56:341

Papayiannis LC, Paraskevopoulos A, Katis NI (2007b) First report of Tomato yellow leaf curl virus infecting common bean (Phaseolus vulgaris) in Greece. Plant Dis 91:465

Paprotka T, Boiteux LS, Fonseca MEN, Resende RO, Jeske H, Faria JC, Ribeiro SG (2010) Genomic diversity of sweet potato geminiviruses in a Brazilian germplasm Bank. Virus Res 149:224–233

Patil BL, Fauquet CM (2010) Differential interaction between cassava mosaic geminiviruses and geminivirus satellites. J Gen Virol 91:1871–1882

Paul S, Ghosh R, Das S, Palit P, Acharyya S, Das A, Mir JI, Chaudhuri S, Ghosh SK, Roy A (2009a) First report of *Tomato leaf curl Joydebpur virus* and associated betasatellite in kenaf (*Hibiscus cannabinus*) plants showing leaf curl symptoms from southern India. Plant Pathol 58:403

Paul S, Ghosh R, Chaudhuri S, Ghosh SK, Roy A (2009b) Biological and molecular variability of the begomoviruses associated with leaf curl disease of Kenaf in India. J Plant Pathol 91:637–647

Phadvibulya V, Boonsirichai K, Adthalungrong A, Srithongchai W (2009) Selection for resistance to yellow vein mosaic virus disease of okra by induced mutation. In: Induced plant mutations in the genomics era. Proceedings of an international joint FAO/IAEA symposium, pp 349–351

Pico B, Diez MJ, Nuez F (1996) Viral diseases causing the greatest economic losses to the tomato crop II. The tomato yellow leaf curl virus-a review. Sci Hortic 67:151–196

Pita JS, Fondong VN, Sangare A, Otim-Nape GW, Ogwal S, Fauquet CM (2001) Recombination, pseudorecombination and synergism of geminiviruses are determinant keys to the epidemic of severe cassava mosaic disease in Uganda. J Gen Virol 82:655–665

Polston JE, Anderson PK (1997) The emergence of whitefly-transmitted geminiviruses of tomato in the western hemisphere. Plant Dis 81:1358–1369

Polston JE, Chellemi DO, Schuster DJ, McGovern RJ, Stansly PA (1996) Spatial and temporal dynamics of tomato mottle geminivirus and *Bemisia tabaci* (Genn.) in Florida tomato fields. Plant Dis 80:1022–1028

Polston JE, Bois D, Ano G, Poliakoff N, Urbino C (1998) Occurrence of a new strain of potato yellow mosaic geminivirus infecting tomato in the Eastern Caribbean. Plant Dis 82:126

Polston JE, McGovern RJ, Brown LG (1999) Introduction of tomato yellow leaf curl virus in Florida and implications for the spread of this and other geminiviruses in tomato. Plant Dis 83:984–988

Polston JE, Rosebrock TR, Sherwood T, Creswell T, Shoemaker PJ (2002) Appearance of Tomato yellow leaf curl virus in North Carolina. Plant Dis 86:173

Qazi J, Mansoor S, Amin I, Awan MY, Briddon RW, Zafar Y (2006) First report of *Mungbean yellow mosaic India virus* on mothbean in Pakistan. Plant Pathol 55:818

Qing L, Xiong Y, Sun XC, Yang SY, Zhou CY (2010) First report of Tobacco curly shoot virus infecting pepper in China. Plant Dis 94:637

Radhakrishnan G, Malathi VG, Varma A (2001a) Novel features of cotton leaf curl disease in India. In: Abstracts of the 3rd international geminivirus symposium, John Innes Centre, Norwich, p 53

Radhakrishnan G, Malathi VG, Varma A (2001b) Are extra DNA components a regular feature of diseases caused by WTGs? In: Abstracts of the 3rd international geminivirus symposium, John Innes Centre, Norwich, p 54

Rafaelo MG, Andrea CM, Dirce FL, Poliane FA, Eduardo CA, Francisco MZ, Márcia RA, Elizabeth PBF (2003) A naturally occurring recombinant DNA-A of a typical bipartite begomovirus does not require the cognate DNA-B to infect *Nicotiana benthamiana* systemically. J Gen Virol 84:715–726

Raj SK, Singh BP (1996) Association of geminivirus infection with yellow green mosaic disease of *Cucumis sativus*: diagnosis by nucleic acid probes. Indian J Exp Biol 34:603–605

Raj SK, Khan MS, Singh R (2005) Natural occurrence of a begomovirus on pigeonpea in India. Plant Pathol 54:809

Raj SK, Khan MS, Snehi SK, Srivastava S, Singh HB (2006a) First report of *Tomato leaf curl Karnataka virus* infecting soybean in India. Plant Pathol 55:817

Raj SK, Khan MS, Snehi SK, Srivastava S, Singh HB (2006b) A yellow mosaic disease of soybean in Northern India is caused by cotton leaf curl Kokhran virus. Plant Dis 90:975–975

Raj SK, Khan MS, Snehi SK, Roy RK (2007) Yellow vein netting of Bimili jute (*Hibiscus cannabinus* L.) in India caused by a strain of *Tomato leaf curl New Delhi virus* containing DNA β. Australas Plant Dis Notes 2:45–47

Raj SK, Snehi SK, Khan MS, Singh R, Khan AA (2008a) Molecular evidence for association of Tomato leaf curl New Delhi virus with leaf curl disease of papaya (Carica papaya L.) in India. Australas Plant Dis Notes 3:152–155

Raj SK, Snehi SK, Kumar S, Khan MS, Pathre U (2008b) First molecular identification of a begomovirus in India that is closely related to *Cassava mosaic virus* and causes mosaic and stunting of Jatropha curcas L. Australas Plant Dis Notes 3:69–72

Raj SK, Snehi SK, Khan MS, Tiwari AK, Rao GP (2010) First report of *Pepper leaf curl Bangladesh virus* strain associated with bitter gourd (*Momordica charantia* L.) yellow mosaic disease in India. Aust Plant Dis Notes 5:14–16

Rajinimala N, Rabindran R (2007) First report of *Indian cassava mosaic virus* on bittergourd (Momordica charantia) in Tamil Nadu, India. Australas Plant Dis Notes 2:81–82

Ramos N, Fernandes JE, Arsenio AF, Mangerico S, Neto E (2002) Control of the *Bemisia tabaci/Tomato yellow leaf curl virus* complex in tomato nurseries in Portugal. EPPO Bull 32:37–38

Ravishankar KV, Aghora TS, Mohan N, Naveen AH, Krishnareddy M (2009) Identification of RAPD marker linked to Mungbean yellow mosaic virus resistance in French bean (*Phaseolus vulgaris* L.). J Hortic Sci 4:167–169

Reddy MK, Venkataravanappa V, Madhuvanthi B, Jalali S (2010) Molecular characterization of begomoviruses associated with papaya leaf curl disease in India. Acta Hortic 851:465–472

Ribeiro SG, de Avila AC, Bezerra IC, Fernandes JJ, Faria JC, Lima MF, Gibbertson RL, Maciel-Zambolim E, Zerbini FM (1998) Widespread occurrence of tomato geminiviruses in Brazil, associated with the new biotype of the whitefly vector. Plant Dis 82:830

Ribeiro SG, Ambrozevicius LP, Avila AC, Bezerra IC, Calegario RF, Fernandes JJ, Lima MF, Mello RN, Rocha H, Zerbini FM (2003) Distribution and genetic diversity of tomato-infecting begomoviruses in Brazil. Arch Virol 148:281–295

Ribeiro SG, Martin DP, Lacorte C, Simoes IC, Orlandini DR, Inoue-Nagata AK (2007) Molecular and biological characterization of Tomato chlorotic mottle virus suggests that recombination underlies the evolution and diversity of Brazilian tomato begomoviruses. Phytopathology 97:702–711

Rochester DE, de Paulo JJ, Fauquet CM, Beachy RN (1994) Complete nucleotide sequence of the geminivirus tomato yellow leaf curl virus, Thailand isolate. J Gen Virol 75:477–485

Rodrigo AV, Jeonggu S, Pongtharin L (2004) Whitefly transmission of sweet potato viruses. Virus Res 100:123–128

Rodríguez-Pardina PE, Hanada K, Laguna IG, Zerbini FM, Ducasse DA (2011) Molecular characterisation and relative incidence of bean- and soybean-infecting begomoviruses in northwestern Argentina. Ann Appl Biol 158:69–78

Rojas A, Kvarnheden A, Valkonen JPT (2000) Geminiviruses infecting tomato crops in Nicaragua. Plant Dis 84:843–846

Rojas A, Kvarnheden A, Marcenaro D, Valkonen JP (2005) Sequence characterization of Tomato leaf curl Sinaloa virus and Tomato severe leaf curl virus: phylogeny of New World begomoviruses and detection of recombination. Arch Virol 150:1281–1289

Rojas MR, Kon T, Natwick ET, Polston JE, Akad F, Gilbertson RL (2007) First report of Tomato yellow leaf curl virus associated with Tomato yellow leaf curl disease in California. Plant Dis 91:1056

Rossel HW, Thottappilly G (1985) Virus diseases of important food crops in tropical Africa. International Institute of Tropical Agriculture, Ibadan

Roy A, Malathi VG (2001) Molecular cloning of cowpea golden mosaic geminivirus and its relationship with mungbean yellow mosaic geminivirus. Trop Agric Res 13:341–352

Rybicki EP, Pietersen G (1999) Plant virus disease problems in the developing world. Adv Virus Res 53:127–175

Sabina I, Munshi AD, Mandal B, Kumar R, Behera TK (2010) Genetics of resistance in Luffa cylindrica Roem. against Tomato leaf curl New Delhi virus. Euphytica 174:83–89

Salati R, Shorey M, Briggs A, Calderon J, Rojas MR, Chen LF, Gilbertson RL, Palmieri M (2010) First report of *Tomato yellow leaf curl virus* infecting tomato, tomatillo, and peppers in Guatemala. Plant Dis 94:482

Samad A, Gupta MK, Shasany AK, Ajayakumar PV, Alam M (2009) Begomovirus related to *Tomato leaf curl Pakistan virus* newly reported in Mentha crops in India. Plant Pathol 58:404

Sanz AI, Fraile A, Garcia-Arenal F, Zhou X, Robinson DJ, Khalid S, Butt T, Harrison BD (2000) Multiple infections, recombination and genome relationship among begomovirus isolates found in cotton and other plants in Pakistan. J Gen Virol 81:1839–1849

Saunders K, Bedford ID, Briddon RW, Markham PG, Wong SM, Stanley JA (2000) Unique virus complex causes Ageratum yellow vein disease. Proc Natl Acad Sci USA 97:6890–6895

Saunders K, Bedford ID, Stanley J (2001a) Pathogenicity of a natural recombinant associated with Ageratum yellow vein disease: Implications for geminivirus evolution and disease aetiology. Virology 282:38–47

Saunders K, Salim N, Mali VR, Malathi VG, Markham PG, Stanley J (2001b) Characterisation of Sri Lankan cassava mosaic virus: evidence for acquisition of a DNA B component by a monopartite begomovirus. In: Abstracts of the 3rd international geminivirus symposium, John Innes Centre, Norwich, p 104

Senanayake DMJB, Mandal B, Lodha S, Varma A (2007) First report of *Chilli leaf curl virus* affecting chilli in India. Plant Pathol 56:343

Shahid MS, Ali L, Wajid S (2009) Cotton leaf curl Rajasthan virus infecting tomato in Pakistan. Pak J Sci Ind R 52:319–321

Sharma SR, Varma A (1976) Cowpea yellow fleck, a whitefly transmitted disease of cowpea. Indian Phytopathol 29:421–423

Sharma P, Ikegami M, Kon T (2010) Identification of the virulence factors and suppressors of post-transcriptional gene silencing encoded by Ageratum yellow vein virus, a monopartite begomovirus. Virus Res 149:19–27

Shelly P, Kushwaha CM, Mishra AK, Singh V, Jain RK, Varma A (2005) Engineering tomato for resistance to tomato leaf curl disease using viral rep gene sequences. Plant Cell Tiss Organ Cult 83:311–318

Shih SL, Green SK, Tsai WS, Lee LM, Wang JT, Tesfaye A (2006) First report of a begomovirus associated with Tomato yellow leaf curl disease in Ethiopia. Plant Dis 90:974

Shih SL, Tsai WS, Green SK, Singh D (2007) First report of *Tomato leaf curl Joydebpur virus* infecting chilli in India. Plant Pathol 56:341

Shih SL, Kumar S, Tsai WS, Lee LM, Green SK (2009) Complete nucleotide sequences of okra isolates of Cotton leaf curl Gezira virus and their associated DNA- beta from Niger. Arch Virol 154:369–372

Shih SL, Tsai WS, Lee LM, Wang JT, Green SK, Kenyon L (2010) First report of *Tomato yellow leaf curl Thailand virus* associated with pepper leaf curl disease in Taiwan. Plant Dis 94:637

Simon B, Cenis JL, Beitia F, Khalid S, Moreno IM, Fraile A, Garcia Arenal F (2003) Genetic structure of field populations of begomoviruses and of their vector *Bemisia tabaci* in Pakistan. Phytopathology 93:1422–1429

Singh SR, Allen DJ (1979) Cowpea pests and diseases, manual series no. 2. International Institute of Tropical Agriculture, Ibadan, p 113

Singh AK, Chattopadhyay B, Pandey PK, Singh AK, Chakraborty S (2007) A new begomovirus species causing leaf curl disease of radish in India. Plant Dis 91:1053

Singh R, Raj SK, Prasad V (2008) Molecular characterization of a strain of *Squash leaf curl China virus* from North India. J Phytopathol 156:222–228

Singh AK, Mishra KK, Chattopadhyay B, Chakraborty S (2009) Biological and molecular characterization of a begomovirus associated with yellow mosaic vein mosaic disease of pumpkin from Northern India. Virus Genes 39:359–370

Sivalingam PN, Varma A (2007a) PCR based diagnosis of begomoviruses associated with Tomato leaf curl disease in India. J Plant Biochem Biot 16:17–22

Sivalingam PN, Varma A (2007b) Non-tomato natural hosts of tomato infecting begomoviruses in north-western India. Indian J Virol 18:20–27

Sivalingam PN, Malathi VG, Varma A (2004) Association of DNA β molecules with mono- and bipartite begomoviruses affecting tomato in India. In: 4th international geminiviruses conference, Cape town, 16/2. W4-1

Sivalingam PN, Malathi VG, Varma A (2010) Molecular diversity of the DNA- beta satellites associated with tomato leaf curl disease in India. Arch Virol 155:757–764

Sivanathan P (1977) Virus diseases of crops in Sri Lanka. Trop Agric Res Ser 10:65–68

Sohrab SS, Mandal B, Pant RP, Varma A (2003) First report of association of Tomato leaf curl New Delhi virus (ToLCNDV) with yellow mosaic disease of *Luffa cylindrical* in India. Plant Dis 87:1148

Sohrab SS, Mandal B, Ali A, Varma A (2006) Molecular diagnosis of emerging begomovirus diseases in cucurbits occurring in northern India. Indian J Virol 17:88–95

Sohrab SS, Mandal B, Ali A, Varma A (2010) Chlorotic curly stunt: a severe begomovirus disease of bottle gourd in Northern India. Indian J Virol 21:56–63

Srivastava N, Chandra R, Saxena S, Bajpai A (2010) PCR based amplification and detection of Papaya leaf curl virus (PaLCuV). Acta Hortic 851:241–245

Stanley J, Saunders K, Pinner MS, Wong SM (1997) Novel defective interfering DNAs associated with Ageratum yellow vein geminivirus infection of *Ageratum conyzoides*. Virology 239:87–96

Storey HH (1931) A new virus disease of the tobacco plant. Nature 128:187–188

Sun HX, Ji YH, Xiong RY, Zhao TM, Yu WG, Zhou YJ (2009) Trend of whitefly-transmitted geminivirus on tomato from Jiangsu Province in 2008. Jiangsu J Agric Sci 25:1278–1281

Swanson MM, Harrison BD (1994) Properties, relationships and distribution of cassava mosaic geminiviruses. Trop Sci 34:15–25

Swanson MM, Varma A, Muniyappa V, Harrison BD (1992) Comparative epitope profiles of the particle proteins of whitefly-transmitted geminiviruses from nine crop legumes in India. Ann Appl Biol 120:425–433

Tahir M, Haider MS (2006) First report of a begomovirus associated with leaf curl disease of bell pepperin Pakistan. Plant Pathol 55:570

Tahir M, Haider MS (2007) First report of tomato leaf curl New Delhi virus infecting bitter gourd in Pakistan. Plant Pathol 54:807

Tahir M, Haider MS, Briddon RW (2010a) Complete nucleotide sequences of a distinct bipartite begomovirus, bitter gourd yellow vein virus, infecting Momordica charantia. Arch Virol 155:1901–1905

Tahir M, Haider MS, Briddon RW (2010b) First report of Squash leaf curl China virus in Pakistan. Australas Plant Dis Notes 5:21–24

Thinlay Penjore U (1996) Survey of leaf curl virus of chilli and tomato in Bhutan. In: Proceedings of the phase I final workshop of the South Asian vegetable research network, Kathmandu, p 216

Thresh JM, Otim-Nape GW, Thankappan M, Muniyappa V (1998) The mosaic diseases of cassava in Africa and India caused by whitefly-borne geminiviruses. Rev Plant Pathol 77:935–945

Tiendrebeogo F, Traore VSE, Barro N, Traore AS, Konate G, Traore O (2008) Characterization of Pepper yellow vein mali virus in Capsicum sp. in Burkina Faso. Plant Pathol J Faisalabad 7:155–161

Tiendrebeogo F, Lefeuvre P, Hoareau M, Villemot J, Konate G, Traore AS, Barro N, Traore VS, Reynaud B, Traore O, Lett JM (2010) Molecular diversity of Cotton leaf curl Gezira virus isolates and their satellite DNAs associated with okra leaf curl disease in Burkina Faso. Virol J 7:48

Tiwari AK, Sharma PK, Khan MS, Snehi SK, Raj SK, Rao GP (2010a) Molecular detection and identification of Tomato leaf curl New Delhi virus isolate causing yellow mosaic disease in bitter gourd (*Momordica charantia*), a medicinally important plant in India. Med Plants 2:117–123

Tiwari N, Singh VB, Sharma PK, Malathi VG (2010b) Tomato leaf curl Joydebpur virus – a monopartite begomovirus causing severe leaf curl in tomato in West Bengal. In: Abstracts of the conference on whitefly and thrips transmitted viruses, New Delhi, p 78

Torres-Pacheco I, Garzon-Tiznado JA, Brown JK, Becerra-Flora A, Rivera-Bustamante RF (1996) Detection and distribution of geminiviruses in Mexico and the southern United States. Phytopathology 86:1186–1192

Tripathi S, Varma A (2003) Identification of sources of resistance Lycopersicon to Tomato leaf curl geminivirus (ToLCV) by agroinoculation. Euphytica 129:43–52

Tsai WS, Shih SL, Green SK, Lee LM, Luther GC, Ratulangi M, Sembel DT, Jan FJ (2009) Identification of a new begomovirus associated with yellow leaf curl diseases of tomato and pepper in Sulawesi. Indonesia Plant Dis 93:321

Usharani KS, Surendranath B, Anilkumar VA, Malathi VG (2001) Host range determinant resides in the DNA B of soybean isolate of Indian mungbean yellow mosaic virus (IMYMV-So). In: Abstracts of the 3rd international geminivirus symposium, John Innes Centre, Norwich, p 57

Usharani KS, Surendernath B, Paul-Khurana SM, Garg ID, Malathi VG (2004) Potato laf curl – a new disease of potato in northern India caused by a strain of Tomato leaf curl New Delhi virus. Plant Pathol 53:235

van Regenmortel MHV, Fauquet CM, Bishop DHL, Carstens EB, Estes MK, Lemon SM, Maniloff J, Mayo MA, McGeoch DJ, Pringle CR, Wickner RB (2000) Virus taxonomy, classification and Nomenclature of viruses. The 7th International Committee on Taxonomy of Viruses. Academic Press, San Diego.USA

Varma PM (1963) Transmission of plant viruses by whiteflies. Bulletin NISI, National Institute of Science in India, pp 11–33

Varma A (1984) Virus diseases of fruits and vegetable crops in Nigeria. FAO Technical Report AG/DP/NIR/72/007, FAO, Rome

Varma A (1990) Changing pattern of plant diseases caused by whitefly transmitted geminiviruses. In: Abstracts of the 8th international congress of virology, Berlin, p 31

Varma A (1993) Integrated management of plant viral diseases. In: Integrated pest management, Ciba foundation symposium, vol 155, Wiley, Chichester, pp 140–157

Varma A (2010) Emergence and spread of the viruses transmitted by whiteflies and thrips in the backdrop of 'Green Revolution'. In: Abstracts of the conference on whitefly and thrips transmitted viruses, New Delhi, p 9

Varma A, Biswas KK (2009) Viral diseases of pulses: emerging concerns. In: Souvenir, international conference on grain legumes. Indian Institute of Pulses Research, Kanpur, pp 63–69

Varma A, Giri BK (1998) Virus diseases of cucurbits in India. In: Nayar NM, More TA (eds.) Cucurbits. Oxford and IBH Publishing House Private Limited, New Delhi

Varma A, Malathi VG (2003) Emerging geminivirus problems: a serious threat to crop production. Ann Appl Biol 142:145–164

Varma A, Mandal B (2003) Other vegetables: amaranthus, chayote, colocasia and xanthosoma, egg plant, lablab, okra, onion, pea and sweet pepper. In: Loebenstine G, Thottappilly G (eds.) Virus and virus-like diseases of major crops in developing countries. Kluwer, Dordrecht

Varma A, Praveen S (2010) Phylogeographic evolution of plant viruses. Nature at work: ongoing saga of evolution. In: Sharma VP (ed.) Publication of the National Academy of Sciences, India. 1st edn. Springer (India) Private Limited, New Delhi

Varma A, Reddy DRR (1984) Golden and green mosaics- two new diseases of cowpea in northern India. Indian Phytopathol 37:409

Varma A, Shelly P (2006) GE tomato resistant to leaf curl disease. ISB news report

Varma A, Dhar AK, Mandal B (1992) MYMV transmission and control in India. In: Green SK, Doo-Hwan Kim (eds.) Mungbean yellow mosaic disease. Proceedings of an international workshop, Asian Vegetable Research and Development Center, Taipei, pp 8–27

Varma A, Malathi VG, Handa A, Aiton M, Harrison BD, Varma JP, Singh RP, Singh M, Srivastava M, Singh J (1993) Occurrence of leaf-curl of cotton and okra in Northern India. In: Abstracts of the 6th international congress of plant pathology, Montreal, pp 17.5.14

Varma A, Puri SN, Raj S, Bhardwaj RP, Kannan A, Jayaswal AP, Srivastava KP, Ajmera BD, Mallik F (1995) Leaf curl disease of cotton in North-West India. Report of the Indian Council of Agricultural Research Committee, New Delhi, pp 17

Varma A, Mandal B, Malathi VG (1998) Putative location of common region and coat protein gene of blackgram isolate of mungbean yellow mosaic geminivirus. J Plant Biochem Biot 7:07–12

Vasudeva RS, Samraj J (1948) A leaf curl disease of tomato. Phytopathology 38:364–369

Wan F, Zhang G, Liu S, Luo C, Chu D, Zhang Y, Zhang L, Jiu M, Lu Z, Cui X et al (2009) Invasive mechanism and management strategy of Bemisia tabaci (Gennadius) biotype B: progress report of 973 Program on invasive alien species in China. Sci China C Life Sci 52:88–95

Were HK, Winter S, Maiss E (2003) Distribution of begomoviruses infecting cassava in Africa. J Plant Pathol 83:145–151

Winter S, Butgereitt A, Thottappilly G (1999) Cowpea golden mosaic virus and related geminiviruses associated with Vigna spp. in Nigeria. Poster presentation, International Virology Congress, Sydney

Xie Y, Peijun Wu, Liu P, Gong H, Zhou X (2010) Characterization of alphasatellites associated with monopartite begomovirus/betasatellite complexes in Yunnan. China Virol J 7:178

Xiong Y, Qing L, Ren F, Li F, Sun XC (2010) First report of Tobacco curly shoot virus on Mirabilis jalapa Linn. in China. J Plant Pathol 92:546

Xu YP, Zhou XP (2007) A new begomovirus associated with leaf curl disease of Euphorbia pulcherrima. J Plant Pathol 89(Suppl 3):S69. 2

Xu YP, Cai XZ, Zhou XP (2007) Tomato leaf curl Guangxi virus is a distinct monopartite begomovirus species. Eur J Plant Pathol 118:287–294

Yadav RK, Shukla RK, Chattopadhyay D (2009) Soybean cultivar resistant to Mungbean Yellow Mosaic India Virus infection induces viral RNA degradation earlier than the susceptible cultivar. Virus Res 144:89–95

Yang CX, Wu ZJ, Xie LH (2009) First report of the occurrence of *Sweet potato leaf curl virus* in tall morningglory (*Ipomoea purpurea*) in China. Plant Dis 93:764

Yassin AM, Nour MA (1965) Tomato leaf curl disease in the Sudan and their relation to tobacco leaf curl. Ann Appl Biol 56:207–217

Yu WG, Zhao TM, Yang ML, Zhao LP, Ji YH, Zhou YJ (2009) PCR detection and sequence analysis of whitefly-transmitted geminivirus in tomato from Anhui and Shandong provinces. Jiangsu J Agric Sci 25:747–751

Zambrano K, Carballo O, Geraud F, Chirinos D, Fernández C, Marys E (2007) First report of Tomato yellow leaf curl virus in Venezuela. Plant Dis 91:768–768

Zhang H, Hu G, Zhou X (2010) Molecular characterization of Tomato leaf curl Hainan virus, a new begomovirus, and evidence for recombination. J Phytopathol 158:829–832

Zhou X, Liu Y, Calvert L, Munoz C, Otim-Nape GW, Robinson DJ, Harrison BD (1997) Evidence that DNA-A of a geminivirus associated with severe cassava mosaic disease in Uganda has arisen by interspecific recombination. J Gen Virol 78:2101–2111

Zhou X, Liu Y, Robinson DJ, Harrison BD (1998) Four DNA- a variants among Pakistani isolates of cotton leaf curl virus and their affinities to DNA-A of geminivirus isolates from okra. J Gen Virol 79:915–923

Zhou XP, Xie Y, Tao X, Zhang Z, Li Z, Fauquet CM (2003) Characterization of DNA β associated with begomoviruses in China and evidence for co-evolution with their cognate viral DNA -A. J Gen Virol 84:237–247

Zoysa IJ (1996) Leaf curl virus of tomato in Sri Lanka. In: Proceedings of the phase I final workshop of the South Asian vegetable research network, Kathmandu, pp 265–269

Chapter 11
Management of *Bemisia tabaci* Whiteflies

A. Rami Horowitz, Yehezkel Antignus, and Dan Gerling

Abstract This review presents and discusses the merits of the methodologies available for implementing integrated pest management (IPM) of *B. tabaci* populations: namely, chemical control with selective insecticides, biological control, crop plant resistance and physical/mechanical methods. Insecticides, by their poisonous nature, are often harmful to natural enemies and therefore, disruptive to overall pest management. However, the more modern materials that are effective for *B. tabaci* control are relatively specific to the target pests, and therefore are less harmful to natural enemies and the environment; consequently, they are also more suitable for integrative combination with other methods. Natural enemies, by themselves, usually do not form a suitable solution of *B. tabaci*- caused problems. However, their occurrence and use greatly reduces the pest's populations. Since viral plant diseases transmitted by *B. tabaci* are not curable, the principal tactics for their management should be based on prevention of transmission by physical-mechanical methods and/or on utilization of host-plant resistance. The correct implementation of natural enemies will help to reduce whitefly numbers, which can then be more readily managed using cultural and, only if necessary, chemical countermeasures.

A.R. Horowitz (✉)
Department of Entomology, Agricultural Research Organization (ARO), Gilat Research Center, M.P. Negev 85280, Israel
e-mail: hrami@volcani.agri.gov.il

Y. Antignus
Agricultural Research Organization (ARO), The Volcani Center, Institute of Plant Protection, P.O.B. 6, Bet Dagan 50250, Israel
e-mail: antignus@volcani.agri.gov.il

D. Gerling
Department of Zoology, Tel Aviv University, Ramat Aviv 69978, Israel
e-mail: dange@tauex.tau.ac.il

W.M.O. Thompson (ed.), *The Whitefly, Bemisia tabaci (Homoptera: Aleyrodidae) Interaction with Geminivirus-Infected Host Plants*,
DOI 10.1007/978-94-007-1524-0_11,

Thus, adopting IPM will alleviate the numerous concerns that accompany the use of chemicals, including those associated with environmental pollution and the widespread resistance that plagues *B. tabaci* management.

11.1 Introduction

The whitefly *Bemisia tabaci* (Gennadius) (Hemiptera: Aleyrodidae) is a pest of many agricultural systems including various vegetable, ornamental and field crops (Byrne and Bellows 1991; Oliveira et al. 2001; Stansly and Naranjo 2010). It directly damages the plants by feeding on phloem sap, and excretes honeydew on the leaves and fruit. The sticky, sugary surface forms a substrate for the growth of black sooty mold fungi that stain the crop and cover the leaves, thus preventing proper photosynthesis. The resulting stickiness and discoloration greatly reduce the value of agricultural crops such as ornamentals, vegetables and cotton. In the latter, the honeydew may cause fiber stickiness that interferes with the spinning process in the textile mills, and greatly reduces the product's value (Hequet et al. 2007).

B. tabaci is a vector of several important families of plant viruses (Jones 2003; Hogenhout et al. 2008) (see also Chap. 1) and in some crops (e.g. tomatoes and cassava) the resulting virus diseases are limiting growth-factors and may cause total crop loss. Most of the important virus diseases transmitted by *B. tabaci* belong to the geminivirus group (family: *Geminiviridae*).

B. tabaci is known for its genetic diversity, which is expressed in a complex of biotypes (Brown et al. 1995; Perring 2001; De Barro et al. 2005) or, as recently suggested, a complex of separate species (Xu et al. 2010; De Barro et al. 2011). The biotypes are largely differentiated based on biochemical or molecular polymorphism, and differ in characteristics such as host plant range, the capacity to cause plant disorders, attraction by natural enemies, expression of resistance, and plant virus-transmission capabilities (e.g. Bedford et al. 1994; Brown et al. 1995; Sanchez-Campos et al. 1999; Perring 2001; Horowitz et al. 2005). Recent reports have suggested that the floral composition of bacterial symbionts might be specific to certain biotypes (Gottlieb et al. 2006; Chiel et al. 2007) and might confer upon them resistance to insecticides (Kontsedalov et al. 2008). The most widespread biotype, B, was recognized in the late 1980s (Costa and Brown 1991; Costa et al. 1993), following extensive outbreaks of *B. tabaci* in the southwestern USA, and has a worldwide distribution. An additional widespread biotype, Q, which probably originated in the Iberian Peninsula (Guirao et al. 1997), has since spread globally (Horowitz et al. 2003; Boykin et al. 2007; Chu et al. 2010).

Management of *B. tabaci* populations and, in particular, management of the viral plant diseases it transmits, is difficult. This is due to the pest's elevated population growth rates, rapid evolution of resistance to insecticides, and the relatively protected location of the individuals on the underside of the leaves. *B. tabaci* is highly polyphagous and is known to develop on over 500 plant species, including a large number of fiber, vegetable, and ornamental crops (Mound and Halsey 1978;

Oliveira et al. 2001). Another remarkable feature is its easy adaptation to changing environmental conditions, especially in subtropical and tropical agroecosystems and in greenhouse-grown crops even in temperate climates (Brown 2007; Castle et al. 2010). Brown (2007) proposed that monoculture cropping together with year-round production practices are mostly responsible for the present whitefly and viral disease outbreaks. Since viral plant diseases transmitted by *B. tabaci* are not curable, the principal strategies for their management are based on prevention of transmission (Antignus 2007), and/or on utilization of host-plant resistance (Lapidot and Friedmann 2002).

At present, the use of insecticides is the main approach employed to manage *B. tabaci* populations. This practice is greatly restricted; however, due to both environmental concerns and the widespread resistance that *B. tabaci* has developed to most of the insecticides in use (Palumbo et al. 2001; Horowitz et al. 2007; Castle et al. 2010). Consequently, increasing importance is being placed upon control by other methods (including cultural, mechanical and biological) as a means of managing pest populations.

This chapter reviews the known measures used for reducing populations of *B. tabaci*, advocating the view that only an encompassing approach, incorporating integrated pest management (IPM) programs, will offer effective and sustainable strategies for managing whiteflies.

11.2 Chemical Control: New Agents, and Insecticide Resistance Management

Several comprehensive reviews of chemical control against *B. tabaci* and insecticide resistance in this pest have been published during the last two decades (e.g., Denholm et al. 1996; Horowitz and Ishaaya 1996; Palumbo et al. 2001; Horowitz et al. 2007; Castle et al. 2010). Here, we summarize the present situation with regard to insecticide efficacy and review information on newer insecticide groups and on novel, recently initiated ones for controlling whiteflies. The problems, associated with insecticide resistance and its management are also considered.

11.2.1 Conventional Insecticides

The most extensively used insecticide classes – organochlorines, organophosphates (OPs), carbamates, and pyrethroids – have generally been the most seriously threatened by resistance; in addition, there is a tendency to ban their use because of their detrimental effect to humankind and the environment.

Early reports (e.g. Sharaf 1986; Dittrich et al. 1990; Horowitz and Ishaaya 1996) presented data on more than 50 conventional insecticides for controlling populations and suppressing virus transmission by *B. tabaci*. The most common conventional insecticides used were carbamates, OPs and pyrethroids. However, many studies have pointed out the limited control achieved by these conventional insecticides in commercial fields and greenhouses, especially at high levels of infestation (e.g. Henneberry and Butler 1992; Henneberry 1993; Palumbo et al. 2001).

Until the mid 1990s, spray mixtures of synergized pyrethroids had been the most effective combination for controlling *B. tabaci* populations, especially in the southwestern USA (Horowitz and Ishaaya 1996; Prabhaker et al. 1998; Palumbo et al. 2001; Castle et al. 2010). These involved combining high levels of pyrethroids with moderate levels of compounds from a different chemical class such as organophosphates and carbamates. The increased efficacy of these mixtures can be attributed to the additive effects of both compounds along with the prevention of a rapid development of resistance (e.g. Denholm et al. 1998). However, the uncontrolled use of these synergized pyrethroids to manage problematic whitefly populations in cotton fields nonetheless resulted in reduced susceptibility in *B. tabaci* populations (Simmons and Dennehy 1996; Dennehy et al. 1997).

It is difficult to achieve comprehensive control of *B. tabaci* with conventional insecticides; the reasons include the under-leaf habitat of immature stages and adults, the presence of older larvae in the lower canopy of the crop, the pest's highly polyphagous nature, and the easy dispersion of adults by wind (Horowitz and Ishaaya 1996; Palumbo et al. 2001).

11.2.2 *Newer Insecticides for Controlling Whiteflies*

Most of the newer insecticides are preferable because of their specificity to target pests, their effectiveness at low rates and their non-persistent characteristics in the environment. Their selectivity renders many of them suitable for IPM programs (Casida and Quistad 1998).

11.2.2.1 Insect Growth Regulators (IGRs)

The need for a greater diversity of insecticides for *B. tabaci* control was met by the development of the non-neurotoxic IGRs (Horowitz and Ishaaya 1996). These potent control agents include in particular buprofezin (Ishaaya 1990) and pyriproxyfen (Ishaaya and Horowitz 1992). Another IGR, novaluron, an inhibitor of chitin synthesis belonging to the benzoylphenyl ureas (BPUs), also has some effect on *B. tabaci* (Ishaaya et al. 2003). Since buprofezin and pyriproxyfen have uniquely different modes of action and desirable biological and environmental profiles, they are considered important components in IPM programs for controlling whiteflies in cotton, vegetables and ornamentals (Horowitz et al. 1999; Ellsworth and Martinez-Carrillo 2001; Horowitz and Ishaaya 2004; Castle et al. 2010).

Buprofezin is a thiadizine-like compound with long residual activity that acts as a chitin synthesis inhibitor. It has both contact and vapor activity, and affects nymphal stages of sucking insects, especially whiteflies (Ishaaya et al.1988; De Cock and Degheele 1998). Its mode of action resembles that of BPUs, although its structure is not analogous. Because of chitin deficiency, the procuticle of the whitefly nymph loses its elasticity and the insect is unable to molt (De Cock and Degheele 1998). Similar to the BPUs, buprofezin is effective on immature stages and not on adults. The compound has a mild effect on natural enemies (De Cock and Degheele 1998). It has been used since the 1990s in glasshouses and fields in many regions of the world with resulting widespread resistance, reaching levels of 10- to 50-fold in greenhouses in the UK, the Netherlands, Spain, and Israel (Cahill et al. 1996; Horowitz et al. 1994; Elbert and Nauen 2000).

Pyriproxyfen is a potent juvenile hormone (JH) mimic affecting the hormonal balance in insects, suppressing embryogenesis, metamorphosis and adult formation (Itaya 1987; Langley 1990; Koehler and Patterson 1991). It has been considered a leading insecticide for controlling whiteflies (Ishaaya and Horowitz 1995; Horowitz et al. 2005; Crowder et al. 2008; Castle et al. 2010), especially biotype B. However, its widespread use has again resulted in resistance build-up in Israel since the 1990s (Horowitz and Ishaaya 1994; Horowitz et al. 1999, 2002, 2005). Recently, cases of strong resistance to pyriproxyfen have been associated with the Q rather than the B biotype (Horowitz et al. 2002, 2003, 2005; Dennehy et al. 2005). In addition, simulation models conducted with the B biotype have demonstrated that resistance to pyriproxyfen in this biotype can be managed through modification of operational and environmental factors that can be controlled by the growers (Crowder et al. 2008).

11.2.2.2 Neonicotinoid Insecticides

The neonicotinoids are among the most effective group of insecticides. They exhibit systemic and translaminar properties, and high residual activity (Elbert et al. 1990, 1998; Takahashi et al. 1992; Horowitz et al. 1998) especially against sucking insects such as whiteflies, aphids and leafhoppers, and various coleopteran pests such as the Colorado potato beetle, *Leptinotarsa decemlineata* (Say). They represent a class of chemicals, nicotine mimics, in which the biochemical target is the nicotinic acetylcholine receptor (nAChR) of both the central and peripheral nervous systems (Bai et al. 1991). Similar to naturally occurring nicotine, the neonicotinoids bind to these receptors, resulting in excitation and paralysis followed by insect death (Yamamoto et al. 1995; Tomizawa and Casida 2003).

Imidacloprid was the first commercial neonicotinoid successfully used for controlling agricultural pests. It has been widely used as a seed dressing and in soil applications, as it is considered a relatively polar material with good xylem mobility.

At present, in many greenhouses and open field crops, pest management has become particularly reliant on neonicotinoids. This has led to a neonicotinoid overuse that, coupled with a strong risk of cross-resistance between various chemicals of this group, threatens their effectiveness (Cahill and Denholm 1999; Li et al. 2001; Horowitz et al. 2004; Prabhaker et al. 2005; Wang et al. 2009). Indeed, increasing numbers of resistance cases to neonicotinoids (e.g. imidacloprid and thiamethoxam) have been reported, especially in the Q biotype of *B. tabaci* (Horowitz et al. 2004; Nauen and Denholm 2005; Roditakis et al. 2009; Luo et al 2010), but also in the B biotype (Wang et al. 2009; Schuster et al. 2010). Thus, it is of utmost importance to develop recommended resistance management strategies for this important group (Elbert et al. 1996; Nauen and Denholm 2005). Similar principles for combating insecticide resistance that apply to all chemical groups are also valid for neonicotinoids. These are based on limiting spatial/temporal exposure to key compounds (e.g. Dennehy and Denholm 1998), and alternating between products without cross-resistance (a difficult task in light of the decline in the introduction of new insecticides). Another important principal is the monitoring and detecting neonicotinoid resistance in order to recommend and implement effective resistance-management strategies (Denholm et al. 1998; Horowitz et al. 2004, 2007; Nauen and Denholm 2005).

11.2.2.3 Diafenthiuron, a Thiourea Derivative

Diafenthiuron is an effective whitefly controlling compound that has been used particularly in Europe and Israel as an alternative to pyriproxyfen for *B. tabaci* control in cotton since 1998 (Horowitz et al. 1999; Palumbo et al. 2001). The compound is a thiourea derivative with both insecticidal and acaricidal activities against some species of hemipterans and phytophagous mites (Steinemann et al.1990; Horowitz and Ishaaya 1996). In the presence of sunlight, diafenthiuron is phytochemically converted to a carbodiimide derivative enhancing its insecticidal activity (Steinemann et al. 1990). It directly affects insect respiration through the inhibition of oxidative phosphorylation and disruption of mitochondrial ATP synthesis (Ruder et al. 1991). Diafenthiuron suppresses the formation of whitefly progeny when adult females are exposed to treated plants (Ishaaya et al. 1993), and is more potent against nymphs than pupae or eggs. It is considered as one of a few whitefly adulticides that is still used effectively (ARH, 2010, personal communication). The compound has low mammalian toxicity with a relatively mild effect on natural enemies and pollinators (Streibert et al. 1988; De Cock et al. 1996); hence, it is considered an important component in IPM programs.

11.2.2.4 Pyridine Insecticides (Pymetrozine)

Pymetrozine, an azomethine pyridine insecticide, is highly specific against sucking insect pests (Flückiger et al. 1992a, b; Fuog et al. 1998). It affects the nerves controlling the salivary pump and causes immediate and irreversible cessation of feeding due to an obstruction of stylet penetration, followed by starvation and insect death (Kayser et al. 1994). The compound is a powerful toxicant against aphids, whiteflies (including *B. tabaci* and *Trialeurodes vaporariorum* [Westwood]) and plant hoppers.

Pymetrozine has systemic and translaminar activities and can be used as a drench or in foliar application (Flückiger et al. 1992a, b). It is also effective in lessening aphid-transmitted diseases caused by persistent viruses (Fuog et al. 1998); its action can be enhanced by the addition of mineral oil (Asjes and Blom-Barnhoorn 2002).

A recent report (Gorman et al. 2010) has unexpectedly demonstrated in *B. tabaci* the existence of cross-resistance between neonicotinoids and pymetrozine (although they differ in structural and functional characteristics); probably reflecting the overexpression of a cytochrome-P450 monooxygenase that is able to metabolize both types of compound. This finding is of utmost importance in insecticide resistance management (IRM) strategies for *B. tabaci* in which both neonicotinoids and pymetrozine should be applied in the same 'window of opportunity' (Gorman et al. 2010).

11.2.3 Novel Insecticides for Controlling *B. tabaci*

11.2.3.1 The Ketoenols: Derivatives of Tetronic Acids (Spiromesifen) and Tetramic Acids (Spirotetramat)

Spiromesifen

Spiromesifen belongs to a new class of pesticides that are derivatives of spirocyclic tetronic acid, which affects mainly whiteflies and mites. It acts as an inhibitor of lipid biosynthesis that interferes with development of the egg and immature stages and reduces adult female fecundity (Bretschneider et al. 2003; Nauen et al. 2005). Spiromesifen acts effectively on the egg and early nymphal stages of *B. tabaci* (both biotypes B and Q) but adults and late nymphal stages are only moderately affected (Prabhaker et al. 2008; Kontsedalov et al. 2009). A recent study has reported that eggs laid by treated *B. tabaci* females are smaller than the normal size and have an abnormally perforated chorion; hence, the females are unable to complete oviposition (Kontsedalov et al. 2009).

Laboratory and field results indicated that spiromesifen had favorable selectivity to natural enemies and pollinators such as *Orius laevigatus* (Fieber), *Eretmocerus mundus* Mercet, and bumblebees (Bielza et al. 2005, 2009). Several reports have pointed to the absence of cross-resistance between spiromesifen and other major commonly used insecticides from different chemical groups, such as neonicotinoids and pyriproxyfen, suggesting the possible use of spiromesifen in IRM programs (Nauen and Konanz 2005; Prabhaker et al. 2008; Kontsedalov et al. 2009).

Spirotetramat

Another insecticide belonging to the ketoenol group is spirotetramat, a novel spirocyclic tetramic acid derivative, and also a lipid biosynthesis inhibitor. Spirotetramat is a systemic insecticide with phloem and xylem mobility for the control of sucking insects, including aphids, whiteflies, psyllids and scales. It is particularly effective against juvenile stages of sucking pests and it significantly reduces fecundity and fertility of *B. tabaci* females (Brück et al. 2009). This insecticide is in its early stage of registration and marketing; hence, more laboratory and field studies are needed to evaluate its efficacy under field conditions.

11.2.3.2 Ryanodine Receptor Insecticides (the Diamides)

Ryanodine receptors are a class of ligand-gated calcium channels controlling the release of calcium from intracellular stores. Ryanodine is a plant alkaloid used as a natural botanical insecticide. Recently, two classes of synthetic agents have been developed for commercial compounds that target insect ryanodine receptors. So far, two insecticides are being studied and registered: Rynaxypyr®, which is more potent against lepidopteran pests; and Cyazypyr™, which targets sucking pests such as whiteflies and aphids as well as other types of insect pests (Sattelle et al. 2008; Lahm et al. 2009)

Cyantraniliprole (Cyazypyr™)

This is a new insecticide whose modes of action are based on the activation of ryanodine receptors causing the release and depletion of intracellular stores of calcium ions. It has very low mammalian toxicity, favorable ecotoxicity profiles and is a potential agent for controlling whiteflies, aphids, thrips, leafminers, caterpillars, and beetle pests. Since it is considered safe to beneficial arthropods, it should be studied in depth as it could be an important component in IPM/IRM programs for controlling *B. tabaci* and other whiteflies in vegetable and ornamental crops.

11.2.4 Drawbacks of Chemical Control for Management of B. tabaci Whiteflies

Although some biological and physical control methods as well as other approaches have been useful in the management of *B. tabaci*, the use of insecticides remains the primary means of control. In some cropping systems, the overuse of insecticides for such control often results in the development of resistance. An extreme example involves the activities of some tomato growers in their effort to control *B. tabaci* and *Tomato yellow leaf curl virus* (TYLCV) (Horowitz et al. 2007). Since whiteflies carrying the virus are able to infect a tomato plant with TYLCV within 4 h of inoculative feeding, insecticides with a quick killing effect on adults are required in order to prevent virus spread. Starting with two applications per week with conventional insecticides the growers reached a situation in which even daily application was ineffective due to buildup of resistance.

While chemical control for managing whitefly adults prior to virus transmission is still common, the use of other control tactics is preferable because of resistance and environmental concerns. Other control countermeasures, such as cultivars resistant to virus and/or vectors and physical barriers to immigration by insect pests, would be highly effective ways of reducing damage by viral diseases. Hence, insecticide applications against the vector should be used as a last measure in combating viral diseases; while the use of other control tactics would assist in moderating development of resistance to insecticides.

11.3 The Use of Natural Enemies for Biological Control of *B. tabaci*

The use of natural enemies for biological control is considered a very effective way for controlling insect pests (Van Driesche and Bellows 1996). Whiteflies feature on top of the list of successful cases in the history of biological control through the use of predators and parasitoids (Gerling 1990). However, *B. tabaci* has been an exception in as much as it often causes severe outbreaks and damage in spite of its extensive enemy fauna. Nonetheless, efforts to control this important pest biologically have not ceased and numerous studies on the enemy fauna as well as on the technology necessary for achieving biological control are continuing. Work on the subject has led to the discovery of numerous natural enemies including the possibility of specific biotype-natural enemy interactions in *B. tabaci* (Kirk et al. 2000). This section is intended to summarize the present state of the art and directs the reader to more extensive contributions on the subject (e.g. Gerling et al. 2001; Arno et al. 2010).

11.3.1 Biological Control

As discussed here, biological control constitutes the utilization of natural enemies for the decimation of pest populations. In our case, it refers to their use in the control of *B. tabaci*; if successful, this measure, especially when coupled with the development of plant varieties that are tolerant or resistant to viral infection can support a sustainable, insecticide-free plant culture. In this section, the natural enemy inventory is surveyed and their efficacy and use discussed.

Natural enemies of whiteflies include fungi, which are the only disease-causing organisms currently known to attack whiteflies, predators and parasitoids. Both generalist and specific natural enemies of whiteflies belonging to the *B. tabaci* complex are known. The former include the disease-causing fungi, most predators and many of the parasitoids. Only some parasitoid species (e.g. *Er. mundus*) are not known to grow on non-*Bemisia* whitefly species.

11.3.1.1 The Fungi

Bemisia species typically grow under arid and semi-arid conditions and therefore do not usually show spontaneous infections by insect-attacking fungi, as known for some citrus-infesting whiteflies. However, they are susceptible to most, if not to all the known whitefly-attacking fungi. Species of *Aschersonia*, *Beauveria*, *Metarhizium*, *Paecilomyces*, and *Verticillium* have all been found infecting *B. tabaci* and their use for its control has been attempted (Ravensberg et al. 1990; Knauf and Wright 1994; Bolckmans et al. 1995; Meekes et al. 2002). Currently, commercial products such as Mycotal® (*Verticillium lecanii*-m), Botanigard® (*Beauveria bassiana*) and PreFeRal® (*Paecilomyces fumosoroseus*) are available on the market; a more extensive list can be found in Stansly and Natwick (2010).

11.3.1.2 The Predators

Reviews of the work done with *B. tabaci* predators and extensive lists of predators, including 38 species of *Araneae* and 123 species of insects, are available (Gerling et al. 2001; Arno et al. 2010). About one third of the insect species listed (44 out of 123) were added during the period 2001–2010, which indicates the very extensive native elements in the different local faunas that are associated with *B. tabaci*.

The extent of naturally occurring predation has been investigated by several authors with varying results. Most notably, Asiimwe et al. (2007) found six different predator species while studying their effect on *B. tabaci* infesting cassava in Uganda, whereas Naranjo et al. (1998) determined that predation was by far the dominant mortality factor of *B. tabaci* populations in Arizona cotton. Other studies on *B. tabaci* predators dealt with their bionomics under laboratory conditions, mainly as a basis for estimating their future utility in pest management [see Stansly and Natwick (2010) for a review]. Recent studies have also indicated the mobility of many predators in nature and the relevance of their population dynamics in non-agricultural fields to their abundance and efficacy as predators. Hence, these have included the types and abundance of prey and the host plants on which they are found. In the zoophytophagous hemipteran species, molecular investigations have shown that predator movements, as related to the host plants, can be followed using molecular tracking techniques (Agusti et al. 2009). This is expected to facilitate the encouragement of the appropriate flora, in particular in the vicinity of *B. tabaci*-infested greenhouses and thus to improve biological control efforts.

Predators are being used to control *B. tabaci* principally under greenhouse conditions. The mite *Amblyseius swirskii* Athias-Henriot is considered one of the most effective natural enemies, being active on most vegetable species except tomatoes, and is used extensively. Several species of zoophytophagous hemipterans are used commercially, mainly *Macrolophus caliginosus* (Wagner) in northern Europe and *Nesidiocoris tenuis* Reuter in the Mediterranean basin. The sensitivity and particular affinity of predators for specific plant species, calls for studying and utilizing a spectrum of natural enemies. Thus, *B. tabaci* on gerbera plants is controlled to a large extent using a species of *Delphastus* (Coccinellidae), whereas the incompatibility of *A. swirskii* with tomato plants is overcome by using both additional predator species and parasitoids.

11.3.1.3 The Parasitoids

The life cycle of parasitoids includes an egg that is deposited in or under the whitefly nymph, a larva that feeds on the whitefly nymph, a pupa lodging within the dead whitefly skin and a free-living adult. All parasitoids of *B. tabaci* are solitary and the ones most used belong to the genera *Encarsia* and *Eretmocerus*. Host ranges vary, with most having a wide range as evidenced by the large number of new associations of the pest and parasitoid species that have been revealed following introduction of *B. tabaci* into new regions. The life cycles of the two genera differ, with *Eretmocerus* having a normal male/female relationship whereas the biparental *Encarsia* is autoparasitic. In autoparasitic species, the female larvae of the parasitoid are primary parasitoids of whitefly nymphs, whereas the males develop at the expense of already developing parasitoids within the whitefly nymph. In these species the males are relatively rare, amounting to ca. 5% compared with 95% females. This autoparasitic habit is usually not considered to hinder the ability of the parasitoid to control the pest. However, at least one case is recorded in which the autoparasitic *En. pergandiella* Howard disrupted successful control of the greenhouse whitefly *T. vaporariorum* by *En. formosa* (Gahan), by selectively laying its male-producing eggs in parasitized whiteflies (Gerling et al. 2001).

Several species are actively used for control of *B. tabaci* in greenhouses whereas others have been introduced and released in order to help control the pest in the field. The former include mainly *En. formosa* and two species of *Eretmocerus*, *Er. eremicus* Rose and Zolnerowich and *Er. mundus*. The latter species is considered specific to *B. tabaci*, whereas the former two are known also to parasitize the greenhouse whitefly and are both reared on and used against both host species. A trade-off thus exists between the higher host specificity of *Er. mundus* and the ease of rearing of the two other species. Field releases of parasitoids have been carried out in numerous countries, most recently in Australia (De Barro and Coombs 2009) and the US (Gould et al. 2008), both of which ended in the establishment of several new parasitoid species, most notably *Er. hayati* Zolnerowich and Rose.

Under natural conditions, different natural enemy species are active together. Therefore, a combination of predators and parasitoids, both those that occur naturally and the introduced organisms, is usually used for the control of *B. tabaci*. Moreover, the same predators (e.g. *N. tenuis* and *A. swirskii*) may prey on additional pests such as spider mites or the lepidopteran pest *Tutta absoluta* (Meyrick) and therefore be both introduced and naturally present in the crops.

11.4 Physical and Cultural Control for Management of *Begomoviruses*

11.4.1 The Significance of Physical and Cultural Control Practices

Four types of damage are caused by the whitefly *B. tabaci* during its feeding on host plants: depletion of the plant nutrients, plant toxication, plant contamination with honeydew and sooty molds and the spreading of viral plant diseases. The latter, which involves plant viruses belonging to several taxonomic groups, is by far the most damaging (Duffus 1987; Byrne et al. 1990). Of the transmitted viruses, Geminiviruses, belonging to the genus *Begomovirus*, form the most significant group in terms of economic damage (Moffat 1999; Polston and Anderson 1997; Rybicki et al. 2000). The genus comprises 196 recognized virus species (ICTV, Virus Taxonomy 2009 Release, http://www.ictvonline.org./VirusTaxonomy.asp?version=2009&bhcp=1) infecting a wide range of economically important crops worldwide (Moffat1999, Polston and Anderson 1997). Begomoviruses are transmitted by *B. tabaci* in a persistent circulative manner. They are retained in the insect for periods ranging from weeks to lifetime. The putative transmission mechanism incorporates two phases: passage of the acquired virus through the insect gut wall into the haemocoel; followed by its passage into the salivary glands, from which the virus is released during feeding into the plant phloem system with the saliva (Cohen and Antignus 1994; Hunter et al. 1998; Rosell et al. 1999). The persistent mode of transmission of geminiviruses enables the use of chemical control to impede disease spread by the whitefly vector through the use of insecticides. However, the high inoculation pressure exerted by the large viruliferous whitefly populations makes this option both inefficient and undesirable, in light of the ecological damage and the development of whitefly resistance to insecticides (Horowitz et al. 1994). Alternatively, IPM techniques seem to offer less disruptive solutions while supporting sustainable agriculture. The underlying principle of cultural control, which is a dominant component of IPM, is the modification of management practices that render the environment less favorable for pest invasion, reproduction, survival and dispersal, thus reducing pest numbers. However, all the measures involve indirect preventive methods, the success of which is often difficult to assess. Many cultural procedures used for virus control are aimed at eradicating or altering one or more of the primary participants in the transmission process (vector, virus infected source plants and/or the crop) or preventing their interaction. While complementary cultural practices may not totally prevent spread, infection may be delayed and the damage level caused by the viral infection may be reduced. The following section describes some of the cultural methods aimed at affecting epidemiological parameters of *B. tabaci* in such a way as to reduce the spread of geminiviruses in protected crops and in the open field (Antignus 1999, 2007, 2010; Hilje et al. 2001).

11.4.2 Phytosanitation

The polyphagous nature of *B. tabaci* (Greathead 1986) increases the risk of virus spread from infected cultivated plants as well as from weeds that serve as natural virus reservoirs. Eradication of potential host plants is relevant in isolated areas, especially in arid regions where natural vegetation is poor. The phytosanitation approach was successfully implemented by imposing a crop-free period of several weeks, thus decreasing the whitefly populations as well as the inoculum sources (Hilje et al. 2001; Ucko et al. 1998).

11.4.3 Plant Barriers

Living barriers can be constructed by planting tall plant species around the field perimeter, in which the primary crop is grown. This approach has been used successfully in some cases, blocking the infiltration of whiteflies into the field and increasing the densities of its predators; however, failures were reported in other cases (Hilje et al. 2001). A variation of this approach is achieved by increasing the density of crop plants per unit area, thus increasing the number of plants escaping infection due to avoidance of contact with the whitefly vector. Increasing planting densities of cassava resulted in a reduction in disease incidence caused by the *African cassava mosaic virus* (ACMV) (Fargette et al. 1990). Another option of using plant barriers is that of the introduction of bait plants into the susceptible crop. It was shown that planting alternate rows of tomatoes and cucumbers delayed spread of TYLCV in the tomato crop thus resulting in a significant yield increase of the latter. Cucumbers can serve as excellent bait plants because they are known as favorite hosts for *B. tabaci* (Al-Musa 1982).

11.4.4 Mechanical Barriers

This approach is based on mechanical interference that prevents contact between the insect vector and the target plant. Any material that is fine enough to block insect infiltration, but not too fine to provide plants with light and adequate ventilation can be used for this purpose. Polypropylene sheets (Agryl®) were effective in reducing disease incidence caused by TYLCV in tomatoes (Cohen and Berlinger 1986). In Israel, when the whitefly population peaks during autumn, zucchini crops are grown under low tunnels covered with Agryl sheets to prevent infection by *Squash leaf curl* begomovirus (SLCV). The Agryl cover is removed at flowering to enable pollination by bees. This procedure, which delays infection, results in significant damage reduction (Antignus et al. 2004a). However, this approach was found inefficient in cases where viral infection affects fruit quality. A delay in infection of watermelons with

Watermelon chlorotic stunt begomovirus (WmCSV) is insufficient to protect plants from the devastating effects of the viral infection (Antignus, unpublished).

Fifty-mesh screens installed as greenhouse walls were found to be highly effective in protection of greenhouse tomatoes from infection by TYLCV (Berlinger et al. 1991). However, these dense screens prevent adequate ventilation, especially during midsummer when temperatures are at peak. The negative effect of the resulting heat stress can be decreased, however, by using different types of cooling systems. The protection efficiency of 50 mesh screens can be dramatically increased by introduction of a UV-absorbing additive into the polyethylene used for the production of the screens. These screens are characterized by a double insect–exclusion mechanism based on both their physical and optical properties. The first UV-absorbing screens (BioNet®) were developed and reported by Antignus et al. (1998). When compared to ordinary 50 mesh screens, the UV-absorbing screens improved inhibition of whitefly penetration and spread of TYLCV by a factor of 4 (Antignus et al. 1998). These results were later confirmed by others (Diaz and Fereres 2007; Ben-Yakir et al. 2008; Legarrea et al. 2010).

11.4.5 Optical Barriers

11.4.5.1 UV-Absorbing Films

Plant attraction to insects follows a chain of events, beginning with insect orientation to the plant from a distance and ending with establishment on plants in order to feed and oviposit. By interfering with various links along this pathway, the contact between the vector and the plant, normally leading to infection, can be prevented. Most insects (Kevan et al. 1991) and mites (McEnrone and Dronka 1966) have photoreceptors sensitive to radiation in the UV range with light signals in this range playing an important role in their ecological behavior. UV-A radiation (320–400 nm) is a necessary stimulus for whiteflies and other insects, affecting their flight orientation and dispersal activity (Coombe 1982; Antignus et al. 2001; Raviv and Antignus 2004; Doukas and Payne 2007).

Photoselective greenhouse cladding materials can serve as mega filters to eliminate parts of the necessary light spectrum, thus inhibiting insect development and virus epidemics (Antignus et al. 1996). UV-absorbing polyethylene films were highly efficient in protection of greenhouse crops against infestation by different insect pests and viral diseases. Tomato crops grown in commercial greenhouses with a UV-absorbing polyethylene roof and 50-mesh screen-covered- walls were approximately 50% less infested with *B. tabaci*, and TYLCV disease incidence was fivefold lower compared with control structures covered with ordinary polyethylene film (Antignus et al. 1996; Raviv and Antignus 2004).

The protective effect of UV-absorbing films can be explained by a putative mechanism described as a "two-compartment effect". A UV-deficient compartment is formed within the covered greenhouse by the UV–absorbing films, while the surrounding environment with its normal level of UV irradiation forms the second compartment. Whiteflies that approach the greenhouse wall from the external environment exhibit a positive UV phototactic behavior and as they lose contact with UV near the walls of the protected greenhouse, they become diverted from their original course, away from the UV-deficient greenhouse walls (Antignus et al. 2001; Antignus 2010). The protective effect of UV-absorbing films was also associated with reduced flight activity in the UV-deficient environment (Antignus et al. 2001; Chyzik et al. 2003; Antignus and Ben-Yakir 2004); and under these conditions the efficiency of virus transmission was diminished (Antignus 2010).

11.4.5.2 Sticky Traps for Monitoring and/or Management Purposes

Mound (1962), suggested that *B. tabaci* is attracted by two groups of wavelengths of transmitted light, the blue/ultra-violet, and the yellow parts of the spectrum. He correlated the reaction to ultra–violet to the induction of migratory behavior, whereas yellow radiation induces vegetative behavior, which may be part of the host selection mechanism. The attraction of whiteflies to yellow has been utilized as an important instrument in sampling and monitoring of whiteflies populations (e.g., Gerling and Horowitz 1984). This vision cue of whiteflies (and other insect pests) was also utilized as a very useful, often overlooked tool for management of pest populations in greenhouses (Van de Veire and Vacante 1984). The so-called "yellow sticky cards" are practically used for mass trapping of winged aphids, whiteflies, thrips, leafminers, fungus gnats, and shore flies. However, beneficial insects such as the whitefly parasitoid, *En. formosa,* can also be caught at times. Yellow sticky traps in various forms can catch large numbers of adult whiteflies. Large yellow-sticky boards or tapes are used in 'hot spots' at a rate of about one per plant. Alternatively, reams of yellow, sticky tape can be draped between posts along plant rows (http://www.gov.on.ca/english/inforas/releases/).

11.4.5.3 Soil Mulches

The use of soil mulch to protect tomato plants from infestation by whiteflies was reported by Avidov (1956), who used sawdust or whitewash spray to mulch the crop seedbeds. Similar results were obtained from straw mulches that not only markedly reduced whitefly population but also delayed the spread of *Cucumber vein yellowing virus* (CVYV) and TYLCV vectored by *B. tabaci* (Cohen 1982). Later on, Cohen and Melamed-Madjar (1978) tested yellow, aluminum, and blue polyethylene film, demonstrating the high efficiency of the yellow polyethylene in delaying infection of tomatoes by TYLCV. Similar protective effects against whiteflies, aphids and their vectored viruses were reported later by others (Suwwan et al. 1988; Csizinszky et al. 1995, 1997; Summers et al. 2005). Yellow and silver polyethylene mulches efficiently protected zucchini plants from the spread of SLCV. Two weeks after planting, disease incidence was lowest (10–20%) in plants grown over yellow and silver mulches, respectively, compared with 50% disease incidence in the un-mulched plots. The landing rate of whiteflies on plants grown over silver and yellow soil mulches was 5–7 folds lower than that on plants grown over bare soil (Antignus et al. 2004a, 2005). Spectrophotometric analysis of light reflection from yellow and silver polyethylene mulches, soil surfaces, and the plant canopy has demonstrated a relatively high level of light reflection from the plastic mulches in a range of 300–700 nm, compared with low levels of light reflection from bare soil. The plant foliage had a distinct reflection peak of 550 nm, considerably higher than that of the reflection from the bare soil. Under these circumstances, the contrast between the soil background and the plant canopy was maximal, enabling insects to detect the crop for landing. In cases where the background of the plant was formed by yellow or silver mulches, the amount of reflected light in the visible range was considerably higher than the reflection of the soil and plant canopy. The poor contrast resulting from the reflection of the plastic interfered with the ability of the insect to detect the plant image and perceive a landing signal (Antignus et al. 2005).

11.4.5.4 Colored Shading Nets

Photoselective shade netting is designed to selectively screen various light spectral components of the solar radiation and/or transform direct light into diffused light. The spectral manipulations offered by this technology are utilized to promote desired physiological responses in ornamental plants, vegetables and fruit trees (Shahak et al. 2008). Recently, it was shown that pepper crops grown under pearl and yellow ChromatiNets® were protected against the aphid–borne non-persistent viruses, *Potato virus Y* (PVY) and *Cucumber mosaic virus* (CMV) (Shahak et al. 2008; Antignus et al. 2009). The immigration of *B. tabaci* to tomato plants grown in 'walk in' tunnels covered with pearl and yellow nets was reduced by 50% compared with their immigration into tunnels covered with conventional black shading net. The spread of TYLCV under these circumstances was lowered by a factor of 2.5 compared with its spread under the black nets (Antignus et al. unpublished).

11.4.6 Breeding for Resistance Against Begomoviruses and the Whitefly Vector

The most promising and efficient approach for managing the transmission of virus diseases involves breeding for resistance.

11.4.6.1 Breeding in Tomato

The devastating effects of TYLCV on tomato crops have stimulated extensive breeding programs worldwide, directed to introduce resistance genes into tomato cultivars. All efforts to identify sources for resistance in the domesticated tomato

(*Solanum lycopersicum* L.) have failed. However, resistance sources against TYLCV were obtained from wild tomatoes. TY-20 (Hazera, Israel) was the first commercial tomato hybrid with tolerance to TYLCV that was produced by crosses with *S. peruvianum* (L.). This cultivar reacted to TYLCV by mild intervenal chlorosis (Pilowski et al. 1989). The tolerance expressed by TY-20 was found to be controlled by five recessive genes (Pilowski and Cohen 1990). The relatively mild symptoms expressed by this cultivar were accompanied by low accumulation of viral DNA in the infected plants (Rom et al. 1992). Improved levels of resistance were derived later from *S. chilense* [LA1969 accessions *L. cheesmanii, L. hirsutum, L. habrochaites* and *L. pimpinellifolium* (Lapidot and Friedman 2002; Vidavski et al. 2008)]. A TYLCV partially dominant resistance gene from *S. chilense* accession LA 1969, designated Ty-1 was mapped on the tomato genome (Zamir et al. 1994). Ty-1 confers a relatively high level of resistance resulting in low accumulation of viral DNA as well as symptom attenuation. *S. chilense* also served as a resistance source to the bipartite begomovirus *Tomato mottle virus* (ToMoV) (Scott and Schuster 1991).

11.4.6.2 Breeding in the Common Bean

The common bean, *Phaseolus vulgaris* L., is an economically important crop that is severely affected by the begomoviruses *Bean golden mosaic virus* (BGMV), *Bean golden yellow mosaic virus* (BGYMV) and TYLCV. Resistance to TYLCV was observed in different commercial varieties but its inheritance has not yet been studied (Lapidot and Friedman 2002). Several resistance sources against BGYMV have been identified and used in breeding programs. Resistant cultivars developed from single resistance sources lost resistance under heavy inoculation pressure. Pyramiding of resistance genes from different sources yielded highly resistant lines (Singh et al. 2000).

11.4.6.3 Breeding in Cassava

Cassava (*Manihot esculenta* L.) is a staple food in many less developed countries. Begomoviruses are a bottleneck and a limiting factor in the cultivation of this important crop. The vegetative propagation of the crop and the lack of virus free propagation material are aggravating the problem. Four distinct begomoviruses were found in cassava plants suffering from Cassava Mosaic Diseases (CMD): *African cassava mosaic virus* (ACMV), *East African cassava mosaic virus* (EACMV), EACMV-Ug, a recombinant virus derived from ACMV and EACMV, and *South African cassava mosaic virus* (SACMV) (Pita et al. 2001). Resistance sources to CMD were not found in domesticated cassava. Introgression of resistance genes was done by inbreeding with related *Manihot* species by interspecific crosses (Hahn et al. 1980). Highly successful Cassava mosaic disease (CMD) pandemic management programs have been implemented in East Africa (Legg 1999). These have been based on the dissemination and multiplication of CMD-free planting material of resistant varieties that are high yielding and have higher dry matter content compared to the traditional cultivars.

11.4.6.4 Breeding in Cotton

Cotton (*Gossypum hirsutum* L.) is affected by *Cotton leaf crumple virus* (CLCrV) which is widespread in the southern USA (Brown and Nelson 1984). The virus appears late in the growing season and causes mild symptoms that do not have a devastating effect on the crop. Resistance seems to be linked to low yields and therefore hinders progress of the breeding programs (Lapidot and Friedman 2002).

Cotton in Pakistan was devastated by a complex of the monopartite begomovirus, *Cotton leaf curl virus* and a satellite DNA (DNA β) that is dependent on the virus for its replication and encapsidation (Briddon et al. 2001). Some highly resistant cotton cultivars have been developed in Pakistan; however, their yield level is unsatisfactory (Briddon and Markham 2000).

11.4.6.5 Breeding Against the Whitefly Vector

Host-plant resistance is an important cornerstone as an IPM component for suppression of *B. tabaci* populations (Berlinger 1986), and may provide a more bio-rational approach for reducing the impact of whitefly transmitted viruses and plant disorders than reliance on pesticides. The phenomenon of plant resistance can be described as a relative reduction in pest population size compared with standard varieties due to genetic characteristics of these plants (De Ponti et al. 1990). Some of the resistance mechanisms against whiteflies are associated with the morphological and chemical characteristics of the plant, such as leaf hairiness, density of canopy, pH level in the plant tissue and production of toxic secondary metabolites produced within the tissue or excreted to the leaf surface (Berlinger 1986). A trial conducted in Sudan on the susceptibility of eight tomato varieties to *B. tabaci* infestation reported low numbers of whitefly eggs, nymphs and adults on the tomato varieties 'Red Cloud' and 'Strain B' (Kisha 1984). Accessions of *S. habrochaites* and *S. peruvianum* were resistant to *Tomato leaf curl virus* TLCV (Muniyappa et al. 1991). The resistance in this case was associated with the presence of exudates from trichome glands on the leaf surface, on which the whiteflies were trapped after landing, preventing feeding and colonization (Channarayappa et al. 1992). Resistance to whiteflies and other insects in wild species of the tomato *Solanum pennellii* (Corr.) D'Arcy was attributed to the presence of sugar esters in the glandular exudate of type IV trichomes; however, the involvement of these compounds in resistance was broadly questioned in other studies (Nombela and Munitz 2010). In later studies, it was found that the *Mi-1* gene, introgressed into varieties of cultivated tomato from its wild relative *S. peruvianum,* confers resistance to root- knot nematodes and also regulates resistance to insects such as the potato aphid, *Macrosiphum euphorbiae* (Thomas) and *B. tabaci* (reviewed by Nombela and Munitz 2010). In addition, tomato varieties carrying the *Mi-1* gene were more resistant to the Q biotype of *B. tabaci* than to the B biotype. Characterization of the locus of the *Mi-1* gene demonstrated the presence of two transcribed genes, *Mi-1* and *Mi-1.2*, with 91% homology. Experiments with transgenic plants harboring the cloned *Mi-1.2* gene indicate that this gene is responsible for the resistance in tomato plants to both B and Q biotypes in 2-month-old tomato plants. Resistance against *B. tabaci* was found in many additional cultivated crops like cotton, beans, cucumbers, zucchini, cassava, corn and others (Reviewed by Nombela and Munitz 2010).

11.4.7 Genetically Engineered Resistance

Genetic engineering opens up the way to broadening and enriching the pool of natural resistance genes against virus diseases. The initial attempts to create transgenes conferring virus resistance were based on the expression of a viral coat protein gene in the plant. The induced protective effects, similar to classical cross protection, were categorized under the term 'coat- protein-mediated' protection, which is part of the pathogen-derived-protection (PDR) strategies (Sanford and Johnson 1985). Later on, constructs that include mutated or truncated virus genes or virus RNA sequences were used for plant transformation in a way that interfered with virus infection or silenced the expression of viral genes. Alternative approaches that have recently been investigated have included the use of geminivirus-induceable toxic proteins to kill virus-infected cells and the expression of DNA-binding proteins that either disrupt geminivirus infections or diminish their harmful effects (Lapidot and Friedman 2002; Prins et al. 2008; Shepherd et al. 2009). In many of the studies describing the effects of transgenes on viral infection, experiments were conducted in model plants with established transformation protocols rather than in the crop itself. Unfortunately, results from model systems do not necessarily reflect the effects of transgenes in crop plants.

11.4.7.1 Replication (*Rep*) Associated Proteins

Truncated geminivirus *Rep* genes were used to introduce resistance into tomatoes against *Tomato yellow leaf curl Sardinia virus* (TYLCSV) (Brunetti et al. 1997). A similar approach was used by Antignus et al. (2004b), who transformed tomato plants with a truncated *Rep* derived from TYLCV-Is mild. In both cases, the expression of the *Rep* genes interfered specifically with the cognate viral infection and not with virus strains with lower identities at the amino acid level.

11.4.7.2 Movement Proteins (*mp*)

Tomato plants transformed with a mutated *Bean dwarf mosaic virus* (BDMV) *mp* gene, showed resistance to *Tomato mottle virus* (ToMoV) – a virus with 93% amino acid sequence identity (von Arnim and Stanely 1992). It seems that transformations with the *mp* genes provide a broader resistance range; however, this approach is not feasible in cases where over-expressions induce toxic effects.

11.4.7.3 Gene Silencing

Gene silencing is a form of genetic regulation used by plants, animals and fungi to control the expression of different genes. Blocking of expression can result either from transcriptional gene silencing by DNA methylation or from degradation of mRNA, known as post-transcriptional gene-silencing (PTGS) (Baulcombe 1996; Prins et al. 2008). PTGS can be triggered by the expression of dsRNAs homologous to virus sequences. PTGS was used successfully against the begomoviruses ACMV, *Mung bean yellow mosaic virus* (MYMV), TYLCV, ToLCV, BGMV *Srilanka cassava mosaic virus* and EACMV (Shephered et al. 2009).

11.4.7.4 Antisense RNA

In vivo base pairing of RNA molecules with sequence complementation to the viral RNA can prevent RNA translation or induce PTGS. This strategy has been successfully exploited to target and selectively suppress the expression of geminivirus genes of TGMV, TYLCV, ToLCV and others; however, this approach failed when tested against other geminiviruses (Shephered et al. 2009).

RNA-based resistance strategies have better a chance of being backed by State legislatives due to their safer status compared with strategies that rely on the expression of foreign proteins. However, the major drawbacks of this approach are associated with their high virus specificity and the activity of certain geminivirus genes known as gene silencing suppressors (Shephered et al. 2009). More work is required to evaluate the practical merits of the different types of transgenic resistance described above. In most cases, the transgenic plants were tested under laboratory conditions, making it hard to predict how these plants will perform under field conditions. It can also be anticipated that the exposure of resistant transgenic plants to natural virus populations in the open field, will initiate an evolutionary process in the virus population through mutation and recombination events and the formation of new virus strains that may threaten resistance stability (Garcia-Andres et al. 2009).

11.5 Concluding Remarks

In this review, we have dealt with the various components of IPM applicable in managing the populations of *B. tabaci*: namely, chemical control with selective insecticides, biological control, crop plant resistance and physical/mechanical methods. It is hoped that the integration of these methods will contribute to improving management of the pest, while replacing the predominant use of insecticides that is currently the main approach employed. Adopting IPM will alleviate the numerous concerns that accompany the use of chemicals, including those associated with environmental pollution and the widespread resistance that plagues *B. tabaci* management. It is encouraging that most modern insecticides that are effective for *B. tabaci* control are relatively specific to the target pests, and therefore are less harmful to natural enemies and the environment; consequently, they should also prove more suitable for integrative combination with other methods.

Since viral plant diseases transmitted by *B. tabaci* are not curable, the principal tactics for their management should be based on prevention of transmission and/or on utilization of host-plant resistance. In addition to important practices like sanitation and crop-free periods, the use of various types of physical and optical obstacles against whiteflies have been studied and implemented. They include screen mesh and mechanical and optical barriers. The efficacy of the latter has been greatly enhanced by including UV-absorbing films and soil mulches that manipulate the behavior of the pest.

Breeding for plant resistance to both the whitefly vector and the begomoviruses has also taken place. New and more tolerant or resistant crop varieties have been successfully implemented and will become commercially available in the near future. Regarding genetic engineering, although great advances have been made in breeding for resistance, its use is prone to be delayed due to ethical and legislative considerations.

Whiteflies top the list of successful cases in the history of biological control using predators and parasitoids. However, using biological control as a sole countermeasure against the viral vectoring activity of *B. tabaci* is not in itself sufficient because of the high efficacy of virus transmission by even a few whitefly adults and the resulting severe damage to high-value crops. Nonetheless, the correct implementation of natural enemies will help to reduce whitefly numbers, which can then be more readily managed using mechanical and, only if necessary, chemical means.

Acknowledgements Our thanks are due to the numerous persons who assisted us by providing information and helping to construct this chapter: Dr. Jan Van der Blom (Coexphal, Spain), Dr. Karel Bolkmans (Koppert, The Nethrelands), Dr. Paul De Barro (CSIRO, Australia), Dr. Isaac Ishaaya (ARO, Israel), Dr. Minho Lee, (National Academy of Agricultural Science, S. Korea), Prof. Bao-Li Qiu (South China Agricultural University), and to Ms. Nomi Paz (Tel Aviv University, Department of Zoology) for the linguistic assistance.

References

Agusti N, Pumariño L, Fernandez M, Gabarra R, Alomar O (2009) Molecular markers for dispersal studies: plant DNA detection within omnivorous predators. In: IOBC/WPRS workgroup "Integrated control in protected crops, Mediterranean climate" Chania, pp 313–318

Al-Musa A (1982) Incidence, economic importance and control of tomato yellow leaf curl in Jordan. Plant Dis 66:561–563

Antignus Y (1999) Cultural control of insect-transmitted viruses. In: Gomez IC (ed.) Current trends in epidemiology and virus control in horticultural crops. Fundacion para la Investigacion Agraria en la Provincia de Almeria, Almeria, pp 79–89

Antignus Y (2007) The management of tomato yellow leaf curl virus in greenhouses and the open field, a strategy of manipulation. In: Czosnek H (ed.) Tomato yellow leaf curl virus disease. Springer, Dordrecht, pp 263–278

Antignus Y (2010) Optical manipulations block the spread of *Bemisia tabaci* in greenhouses and the open field. In: Stansly PA, Naranjo SE (eds.) *Bemisia*: bionomics and management of a global pest. Springer, Dordrecht, pp 349–356

Antignus Y, Ben-Yakir D (2004) Greenhouse photoselective cladding materials serve as an IPM tool to control the spread of insect pests and their vectored viruses. In: Horowitz AR, Ishaaya I (eds.) Insect Pest Management. Springer, Berlin, pp 319–335

Antignus Y, Mor N, Ben-Joseph R, Lapidot M, Cohen S (1996) UV-absorbing plastic sheets protect crops from insect pests and from virus diseases vectored by insects. Environ Entomol 25:919–924

Antignus Y, Lapidot M, Hadar D, Messika Y, Cohen S (1998) UV absorbing screens serve as optical barriers to protect vegetable crops from virus diseases and insect pests. J Econ Entomol 91:1401–1405

Antignus Y, Nestel D, Cohen S, Lapidot M (2001) Ultraviolet-deficient greenhouse environment affects whitefly attraction and flight behavior. Environ Entomol 30:394–399

Antignus Y, Lachman O, Pearlsman M, Koren A, Matan E, Tregerman M, Ucko O, Messika Y, Omer S, Unis H (2004a) Development of an IPM system to reduce the damage of squash leaf curl begomovirus in zucchini squash crops. In: Abstract compendium, 2nd European whitefly symposium, Cavtat

Antignus Y, Vunsh R, Lachman O, Pearlsman M, Maslenin L, Hananya U, Rosner A (2004b) Truncated rep gene originated from tomato yellow leaf curl virus-Israel [Mild] confers strain-specific resistance in transgenic tomato. Ann Appl Biol 144:39–44

Antignus Y, Lachman O, Pearlsman M (2005) Light manipulation by soil mulches protects crops from the spread of Begomoviruses. In: Abstracts of the 9th international plant virus epidemiology symposium, Lima

Antignus Y, Ben-Yakir D, Offir Y, Messika Y, Dombrovsky A, Chen M, Ganot L, Yehezkel H, Ganz S, Shahak Y (2009) Colored shade nets form optical barrier protecting pepper and tomato crops against aphid-borne non-persistent viruses. Sade Va'Yerek 12:60–62 (in Hebrew)

Arno J, Gabarra R, Liu TX, Simmons AM, Gerling D (2010) Natural enemies of *Bemisia tabaci*: predators and parasitoids. In: Stansly PA, Naranjo SE (eds.) *Bemisia*: bionomics and management of a global pest. Springer, Dordrecht, pp 385–421

Asiimwe P, Ecaat JS, Otim M, Gerling D, M. Guershon Kyamanywa S and Legg J (2007) Mortality factors affecting populations of Bemisia tabaci on cassava. In: Uganda 4th international Bemisia workshop international whitefly genomics workshop, Duck Key, 3–8 Dec 2006. J Insect Sci. http://www.insectscience.org/

Asjes CJ, Blom-Barnhoorn GL (2002) Control of aphid vector spread of lily symptomless virus and lily mottle virus by mineral oil/insecticide sprays in Lilium. Acta Hortic 70:277–281

Avidov Z (1956) Bionomics of the tobacco whitefly (*Bemisia tabaci* Gennad.) in Israel. Ktavim 7:25–41

Bai D, Lummis SCR, Leicht W, Breer H, Sattelle DB (1991) Actions of imidacloprid and a related nitromethylene on cholinergic receptors of an identified insect motor neurone. Pestic Sci 33:197–204

Baulcombe DC (1996) Mechanisms of pathogen derived resistance to viruses in transgenic plants. Plant Cell 8:1833–1844

Bedford ID, Briddon RW, Brown JK, Rosell RC, Markham PG (1994) Geminivirus transmission and biological characterisation of *Bemisia tabaci* (Gennadius) biotypes from different geographic regions. Ann Appl Biol 125:311–325

Ben-Yakir D, Hadar MD, Offir Y, Chen M, Tregerman M (2008) Protecting crops from pests using OptiNet® and ChromatiNet® shading nets. Acta Hortic 770:205–212

Berlinger MJ (1986) Host plant resistance to *Bemisia tabaci*. Agric Ecosyst Environ 17:69–82

Berlinger MJ, Dahan R, Mordechi S, Liper A, Katz J, Levav N (1991) The use of nets to prevent the penetration of *Bemisia tabaci* into greenhouse. Hassadeh 71:1579–1583 (in Hebrew)

Bielza P, Contreras J, Quinto V, Izquierdo J, Mansanet V, Elbert A (2005) Effects of Oberon® 240 SC on bumblebees pollinating greenhouse tomatoes. Pflanzenschutz-Nachrichten Bayer 58:469–484

Bielza P, Fernández E, Grávalos C, Izquierdo J (2009) Testing for non-target effects of spiromesifen on *Eretmocerus mundus* and *Orius laevigatus* under greenhouse conditions. Biocontrol 54:229–236

Bolckmans K, Sterk G, Eyal J, Sels B, Stepman W (1995) PreFeRal (*Paecilomyces fumosoroseus* strain Apopka 97), a new microbial insecticide for the biological control of whiteflies in greenhouses. Med Fac Landbouww Univ Gent 60(3a):707–711

Boykin LM, Shatters RG Jr, Rosell RC, McKenzie CL, Bagnall RN, De Barro P, Frohlich DR (2007) Global relationships of *Bemisia tabaci* (Hemiptera: Aleyrodidae) revealed using Bayesian analysis of mitochondrial COI DNA sequences. Mol Phylogenet Evol 44:1306–1319

Bretschneider T, Benet-Buchholz J, Fischer R, Nauen R (2003) Spirodiclofen and spiromesifen – novel acaricidal and insecticidal tetronic acid derivatives with a new mode of action. Chimia 57:697–701

Briddon RW, Markham PG (2000) Cotton leaf curl virus disease. Virus Res 71:151–159

Briddon RW, Mansour S, Bedford ID, Pinner MS, Saunders K, Stanely J, Zafar Y, Malik KA, Markham PG (2001) Identification of DNA components required for induction of cotton leaf curl disease. Virology 285:234–243

Brown JK (2007) The B*emisia tabaci* complex: genetic and phenotypic variation and relevance to TYLCV –vector interactions. In: Czosnek H (ed.) Tomato yellow leaf curl virus disease. Springer, Dordrecht, pp 25–56

Brown JK, Nelson MR (1984) Geminate particles associated with cotton leaf crumple disease in Arizona. Phytopathology 74:987–990

Brown JK, Frohlich DR, Rosell RC (1995) The sweetpotato or silverleaf whiteflies: biotypes of *Bemisia tabaci* or a species complex? Annu Rev Entomol 40:511–534

Brück E, Elbert A, Fischer R et al (2009) Movento®, an innovative ambimobile insecticide for sucking insect pest control in agriculture: biological profile and field performance. Crop Prot 28:838–844

Brunetti A, Tavazza E, Noris E, Tavazza P, Caciagli P, Ancora G, Crespi S, Accoto GP (1997) High expression of truncated viral rep protein confers resistance to tomato yellow leaf curl virus in transgenic tomato plants. Mol Plant Microbe Interact 10:571–579

Byrne DN, Bellows TS Jr (1991) Whitefly biology. Annu Rev Entomol 36:431–457

Byrne DN, Bellows TS, Parella MP (1990) Whiteflies in agricultural systems. In: Gerling D (ed.) Whiteflies: their bionomics, pest status and management. Intercept, Andover, pp 227–261

Cahill M, Denholm I (1999) Managing resistance to chloronicotinyl insecticides: rhetoric or reality? In: Yamamoto I, Casida J (eds.) Neonicotinoid insecticides and the nicotinic acetylcholine receptor. Springer, Tokyo, pp 253–270

Cahill M, Jarvis W, Gorman K, Denholm I (1996) Resolution of baseline responses and documentation of resistance to buprofezin in *Bemisia tabaci* (Homoptera: Aleyrodidae). Bull Entomol Res 86:117–122

Casida JE, Quistad GB (1998) Golden age of insecticide research: past, present, or future. Annu Rev Entomol 43:1–16

Castle SJ, Palumbo JC, Prabhaker N, Horowitz AR, Denholm I (2010) Ecological determinants of *Bemisia tabaci* resistance to insecticides. In: Stansly PA, Naranjo SE (eds.) *Bemisia*: bionomics and management of a global pest. Springer, Dordrecht

Channarayappa C, Shivasankar G, Muniyappa V, Frist RH (1992) Resistance of Lycopersicon species to *Bemisia tabaci*, a tomato leaf curl vector. Can J Bot 70:2184–2192

Chiel E, Gottlieb Y, Zchori-Fein E, Mozes-Daube N, Katzir N, Inbar M, Ghanim M (2007) Biotype-dependent secondary symbiont communities in sympatric populations of *Bemisia tabaci*. Bull Entomol Res 97:407–413

Chu D, Wan FH, Zhang YJ, Brown JK (2010) Change in the biotype composition of *Bemisia tabaci* in Shandong Province of China from 2005 to 2008. Environ Entomol 39:1028–1036

Chyzik R, Dobrinin S, Antignus Y (2003) Effect of a UV-deficient environment on the biology and flight activity of *Myzus persicae* and its hymenopterous parasite *Aphidius matricariae*. Phytoparasitica 31:467–477

Cohen S (1982) Control of whitefly vectors of viruses by color mulches. In: Harris KF, Maramorosch K (eds.) Pathogens, vectors and plant diseases, approaches to control. Academic, New York, pp 45–56

Cohen S, Antignus Y (1994) Tomato yellow leaf curl virus, a whitefly-borne geminivirus of tomatoes. In: Harris KS (ed.) Advances in disease vector research, vol 10. Springer-Verlag, New York, pp 259–288

Cohen S, Berlinger MJ (1986) Transmission and cultural control of whitefly-borne viruses. Agric Ecosyst Environ 17:89–97

Cohen S, Melamed-Madjar V (1978) Prevention by soil mulching of the spread of tomato yellow leaf curl virus transmitted by *Bemisia tabaci* (Gennadius) (Hemiptera: Aleyrodidae) in Israel. Bull Entomol Res 68:465–470

Coombe PE (1982) Visual behavior of the greenhouse whitefly, *Trialeurodes vaporariorum*. Physiol Entomol 7:243–251

Costa HS, Brown JK (1991) Variation in biological characteristics and esterase patterns among populations of *Bemisia tabaci* (Genn.) and the association of one population with silverleaf symptom induction. Entomol Exp Appl 61:211–219

Costa HS, Brown JK, Sivasupramaniam S, Bird J (1993) Regional distribution, insecticide resistance, and reciprocal crosses between the 'A' and 'B' biotypes of *Bemisia tabaci*. Insect Sci Appl 14:255–266

Crowder DW, Ellsworth PC, Tabashnik BE, Carriére Y (2008) Effects of operational and environmental factors on evolution of resistance to pyriproxyfen in the sweetpotato whitefly (Hemiptera: Aleyrodidae). Environ Entomol 37:1514–1524

Csizinszky AA, Schuster DJ, Kring JB (1995) Color mulches influence yield and insect pest populations in tomatoes. J Am Soc Hortic Sci 120:778–784

Csizinszky AA, Schuster DJ, Kring JB (1997) Evaluation of colored mulches and oil sprays for yield and for the control of silverleaf whitefly, *Bemisia argentifolii* (Bellows and Perring) on tomatoes. Crop Prot 16:475–481

De Barro PJ, Coombs MT (2009) Post-release evaluation of *Eretmocerus hayati* Zolnerowich and Rose in Australia. Bull Entomol Res 99:193–206

De Barro PJ, Trueman JWH, Frohlich DR (2005) *Bemisia argentifolii* is a race of *B. tabaci* (Hemiptera: Aleyrodidae): the molecular genetic differentiation of *B. tabaci* populations around the world. Bull Entomol Res 95:193–203

De Barro PJ, Liu SS, Boykin LM, Dinsdale A (2011) *Bemisia tabaci:* a statement of species status. Annu Rev Entomol 56:1–19

De Cock A, Degheele D (1998) Buprofezin: a novel chitin synthesis inhibitor affecting specifically planthoppers, whiteflies and scale insects. In: Ishaaya I, Degheele D (eds.) Insecticides with novel modes of action: mechanism and application. Springer, Berlin/Heidelberg/New York, pp 74–91

De Cock A, de Clercq P, Tirry L, Degheele D (1996) Toxicity of diafenthiuron and imidacloprid to the predatory bug *Podisus maculiventris* (Heteroptera: Pentatomidae). Environ Entomol 25:476–480

De Ponti OMB, Romanow LR, Berlinger MJ (1990) Whitefly plant relationships: plant resistance. In: Gerling D (ed.) Whiteflies: their bionomics, pest status and management. Intercept Ltd, Andover, pp 91–106

Denholm I, Cahill M, Byrne FJ, Devonshire AL (1996) Progress with documenting and combating insecticide resistance in *Bemisia*. In: Gerling D, Mayer RT (eds.) *Bemisia*: 1995 taxonomy, biology, damage, control and management. Intercept Ltd, Andover, pp 577–603

Denholm I, Cahill M, Dennehy TJ, Horowitz AR (1998) Challenges with managing insecticide resistance in agricultural pests, exemplified by the whitefly, *Bemisia tabaci*. Phil Trans R Soc Series B 353:1757–1767

Dennehy TJ, Denholm I (1998) Goals, achievements and future challenges of the Arizona whitefly resistance management program. In: Dugger CP, Richter DA (eds.) Proceedings of the beltwide cotton production research conference, San Diego. National Cotton Council of America, Memphis, 5–9 Jan 1998, pp 68–72

Dennehy TJ, Williams L III, Li X, Wigert M, Birdwell E (1997) Status of whitefly resistance to insecticides in Arizona cotton. In: Silvertooth JC (ed.) Cotton, a college of agriculture report, series P-108, University of Arizona, College of Agriculture, Tucson, pp 232–253

Dennehy TJ, DeGain BA, Harpold VS, Brown JK, Morin S, Fabrick JA, Byrne FJ, Nichols RL (2005) New challenges to management of whitefly resistance to insecticides in Arizona. In: Byrne DN, Baciewicz P (eds.) 2005 Vegetable report, series P-144, College of Agriculture and Life Sciences, University of Arizona, pp 1–31. http://cals.arizona.edu/pubs/crops/az1382/az1382_2.pdf

Diaz BM, Fereres A (2007) Ultraviolet-blocking materials as a physical barrier to control insect pests and pathogens in protected crops. Pest Tech 1:85–95

Dittrich V, Ernst GH, Ruesh O, Uk S (1990) Resistance mechanisms in sweetpotato whitefly (Homoptera: Aleyrodidae) populations from Sudan, Turkey, Guatemala, and Nicaragua. J Econ Entomol 83:1665–1670

Doukas D, Payne C (2007) Greenhouse whitefly (Homoptera: Aleyrodidae) dispersal under different UV-light environments. J Econ Entomol 100:380–397

Duffus J (1987) Whitefly transmission of plant viruses. In: Kerry FH (ed.) Current topics in vector research. Springer, New York, pp 74–91

Elbert A, Nauen R (2000) Resistance in *Bemisia tabaci* (Homoptera: Aleyrodidae) to insecticide in southern Spain with special reference to neonicotinoids. Pest Manag Sci 56:60–64

Elbert A, Overbeck H, Iwaya K, Tsuboi S (1990) Imidacloprid, a novel systemic nitromethylene analogue insecticide for crop protection. In: Proceedings of 1990 Brighton crop protection conference – pests and diseases, Brighton, pp 21–28

Elbert A, Nauen R, Cahill M, Devonshire AL, Scarr AW, Sone S, Steffens R (1996) Resistance management with chloronicotinyl insecticides using imidacloprid as an example. Pflanzen-Nachrich Bayer 49:5–53

Elbert A, Nauen R, Leicht W (1998) Imidacloprid, a novel chloronicotinyl insecticide: biological activity and agricultural importance. In: Ishaaya I, Degheele D (eds.) Insecticides with novel modes of action: mechanism and application. Springer, Berlin/Heidelberg/New York, pp 50–73

Ellsworth PC, Martinez-Carrillo JL (2001) IPM for *Bemisia tabaci*: a case study from North America. Crop Prot 20:853–869

Fargette D, Fauquet C, Grenier I, Thresh MJ (1990) The spread of African cassava virus into the within cassava fields. J Phytopathol 130:289–302

Flückiger CR, Kristinsson H, Senn R, Rindlisbacher A, Buholzer H, Voss G (1992a) CGA 215′944 – a novel agent to control aphids and whiteflies. In: Brighton crop protection conference – pests and diseases, vol 1, Brighton, pp 43–50

Flückiger CR, Senn R, Buholzer H (1992b) CGA 215′944 – opportunities for use in vegetables. In: Proceedings of 1992 Brighton crop protection conference – pests and diseases, vol 3, Brighton, pp 1187–1192

Fuog D, Fergusson SJ, Flückiger C (1998) Pymetrozine: a novel insecticide affecting aphids and whiteflies. In: Ishaaya I, Degheele D (eds.) Insecticides with novel modes of action: mechanism and application. Springer, Berlin/Heidelberg/New York, pp 40–49

Garcia-Andres S, Tomas DM, Navas-Castillo J, Moriones E (2009) Resistance driven-selection of begomoviruses associated with the tomato yellow leaf curl disease. Virus Res 146:66–72

Gerling D (1990) Whiteflies their bionomics, pest status and management. Intercept Ltd, Andover

Gerling D, Horowitz AR (1984) Yellow traps for evaluating the population levels and dispersal patterns of *Bemisia tabaci* (Gennadius) (Homoptera: Aleyrodidae). Ann Entomol Soc Am 77:753–759

Gerling D, Alomar O, Arno J (2001) Biological control of *Bemisia tabaci* using predators and parasitoids. Crop Prot 20:779–799

Gorman K, Slater R, Blande J, Clarke A, Wren J, McCaffery A, Denholm I (2010) Cross-resistance relationships between neonicotinoids and pymetrozine in *Bemisia tabaci* (Hemiptera: Aleyrodidae). Pest Manag Sci 66:1186–1190

Gottlieb Y, Ghanim M, Chiel E, Gerling D, Portnoy V, Steinberg S, Tzuri G, Horowitz AR, Belausov E, Mozes-Daube N, Kontsedalov S, Gershon M, Gal S, Katzir N, Zchori-Fein E (2006) Identification and localization of a Rickettsia sp in Bemisia tabaci (Homoptera: Aleyrodidae). Appl Environ Microbiol 72:3646–3652

Gould JR, Hoelmer KA, Goolsby JA (2008) Classical biological control of *Bemisia tabaci* in the United States: a review of interagency research and implementation (co-editors). In: Hokkanen HMT (ed.) Progress in biological control, vol 4. Springer, Dordrecht

Greathead AH (1986) Host plants. In: Cock MJW (ed.) *Bemisia tabaci* – a literature survey. CAB International Institute of Biological Control, Silwood Park, pp 17–26

Guirao P, Beitia F, Cenis JL (1997) Biotype determination of Spanish populations of *Bemisia tabaci* (Hemiptera: Aleyrodidae). Bull Entomol Res 87:587–593

Hahn SK, Terry ER, Leuschner K (1980) Breeding cassava for resistance to cassava mosaic disease. Euphytica 29:673–683

Henneberry TJ (1993) Sweetpotato whitefly - current status and national research and action plan. In: Dugger CP, Richter DA (eds.) Proceedings beltwide cotton production conferences, New Orleans, 10–14 Jan, pp 663–666

Henneberry TJ, Butler GD Jr (1992) Whiteflies as a factor in cotton production with specific reference to *Bemisia tabaci* (Gennadius). In: Dugger CP, Richter DA (eds.) Proceedings beltwide cotton production conferences, Nashville, 6–10 Jan 1992, pp 674-683

Hequet E, Henneberry TJ, Nichols RL (eds.) (2007) Sticky cotton: causes, effects, and prevention. USDA-ARS Technical Bulletin No 1915

Hilje L, Costa HS, Stansly PA (2001) Cultural practices for managing *Bemisia tabaci* and associated viral diseases. Crop Prot 20:801–812

Hogenhout SA, Ammar ED, Whitfield AE, Redinbaugh MG (2008) Insect vector interactions with persistently transmitted viruses. Annu Rev Phytopathol 46:327–359

Horowitz AR, Ishaaya I (1994) Managing resistance to insect growth regulators in the sweetpotato whitefly (Homoptera: Aleyrodidae). J Econ Entomol 87:866–871

Horowitz AR, Ishaaya I (1996) Chemical control of *Bemisia tabaci* – management and application. In: Gerling D, Mayer RT (eds.) *Bemisia*: 1995 taxonomy, biology, damage, control and management. Intercept Ltd, Andover, pp 537–556

Horowitz AR, Ishaaya I (2004) Biorational insecticides - mechanisms, selectivity and importance in pest management. In: Horowitz AR, Ishaaya I (eds.) Insect pest management. Springer, Berlin, pp 1–28

Horowitz AR, Forer G, Ishaaya I (1994) Managing resistance in *Bemisia tabaci* in Israel with emphasis on cotton. Pestic Sci 42:113–122

Horowitz AR, Mendelson Z, Weintraub PG, Ishaaya I (1998) Comparative toxicity of foliar and systemic applications of two chloronicotinyl insecticides, acetamiprid and imidacloprid, against the cotton whitefly, *Bemisia tabaci*. Bull Entomol Res 88:437–442

Horowitz AR, Mendelson Z, Cahill M, Denholm I, Ishaaya I (1999) Managing resistance to the insect growth regulator pyriproxyfen in *Bemisia tabaci*. Pestic Sci 55:272–276

Horowitz AR, Kontsedalov S, Denholm I, Ishaaya I (2002) Dynamics of insecticide resistance in *Bemisia tabaci* – a case study with an insect growth regulator. Pest Manag Sci 58:1096–1100

Horowitz AR, Denholm I, Gorman K, Cenis JL, Kontsedalov S, Ishaaya I (2003) Biotype Q of *Bemisia tabaci* identified in Israel. Phytoparasitica 31:94–98

Horowitz AR, Kontsedalov S, Ishaaya I (2004) Dynamics of resistance to the neonicotinoids acetamiprid and thiamethoxam in *Bemisia tabaci* (Homoptera: Aleyrodidae). J Econ Entomol 97:2051–2056

Horowitz AR, Kontsedalov S, Khasdan V, Ishaaya I (2005) Biotypes B and Q of *Bemisia tabaci* and their relevance to neonicotinoid and pyriproxyfen resistance. Arch Insect Biochem Physiol 58:216–225

Horowitz AR, Denholm I, Morin S (2007) Resistance to insecticides in the TYLCV vector, *Bemisia tabaci*. In: Czosnek H (ed.) Tomato yellow leaf curl virus disease. Springer, Dordrecht, pp 305–325

Hunter WB, Hiebert E, Tsai JH, Polston JE (1998) Location of geminiviruses in the whitefly *Bemisia tabaci* (Homoptera: Aleurodidae). Plant Dis 82:1147–1157

Ishaaya I (1990) Buprofezin and other IGRs for controlling cotton pests. Pestic Outlook 1(2):30–33

Ishaaya I, Horowitz AR (1992) A novel phenoxy juvenile hormone analog (pyriproxyfen) suppresses embryogenesis and adult emergence of the sweetpotato whitefly (Homoptera: Aleyrodidae). J Econ Entomol 85:2113–2117

Ishaaya I, Horowitz AR (1995) Pyriproxyfen, a novel insect growth regulator for controlling whiteflies: mechanism and resistance management. Pestic Sci 43:227–232

Ishaaya I, Mendelson Z, Melamed-Madjar V (1988) Effect of buprofezin on embryogenesis and progeny formation of sweetpotato whitefly (Homoptera: Aleyrodidae). J Econ Entomol 81:781–784

Ishaaya I, Mendelson Z, Horowitz AR (1993) Toxicity and growth-suppression exerted by diafenthiuron in the sweetpotato whitefly, *Bemisia tabaci*. Phytoparasitica 21:199–204

Ishaaya I, Kontsedalov S, Horowitz AR (2003) Novaluron (Rimon), a novel IGR: potency and cross-resistance. Arch Insect Biochem Physiol 54:157–164

Itaya W (1987) Insect juvenile hormone analogue as an insect growth regulator. Sumitomo Pyrethroid World 8:2–4

Jones DR (2003) Plant viruses transmitted by whiteflies. Eur J Plant Pathol 109:195–219

Kayser H, Kaufmann L, Schürmann F (1994) Pymetrozine (CGA 215'944): a novel compound for aphid and whitefly control. In: An overview of its mode of action. Proceedings of 1994 Brighton crop protection conference – pests and diseases, vol 2, Brighton, pp 737–742

Kevan PG, Straver WA, Offer O, Laverty TM (1991) Pollination of greenhouse tomatoes by bumblebees in Ontario. Proc Entomol Soc Ont 122:15–19

Kirk AA, Lacey LA, Brown JK, Ciomperlik MA, Goolsby JA, Vacek DC, Wendel LE, Napompeth B (2000) Variation in the *Bemisia tabaci* species complex (Hemiptera: Aleyrodidae) and its natural enemies leading to successful biological control of Bemisia biotype B in the USA. Bull Entomol Res 90:317–327

Kisha JSA (1984) Whitefly, *Bemisia tabaci*, infestations on tomato varieties and a wild Lycopersicon species. Ann Appl Biol 104(Supplement, Tests of Agrichemicals and Cultivars 5):124–125

Knauf TA, Wright JE (1994) *Beauveria bassiana* (ATCC 74040): control of insect pests in field crops and ornamentals. In: Brighton crop protection conference – pest diseases, vol 3, Brighton, pp 1103–1108

Koehler PG, Patterson RJ (1991) Incorporation of pyriproxyfen in a German cockroach (Dictyoptera: Blattellidae) management program. J Econ Entomol 84:917–921

Kontsedalov S, Zchori-Fein E, Chiel E, Gottlieb Y, Inbar M, Ghanim M (2008) The presence of *Rickettsia* is associated with increased susceptibility of *Bemisia tabaci* (Homoptera: Aleyrodidae) to insecticides. Pest Manag Sci 64:789–792

Kontsedalov S, Gottlieb Y, Ishaaya I, Nauen R, Horowitz AR, Ghanim M (2009) Toxicity of spiromesifen to the developmental stages of *Bemisia tabaci* biotype B. Pest Manag Sci 65:5–13

Lahm GP, Cordova D, Barry JD (2009) New and selective ryanodine receptor activators for insect control. Bioorg Med Chem 17:4127–4133

Langley P (1990) Control of the tsetse fly using a juvenile hormone mimic, pyriproxyfen. Sumitomo Pyrethroid World 15:2–5

Lapidot M, Friedmann M (2002) Breeding for resistance to whitefly-transmitted geminiviruses. Ann Appl Biol 140:109–127

Legarrea S, Karnieli A, Fereras A, Weintraub PG (2010) Comparison of UV-absorbing nets in pepper crops, spectral properties, effects on plants and pest control. Photochem Photobiol 86:324–330

Legg JP (1999) Emergence, spread and strategies for controlling the pandemic of cassava mosaic virus disease in East and Central Africa. Crop Prot 18:627–637

Li AY, Dennehy TJ, Li S, Wigert ME, Zarborac M, Nichols RL (2001) Sustaining Arizona's fragile success in whitefly resistance management. In: Dugger CP, Richter DA (eds.) Proceedings of the beltwide cotton production research conference, Anaheim. National Cotton Council of America, Memphis, 9–13 Jan 2001, pp 1108–1114

Luo C, Jones CM, Devine G, Zhang F, Denholm I, Gorman K (2010) Insecticide resistance in Bemisia tabaci biotype Q (Hemiptera: Aleyrodidae) from China. Crop Prot 29:429–434

McEnrone WD, Dronka K (1966) Color vision in the adult female two-spotted spider mite. Science 154:782–784

Meekes ETM, Fransen JJ, van Lenteren JC (2002) Pathogenicity of *Aschersonia* spp. against whiteflies *Bemisia argentifolii* and *Trialeurodes vaporariorum*. J Invertebr Pathol 81:1–11

Moffat AS (1999) Geminiviruses emerge as serious crop threat. Science 286:1835

Mound L (1962) Studies on the olfaction colour sensitivity of *Bemisia tabaci* Genn. Aleyrodidae. Entomol Exp Appl 99–104

Mound LA, Halsey SH (1978) Whitefly of the world: a systematic catalogue of the Aleyrodidae (Homoptera) with host plant and natural enemy data. British Museum (Natural History), Chichester

Muniyappa V, Jalikop SH, Saikia AK, Chennarayappa SG, Bhat AI, Ramappa HK (1991) Reaction of Lycopersicon cultivars and wild accessions to tomato leaf curl virus. Euphytica 56:37–41

Naranjo SE, Ellsworth PC, Diehl JW (1998) Whitefly management in Arizona: contribution of natural enemies to whitefly mortality. P-112, University of Arizona, Tucson, pp 324–329

Nauen R, Denholm I (2005) Resistance of insect pests to neonicotinoid insecticides: current status and future prospects. Arch Insect Biochem Physiol 58:200–215

Nauen R, Konanz S (2005) Spiromesifen as a new chemical option for resistance management in whiteflies and spider mites. Pflanzenschutz-Nachrichten Bayer 58:485–502

Nauen R, Schnorbach HJ, Elbert A (2005) The biological profile of spiromesifen (Oberon) – a new tetronic acid insecticide/acaricide. Pflanzenschutz-Nachrichten Bayer 58:417–440

Nombela G, Muniz M (2010) Host plant resistance for the management of *Bemisia tabaci*: a multi-crop survey with emphasis on tomato. In: Stansly PA, Naranjo SE (eds.) *Bemisia*: bionomics and management of a global pest. Springer, Dordrecht, pp 357–383

Oliveira MRV, Henneberry TJ, Anderson P (2001) History, current status, and collaborative research projects for *Bemisia tabaci*. Crop Prot 20:709–723

Palumbo JC, Horowitz AR, Prabhaker N (2001) Insecticidal control and resistance management for *Bemisia tabaci*. Crop Prot 20:739–765

Perring TM (2001) The *Bemisia tabaci* species complex. Crop Prot 20:725–737

Pilowski M, Cohen S (1990) Tolerance in tomato yellow leaf curl virus derived from *Lycopersicon peruvianum*. Plant Dis 74:248–250

Pilowski M, Cohen S, Ben-Joseph R, Shlomo A, Chen L, Nahon S, Krikun J (1989) TY-20, a tomato cultivar tolerant to tomato yellow leaf curl virus. Hassadeh 69:1212–1215 (in Hebrew)

Pita JS, Fondong A, Sangare A, Kokora RNN, Fauquet CM (2001) Genomic and biological diversity of the African cassava geminiviruses. Euphytica 120:115–125

Polston JE, Anderson PK (1997) The emergence of whitefly-transmitted geminiviruses in tomato in the Western Hemisphere. Plant Dis 81:1358–1369

Prabhaker N, Toscano NC, Henneberry TJ (1998) Evaluation of insecticide rotations and mixtures as resistance management strategies for *Bemisia argentifolii* (Homoptera: Aleyrodidae). J Econ Entomol 91:820–826

Prabhaker N, Castle SJ, Toscano N, Henneberry TJ (2005) Assessment of cross-resistance potential among neonicotinoid insecticides in *Bemisia tabaci* (Hemiptera: Aleyrodidae). Bull Entomol Res 95:535–543

Prabhaker N, Castle SJ, Buckelew L, Toscano NC (2008) Baseline susceptibility of *Bemisia tabaci* B biotype (Hemiptera: Aleyrodidae) populations from California and Arizona to spiromesifen. J Econ Entomol 101:174–181

Prins M, Laimer M, Noris A, Schubert J, Wassenger M, Tepper M (2008) Strategies for antiviral resistance in transgenic plants. Mol Plant Pathol 9:73–83

Ravensberg WJ, Malais M, van der Schaaf DA (1990) Application of *Verticillium lecanii* in tomatoes and cucumber to control whitefly and thrips. Bull IOBC 13(5):173–178

Raviv M, Antignus Y (2004) UV Radiation effects on pathogens and insect pests of greenhouse-grown crops. Photochem Photobiol 79:219–226

Roditakis E, Grispou M, Morou E, Kristoffersen JB, Roditakis N, Nauen R, Vontas J, Tsagkarakou A (2009) Current status of insecticide resistance in Q biotype *Bemisia tabaci* populations from Crete. Pest Manag Sci 65:313–322

Rom M, Antignus Y, Gideoni D, Pilowski M, Cohen S (1992) Comparative study of tomato yellow leaf curl virus (TYLCV) DNA accumulation in tolerant and susceptible tomato lines. Plant Dis 77:253–257

Rosell RC, Torres-Jerez I, Brown KJ (1999) Tracing geminivirus-whitefly transmission pathway by polymerase chain reaction in whitefly extracts, saliva, hemolymph and honeydew. Phytopathology 89:239–246

Ruder FJ, Guyer W, Benson JA, Kayser H (1991) The thiourea insecticide/acaricide diafenthiuron has a novel mode of action: inhibition of mitochondrial respiration by its carbodiimide product. Pestic Biochem Phys 41:207–219

Rybicki EP, Briddon RW, Brown JK, Fauquet CM, Maxwell DP, Stanely J, Harrison BD, Markham PG, Bisaro DM, Robinson D (2000) Family geminiviridae. In: van Regenmortel MHV, Fauquet CM, Bishop DHYL, Carstens E, Estes M, Lemon S, Maniloff J, Mayo MA, McGeoch D, Pringle C, Wickner R (eds.) Virus taxonomy, seventh report of the international committee on taxonomy of viruses. Academic, New York, pp 285–297

Sanchez-Campos S, Navas-Castillo J, Camero R, Soria C, Diaz JA, Moriones E (1999) Displacement of *tomato yellow leaf curl virus* (TYLCV-Sr) by TYLCV-Is in tomato epidemics in Spain. Phytopathology 89:1038–1043

Sanford JC, Johnston SA (1985) The concept of pathogen derived resistance. J Theor Biol 113:395–405

Sattelle DB, Cordova D, Cheek TR (2008) Insect ryanodine receptors: molecular targets for novel pest control chemicals. Invertebr Neurosci 8:107–119

Schuster DJ, Mann RS, Toapanta M, Cordero R, Thompson S, Cyman S, Shurtleff A, Morris RF II (2010) Monitoring neonicotinoid resistance in biotype B of *Bemisia tabaci* in Florida. Pest Manag Sci 66:186–195

Scott JW, Schuster D (1991) Screening of accessions for resistance to the Florida tomato geminivirus. Tomato Genet Coop Rep 41:48–50

Shahak Y, Gal E, Offir Y, Ben-Yakir D (2008) Photoselective shade netting integrated with greenhouse technologies for improved performance of vegetable and ornamental crops. Acta Hortic 797:75–80

Sharaf N (1986) Chemical control of *Bemisia tabaci*. Agric Ecosyst Environ 17:111–127

Shepherd DN, Martin DP, Thomson JA (2009) Transgenic strategies for developing crops resistant to geminiviruses. Plant Sci 176:1–11

Simmons AL, Dennehy TJ (1996) Contrasts of three insecticide resistance monitoring methods for whitefly. In: Dugger CP, Richter, DA (eds.) Proceedings of the beltwide cotton production research conference, Nashville, National Cotton Council of America, Memphis, 9–12 Jan 1996, pp 748–752

Singh SP, Morales FJ, Miklas PN, Teran H (2000) Selection for bean golden resistance in intra- and interracial bean populations. Crop Sci 40:1565–1572

Stansly PA, Naranjo SE (eds.) (2010) *Bemisia*: bionomics and management of a global pest. Springer, Dordrecht

Stansly PA, Natwick A (2010) Integrated systems for managing *Bemisia tabaci* in protected and open field agriculture. In: Stansly PA, Naranjo SE (eds.) *Bemisia*: bionomics and management of a global pest. Springer, Dordrecht, pp 467–497

Steinemann A, Stamm E, Frei B (1990) Chemodynamics in research and development of new plant protection agents. Pestic Outlook 1(3):3–7

Streibert HP, Drabek J, Rindlisbacher A (1988) CGA 106630 - a new type of acaricide/insecticide for the control of the sucking pest complex in cotton and other crops. In: Proceedings Brighton crop protection conference - pests and diseases, Brighton, pp 25-33

Summers CG, Mitchell JP, Stapleton JJ (2005) Mulches reduce aphid-borne viruses and whiteflies in cantaloupe. Calif Agr 59:90–94

Suwwan MA, Akkawi M, Al-Musa AM, Mansour A (1988) Tomato performance and incidence of tomato yellow leaf curl (TYLC) virus as affected by type of mulch. Scientia Hortic 37:39–45

Takahashi H, Mitsui J, Takausa N, Matsud M, Yoneda H, Suszuki J, Ishimitsi K, Kishimoto T (1992) NI-25, a new type of systemic and broad spectrum insecticide. In: Proceedings of 1992 Brighton crop protection conference – pests and diseases, vol 1,Brighton, pp 88–96

Tomizawa M, Casida JE (2003) Selective toxicity of neonicotinoids attributable to specificity of insect and mammalian nicotinic receptors. Annu Rev Entomol 48:339–364

Ucko O, Cohen S, Ben-Joseph R (1998) Prevention of virus epidemics by a crop free period in the Arava region of Israel. Phytoparasitica 26:313–321

van de Veire M, Vacante V (1984) Greenhouse whitefly control through the combined use of the colour attraction system with the parasite wasp *Encarsia formosa* (Hym.: Aphelinidae). Entomophaga 29:303–310

Van Driesche R, Bellows TS Jr (1996) Biological control. Chapman & Hall, New York

Vidavski F, Czosnek H, Gazit S, Levy D, Lapidot M (2008) Pyramiding of genes conferring resistance to tomato yellow leaf curl virus from different wild tomato species. Plant Breed 127:734–746

von Arnim A, Stanely J (1992) Inhibition of African cassava mosaic virus systemic infection by a movement protein from the related geminivirus tomato golden mosaic virus. Virology 187:555–564

Wang Z, Yao M, Wu Y (2009) Cross-resistance, inheritance and biochemical mechanisms of imidacloprid resistance in B-biotype *Bemisia tabaci*. Pest Manag Sci 65:1189–1194

Xu J, De Barro PJ, Liu SS (2010) Reproductive incompatibility among genetic groups of *Bemisia tabaci* supports the proposition that the whitefly is a cryptic species complex. Bull Entomol Res 100:359–366

Yamamoto I, Yabuta G, Tomizawa M, Saito T, Miyamoto T, Kagabu S (1995) Molecular mechanism of selective toxicity of nicotinoids and neonicotinoids. J Pestic Sci 20:33–40

Zamir D, Ekstein M, Micelson I, Zakay Y, Navot N, Zeidan M, Sarfatti M, Eshed Y, Harel E, Plebam T, Van-Oss H, Keidar N, Rabinowitch HD, Czosnek H (1994) Mapping and introregression of a tomato yellow leaf curl virus tolerance gene, Ty-1. Theor Appl Genet 88:141–146

Chapter 12
Development of Integrated Pest Management (IPM) Strategies for Whitefly (*Bemisia tabaci*)-Transmissible Geminiviruses

Robert L. Gilbertson, Maria Rojas, and Eric Natwick

Abstract Worldwide outbreaks of *Bemisia tabaci* whiteflies, especially biotype B, have facilitated the emergence of whitefly-transmitted geminiviruses (WTG). These viruses cause economically important diseases of vegetable and fiber crops, especially in tropical and subtropical regions of the world. Because small populations of whiteflies can efficiently spread WTGs, management of these diseases is more challenging than for whiteflies alone. In this chapter, we discuss (1) why WTGs have emerged worldwide, (2) key aspects of the biology of WTGs and *B. tabaci*, and (3) how these aspects shape the development of an integrated pest management (IPM) approach for these diseases. The generalized IPM package involves strategies for (1) before the growing season, such as the use of virus- and whitefly-free transplants and propagative stock, and resistant varieties; (2) during the growing season, such as whitefly population suppression, roguing virus-infected plants, floating row covers and reflective mulches; and (3) after the growing season, such as region-wide sanitation, weed management and implementation of a host-free period. Different combinations of strategies will be used depending on the crop, cropping system, and properties of the virus and the whitefly vector. This is illustrated with two case studies: IPM for WTGs in an annual (tomato) and a perennial (cassava) crop.

R.L. Gilbertson (✉) • M. Rojas
Department of Plant Pathology, University of California-Davis, Davis, CA 95616, USA
e-mail: rlgilbertson@ucdavis.edu; mjrojas@ucdavis.edu

E. Natwick
UC Desert Research and Extension Center, University of California Cooperative Extension, 1050 E. Holton Road, Holtville, CA 92250, USA
e-mail: etnatwick@ucdavis.edu

W.M.O. Thompson (ed.), *The Whitefly, Bemisia tabaci (Homoptera: Aleyrodidae) Interaction with Geminivirus-Infected Host Plants*,
DOI 10.1007/978-94-007-1524-0_12,

12.1 The Problem, the Players and Their Interactions

High populations of sweet potato whiteflies, *Bemisia tabaci* (Gennadius), can cause considerable damage to crop plants by direct feeding damage, production of honeydew (e.g., sticky cotton) and the injection of toxic substances that induce physiologic abnormalities such as irregular ripening in tomato and silverleaf in squash (Perring 2001; Brown et al. 1995). However, whiteflies can also cause major damage to plants through the vectoring of various plant-infecting viruses. *B. tabaci* is the vector of more than 100 different plant viruses, including those in the genera *Begomovirus* (*Geminiviridae*), *Crinivirus* (*Closteroviridae*) and *Ipomovirus* (*Potyviridae*) (Jones 2003). Indeed, it can be argued that whiteflies cause more economic damage to plants through their role as vectors of viruses than through the direct damage that they cause.

12.1.1 Geminiviruses and Geminivirus Diseases

Geminiviruses are one of only two types of plant-infecting viruses with a single-stranded DNA genome (the other being the family *Nanoviridae*). Members of the geminivirus family all have small circular single-stranded DNA genomes encapsidated within small twinned icosahedral virions (18 × 30 nm), from which the viruses derive their name (the Latin word gemini means twin) (Rojas et al. 2005). These viruses evolved from circular single-stranded DNA plasmids in bacteria or primitive eukaryotic organisms. In their plant hosts, geminiviruses replicate primarily in the nuclei of phloem and phloem-associated cells, and are transmitted plant-to-plant by phloem-feeding insects (leafhoppers and whiteflies). Geminiviruses are distributed worldwide in weeds and indigenous plants, especially in tropical and subtropical regions of the world. Thus, geminiviruses are believed to be an ancient group of plant viruses that appeared prior to continental drift. Over time, geminiviruses have undergone considerable diversification, and four genera are recognized based on genome structure, type of host plant infected and the type of insect vector (Rojas et al. 2005). In this chapter, we will focus on whitefly-transmitted geminiviruses (WTG), which comprise the genus *Begomovirus*.

Long before geminiviruses were characterized, the diseases they caused in food and fiber crops were well known for causing damage to crops and economic losses. These diseases included African cassava mosaic, bean golden mosaic, cotton leaf curl, squash leaf curl and tomato yellow leaf curl (Varma and Malathi 2003). Since their characterization in the late 1970s-early 1980s, geminiviruses have emerged from relative obscurity to currently being the family of plant-infecting viruses with the largest number of recognized species (Fauquet et al. 2008). The majority of these species are begomoviruses. Key factors that have driven the emergence of WTGs include worldwide distribution of weed-infecting begomoviruses and the rapid spread of *B. tabaci* biotype B throughout tropical and subtropical regions (e.g.,

Africa, Asia [including China, India, and Pakistan], Australia, the Caribbean Basin, Central America, the Mediterranean Basin, Mexico, South America, Southern Europe, and Southern North America). Together, this has driven the process of local or parallel evolution, in which the whiteflies acquire an indigenous progenitor begomovirus from a weed or other host plant, and introduce it into a cultivated crop plant, especially those that are not native to the region. Eventually, through genetic mechanisms such as mutation, reassortment (pseudorecombination) and recombination, new viral forms arise, some of which adapt to and infect the crop plants (Rojas et al. 2005; Seal et al. 2006). Through this process, many crop plants (e.g., cucurbits, common bean, pepper and tomato) have become hosts of a diversity of begomovirus species in different geographical regions of the world. For example, tomatoes are infected by >60 begomovirus species (Czosnek and Laterrot 1997; Fauquet et al. 2008; Polston and Anderson 1997).

12.1.2 *Virus Identification and Biology*

The first step in managing a viral disease is to identify the virus(es) involved. Fortunately, considerable advances have been made in the capacity to detect and characterize WTGs. The use of the PCR and DNA sequencing revolutionized the identification and characterization of these viruses (e.g., Rojas et al. 1993; Wyatt and Brown 1996), and DNA sequences of hundreds of geminiviruses are available in sequence repositories (e.g., GenBank). This has allowed for the development of degenerate DNA primers for rapid PCR-based detection of geminiviruses. In addition, tissue blotting methods have been developed that allow for rapid and simple sample preparation and long-distance transport for laboratory processing (e.g., Zhou et al. 2008). Together, this has made PCR and sequencing the method of choice for geminivirus identification. As advances have also been made in our understanding of the biology of the virus and whitefly vector, identification of the virus also provides potential strategies for disease management.

12.1.3 *Biology of B. tabaci and the Virus-Vector Interaction*

B. tabaci is a complex species that is composed of multiple biotypes and/or species (Perring 2001; Brown et al. 1995). These whiteflies present numerous problems for crop production, which complicates selection of management strategies. In particular, *B. tabaci* biotype B has emerged as an important pest of agricultural and horticultural crops throughout the world (Byrne et al. 1990; de Barro 1995; Perring 2001; Brown et al. 1995), and it colonizes >600 plant species (Oliveira et al. 2001). Not only is this whitefly polyphagous, it causes three types of damage or injury to plants: direct damage, indirect damage and virus transmission (Brown et al. 1995; Perring 2001). Direct feeding removes sap from the phloem system, resulting in wilting and reduced growth and yield. It also may cause leaf chlorosis, premature dropping of leaves and even death of plants. In some plant species, feeding of *B. tabaci* biotype B induces physiological or toxicogenic disorders (e.g., silverleaf of squash and irregular ripening of tomato). Indirect injury is caused from deposition of excrement, called 'honeydew,' on plant surfaces resulting in loss of quality. Honeydew can also cause a reduction in photosynthesis and quality from discoloration caused by growth of 'sooty molds'.

However, some of the most severe damage that *B. tabaci* does to crops is through the transmission of plant viruses, including WTGs (Jones 2003). The mode of whitefly transmission of WTGs is persistent and non-propagative, as most evidence suggests that the virus is not passed to the offspring via the eggs (i.e., transovarially transmitted). In general, the virus can be acquired and transmitted in as little as 5 min (Czosnek et al. 2001; Ghanim et al. 2001; Rosell et al. 1999), although longer feeding times will result in greater efficiency of transmission. There is also a latent period of ~8 h before the adult becomes viruliferous, but then it is believed to carry the virus for life (Czosnek et al. 2001). Thus, an individual viruliferous adult can infect multiple plants, and the virus can be carried long distances in whiteflies associated with plant materials. This mode of virus transmission makes whiteflies efficient vectors, and small populations can spread WTGs rapidly, resulting in high rates of infection in relatively short time periods. The complete life cycle of the whitefly can be completed in as little as 18 days when conditions are favorable (warm temperatures), whereas it may take as long as 2 months when conditions are less favorable (Perring 2001). Finally, there is also evidence that *B. tabaci* whiteflies have developed a mutualistic relationship with certain begomoviruses, such that these whiteflies, but not indigenous ones, have greater rates of reproduction and survival on virus-infected plants (Colvin et al. 2004; Jiu et al. 2007).

12.1.4 IPM for Whitefly-Transmitted Geminiviruses

Because it is not possible to cure plants of infections by WTGs, efforts must be taken to keep plants from becoming infected or to manage the rate, timing or severity of the infection to protect crop health. Typically, growers emphasize whitefly management with insecticides to control WTGs; however, in most cases this does not provide adequate protection. What has been more successful for management of WTGs has been the use of an integrated pest management (IPM) approach (Jones 2004), in which multiple strategies are employed, targeting different levels of the plant-WTG-whitefly interaction. The first widely recognized concept of IPM stressed a combination of chemical, biological, and other control methods for insect pest management (Stern et al. 1959). A key tool in this IPM concept was the economic injury level (EIL), the point between cost and benefit of using a control measure, usually a pesticide. The EIL works well for insect pests, including whiteflies, where there is a direct correlation between insect population density and crop damage (e.g., Gusmao et al. 2006).

The problem in using EILs for WTGs (or other insect-vectored viruses) is that populations of whiteflies well below the threshold can transmit sufficient virus to result in economic losses. Thus, the development of an IPM package for WTGs is more complicated because of the virus-vector interaction. Such a program should involve monitoring whitefly populations and, possibly, management with insecticides, but this strategy alone is usually not sufficient. For WTGs, management decisions may be based on an "action threshold," the population level at which a control measure should be implemented to avoid significant damage to a crop (Ellsworth et al. 1994; Naranjo and Flint 1995; Dik and Albajes 1999; Gould and Naranjo 1999; Chu et al. 2007). The action threshold is often more practical than an economic threshold based on EIL, and may require less rigorous criteria for making rational management decisions. Therefore, it need only demonstrate a potential for a significant crop loss, such as when an insect-transmitted virus is involved (Nichols et al. 1994; Palumbo et al. 1994; Naranjo et al. 1998). However, we are not aware of any insect vector action thresholds established for management of insect-transmitted viruses.

12.2 Generalized Scheme for IPM of Whitefly-Transmitted Geminiviruses

There are a number of very effective strategies that, when combined into an IPM package, can provide effective management of diseases caused by WTGs. The specific strategies used for the IPM package in a given agroecosystem will be dependent on knowledge of the crop plant, the cropping system and the climatic conditions, and the biology of the virus and the vector. However, it is possible to present a generalized scheme for IPM of whitefly-transmitted viruses. This scheme is divided into three parts: (1) before the growing season, (2) during the growing season and (3) after the growing season.

12.2.1 Before the Growing Season

Advance preparation in terms of the cultivar of the crop, the source of the planting material (seed, transplants or propagative material), and field location are very important. Ideally, this will be implemented within the framework of a regional system-wide approach for management of whiteflies and WTGs.

12.2.1.1 Cultivar and Seed Selection

Selection of the proper cultivar is important for many reasons, but in the case of IPM of WTGs, the key points relate to availability of virus resistance or tolerance and certain horticultural aspects (e.g., earliness). Host plant resistance to whiteflies and/or WTGs provides an ideal pest management tool, with little or no environmental impact. Unfortunately, host plant resistance is not available for many whitefly-transmitted begomoviruses, and there are even fewer examples of resistance to the vector.

However, resistance to a number of WTGs has been identified in wild species or other sources. For example, tomato varieties with resistance to TYLCV are commercially available (Lapidot and Friedmann 2002), as are common bean varieties with resistance to golden mosaic caused by *Bean golden yellow mosaic virus* (Morales 2001) and dwarf mosaic caused by *Bean dwarf mosaic virus* (Seo et al. 2004), cassava varieties with resistance to the complex of begomoviruses causing African cassava mosaic disease (Legg and Fauquet 2004) and cotton varieties with resistance to the begomovirus/betasatellite complex that causes cotton leaf curl disease (Briddon 2003; Tarr 1951). For cucurbits and peppers, resistant varieties are not currently available, although potential sources of resistance have been identified.

It is important for these varieties to be evaluated in different regions where WTGs are economically important, and for this information to be made available to growers. Thus, even when a resistant variety is available, it may not be adapted for production in a certain area or it may not meet market standards. For example, if TYLCV were to spread from southern California (Rojas et al. 2007) into major processing tomato areas of California, the currently available TYLCV-resistant fresh market varieties would not be commercially acceptable. In Guatemala, a hybrid variety that was highly resistant to the prevalent WTGs was not commercially acceptable because of the fruit shape and firmness. Another problem is the availability of seeds of resistant varieties. Many WTGs are important in developing countries where subsistence farmers are involved in vegetable production. A big challenge is making seed of these varieties available to growers, particularly in the case of hybrids. Finally, it is also important to note that there is evidence that resistance in certain crops, e.g., cotton and cassava, has been overcome by new forms of WTGs in certain geographical locations. This further highlights the need for long-term cultivar testing in multiple environments.

12.2.1.2 Planting Virus- and Whitefly-Free Transplants and Propagative Stock

In general, the earlier that a plant is infected with a virus, the more severe the disease symptoms and the greater the yield loss. Therefore, it is critical to establish new plantings with virus-free and whitefly-free transplants (e.g., cucurbits, peppers and tomatoes) or propagative stock (e.g., cassava and sweet potatoes) (Fig. 12.1a–d). The first step is to keep transplant propagation facilities free of whiteflies. Greenhouses should have induced positive airflow, double door airlock entrances and roofs covered with UV absorbing films. All vents and other openings should be covered with whitefly-resistant, fine-mesh screening of 0.25×0.8 mm openings or less. Sanitation within and around the propagation facilities is also important. Potential whitefly or virus host plants must be eliminated in and around the facility, with discarded plant materials sealed in whitefly-proof containers or destroyed.

Fig. 12.1 Production of propagative materials (e.g., cuttings, transplants, etc.) that are free of whiteflies and whitefly-transmitted geminiviruses is an essential component of an effective integrated pest management (IPM) package. (**a**) production of vegetable (e.g., melon, pepper and tomato) transplants in specialized greenhouse operations, (**b**) production of vegetable transplants in small-scale screenhouses, and (**c**) and (**d**) production of transplants in the field

In addition, systemic neonicotinoid class insecticides (e.g., imidacloprid or thiamethoxam) applied as soil drenches, along with foliar insecticide sprays, can be used in greenhouse operations to suppress whitefly populations. This is essential in certain situations, such as the slat-house operations used in Southern California's Imperial Valley for tomato and pepper transplant production, to prevent TYLCV infection (Rojas et al. 2007). Monitoring of whitefly adults with yellow sticky traps can be used to know when foliar insecticides need to be applied (Gillespe and Quiring 1987). One approach is to place one trap per 80 plants or at least one per 6 m^2, at the beginning of the transplant production, to be used as a monitoring and control measure. Finally, it is highly advisable that transplants be treated with an insecticide at least 7 days before shipping. If specialized transplant production facilities are not available, field-produced transplants should be grown in isolated seedbeds and treated with systemic insecticide. In situations where whitefly and virus pressure is high, seedlings produced in the field can be protected from whiteflies with floating row covers (see below).

Finally, it is also important that transplants are produced locally, because WTGs can be spread long distance in transplants or in whiteflies associated with transplants (Jones 2003). Furthermore, because symptoms may not develop until 10–14 days after inoculation, visual inspection is not sufficient to assure transplants are virus-free. Thus, it is critical not to import transplants produced in areas where

WTGs are prevalent. The long-distance transport of infected plant materials (probably transplants) have been responsible for the introduction of the Old World TYLCV into the New World (Polston and Anderson 1997; Salati et al. 2002), the New World *Squash leaf curl virus* into the Old World (Antignus et al. 2003) and *Cucurbit leaf crumple virus* (CuLCrV) from the Western United States into Florida (Akad et al. 2008).

12.2.1.3 Location and Time of Planting

If possible, new plantings should be established following a host-free period or during periods when virus and whitefly pressure are low. If it is not possible to institute a regional host-free period, new plantings should be delayed until old plantings are removed. If this is not practical, new plantings should not be established near or adjacent to old established crops infested with whiteflies and begomoviruses, as this will result in early infections and the potential for substantial yield losses. If there is good information available on the seasonal patterns of whitefly populations and virus pressure, planting times can be modified to avoid periods of high pressure. If multiple staggered plantings are planned, barrier crops can be planted prior to the establishment of the plantings. In addition, field locations should be planned such that later plantings are established upwind of earlier planting and in blocks such that minimal area of the field is exposed to wind. However, under heavy virus pressure, these approaches alone are unlikely to substantially reduce virus infection in the field.

12.2.2 During the Growing Season

12.2.2.1 Whitefly Management

Suppression of whitefly populations with insecticides, especially in areas with histories of whitefly outbreaks, can be an important component of a successful IPM package for WTGs. Insecticides are most commonly applied as foliar sprays or injected into the soil, but may also be applied via chemigation through drip irrigation. Soil applications are typically systemic insecticides, mostly in the neonicotinoid chemical class. The prophylactic use of soil applied systemic insecticides have been documented to slow, reduce or delay virus transmission by whiteflies; however, the use of insecticides alone often does not deliver sufficient protection from WTGs to prevent economically important crop damage.

In general, it is advisable to apply a neonicotinoid insecticide as a drench at the time of transplanting. This is best accomplished by adding the insecticide into the transplant water, as this will extend upon the treatment applied in the transplant greenhouse or seedbed. Together, these treatments should provide protection from whiteflies for 45–60 days (Kerns and Palumbo 1995). However, more recent field insecticide efficacy research in the southwestern United States has shown that *B. tabaci* nymphal populations begin to increase between 30 and 40 days after injection of imidacloprid into soil at planting. In such cases, foliar applications of materials such as Courier (buprofezin), an insect growth regulator (IGR) that inhibits chitin synthesis (Ishaaya et al. 1988), and Oberon (spiromesifen), an insecticide that inhibits lipid biosynthesis (Nauen et al. 2003), can be used. These insecticides prevent whitefly nymphal development and can provide suppression that is equal to soil-applied neonicotinoid treatments. However, the IGRs are not efficacious against whitefly adults (Stansly and Natwick 2010). For control of adults, pyrethroid insecticides and endosulfan are commonly used in vegetable crops and cotton (Chu et al. 1998). Fulfill (pymetrozine) is an insecticide with a unique mode of action, paralyzing the muscles of homopterous insects used during feeding, and thereby preventing ingestion of plant sap and transmission of viruses (Polston and Sherwood 2003). These treatments can effectively manage whitefly populations and spread of WTGs within a field, but plants can still get infected by virus from viruliferous whiteflies emigrating from outside of the field.

Monitoring needs and requirements for whitefly populations vary depending on many factors such as type of crop, structure of the crop canopy, and the potential for economic damage due to virus diseases. Prior to registration of IGRs for whitefly control, insecticide treatment decisions were based only on adult sampling. Now, both whitefly adults and nymphs may need to be monitored on leaves of many vegetable crops from emergence of seedlings until harvest. Whitefly adults are monitored for timing applications of adulticides, and nymphs for timing IGR applications. Foliar sprays are usually applied based on an action threshold (Nichols et al. 1994; Palumbo et al. 1994; Naranjo et al. 1998), and nymphal action thresholds have been developed for applications of IGRs (Ellsworth and Martinez-Carrillo 2001). Whitefly adults are highly mobile and inter-crop movement may be of concern due to virus disease. Thus, monitoring adult populations, their movement and the percentage of adults carrying a virus is important for an area-wide whitefly management program or for virus disease management programs (e.g., Salati et al. 2002). Factors that complicate whitefly control with foliar insecticides are: (1) adults and nymphs are found mostly on the abaxial leaf surface often escaping contact with spray droplets, and (2) *B. tabaci* and some other whitefly species have developed resistance to many insecticides. Therefore, it is critical to rotate chemical classes with different modes of action for insecticide resistance management (IRM), and to continue to identify new insecticides with different modes of activity (Castle et al. 2009).

Many annual cropping systems rely heavily on insecticides when whitefly populations are high and pose a major threat to production. Although insecticides often temporarily reduce whitefly populations below EIL, this generally does not protect plants from WTGs due to the persistent nature of virus transmission and the migration of viruliferous whiteflies into fields. Rubenstein et al. (1999) showed that viruliferous adult whiteflies had enough time into inoculate TYLCV to imidacloprid-treated plants before they died; therefore, transmission efficiency was similar for plants treated with imidacloprid and those not treated. However, Ahmed et al. (2001) showed that imidacloprid applied immediately after planting and 6 weeks later, protected tomato plants against TYLCV up to 12 weeks after sowing, and similar results were reported for thiamethoxam (Mason et al. 2000). In most cases, suppression of whitefly populations with insecticide applications alone is unlikely to provide adequate management of WTGs.

12.2.2.2 Roguing

This strategy involves the physical removal of virus-infected plants over the course of the growing season. Roguing needs to be done soon after plots are established, and is most helpful if the incidence of the virus is low (<5%). After roguing, it is also important that there is a minimal level of virus spread in the field, as well as a limited amount of introduction of virus from outside the field. If whitefly populations are high, plants should be treated with an insecticide to kill whitefly adults prior to roguing. If nymphs are present, rogued plants should be removed in plastic bags and disposed of well away from production fields. Ideally, fields should be monitored weekly and symptomatic plants removed. At least, fields should be checked twice, 7–14 and 21–28 days after transplanting, so that the diagnostic symptoms have had a chance to develop.

12.2.2.3 Exclusion: Protected Culture in Greenhouses and Screen Houses

Vegetables can be protected from whitefly damage and virus infection by physical means, i.e., preventing the insects from contacting susceptible plants. In the most extreme case, the entire crop is grown in a greenhouse or screen house, and plants are protected from whiteflies for the entire production cycle (Fig. 12.2a–c). When these structures are kept free of whiteflies (e.g., through the use of glass, plastic or screening; vents covered with screening and double doors with positive pressure), excellent management of whiteflies and WTGs can be achieved. For example, this approach has been successfully used to protect tomatoes from TYLCV in Israel (Ausher 1997; Berlinger et al. 2002). In the Baja Peninsula of Mexico, tomatoes and other vegetables are produced in large screen houses (mallas) to protect the crops from a number of insect pests and diseases, including WTGs. In Central America, where these viruses can limit field production of tomatoes and peppers, protected culture is becoming more common (Fig. 12.2a–c). However, as this type of production is expensive and labor intensive, it tends to be targeted for the export market rather than local consumption.

12.2.2.4 Exclusion: Protection of Plants in the Field with Floating Row Covers

The other commonly used method of exclusion is the covering of young plants, either emerging from seeds or that have been transplanted, with protective netting (Fig. 12.2d). This netting is a spun-bonded polyester material (commercially

Fig. 12.2 Production of vegetable crops (e.g., cucumbers, peppers and tomatoes) in protected culture can be an effective component of an integrated pest management (IPM) package for whitefly-transmitted geminiviruses. (**a**) greenhouses for production of vegetable crops, (**b**) production of peppers in greenhouses, (**c**) production of tomatoes in greenhouses and (**d**) use of 'floating row covers' for protection of vegetables from whiteflies and whitefly-transmitted geminiviruses after transplanting (generally provides about 30 days of protection until the nets need to be removed)

available as Agribon or Agril) and is placed directly over the rows of emerging seedlings or transplants. The covers are typically placed over the plants without any type of support, such that it is a 'floating' row cover and moves with the growth of the plants. In other cases, semi-circular lengths of wire or piping are used to provide support and keep the netting from directly contacting the leaves. These materials allow passage of adequate amounts of light for normal plant growth, although there have been some reports that the microclimate formed under the row covers can favor development of foliar diseases caused by bacteria and fungi. In general, the row covers are left on for ~30 days or until pollination, e.g., in the case of cucurbits.

It is well established that the use of row covers can protect plants from whiteflies and reduce the spread of WTGs in crops like cucurbits, pepper and tomato (Natwick and Durazo 1985; Natwick and Laemmlen 1993; Orozco-Santos et al. 1995; Webb and Linda 1992). In cases of severe whitefly and virus pressure, this protection can make the difference in whether a marketable crop is produced. This approach has been shown to slow the spread of TYLCV in Israel (Cohen and Berlinger 1986), and is being used in Guatemala to protect tomato and peppers from infection with various WTGs. Row covers have also been successfully used in Guatemala to protect melons from WTGs and the whitefly-transmitted crinivirus, *Cucurbit yellow stunting disorder virus* (CYSDV). However, the use of floating row covers is expensive, and should be used to protect plants during periods where whitefly and virus pressure are known to be high. Small-scale farmers can use row covers to protect seed beds or to produce small tunnels in which seedlings can be protected from whiteflies during this critical stage of growth (Hilje et al. 2001). Of course, once the row covers are removed, the crop is again exposed to viruliferous whiteflies and high rates of infection can still be experienced, albeit only in the new growth. In these situations, it is important to use early maturing or resistant varieties and/or include whitefly management strategies (i.e., insecticides).

12.2.2.5 Barrier Crops and Mulches

There are a number of other cultural practices that can be used to protect crops from whiteflies and, thus, slow the spread of WTGs. Physical barriers can be designed to prevent the movement of whiteflies into fields of susceptible crops. Barriers may be non-living, such as plastic (yellow plastic with sticky material to trap insects) or screening; or living, such as the planting of a tall-growing plant species (non-hosts of the whitefly and WTGs) in between fields of susceptible crops. The best barrier plants for WTGs are monocots such as corn, sorghum and elephant grass. However, there is little evidence that barriers effectively reduce whitefly migration or virus spread, because whiteflies can fly or be wind-carried over barriers, and transmit WTGs for long periods of time, due to the persistent nature of transmission (Hilje et al. 2001). Thus, barriers are generally not an essential component of the IPM package for WTGs.

Mulches are designed to prevent insects from recognizing and landing on a crop that is susceptible to virus infection. Like barriers, mulches can be non-living (plastic or some other material) or living (plants grown among the susceptible crop). In terms of non-living mulches, the most effective materials are colored or UV-reflective plastic. These have been reported to have some success in reducing whitefly population densities as well as the incidence of WTGs (Antignus 2000). In Florida, a UV reflective mulch treatment reduced the incidence of CuLCrV in zucchini squash (Nyoike et al. 2008); and the incidence of *Tomato mottle virus* (ToMoV) and TYLCV in Florida and Jordan, respectively (Csizinszky et al. 1995; Suwwan et al. 1988). These mulches can also result in improved crop growth. On the other hand, mulches are expensive, labor-intensive and can be deleterious to the environment. Thus, mulches are more typically used in a high value crop, such as fresh market tomatoes.

Living mulches involve planting low-growing ground cover-type plants, which are non-hosts for whiteflies and WTGs, among a susceptible crop. These living mulches reduce whitefly populations by causing the insects to leave the field due to the presence of the non-host plants (Hilje et al. 2001). In studies carried out in Costa Rica, reflective polyester mulches and living mulches of perennial peanuts and coriander reduced the number of incoming adult whiteflies and the spread of the begomovirus *Tomato yellow mottle virus* (Hilje and Stansly 2008). In Florida, a living mulch of buckwheat reduced the incidence of CuLCrV in zucchini squash, especially when combined with imidacloprid treatment (Nyoike et al. 2008). However, living mulches alone only reduce virus incidence and may not prevent economic losses when virus pressure is high; thus, this method should be used in combination with other approaches (e.g., virus-free transplants, insecticides and crop-free periods) to achieve higher levels of protection. On the other hand, living mulches are less expensive and more environmentally friendly than plastic mulches and may be more readily adopted by small farmers, especially if the living mulch is a crop plant, such as coriander (Hilje and Stansly 2008).

12.2.3 After the Growing Season

12.2.3.1 Sanitation

Field sanitation is a critical component of an effective IPM program for WTGs because old plants are often the most important sources of whiteflies and WTGs for newly planted crops. Moreover, as these plants senesce, they become a less preferred source of nutrition, and whiteflies will move to later-planted crops, where they can initiate new infections. Thus, it is absolutely essential that, after harvest, old plants are promptly removed and destroyed, preferably within a week after harvest. This can be achieved by physical removal, application of desiccating herbicides and oils, tillage or burning. In Florida, it is recommended that vegetable crops be destroyed within 5 days of harvest, and crop residues burned down with a combination of a contact herbicide/oil and a non-ionic adjuvant (Schuster et al. 2007). It is also essential to uproot plants, such that new growth (e.g., sprouts) will not develop from old plants remaining in the field; this young growth will often have high titers of virus and whiteflies, and can serve as an important inoculum source for newly planted fields (Fig. 12.3b,c).

In cases where harvested plants have high populations of whiteflies, it may be advisable to destroy the whiteflies as well as the plants. For example, this could involve the application of a contact insecticide followed by deep plowing of the field to destroy the plants (Fig. 12.3b, c, d). To minimize migration of whitefly adults, avoid crop destruction activities in the field during times when prevailing winds would facilitate movement of whiteflies to adjacent fields of susceptible crops. It may also be helpful to destroy crops within a field, block by block, as harvest is completed rather than waiting and destroying the entire field at one time.

An excellent example of the importance of sanitation of old harvested plants in management of WTGs is the role of ratoon cotton plants in cotton leaf curl disease in the Sudan (Tarr 1951). In the early to mid 1900s, large acreages (~200,000 acres) of irrigated Egyptian cotton (*Gossypium barbadense* L.) were grown in Gezira, Sudan. Beginning around 1930, cotton leaf curl disease emerged as a major constraint on production. It was established that the virus was transmitted by whiteflies and that it had a narrow host range (mostly plants in the family *Malvaceae*). Furthermore, it was found that a major source of inoculum, was ratoon cotton

Fig. 12.3 Sanitation and host-free periods play an important role in integrated pest management (IPM) packages for whitefly-transmitted geminiviruses. (**a**) weed management, both in and around fields, can reduce potential sources of whiteflies and whitefly-transmitted geminiviruses; weeds (*Cleome viscosa*) growing among tomatoes in the Dominican Republic, (**b** and **c**) sprouts emerging from infected plants following harvest can have extremely high titers of whitefly-transmitted geminiviruses and serve as sources of inoculum for newly planted fields, such as this example of *Tomato yellow leaf curl virus*-infected sprouts emerging from a harvested tomato plant in the Dominican Republic, (**d**) prompt removal and destruction of harvested plants is essential for an effective IPM package for whitefly-transmitted geminiviruses; this processing tomato field was promptly plowed following harvest, and (**e** and **f**) the successful re-introduction of tomato culture in Mali, West Africa in an area heavily impacted by whitefly-transmitted geminiviruses following implementation of region-wide sanitation, a host-free period and the planting of early maturing tolerant hybrid tomato varieties

'plants' that sprouted from the cotton 'tree' stumps left in the ground after harvest. A root pulling apparatus was devised for removing these stumps and, following the widespread implementation of this approach, the incidence of the cotton leaf curl disease in Gezira was reduced substantially.

Finally, because of the mobile nature of whiteflies, it is imperative that effective field sanitation be practiced on a regional level. Obviously, if all farmers are not participating in the sanitation program, the effectiveness is reduced, as old plants in fields where sanitation was not practiced will serve as inoculum sources for newly planted fields. Implementation of an effective area-wide sanitation program typically requires the involvement of government (extension and regulatory agencies) and industry (for implementation), at least until it becomes a standard practice for local farmers.

12.2.3.2 Weed Management

Weeds can harbor whiteflies and be a source of WTGs (Fig. 12.3a). As previously mentioned, weeds are commonly infected with indigenous begomoviruses, which are generally different species than those infecting crop plants. Moreover, although most of the crop-infecting begomoviruses evolved from weed-infecting progenitors, the changes in the virus that allowed for adaptation to the crop plant make it less fit in the original weed host. Thus, relatively few studies have demonstrated that a particular weed host plays a major role in the epidemiology of crop-infecting begomoviruses.

However, the finding that TYLCV induces symptomless, low virus titer infections in a range of weed species may explain why so few weed hosts of begomoviruses have been identified (Salati et al. 2002). This finding also has raised new concerns about the role of weeds in the epidemiology of diseases caused by WTGs. An unresolved question is how efficiently whiteflies acquire WTGs from symptomless weed hosts. In the Dominican Republic, a 3 month whitefly host-free period has provided effective management of TYLCV, despite the widespread distribution of symptomless weed hosts. This suggests that the virus is not efficiently acquired from these hosts. This notion is also consistent with emerging evidence that there is a strong positive correlation between virus titer in the acquisition host and the capacity of insect vector to acquire and transmit the virus (Chen and Gilbertson 2009; Froissart et al. 2010).

Thus, there is no evidence that weed control alone will be an effective management strategy for WTGs. Furthermore, such programs are expensive and very difficult to implement in subtropical and tropical regions where weeds are present in great abundance throughout the year. In a situation where one of two weed species are known to play a key role in a disease, such as was the case for TYLCV in the Jordan Valley (Cohen et al. 1988), it may be feasible to conduct a program of weed eradication. Similarly, if certain weeds are preferred whitefly hosts, these could be targeted for eradication or insecticide sprays. In general, growers should be encouraged to manage weeds in fields as a desirable agronomic practice and, if possible, around fields (Fig. 12.3a).

12.2.3.3 Host-Free Period

In temperate regions, the winter season provides a natural host-free period. However, in tropical and sub-tropical areas crops can be grown all year, and the only way to break this pattern of continuous cropping is by having a defined period of time where the susceptible crop(s) is not grown. In terms of WTGs, this approach can be very successful because individual viruses tend to have a relatively narrow host range and the most important inoculum source is often the susceptible crop plant itself. Furthermore, even though *B. tabaci* (especially biotype B) has a wide host range, the removal of one or more key preferred crop hosts, such as beans, cucurbits, or tomatoes, can also lead to substantial reductions in whitefly populations, which further adds to the efficacy of the host-free period. Indeed, one option is to actually have a whitefly host-free period, which in effect targets both the insect vector and the virus. In addition, during a host-free period, volunteer crop plants or weed hosts of the virus or whitefly vector also need to be destroyed. A successful host-free period requires a community or regional approach where all growers agree to terminate production of a particular crop within a geographically defined area, and to delay planting the crop until an established planting date.

The host-free period is best suited to annual crops that are harvested over a short period, like vegetables or cotton. It is more difficult with perennial or semi-perennial crops, such as cassava and okra, which are continuously grown and harvested. The length and time of year for a host-free period will depend of the particular crop(s) and cropping systems used in a region. However, the host-free period should allow enough time for one or two generations of the whitefly vector (recall that whiteflies generally do not pass the virus to the offspring via eggs [transovarial transmission]). The time of year when the host-free period is imposed may be when whitefly populations are low, or a time where the susceptible crop(s) are traditionally not grown. These very important decisions about a host-free period must be made in conjunction with the growers in the region. For example, the host-free period for tomato-infecting begomoviruses in Mali is June-August, which is the rainy season, a period when whitefly populations are low and there is less tomato production (Fig. 12.3e, f). The host-free period can be legally enforced, such as the one in the Dominican Republic; or voluntary, such as the one implemented in Mali, although this required the support of the village chiefs, who play a key role in local agricultural decisions.

The importance of a host-free period for managing WTGs was observed for cotton leaf curl disease in the Gezira region of the Sudan (Tarr 1951). Here, in this production system, the cotton harvest is completed in March and a 'dead' or cotton-free period of 2–3 months was implemented before the next crop was planted. However, because of the infected ratoon cotton that would appear following rains in July, the cotton-free period did not help manage the disease, even with government mandated programs banning the cultivation of suspected alternate hosts (e.g., the crop plant *Hibiscus esculentus* [bamia], and the malvaceous ornamental shrub *Malvaviscus arboreous*). Once the importance of the ratoon cotton as an inoculum source was recognized and these plants were eliminated by regional sanitation measures, the cotton-free period became a much more effective strategy. This example shows how an effective host-free period goes hand-in-hand with effective sanitation.

12.3 Putting It All Together: Case Studies of IPM for Whitefly-Transmitted Viruses

The development of an effective IPM program for WTGs is dependent on knowledge of the host, the virus, the whitefly vector and the environment/cropping system (Jones 2004). Also, the IPM package that can be developed will also have to be compatible with the type of farmer and farming system involved. In some cases, such as subsistence farmers in developing countries, certain practices will not be practical, such as whitefly control with insecticides and floating row covers. Two case studies will be presented that bring together a number of management strategies into an IPM package. These reflect different situations: an annual vs. perennial host, different viruses and virus biology and different cropping practices.

12.3.1 Case Study 1: Management of a Whitefly-Transmitted Geminivirus in an Annual Crop: IPM of Tomato-Infecting Begomoviruses

12.3.1.1 The Host: Cultivated Tomato

Tomato (*Solanum lycopersicum* L.) is second to potato as the most important vegetable crop in the world. A member of the nightshade family (*Solanaceae*), it is a herbaceous perennial plant with a viney growth habit, although the cultivated tomato is most commonly grown as an annual crop. Tomato is an extremely versatile crop that can be consumed fresh or processed (paste, sauce, catsup), and it is often an important cash crop for small farmers in developing countries. Tomato is a New World crop, with the center of origin in the Andes Mountains of South America (parts of Peru, Ecuador and Chile). Numerous wild species of tomato have been identified in this region, and these represent a valuable genetic resource, including sources of genes that confer resistance to WTGs. The tomato was first domesticated in Mexico, where it was consumed by the Aztecs and other indigenous people. The Spanish introduced tomatoes to Europe in the sixteenth century, and subsequently to the Caribbean and to Asia (Philippines). Tomatoes spread throughout Europe and were introduced back into the New World (e.g., North America). Today, tomatoes are grown worldwide, with production in 144 countries, and production has increased substantially over the past 30 years.

12.3.1.2 Partners in Crime: The Worldwide Emergence of *Bemisia tabaci* Biotype B and Tomato-Infecting Begomoviruses

The spread of the tomato plant throughout the world was an important development in terms of the diet of human beings. However, this also exposed the tomato plant to a wide range of insect pests and diseases, most of which were new (i.e., not found in the centers of origin or domestication). In some cases, this resulted in the emergence of new tomato pests and pathogens. Indeed, there is perhaps no better example of this phenomenon than the emergence of *B. tabaci* biotype B and WTGs (Seal et al. 2006). The rapid expansion of irrigated tomato production in tropical and subtropical regions of the world contributed to the explosion of whitefly populations, especially in arid and semi-arid regions. It was in this environment, probably somewhere in western Asia, where the polyphagous *B. tabaci* biotype B originated (Brown et al. 1995). The insect spread quickly, mostly by human activities, and displaced indigenous whiteflies in many areas. This has led to the emergence of biotype B as a pest of world-wide importance, and one that is very difficult to manage (Perring 2001; Brown et al. 1995; Stansly and Natwick 2010).

A pattern emerged in which diseases caused by WTGs would appear in tomato crops ~3–5 years after outbreaks of *B. tabaci* biotype B. Moreover, characterization of the viruses involved in geographically distinct regions (e.g., Asia, Africa, Mexico and Central America, Brazil, etc.) revealed that most were new begomovirus species that were unique to these regions (e.g., Chowda Reddy et al. 2005; Ribeiro et al. 2003; Zhou et al. 2008). This can be attributed to the previously discussed phenomenon of local evolution. Thus, through this complex plant-virus-vector interaction, many new tomato-infecting begomoviruses have emerged worldwide (as mentioned previously, more than 60 species have been described). In contrast to the diversity of the WTGs infecting tomatoes worldwide, the symptoms induced by these viruses are similar, consisting of some combination of stunted and distorted growth and leaf distortion, upcurling, chlorosis, and vein purpling (Fig. 12.4a, b). Collectively, the diseases caused by these viruses are called tomato yellow leaf curl (TYLC[D], Fig. 12.4a) or tomato leaf curl (TLC[D], Fig. 12.4b). In addition, the Israeli strain of TYLCV (TYLCV-IL) has been spread throughout the world by human activities, further complicating the identification and management of tomato-infecting begomoviruses (Polston and Anderson 1997). Today, tomato-infecting begomoviruses cause substantial losses in many tomato-producing regions and, in some cases, the diseases are so severe that tomatoes cannot be cultivated in some areas during certain times.

12.3.1.3 Biological Considerations

An IPM strategy for tomato-infecting begomoviruses requires knowledge of the virus, the vector and the cropping system (Holt et al. 1999). Although there is considerable genetic diversity in the begomoviruses that infect tomato, they all share certain basic properties. The viruses are not seed borne or mechanically transmitted

Fig. 12.4 Symptoms of begomovirus diseases. (**a**) tomato yellow leaf curl in the Dominican Republic, (**b**) tomato leaf curl in Mali and (**c**) African cassava mosaic in Mali

(i.e., by touch), which reflects the phloem limitation of most of these viruses. The host range of most tomato-infecting begomoviruses is narrow, with tomato and a few other solanaceous plants being the most important hosts, and the hosts in which the typical disease symptoms are observed. TYLCV causes symptomless infections in weeds and other plants (Salati et al. 2002), which can serve as reservoir hosts in the absence of tomato. It is likely that other tomato-infecting begomoviruses also induce symptomless infections in weeds. All WTGs are transmitted by *B. tabaci* in a persistent manner, but most are not transovarially transmitted. On the other hand, it is important to identify the begomovirus(es) involved in a disease outbreak in a particular geographic region, because the efficacy of host plant resistance genes varies for different WTGs. Thus, knowledge of the precise WTG involved is important for selection of the appropriate resistant varieties (see below). It is also important to know something of the bionomics of the whitefly vector, such as the prevalent biotype(es), when are populations high and low, the main host plant species of the whitefly and if the population has pesticide resistance.

12.3.1.4 Knowledge of Local Tomato-Growing Practices

In addition to the information about the biology of the virus and vector, there is also a need for a thorough knowledge of the tomato growing practices (e.g., type of tomatoes that are grown [i.e., fresh market or processing], the varieties that are planted, time of planting, and means of harvest and storage). It is critical to know how many months of the year a crop is grown and the number of plantings crops are grown, the means of watering (rain fed or irrigated) and sanitation practices.

12.3.1.5 IPM Package for Tomato-Infecting Geminiviruses: Preplant Activities

Utilize Virus- and Whitefly-Free Transplants

Tomato-infecting begomoviruses are not seed-transmitted, so transplants will not become infected via contaminated or infected seed. However, whiteflies can transmit the virus to plants in the seedling stage, and establishing fields with virus-infected transplants will lead to rapid spread of the virus within the field. Furthermore, infection of plants at such an early stage of growth will lead to the greatest economic losses. Therefore, an essential component of the IPM package is the use of virus- and whitefly-free transplants.

Tomato transplants are typically produced in greenhouses or screenhouses or in seedbeds in the open field (Fig. 12.1a–d). Production of transplant seedbeds in the field is common in developing countries (Fig. 12.1c, d). Here, it is essential to establish these seedbeds away from established fields or promptly following a tomato-free period. If possible, the seed or seedlings should be treated with a systemic insecticide, such as imidacloprid or thiamethoxam. If the seedbeds are established in the presence of high populations of whiteflies, protection of the seedlings with floating row covers may be necessary.

In developed countries, it has become common practice to produce transplants in greenhouses, usually in specialized operations (Fig. 12.1a). While this is done primarily for producing vigorous and uniform transplants, it also allows for production of virus- and whitefly-free transplants. In the case of WTGs, this means keeping whiteflies physically separated from transplants. In addition, a systemic insecticide treatment is also commonly applied. It is also important to maintain a high level of sanitation in greenhouses and to maintain good weed control, as weeds can harbor both whiteflies and viruses. Under these conditions, virus- and whitefly-free tomato transplants can be efficiently produced, and these can be an important tool in the effective management of WTGs, as they eliminate a key source of primary inoculum.

Resistant Varieties

Tomato varieties are selected primarily on the basis of horticultural traits rather than disease resistance, unless a major disease problem emerges. Resistance genes for WTGs have been identified in wild tomato species, and some of these have been introgressed into commercial tomato cultivars. Some of the best characterized of these genes include *Ty-1*, *Ty-3*, *Ty-4* and *Ty-5* loci from *S. chilense* and the *Ty-2* locus from *S. habrochaites* (Hanson et al. 2006; Ji et al. 2007). Molecular markers have been developed for some of these genes, which can facilitate their introduction into preferred tomato cultivars (Ji and Scott 2005).

It is important to note that these genes are not equally effective in their capacity to confer resistance versus different begomovirus species. For example, *Ty-2* provides effective resistance against tomato-infecting begomoviruses from Asia, but not versus those in West Africa and Mexico/Central America. The TYLCV-resistant variety, Gempride, is highly resistant to TYLCV-IL isolates, but is susceptible to infection by the bipartite begomovirus *Pepper huasteco yellow vein virus* from Mexico. Therefore it is important to first screen varieties/germplasm with reported begomovirus resistance in target geographical regions to determine if the resistance will work against the local tomato-infecting begomoviruses. The combination of *Ty-1*, *Ty-2* and *Ty-3*, provides effective resistance versus TYLCV and many other tomato-infecting begomoviruses; thus, breeding programs need to emphasize pyramiding multiple resistance genes (Mejia et al. 2005).

In the development of an effective IPM program for WTGs, it is critical to be aware of whether varieties with resistance to WTGs are available and are these of the desired type (e.g., determinant or indeterminant and fruit type and color). In addition, it is also important to know whether these varieties have been evaluated in the region, and if seed is available and affordable. The most progress in developing begomovirus-resistant tomatoes has been with fresh market tomato varieties with TYLCV resistance. There are fewer processing tomato varieties available with begomovirus resistance, but breeding efforts are ongoing and commercial varieties should be available in the near future. Having identified a variety that is resistant to the local begomoviruses, the next challenges are getting the information to farmers and availability of seed. Most of the begomovirus-resistant tomato varieties are hybrids, which are more expensive than open-pollinated varieties and have to be purchased on an annual basis. For subsistence farmers in developing countries, this is a problem in terms of cost and seed availability. The cost issue can be addressed by selling small quantities of seeds; however, in many regions, the seeds are simply not available. This will continue to be a problem until better extension programs are developed for trialing and publicizing new varieties and seed dissemination networks are established.

In situations where an effective host-free period exists, it is possible to grow early maturing susceptible or tolerant varieties in the beginning of the growing season. In this case, the plants have sufficient time to flower and set fruit before begomovirus pressure builds-up. Host-free periods have provided sufficient cleansing of tomato-growing regions of WTGs to allow for successful planting of susceptible early maturing hybrid processing tomato varieties in the Dominican Republic and Mali (Gilbertson et al. 2007; Zhou et al. 2008; Fig. 12.3e, f). However, as the incidence of the virus increases, which generally occurs in 6–8 weeks, it often becomes necessary to plant resistant varieties. For example, in the Dominican Republic it is necessary to plant resistant varieties later in the growing season due to the intense virus pressure. In Mali, where a second planting is done, virus and whitefly pressure may not reach levels that necessitate planting a resistant variety (Fig. 12.3e, f).

Cultural Practices

A number of cultural practices have been identified that can also be considered in the IPM package. However, these usually have only a limited beneficial effect, and are not a substitute for virus- and whitefly-free transplants or resistant varieties. If possible, new tomato plantings should be established at times when whitefly populations are not at their peak. This varies from region to region, and should be established experimentally. Ideally, there also would be some regional coordination regarding planting date or a host-free period in regions with no winter season. New plantings should not be established near old plantings. If sequential plantings are done, it is important to put the late-planted fields upwind of the early-planted fields. Finally, maintaining optimal health of the tomato crop is also important (e.g., adequate water and fertilizer), as viral infection tends to develop slower in healthy vs. stressed plants.

During the Growing Season

Row Covers

In situations where a tomato crop is being transplanted into the field in the presence of viruliferous whiteflies, plants can be physically protected with 'floating row covers'. These materials are placed over the rows of plants, leaving the ground in between rows uncovered (Fig. 12.2d). The covers can only be left over plants for ~30 days and, if viruliferous whiteflies are still present, these plants will become infected. However, given the nature of systemic spread of viruses, only the new growth emerging after the removal of the cover will be diseased, and some tomato production will be achieved by the lower part of the plant. Thus, floating row covers can delay viral infection and assure at least some yield. On the negative side, the cost of the netting and labor to put it over the plants is considerable, and there is some evidence that the netting can increase the incidence of certain bacterial and fungal diseases of tomato.

Roguing

Roguing is a strategy that can be used with tomato-infecting begomoviruses. However, it is generally effective only if done soon after plots are established, when the incidence of the virus is low (<5%) and virus pressure is not high.

Whitefly Monitoring and Management

It is difficult, if not impossible, to control WTGs through the use of insecticides alone. This approach is also costly and not good for the health of farmers or the environment. However, in intensive production systems, such as fresh market tomato production in Florida and certain greenhouse production systems, intensive use of insecticides can reduce the spread of tomato-infecting begomoviruses such as TYLCV (Berlinger et al. 1986; Mason et al. 2000; Schuster et al. 2007). However, in most production systems, this is not economical or practical.

It is important to monitor whitefly populations to understand the population dynamics on a regional basis, especially to detect the build-up of populations early in the crop production cycle. This can be done by monitoring adult populations with yellow sticky cards or with the leaf turn method, in which adults and/or nymphs are directly counted on the undersurfaces of leaves. However, as mentioned earlier, with WTGs, the challenge is developing threshold populations that can trigger pesticide applications that will slow the spread of the virus.

After the Growing Season

Sanitation

Because weeds can serve as hosts for whiteflies and WTGs, weed management, at least with in and possibly around tomato fields, is also recommended as part of a sanitation program (Fig. 12.3a). Following harvest, it is imperative that tomato plants are promptly removed or and destroyed. It is critical that plants are uprooted, as rooted plants infected with virus can produce sprouts, which will have high titers of WTGs (Fig. 12.3b, c). This new growth will also be sought out by whiteflies, particularly in fields of dead or very old plants. At least, the harvested plants should be uprooted, removed from the field, and destroyed or used for composting. If possible, the harvested plants can be tilled or plowed into the soil (Fig. 12.3d). Like other components of the IPM program, sanitation is most effective if all farmers in a region are willing to use the practice. Dissemination of the information about the importance of good sanitation, and implementing region-wide practices is best accomplished through an effective extension system.

Regional Tomato-Free Period

Most tomato-infecting begomoviruses have a relatively narrow host range, the life cycle of *B. tabaci* is 18–60 days (depending on the temperature) and most of the viruses are not transmitted to the progeny through the egg (a possible exception is TYLCV; Czosnek et al. 2001). Therefore, by having a tomato-free period of 1–3 months, the main source of primary inoculum (tomato) is removed, effectively 'cleansing' the virus from whiteflies in the area. In Israel, the incidence of TYLCV was reduced with a vegetable crop-free period in June and July, coinciding with low whitefly populations (Ucko et al. 1998). Host-free periods also have been successfully used for management of TYLCV in Cyprus and in the Dominican Republic, and for a complex of tomato-infecting begomoviruses in Mali. These host-free periods result in a substantial delay in the appearance of the virus, such that an acceptable crop of tomatoes can be obtained, especially when early maturing tomato varieties are planted (Fig. 12.3e, f). In the case of the Dominican Republic, a PCR-based test was used to demonstrate the efficacy of the host-free period; whiteflies collected during the peak of the tomato-growing season were all carrying TYLCV, whereas whiteflies collected from weeds and other hosts during the host-free period had little or no virus (Salati et al. 2002; Gilbertson et al. 2007). After the implementation of a whitefly host-free period and the introduction of early maturing hybrid varieties, tomato production in the Dominican Republic far exceeded levels before the introduction of TYLCV.

The tomato- or host-free period has proven to be a highly effective tool for management of tomato-infecting begomoviruses (Fig. 12.3e, f). However, for a tomato-free period to be successful it must be a regional program and all farmers must agree to participate. The crops to be included depend on the knowledge of the biology of the virus and the agroecosystem; however, by including other whitefly hosts (e.g., beans, cucurbits, and other solanaceous crops) the crop-free period can also result in a substantial drop in the whitefly population. As previously mentioned, the free period can be either voluntary of legally imposed. Either way, the use of a tomato- or crop-free period should be seriously considered in areas where disease pressure is high and there is a need for a sustainable management approach that is not based on insecticide sprays.

Summary

Cultivated tomato has been a good host for the evolution of new WTGs, and this has been facilitated by the worldwide dissemination of the polyphagous *B. tabaci* biotype B. In many cases, these viruses cause diseases of considerable economic importance, particularly in tropical and subtropical regions. Many components of the IPM package for these viruses do not require specific knowledge of the begomovirus(es) involved, but identification of the WTGs involved in a region may influence the selection of resistant varieties. It is critical to start with virus- and whitefly-free transplants and, if possible, resistant varieties. Ideally, transplants are planted during a period of low virus pressure, such as following a host-free period or away from established fields. If viruliferous whiteflies are present then additional measures may be taken such as floating row covers or management of whitefly populations with insecticides. Roguing infected plants early in the season may slow down spread of the virus as does the use of reflective mulches. Following harvest, it is critical to uproot and destroy old plants, through removal or tillage. In tropical and subtropical regions, the implementation of a 1–3 month tomato- or whitefly host-free period can substantially reduce virus and whitefly pressure for the next crop. The free period provides an effective approach that is not based on pesticides, but does require regional cooperation. By implementing this IPM package (Table 12.1), there is a high probability that effective management of any tomato-infecting begomovirus can be accomplished.

12.3.2 *Case Study 2: Management of a Whitefly-Transmitted Geminivirus in a Perennial Crop: IPM of African Cassava Mosaic Disease*

12.3.2.1 The Cassava Plant and Its' Cultivation

Cassava (*Manihot esculenta*) is an important source of carbohydrates for people in tropical and subtropical countries, particularly in Africa. Worldwide, cassava represents the third largest source of carbohydrates for the human diet. The cassava plant is a perennial woody shrub (member of the spurge family, the Euphorbeaceae) that is grown for its tuberous roots, which contain large quantities of edible starch.

Table 12.1 Integrated pest management strategy for a whitefly-transmitted geminivirus disease of an annual crop (tomato)

Period of implementation	Management strategy
Before the growing season	Use of whitefly- and virus-free transplants
	Use of resistant cultivars
	Modification of planting dates
	Avoid planting new fields near older fields
	Plant new fields up-wind of old fields
During the growing season	Roguing of plants showing symptoms
	Monitor for whitefly populations using established means of sampling
	Apply insecticides only when necessary
	Rotate insecticides to minimize development of resistance (e.g., no more than two uses of any material per season) and use recommended rates (avoid using sub-lethal rates)
After the growing season	Prompt removal of old crops following harvest
	Consider region-wide management of planting dates and locations and implementation of a host-free period

Cassava is a New World crop, having been domesticated in Brazil. However, the majority of cassava is produced in Africa, which relates to the fact that it gives some of the highest yields of food energy per cultivated area, and produces well in poor soils and with minimal amounts of water. Cassava is propagated vegetatively, through the planting of ~15 cm stem sections.

12.3.2.2 Cassava Mosaic Disease (CMD)

This is one of the most damaging diseases of cassava (Fargette et al. 2006; Thresh and Cooter 2005). It can cause substantial yield reductions and is very difficult to manage. CMD is characterized by a striking light-dark green mosaic of leaves, various degrees of leaf and stem distortion and reduced numbers and weight of tubers (Fig. 12.4c). CMD is caused by a complex of whitefly-transmitted begomoviruses, including *African cassava mosaic virus* [ACMV] and *East African cassava mosaic virus* [EACMV]. The severity of the symptoms depends on the cultivar of cassava and the virus(es) involved.

The disease was first described in Africa (Tanzania) in the late 1800s, and it is now found throughout Africa and in some parts of Asia (e.g., India). It has been suggested that WTGs infecting this crop originated in East Africa (Fargette et al. 2006). Subsequently, the virus spread within Central-West Africa, causing losses to cassava production in countries such as Uganda, Zaire, Gabon, Zimbabwe, Botswana and South Africa (Legg and Fauquet 2004; Thresh and Cooter 2005). In the early 1990s, CMD became an even more important constraint, as epidemics of a severe type of CMD developed in East Africa, first in Uganda and then in other countries.

These epidemics have been associated with new highly virulent forms of the viruses, such as a recombinant strain of EACMV (EACMV-UG), and mixed infections of EACMV-like viruses and ACMV. Finally, there is evidence of a synergistic interaction between cassava-adapted whitefly populations and the more virulent forms of the disease, such that these whiteflies show increased rates of colonization and fecundity on plants with severe CMD (Colvin et al. 2004). These epidemics have resulted in extensive yield losses (e.g., estimated losses of $60 million in Uganda in the 1990s; Legg and Fauquet 2004).

12.3.2.3 Biological Considerations

An effective IPM strategy for CMD must consider the perennial nature of the crop and the fact that it is vegetatively propagated (Thresh and Cooter 2005). Also, the viruses that cause CMD have a relatively narrow host range, infecting only members of the plant family Euphorbiaceae; including cassava, castor bean (*Ricinus communis*) and certain wild hosts and weeds. Thus, infected cuttings serve as the most important means of carry over between crops, spread from one geographical area to another, and a key source of initial inoculum. Viruliferous whiteflies can also ‘hitch a ride’ on other plants, providing another means of long-distance spread. The whitefly vector serves as a very effective means of short distance spread within and between fields, and this is a function of whitefly population and CMD symptoms. Thus, although a single whitefly can transmit the disease, greater rates of spread occur with high populations (e.g., >10 whiteflies/plant). In a production area, wind also has a significant influence on whitefly spread, and the disease spreads faster to fields downwind and/or to nearby established fields with high incidences of infected plants compared with fields upwind and further away from fields with many infected plants. In addition, infections often are more prevalent in the outer parts of the field exposed to prevailing winds.

12.3.2.4 IPM Package for Cassava Mosaic Disease: Preplant Activities

Disease-Free Cuttings

For this disease, pre-plant strategies are critical (Guthrie 1990; Thresh and Cooter 2005). One of the key strategies is the planting of virus-free cuttings, which can be accomplished by simply taking cuttings from plants of a desirable cultivar (susceptible or resistant) that are not showing symptoms. These cuttings can be replanted, confirmed virus-free and then the plants used as a source of disease-free cuttings. A procedure for producing and distributing disease-free cuttings (referred to as ‘sanitation’) is described by Guthrie (1990). However, providing sufficient amounts of virus-free cuttings will always be a challenge, and there is a need to develop this technology on regional or even on-farm levels. In the case of materials coming out of breeding programs, it is often easier to develop a program for providing disease-free cuttings because of the involvement of organizations with the necessary resources. Virus-free cassava cutting programs have been initiated by the International Institute of Tropical Agriculture (IITA) in Nigeria, where a system of meristem tip culture and tissue culture-based propagation has been used to generate virus-free germ plasm that can be provided to national programs throughout Africa.

Regardless, the planting of virus-free cuttings is critical as it eliminates the key source of primary inoculum for a field. The use of cuttings derived from plants not showing disease symptoms is adequate, particularly if the symptom-free cuttings are carefully multiplied and confirmed to be free of virus. Establishing an isolated field with virus-free cuttings may provide adequate disease management. However, in areas where cassava is extensively cultivated, viruliferous whiteflies from neighboring fields will introduce the virus into fields established with clean stock. In these scenarios, it is desirable to have a regional management approach, where all farmers are educated about the disease and then coordinated in terms of planting virus-free stock and planting times or locations. However, for this approach to be effective, it is necessary to have an effective regional extension program, something that is often lacking in developing countries.

Resistant Varieties

Another highly desirable pre-plant management strategy is the use of resistant cassava cultivars. Cassava germ plasm with high levels of resistance has been generated via traditional breeding as well as transgenic approaches. High levels of resistance to CMD have been identified in wild cassava relatives (e.g., *M. glaziovii*) and in landraces from Nigeria. The resistance from *M. glaziovii* is multigenic and operates through multiple mechanisms (Fargette et al. 1996; Thresh and Cooter 2005). Through a program of inter-specific crosses and back-crossing, this resistance has been introgressed into commercially acceptable cassava varieties (e.g., TMS lines, such as TMS 30572) that have been distributed to farmers in cassava-growing regions of Africa over the past 20 years. The CMD resistance identified in the Nigerian landraces is conferred by a single dominant gene (CMD2), and this resistance has been combined with that from *M. glaziovii* to generate materials with high levels of resistance and excellent yield potential. Germplasm with this combined resistance is now being released to national programs by IITA. Transgenic cassava with high levels of resistance has also been generated through transformation with various forms of viral genes, especially the Replication-associated protein (Rep) gene (Legg and Fauquet 2004). The mechanism of this resistance appears to involve gene silencing (RNA interference), and evidence suggests it may provide broad spectrum resistance to a range of cassava mosaic begomoviruses. However, the planting of transgenic crops requires regulatory approval and, in most countries with CMD, a regulatory framework for field trials and release of transgenic crops is not in place.

The planting of resistant cultivars is an effective means of managing CMD, including the severe CMD outbreaks in Uganda. Efforts are underway to introduce these varieties into other countries, although there have been some problems with lack of infrastructure to multiply and distribute planting material, and with some of the resistant genotypes being less desirable in terms of taste, texture and agronomic qualities. Many local cassava cultivars (landraces) have been selected over a long period of time and often have locally preferred horticultural properties. Thus, it is important for breeders to address this, perhaps through a more decentralized participatory breeding program (Dixon et al. 2010). In situations where availability of resistant cultivars is limited, planting a mixture of varieties (including some with resistance) can reduce disease incidence in susceptible varieties. Regardless, the continued deployment of resistant cultivars to growers is critical, as the use of CMD-resistant disease-free cuttings provides the most effective control.

Cultural Practices

A number of cultural practices can also be considered, although these may have only limited beneficial effect or are too difficult for farmers to utilize (Thresh and Cooter 2005). If possible, new plantings should be established at times when whitefly populations are low. This varies from region to region and should be established experimentally. Ideally, there also would be some regional coordination regarding planting date. New plantings should not be established near old plantings with CMD. Also, with sequential plantings it is important to put later-planted fields upwind of early planted fields. Elongated plots that are exposed to prevailing wind should be avoided, as this is where the highest infection rates tend to occur. Intercropping of cassava with other crops such as banana, sweet potato and legumes, can reduce virus spread through reduction of whitefly populations. Finally, cassava should be grown under favorable conditions, as CMD spreads slower in fields with healthy plants.

12.3.2.5 During the Growing Season

Roguing

The physical removal of virus-infected plants over the course of the growing season can be useful, particularly when disease incidences are relatively low (<5%). Thus, roguing will be most effective when used in combination with pre-plant measures, such as planting disease-free cuttings. Roguing is also an important method for the amplification of sources of disease-free cuttings. Roguing needs to be done soon after plots are established, and fields should be monitored once or twice shortly after planting as cuttings show symptoms in newly emerging leaves. Ideally, roguing would be implemented on a regional basis.

Whitefly Management

Although there is a correlation between numbers of whiteflies and rate of spread of CMD, managing the disease with insecticide sprays has not been effective, nor is it practical. This relates to the fact that cassava is grown on small plots by subsistence farmers who often lack the understanding of CMD, the training and equipment to apply pesticides, and the resources to purchase the appropriate insecticides. However, it is possible that suppression of whitefly populations, either via biological control or with natural or synthetic insecticides, may help slow the spread of the CMD in certain situations.

After the Growing Season

Sanitation

Cassava is the main host of the viruses that cause CMD. Thus, it is critical to destroy cassava plants promptly after harvest, as well as any other known host plants. This should be done within and around fields (for reservoir hosts) following harvest and, if necessary, before establishing new plantings. Ideally, the planting and harvest times in defined regions or localities could be coordinated to avoid periods of high disease pressure (e.g., high populations of viruliferous whiteflies) and to possibly allow a period with minimal plantings of cassava to help cleanse the agroecosystem of the virus (i.e., akin to a host-free period).

Summary

CMD represents the most important biotic constraint for cassava production, particularly in Africa, but also in Asia. Recent advances in the understanding of the biology of the viruses involved and the whitefly vector, and in identifying and/or generating cassava cultivars with disease resistance has allowed for development

Table 12.2 Integrated pest management strategy for a whitefly-transmitted geminivirus disease of a perennial crop (mosaic disease of cassava)

Period of implementation	Management strategy
Before the growing season	Use of virus-free propagative material
	Use of resistant cultivars
	Modification of planting dates
	Avoid planting new fields near older fields
During the growing season	Roguing of plants showing mosaic symptoms
	Monitor for whitefly populations using established means of sampling
	Apply insecticides only when necessary
	Rotate insecticides to minimize development of resistance (e.g., no more than two uses of any material per season) and use recommended rates (avoid using sub-lethal rates)
After the growing season	Prompt removal of old crops following harvest
	Consider region-wide management of planting dates and locations

of improved management strategies. These include the planting of disease free cuttings and the use of resistant or tolerant cultivars. Certain cultural practices can also help reduce the incidence or spread of disease, as can a systematic roguing program, but only when combined with the use of disease-free cuttings. Extensive sanitation, in the form of prompt removal and destruction of cassava plants following harvest, together with removal of other hosts of the virus, can reduce inoculum pressure for subsequent plantings. This IPM program (or components of it) (Table 12.2) has provided effective disease management, but it has not been widely implemented and CMD continues to cause significant losses in many countries of Africa. This relates to lack of understanding of the disease by farmers, lack of availability of disease-free cuttings, preference for local susceptible varieties instead of resistant genotypes, the perennial nature of cassava (i.e., a constant source of inoculum and difficulties in having cassava-free periods) and lack of extension programs to deliver the IPM package to farmers. It will take a major effort to develop regional coordinated efforts to implement relevant IPM packages. These packages may differ depending on the region or locality, and may require decentralized and participatory breeding efforts to generate resistant varieties that provide the desired horticultural properties that are preferred by local growers and consumers.

References

Ahmed NE, Kanan HO, Sugimoto Y, Ma YQ, Inanago S (2001) Effect of imidacloprid on incidence of *Tomato yellow leaf curl virus*. Plant Dis 85:84–87

Akad F, Webb S, Nyoike TW, Liburd OE, Turechek W, Adkins S, Polston JE (2008) Detection of *Cucurbit leaf crumple virus* in Florida. Plant Dis 92:648

Antignus Y (2000) Manipulation of wavelength dependent behavior of insects:an IPM tool to impede epidemics and restrict spread of insect-borne viruses. Virus Res 71:213–220

Antignus Y, Lachman O, Pearlsman M, Omer S, Unis H, Messila Y, Ucko O, Koren A (2003) Squash leaf curl geminivirus a new illegal immigrant from the Western Hemisphere, a threat to cucurbit crops in Israel. Phytoparasitica 31:415

Ausher R (1997) Implementation of integrated pest management in Israel. Phytoparasitica 25:119–141

Berlinger MJ, Dahan R, Mordechi S (1986) The prevention of *Tomato yellow leaf curl virus* by controlling its vector, *Bemisia tabaci*. Hassadeh 66:686–689

Berlinger MJ, Taylor RAJ, Lebiush-Mordechi J, Shalhevet S, Spharim I (2002) Efficiency of insect exclusion screens for preventing whitefly transmission of tomato yellow leaf curl virus of tomatoes in Israel. Bull Entomol Res 92:367–373

Briddon RW (2003) Cotton leaf curl disease, a multicomponent begomovirus complex. Mol Plant Pathol 4:427–434

Brown JK, Frohlich DR, Rosell RC (1995) The sweetpotato or silverleaf whiteflies: biotypes of *Bemisia tabaci* or a species complex? Annu Rev Entomol 40:511–534

Byrne DN, Bellows TS Jr, Parrella MP (1990) Whiteflies in agricultural systems. In: Gerling D (ed.) Whiteflies: their bionomics, pest status and management. Intercept, Great Britain

Castle S, Palumbo J, Prabhaker N (2009) Newer insecticides for plant virus disease management. Virus Res 141:131–139

Chen LF, Gilbertson RL (2009) Curtovirus-cucurbit interaction: acquisition host plays a role in leafhopper transmission in a host-dependent manner. Phytopathology 99:101–108

Chowda Reddy RV, Colvin J, Muniyappa V, Seal S (2005) Diversity and distribution of begomoviruses infecting tomato in India. Arch Virol 150:845–867

Chu CC, Natwick ET, Perkins HH, Brushwood DE, Henneberry TJ, Castle SJ, Cohen AA, Boykin MA (1998) Upland cotton susceptibility to *Bemisia argentifolii* (Homoptera: Aleyrodidae) infestations. J Cotton Sci 2:1–9

Chu CC, Barnes E, Natwick ET, Chen TY, Ritter D, Henneberry TJ (2007) Trap catches of sweetpotato whitefly (Homoptera: Aleyrodidae) in the Imperial Valley, California, from 1996 to 2002. Insect Sci 14:165–170

Cohen S, Berlinger M (1986) Transmission and cultural control of whitefly-borne viruses. Agric Ecosyst Environ 17:89–97

Cohen S, Kern J, Harpaz I, Ben-Joseph R (1988) Epidemiological studies of the tomato yellow leaf curl virus (TYLCV) in the Jordan Valley, Israel. Phytoparasitica 16:259–270

Colvin J, Omongo C, Maruthi M, Otim-Nape G, Thresh J (2004) Dual begomovirus infections and high *Bemisia tabaci* populations: two factors driving the spread of a cassava mosaic disease pandemic. Plant Pathol 53:577–584

Csizinszky AA, Schuster DJ, Kring JB (1995) Color mulches influence yield and insect pest populations in tomatoes. J Am Soc Hortic Sci 120:778–784

Czosnek H, Laterrot H (1997) A worldwide survey of tomato yellow leaf curl viruses. Arch Virol 142:1391–1406

Czosnek H, Ghanim M, Morin S, Rubenstein G, Fridman V, Zeidan M (2001) Whiteflies: vectors and victims (?) of geminiviruses. Adv Virus Res 57:291–322

De Barro PJ (1995) *Bemisia tabaci* biotype B: a review of its biology density and control. CSIOR Division of Entomology, Technical paper no. 33, 59 pp

Dik AJ, Albajes R (1999) Principles of epidemiology, population biology, damage relationships, and integrated control of diseases and pests. In: Albajes R, Lodovica-Gullino M, van Lenteren JC, Elad Y (eds.) Integrated pest and disease management in greenhouse crops. Springer, Heidelberg

Dixon AGO, Ogbe FO, Okechukwu RU (2010) Cassava mosaic disease in Sub-Saharan Africa: a feasible solution for an unsolved problem. Outlook Agric 39:89–94

Ellsworth PC, Martinez-Carrillo JL (2001) IPM for *Bemisia tabaci*: a case study from North America. Crop Prot 20:853–869

Ellsworth PC, Diehl JW, Dennehy TJ, Naranjo SE (1994) Sampling sweet potato whiteflies in cotton. University of Arizona, IPM Series No. 2

Fargette D, Colon LT, Bouveau R, Fauquet C (1996) Components of resistance of cassava to African cassava mosaic virus. Eur J Plant Pathol 102:645–654

Fargette D, Konate G, Fauquet C, Muller E, Peterschmitt M, Thresh JM (2006) Molecular ecology and emergence of tropical plant viruses. Ann Rev Phytopathol 44:235–260

Fauquet CM, Briddon RW, Brown JK, Moriones E, Stanley J, Zerbini M, Zhou X (2008) Geminivirus strain demarcation and nomenclature. Arch Virol 153:783–821

Froissart R, Doumayrou J, Vuillaume F, Alizon S, Michalakis Y (2010) The virulence-transmission trade-off in vector-borne plant viruses: a review of (non-) existing studies. Philos Trans R Soc B 365:1907–1918

Ghanim M, Morin S, Czosnek H (2001) Rate of *Tomato yellow leaf curl virus* translocation in the circulative transmission pathway of its vector, the whitefly *Bemisia tabaci*. Phytopathology 91:188–196

Gilbertson RL, Rojas MR, Kon T, Jaquez J (2007) Introduction of *Tomato yellow leaf curl virus* into the Dominican Republic: the development of a successful integrated pest management strategy. In: Czosnek H (ed.) Tomato yellow leaf curl virus disease. Springer, Heidelberg, pp 279–303

Gillespe DR, Quiring D (1987) Yellow sticky traps for detecting and monitoring greenhouse whitefly (Homoptera: Aleyrodidae) adults on greenhouse tomato crops. J Econ Entomol 80:675–679

Gould JR, Naranjo SE (1999) Distribution and sampling of *Bemisia argentifolii* (Homoptera: Aleyrodidae) and *Eretmocerus eremicus* (Hymenoptera: Aphelinidae) on cantaloupe vines. J Econ Entomol 92:402–408

Gusmao MR, Picanco MC, Guedes RNC, Galvan TL, Pereira EJG (2006) Economic injury level and sequential sampling plan for *Bemisia tabaci* in outdoor tomato. J Appl Entomol 130:160–166

Guthrie J (1990) Controlling African cassava mosaic disease. CTA, Wageningen

Hanson P, Green SK, Kuo G (2006) *Ty-2*, a gene on chromosome 11 conditioning geminivirus resistance in tomato. Rep Tomato Genet Co-op 56:17–18

Hilje L, Stansly PA (2008) Living mulch ground covers for management of *Bemisia tabaci* (Gennadius) (Homoptera: Aleyrodidae) and *Tomato yellow mottle virus* (ToYMoV) in Costa Rica. Crop Prot 27:10–16

Hilje L, Costa HC, Stansly PA (2001) Cultural practices for managing *Bemisia tabaci* and associated viral diseases. Crop Prot 20:801–812

Holt J, Colvin J, Muniyappa V (1999) Identifying control strategies for tomato leaf curl virus disease using an epidemiological model. J Appl Ecol 36:625–633

Ishaaya I, Mendelson Z, Melamed-Madjar V (1988) Effect of buprofezin on embryo genesis and progeny formation of sweet potato whitefly (Homoptera: Aleyrodidae). J Econ Entomol 81:781–784

Ji Y, Scott JW (2005) Identification of RAPD markers linked to *Lycopersicon chilense* derived resistance genes on chromosome 6 of tomato. Acta Hortic 695:407–411

Ji Y, Schuster DJ, Scott DJ (2007) Ty-3, a begomovirus resistance locus near the tomato yellow leaf curl virus resistance locus Ty-1 on chromosome 6 of tomato. Mol Breed 20:271–284

Jiu M, Zhou X-P, Tong L, Xu J, Yang X et al (2007) Vector-virus mutualism accelerates population increase of an invasive whitefly. PLoS ONE 2(1):e182. doi:10.1371/journal.pone.0000182

Jones DR (2003) Plant viruses transmitted by whiteflies. Eur J Plant Pathol 109:195–219

Jones RAC (2004) Using epidemiological information to develop effective integrated virus disease management strategies. Virus Res 100:5–30

Kerns DL, Palumbo JC (1995) Using admire on desert vegetable crops. Univ Arizona Coop Ext Pub. #195017

Lapidot M, Friedmann M (2002) Breeding for resistance to whitefly-transmittted geminiviruses. Ann Appl Biol 140:109–127

Legg JP, Fauquet CM (2004) Cassava mosaic geminiviruses in Africa. Plant Mol Biol 56:585–599

Mason G, Rancati M, Bosco D (2000) The effect of thiamethoxam, a second generation neonicotinoid insecticide, in preventing transmission of tomato yellow leaf curl geminivirus (TYLCV) by the whitefly *Bemisia tabaci* (Gennadius). Crop Prot 19:473–479

Mejia L, Teni RE, Vidavski F, Czosnek H, Lapidot M, Nakhla MK, Maxwell DP (2005) Evaluation of tomato germplasm and selection of breeding lines for resistance to begomoviruses in Guatemala. Acta Hortic 695:251–255

Morales FJ (2001) Conventional breeding for resistance to *Bemisia tabaci*-transmitted geminiviruses. Crop Prot 20:825–834

Naranjo SE, Flint HM (1995) Spatial distribution of adult *Bemisia tabaci* (Homoptera: Aleyrodidae) in cotton, and development and validation of fixed-precision sampling plans for estimating population density. Environ Entomol 24:261–270

Naranjo SE, Ellsworth PC, Chu CC, Henneberry TJ, Riley DG, Watson TF, Nichols RL (1998) Action thresholds for the management of *Bemisia tabaci* (Homoptera: Aleyrodidae) in cotton. J Econ Entomol 91:1415–1426

Natwick ET, Durazo A III (1985) Polyester covers protect vegetables from whiteflies and virus disease. Calif Agric 39:21–22

Natwick ET, Laemmlen FF (1993) Protection from phytophagous insects and virus vectors in honeydew melons using rowcovers. Fla Entomol 76:120–126

Nauen R, Bretschneider T, Elbert A, Fisher R, Tiemann R (2003) Spirodiclofen and spiromesifen. Pestic Outlook 14:243–245

Nichols RL, Chu CC, Ellsworth PC, Henneberry TJ, Naranjo SJ, Riley DG, Toscano NC, Watson TF (1994) Determining an action threshold to prevent whitefly outbreaks. Phytoparasitica 22:349

Nyoike TW, Liburd OE, Webb SE (2008) Suppression of whiteflies, *Bemisia tabaci* (Hemiptera: Aleyrodidae) and incidence of *Cucurbit leaf crumple virus*, a whitefly-transmitted virus of zucchini squash new to Florida, with mulches and imidacloprid. Fla Entomol 91:460–465

Oliveira MRV, Henneberry TJ, Anderson P (2001) History, current status, and collaborative research projects for *Bemisia tabaci*. Crop Prot 20:709–723

Orozco-Santos M, Perez-Zamora O, Lopez-Arriaga M (1995) Floating row cover and transparent mulch to reduce insect population, virus diseases and increase yield in cantaloupe. Fla Entomol 78:493–501

Palumbo JC, Tonhasca Jr A, Byrne DN (1994) Sampling plans and action thresholds for whiteflies on spring melons. University of Arizona IPM Series Number 1

Perring TM (2001) The *Bemisia tabaci* species complex. Crop Prot 20:725–737

Polston JE, Anderson PK (1997) The emergence of whitefly-transmitted geminiviruses in tomato in the Western Hemisphere. Plant Dis 81:1358–1369

Polston JE, Sherwood T (2003) Pymetrozine interferes with transmission of *Tomato yellow leaf curl virus* by the whitefly *Bemisia tabaci*. Phytoparasitica 31:490–498

Ribeiro SG, Ambrozevicius AC, Avilla AC, Bezerra IC, Calegario RF, Fernandes JJ, Lima MF, de Mello RN, Rocha RF, Zerbini FM (2003) Distribution and genetic diversity of tomato-infecting begomoviruses in Brazil. Arch Virol 148:281–295

Rojas MR, Gilbertson RL, Russell DR, Maxwell DP (1993) Use of degenerate primers in the polymerase chain reaction to detect whitefly-transmitted geminiviruses. Plant Dis 77:340–347

Rojas MR, Hagen C, Lucas WJ, Gilbertson RL (2005) Exploiting chinks in the plant's armor: evolution and emergence of geminiviruses. Ann Rev Phytopathol 43:361–394

Rojas MR, Kon T, Natwick ET, Polston JE, Akad F, Gilbertson RL (2007) First report of *Tomato yellow leaf curl virus* associated with tomato yellow leaf curl disease in California. Plant Dis 91:1056

Rosell RC, Torres-Jerez I, Brown JK (1999) Tracing the geminivirus-whitefly transmission pathway by polymerase chain reaction in whitefly extracts, saliva, hemolymph, and honeydew. Phytopathology 89:239–246

Rubenstein G, Morin S, Czosnek H (1999) Transmission of *Tomato yellow leaf curl geminivirus* to imidacloprid treated tomato plants by the whitefly *Bemisia tabaci* (Homoptera: Aleyrodidae). J Econ Entomol 92:658–662

Salati R, Nahkla MK, Rojas MR, Guzman P, Jaquez J, Maxwell DP, Gilbertson RL (2002) *Tomato yellow leaf curl virus* in the Dominican Republic: characterization of an infectious clone, virus monitoring in whiteflies, and identification of reservoir hosts. Phytopathology 92:487–496

Schuster DJ, Stansly PA, Polston JE, Gilreath PR, McAvoy E (2007) Management of whiteflies, whitefly-vectored plant virus, and insecticide resistance for vegetable production in Southern Florida. ENY-735 (IN695), IFAS Extension, University of Florida, Gainesville

Seal SE, van den Bosch F, Jeger MJ (2006) Factors influencing begomovirus evolution and their increasing global significance: implications for sustainable control. Crit Rev Plant Sci 25:23–46

Seo JS, Gepts P, Gilbertson RL (2004) Genetics of resistance to the geminivirus, *Bean dwarf mosaic virus*, and the role of the hypersensitive response. Theor Appl Genet 108:786–793

Stansly PA, Natwick ET (2010) Integrated systems for managing *Bemisia tabaci* in protected and open field agriculture. In: Stansly PA, Naranjo SE (eds.) *Bemisia*: bionomics and management of a global pest. Springer, Heidelberg

Stern VM, Smith RF, van den Bosch R, Hagen KS (1959) The integrated control concept. Hilgardia 29:81–101

Suwwan MA, Akkawi M, Al-Musa AM, Mansour A (1988) Tomato performance and incidence of tomato yellow leaf curl (TYLC) virus as affected by type of mulch. Sci Hortic 37:39–45

Tarr SAJ (1951) Leaf curl disease of cotton. The Commonwealth Mycological Institute, Kew

Thresh JM, Cooter RJ (2005) Strategies for controlling cassava mosaic virus disease in Africa. Plant Pathol 54:587–614

Ucko OS, Cohen S, Ben-Joseph R (1998) Prevention of virus epidemics by a crop-free period in the Arava region of Israel. Phytoparasitica 26:313–321

Varma A, Malathi VG (2003) Emerging geminivirus problem: a serious threat to crop production. Ann Appl Biol 142:145–162

Webb SE, Linda SB (1992) Evaluation of spun-bounded polyethylene row covers as a method of excluding insects and viruses affecting fall-grown squash in Florida. J Econ Entomol 85:2344–2352

Wyatt SD, Brown JK (1996) Detection of subgroup III geminivirus isolates in leaf extracts by degenerate primers and polymerase chain reaction. Phytopathology 86:1288–1293

Zhou Y-C, Noussourou M, Kon T, Rojas MR, Jiang H, Chen L-F, Gamby K, Foster R, Gilbertson RL (2008) Evidence for local evolution of tomato-infecting begomovirus species in West Africa: characterization of tomato leaf curl Mali virus and tomato yellow leaf crumple virus from Mali. Arch Virol 153:693–706

Chapter 13
Remote Sensing for Detecting and Mapping Whitefly (*Bemisia tabaci*) Infestations

Chenghai Yang and James H. Everitt

Abstract Remote sensing technology has long been used for detecting insect infestations on agricultural crops. With recent advances in remote sensing sensors and other spatial information technologies such as Global Position Systems (GPS) and Geographic Information Systems (GIS), remote sensing is finding more and more practical applications for the detection and management of insect pests, including sweetpotato whitefly, *Bemisia tabaci* (Gennadius). This chapter begins with an extended overview of remote sensing principles and systems that can be used for entomological studies. Properties and behavior of electromagnetic energy, major divisions of the electromagnetic spectrum (i.e., ultraviolet, visible, infrared and microwave), and the interactions between radiation and ground targets are discussed. Major types of remote sensing systems are described, including ground-based spectroradiometers, aerial photographic cameras, airborne digital multispectral and hyperspectral imaging systems, and moderate and high resolution satellite imaging systems. The second part of the chapter provides a brief review on the use of remote sensing for detecting whitefly infestations and presents an application example to illustrate how remote sensing can be integrated with GPS and GIS technologies for detecting and mapping whitefly infestations in cotton fields. The methodologies for ground reflectance and airborne image acquisition and for the integration of image data with GPS and GIS are discussed.

C. Yang (✉) • J.H. Everitt
United States Department of Agriculture, Agricultural Research Service,
Kika de la Garza Subtropical Agricultural Research Center, Weslaco,
TX 78596, USA
e-mail: chenghai.yang@ars.usda.gov; james.everitt@ars.usda.gov

W.M.O. Thompson (ed.), *The Whitefly, Bemisia tabaci (Homoptera: Aleyrodidae) Interaction with Geminivirus-Infected Host Plants*,
DOI 10.1007/978-94-007-1524-0_13,

13.1 Overview of Remote Sensing Technology

Remote sensing is the science or technology of acquiring information about the earth's surface without physically touching it. It uses sensors to measure and record the reflected and emitted electromagnetic radiation from the target area in the field of view of the sensor instrument. The detecting and recording instruments are generally referred to as remote sensors, including photographic cameras, video and digital cameras, electro-mechanical scanners, and radar systems. Remote sensors are typically carried on aircraft and earth-orbiting satellites, but some sensors can be handheld or mounted on ground-based vehicles. In this section, properties and behavior of electromagnetic energy, major divisions of the electromagnetic spectrum (i.e., ultraviolet (UV), visible, infrared and microwave), and the interactions with targets are first discussed. Main types of remote sensing systems are then described, including ground-based spectroradiometers, aerial photographic cameras, airborne digital multispectral and hyperspectral imaging systems, and satellite multispectral imaging systems.

13.2 Electromagnetic Spectrum

The knowledge of electromagnetic energy is very important for the understanding of remote sensing data collection. Various types of natural and artificial electromagnetic energy exist in the universe. The reason we are familiar with visible light is that we can see it with our naked eyes, but other important forms of electromagnetic energy include invisible gamma, X-ray, UV, infrared, microwave and radio energy. Practically all forms of natural electromagnetic energy entering into the earth atmosphere and earth surface are produced by the sun. Today a wide variety of electromagnetic energy is artificially produced on earth and no fundamental physical differences exist between natural and artificial electromagnetic energy. Radiation is the only method by which solar energy can cross millions of kilometers of free space and reach the earth, and it is the method of energy transfer with which we are primarily concerned in remote sensing. Electromagnetic radiation travels in straight path at the speed of light across empty space and only slightly slower in the atmosphere. As this radiation approaches the earth, it passes through the atmosphere and then finally reaches the earth's surface. Some radiation is reflected upward from the surface and some is absorbed and reradiated or emitted from the surface as thermal energy. It is the reflected radiation that forms the basis for photographs and most of the remote sensing images. The thermal, or emitted, energy can also be used to form images. Besides, man-made radiation, such as that generated by imaging radars, is also commonly used in remote sensing.

According to the basics of wave theory, all electromagnetic radiation has predictable properties and behaviors. Electromagnetic radiation consists of an electrical field which varies in magnitude in a direction perpendicular to the direction in which the radiation is traveling, and a magnetic field oriented at right angles to the electrical field. Both these fields travel at the speed of light. Two characteristics of electromagnetic radiation are particularly important for understanding remote sensing. These are the wavelength and frequency. The wavelength is the length of one wave cycle, which can be measured as the distance between successive wave crests. Wavelength is usually represented by the Greek letter lambda (λ). Wavelength is measured in meters (m) or some factor of meters such as centimeters (cm, 10^{-2} m), micrometers (μm, 10^{-6} m) or nanometers (nm, 10^{-9} m). Frequency refers to the number of cycles of a wave passing a fixed point per unit of time. Frequency is normally represented by the Greek letter mu (ν) and measured in hertz (Hz), equivalent to one cycle per second, and various multiples of hertz. Wavelength and frequency are related with the speed of light (c) by the formula $c = \lambda\nu$. Since the speed of light is essentially a constant (3×10^8 m/s), wavelength and frequency are inversely related to each other. The shorter the wavelength, the higher the frequency. Therefore, the characteristics of electromagnetic energy can be specified using either frequency or wavelength. Different disciplines and different applications follow varied conventions. A common practice in the field of remote sensing is to define the regions of the electromagnetic spectrum on the basis of wavelength.

13.3 Major Regions of the Electromagnetic Spectrum

The electromagnetic spectrum ranges from the shorter wavelengths (including gamma and X-rays) to the longer wavelengths (including microwaves and broadcast radio waves). In remote sensing, major regions of the electromagnetic spectrum are defined for convenience and by traditions. Table 13.1 gives the major spectral regions of the electromagnetic spectrum (Avery and Berlin 1992).

For practical purposes, the UV, visible, infrared and microwave regions of the spectrum are useful for remote sensing. Current remote sensors can only selectively measure the electromagnetic radiation in one or some of the spectral regions. There has been no single sensor that is capable of detecting radiation from all the spectral regions. The sun is the natural source of UV radiation, but wavelengths less than 0.3 μm are unable to pass through the atmosphere. Thus only the near UV (0.3–0.4 μm) is available for remote sensing. Although humans cannot see UV radiation directly, bees and certain other insects are visually sensitive to near UV. In addition, UV radiation can induce fluorescence (emission of visible light) in some materials, primarily rocks and minerals, and therefore it is useful for some geological remote sensing applications. However, UV radiation can be easily scattered by the atmosphere, so it is not widely used in the field of remote sensing. The visible spectrum is the region our eyes can detect. Although it constitutes a very small portion of the spectrum, it is very important in remote sensing. Visible light covers a spectral range from approximately 0.4–0.7 μm. It can be subdivided into three equal-wavelength bands that represents the three additive primary colors, blue (0.4–0.5 μm), green (0.5–0.6 μm), and red (0.6–0.7 μm). No single primary color can be formed

Table 13.1 Major spectral regions of the electromagnetic spectrum (Adapted from Avery and Berlin 1992)

Spectral region	Wavelength (μm)
Gamma and X-rays	<0.01
Ultraviolet	0.01–0.4
Visible	0.4–0.7
Infrared	
Near-infrared	0.7–1.5
Mid-infrared	1.5–5.6
Far infrared	5.6–1,000
Microwaveand radio	>1,000

from the mixture of the other two, and all other colors can be formed by mixing the three primaries in appropriate portions. Equal portions of the three additive primaries combine to form white light. The color of an object is defined by the color of light it reflects. A blue object looks blue because it reflects blue light. Intermediate colors are formed when an object reflects two or more of the additive primaries. For example, a yellow object reflects red and green light, but absorbs blue light; a purple object reflects red and blue light, but absorbs green light; and a cyan object reflects blue and green light, but absorbs red light. Moreover, a white object reflects all three primary colors, while a black object absorbs all three primaries. Most of the colors we see are the result of the preferential reflection and absorption of wavelengths that make up white light. For example, the chlorophyll of healthy plants absorbs more of blue and red wavelengths of white light and reflects relatively more of the green wavelengths to our eyes. Thus most of the plants we see are green. We perceive fresh snow as white because snow reflects all wavelengths of visible spectrum equally well.

The infrared region covers the wavelength range from approximately 0.7–1,000 μm. Because of atmospheric attenuation, the infrared region beyond about 15 μm is generally not available for remote sensing studies. That is why some text books define the infrared region from 0.7 to 15 μm. This region of the spectrum is very large relative to the visible region. Because of its broad range, it encompasses radiation with quite varied properties. In physics, the infrared region is subdivided into near-infrared (NIR) (0.7–1.5 μm), middle infrared (1.5–5.6 μm), and far infrared (5.6–1,000 μm). In remote sensing, the infrared region is usually divided into the reflected infrared region and the emitted or thermal infrared region. The reflected infrared region with a wavelength span from 0.7 to about 3 μm represents reflected solar radiation, which behaves like visible light. The thermal IR region is quite different from the visible and reflected IR portions, as this energy is essentially the radiation that is emitted from the earth's surface in the form of heat. The thermal infrared covers wavelengths from approximately 3.0 μm to 1,000 μm, even though wavelengths over 15 μm are not usable in remote sensing.

The portion of the spectrum of more recent interest to remote sensing is the microwave region from about 1 mm to 1 m. This covers the longest wavelengths used for remote sensing. The shorter wavelengths have properties similar to the thermal infrared region while the longer wavelengths approach the wavelengths used for radio broadcasts. At proper wavelengths (>9 mm), microwave radiation can pass through clouds, precipitation, tree canopies, and dry surficial deposits. There are two types of sensors that operate in the microwave region. A passive microwave sensor detects natural microwave radiation that is emitted from the earth's surface, whereas an active microwave sensor or radar propagates an artificial microwave signal to the surface and detects the reflected signal.

13.4 Interactions Between Electromagnetic Radiation and Targets

Radiation used for remote sensing must pass through the earth's atmosphere before reaching the surface. Particles and gases in the atmosphere can affect the incoming light and radiation. These effects are caused by the mechanisms of scattering, absorption and refraction. Because of the barriers formed by these particles and gases, the earth's atmosphere is not completely transparent to electromagnetic radiation. It selectively transmits radiation of certain wavelengths. Those wavelengths that are relatively easily transmitted through the atmosphere are referred to as atmospheric windows. Positions, extents and effectiveness of atmosphere windows are determined by the absorption spectra of atmosphere gases. Atmospheric windows define the wavelengths that can be used for remote sensing. Absorption by the atmosphere has its maximum influence on wavelengths shorter than 0.3 μm and a minimum impact on wavelengths greater than 6 mm. Between the two wavelengths, there are a number of important windows that coincide with significant levels of solar radiation and heat energy emitted by the earth. These windows include UV and visible to NIR (0.3–1.1 μm), mid-infrared (1.5–1.8 μm and 2.0–2.4 μm), thermal (3–5 μm and 8–14 μm) and microwave (>6 mm).

Radiation that is not absorbed or scattered in the atmosphere can reach and interact with the earth's surface. When energy strikes the surface, there are three forms of interaction, including absorption, transmission, and reflection. The total incident energy will interact with the surface in one or more of these three ways. The proportions accounted for by each process depend on the wavelength of the energy, the angle of the illumination, and the material and condition of the target. Absorption occurs when radiation is absorbed into the target while transmission occurs when radiation passes through a target. Reflection occurs when radiation bounces off the target and is redirected. In remote sensing, we are most interested in measuring the radiation reflected from targets. Specular reflection and diffuse reflection represent the two extreme ends of the way in which energy is reflected from a target. If a surface is smooth relative to the wavelengths, specular or mirror-like reflection occurs where all or almost all of the energy is directed away from the surface in a single direction. Diffuse reflection occurs if the surface is rough and the energy is reflected almost uniformly in all directions. Most natural surface features lie somewhere between perfectly specular or perfectly diffuse reflectors. If the wavelengths are

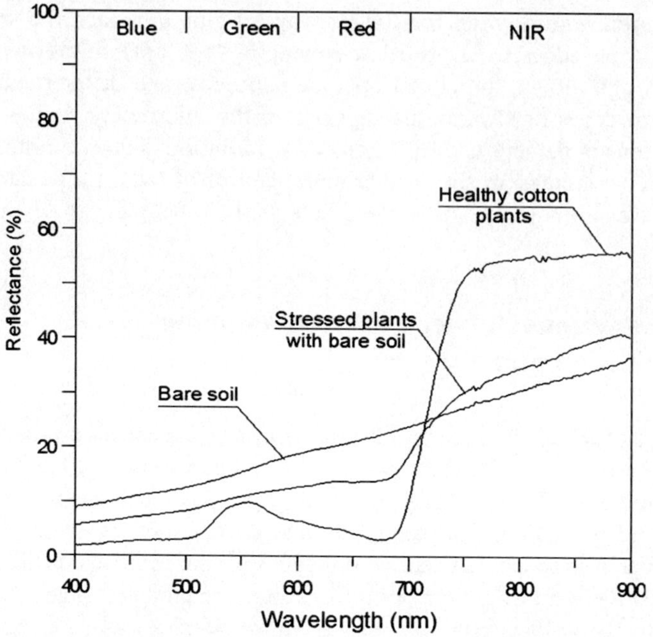

Fig. 13.1 Representative spectra for healthy cotton plants, stressed plants with bare soil exposure, and bare soil

much smaller than the surface variations or the particle sizes that make up the surface, diffuse reflection will dominate. For visible and NIR radiation, many natural surfaces may behave as diffuse reflectors, including uniform grassy surfaces. Many applications of remote sensing depend on the understanding of the spectral behavior of living plant leaves. Let's take plant leaves as an example to illustrate how radiation in the visible to NIR region of the spectrum interacts with them.

In the visible portion of the spectrum, chlorophyll, the green pigment that is responsible for the green color of living vegetation, controls much of the spectral response of the living leave. Chlorophyll strongly absorbs blue and red light for use in photosynthesis, but reflects green light. Leaves appear very green to us in the summer when chlorophyll content is at its maximum. In autumn, there is less chlorophyll in the leaves, so there is less absorption and proportionately more reflection of the red wavelengths, making the leaves appear red or yellow, which is a combination of red and green. In the NIR region of the spectrum, reflection of the leaf is controlled not by chlorophyll but by the structure of the spongy mesophyll tissue. The internal structure of healthy leaves act as excellent diffuse reflectors of NIR light. If our eyes were sensitive to NIR, living plants would appear very bright to us. In fact, measuring the NIR reflectance is one way to determine the health status of vegetation.

Figure 13.1 gives typical reflectance spectra for healthy cotton plants, stressed cotton plants, and bare soil. For the healthy plant canopy, the spectrum has low blue and red reflectance and high green reflectance. Toward the red edge of the visible spectrum (about 700 nm), as the absorption of the red light by chlorophyll begins to decline, reflectance rises sharply and gradually flattens out in the NIR region. The reflectance curve for bare soil is close to a straight line and soil reflectance increases with wavelength gradually in the visible to NIR region of the spectrum. Certainly, spectral response depends on the types of plants and soils. For the given cotton plants and soil shown in Fig. 13.1, the soil has higher reflectance than the plants in the visible region, whereas the plants have higher reflectance in the NIR region. The reflectance curve for stressed plants falls somewhere between the reflectance curves for healthy plants and bare soil. The reflectance for the stressed plants is higher in the visible region and lower in the NIR region than the reflectance for normal plants. Since the stressed plants show the yellowing or brown tone, there is less absorption and more reflection in the blue and red regions. These deviations in reflectance are also attributed by the reduced canopy of the stressed plants and large soil exposure within the field of view of the spectroradiometer. These spectral behaviors are the basis for distinguishing healthy plants from plants that may have been affected by water stress, nutrient deficiency, insect damage, and disease infestation.

It can be seen from this example that different targets have different spectral responses. By measuring the energy that is reflected (or emitted) by a target over different wavelengths, we can determine a spectral response for that target. By comparing the response patterns of different targets, we may be able to distinguish between them. However, if we only compare them at one or two wavelengths, we might not be able to separate them. For example, soil and vegetation in this example may reflect somewhat similarly around 720 nm, but are clearly separable in other wavelengths. Spectral response can be quite variable, even for the same target type, and can also vary with time and location. Understanding the factors that influence the spectral response of the features of interest is critical to correctly interpret the interaction of electromagnetic radiation with the target.

13.5 Remote Sensors

Remote sensors include all the instruments that detect and measure reflected and emitted electromagnetic radiation from a distance. These instruments fall into two broad categories: non-imaging such as a spectroradiometer and imaging such as a camera. According to the types of sensor-carrying platforms, remote sensors can be ground-based, airborne and spaceborne. Both non-imaging and imaging sensors can be carried in all three types of platforms, though non-imaging sensors are primarily used for ground-based applications. Portable non-imaging remote sensing instruments include radiometers and spectroradiometers. The types of radiometers can be single-band radiometers, which measure radiation intensity integrated through one broad waveband, and multispectral radiometers, which measure radiation intensity in more than one broad waveband. Spectroradiometers measure radiation intensity over a continuous range of wavelengths by simultaneously sampling a large number of narrow spectral bands.

Fig. 13.2 A FieldSpec HandHeld portable spectroradiometer (Analytical Spectral Devices, Denver) (*left*), a reference panel (*middle*), and a laptop computer

Figure 13.2 shows a FieldSpec HandHeld portable spectroradiometer (Analytical Spectral Devices, Inc., Denver, CO), which acquires a continuous spectrum by measuring radiation intensity in 512 bands between 325 and 1,075 nm. The spectra shown in Fig. 13.1 were obtained using this instrument. The FieldSpec 3 portable spectrometer from the same company can take measurements from 350 to 2,500 nm with sampling intervals of 1.4 nm at 350–1,000 nm and 2 nm at 1,000–2,500 nm.

Imaging sensors are designed to provide views of a target area from vertical (nadir) or oblique (slanted) perspectives. Usually remote sensing images are captured from vertical or downward directions. Imaging systems can be divided into four main groups: (1) film-based photographic camera, (2) electro-optical sensor, (3) passive microwave, and (4) active microwave or radar. Passive remote sensors, represented by the first three groups, detect radiation emanating naturally from the surface, including reflected sunlight or emitted thermal infrared and microwave energy. Active remote sensors, represented by imaging radar, provide their own energy source to a target and record the reflected radiation from the target. Advantages for active sensors include the ability to obtain measurements anytime, regardless of the time of day or season.

The key components of an aerial photographic camera are the lens and the film. The lens renders a sharply defined image of a given scene, while the film acts as both the radiation detector and the recording medium. Photographic cameras can detect wavelengths ranging from 0.3–0.9 μm, which includes the near UV band, the visible band and the shorter wavelengths of the NIR band.

Electro-optical sensors use mirrors and/or lenses to collect and focus incoming radiation onto different types of photo detectors. Photo detectors are devices formed from substances known to respond to energy over a defined wavelength interval, generating a weak electrical signal with strength related to the radiances of the features in the field view of the sensor. The electrical signal is amplified and then recorded in analog and/or digital format on magnetic media. Subsequently, the stored signals can be viewed on a television screen or downloaded to a computer for viewing and processing. Detectors have been designed with sensitivities for many of the spectral intervals of interest in remote sensing, including regions of the near UV, visible to NIR, mid-infrared, and thermal infrared. The total span of the wavelengths that can be detected by electro-optical detectors extends from 0.3 to 15 μm. The detectors sensitive in these different regions of the spectrum are formed from different materials. The sensors operating in the visible to mid-infrared region can only be used under sunlit conditions. In contrast, thermal sensors exhibit a day or night capability.

Microwave remote sensing encompasses both active and passive forms of sensing. Passive microwave sensors detect low-level microwave radiation emitted from the earth's surface. These devices normally operate at wavelengths between 1.5 mm and 30 cm. Active microwave sensors, or imaging radars, operate within very narrow wavelength bands from 8 mm to 68 cm. Microwave systems use data recording and image display methods similar to electro-optical imaging systems. Both passive and active microwave systems have the important advantage of an all-weather capability at wavelengths exceeding about 3 cm. Moreover, these systems can be used at any time of the day.

13.6 Aerial Photographic Systems

Aerial photography is the oldest and simplest form of remote sensing. Although the status of aerial photography is challenged by continuing innovations in digital imaging technology, it remains a practical and useful remote sensing tool being used today. An aerial photographic camera is generally a framing camera. It instantaneously captures an image from the field of view with each exposure and save it onto a frame on a roll of film. Although handheld cameras with 35-mm film has been used in some remote sensing applications, most of the mapping cameras in use today incorporate an image format measuring 23 cm by 23 cm (9 in. by 9 in.). Cameras with a 70-mm image frame are also used.

Photographic films are sensitive to light from 0.3 to 0.9 μm in wavelength covering the UV, visible, and NIR. Panchromatic films are sensitive to the UV and the visible portions of the spectrum. The two major groups of films are black-and-white and color. Black-and-white films can be panchromatic or infrared. Black-and-white films use one single light-sensitive layer, whereas color films use three light-sensitive layers. UV photography also uses panchromatic film, but a filter is used with the camera to absorb and block the visible energy from reaching the film. As a result, only the UV reflectance is recorded. UV photography is not widely used, because of the atmospheric scattering and absorption that occurs in this region of the spectrum. Black-and-white infrared photography uses film sensitive to the entire 0.3–0.9 μm wavelength range and is useful for detecting differences in vegetation cover, due to its sensitivity to infrared reflectance. The two types of color films are normal color and color infrared (CIR). Normal color film is sensitive to blue, green, and red light – the same as our eyes. The normal color photos appear to us the same way that our eyes see the environment as the colors resemble those which would appear to us (i.e. trees appear green). Unlike normal color film, CIR film is sensitive to green, red, and NIR radiation, which are processed to appear as blue, green, and red, respectively. In a CIR photograph, targets with high NIR reflectance appear red, those with a high red reflectance appear green, and those with a high green reflectance appear blue, thus giving us an unnatural or false presentation of the targets relative to the color we normally perceive them to be. Because of this, a CIR photograph is sometimes called a false color composite or photograph.

Like any remote sensors, aerial photographic cameras can be carried in ground-based, airborne, or spaceborne platforms. The ground coverage of a camera depends on several factors, including the focal length of the lens, platform altitude, and film size. The focal length effectively controls the angular field of view of the lens and determines the imaging area. The longer the focal length, the smaller the area covered on the ground, but with greater spatial detail. The area covered also depends on the altitude of the platform. At high altitudes, a camera will sense a larger area on the ground than at lower altitudes, but with reduced detail. Aerial photos can provide finer spatial detail than most digital images. This is one of the reasons that aerial photography is still used in some applications today. However, new digital cameras can provide comparable spatial detail and aerial photos will eventually be replaced by digital images.

Vertical photographs taken with single-lens frame cameras are commonly used for remote sensing and mapping purposes. These cameras are specifically built for capturing a rapid sequence of photographs while limiting geometric distortion. They are often linked with navigation systems onboard the aircraft platform, to allow for accurate geographic coordinates to be instantly assigned to each photograph. Most camera systems also include mechanisms which compensate for the effect of the aircraft motion relative to the ground, in order to limit distortion as much as possible. When obtaining vertical aerial photographs, the aircraft normally flies in a series of lines, each called a flight line. Photos are taken in rapid succession looking straight down at the ground, often with a 50–60% overlap between successive photos. The overlap ensures total coverage along a flight line and also facilitates stereoscopic viewing. Aerial photographs are most useful when fine spatial detail is more critical than spectral information, as their spectral resolution is generally coarse when compared to data captured with electronic sensing devices. The geometry of vertical photographs is well understood and it is possible to make very accurate measurements from them for a variety of different applications. The science of making measurements from photographs is called photogrammetry and has been performed extensively since the very beginnings of aerial photography. Photos are most often interpreted manually by a human analyst. They can also be scanned to create a digital image and then analyzed in a digital computer environment.

13.7 Airborne Electro-Optical Imaging Systems

Unlike photographic cameras, electro-optical sensors use non-film detectors to convert the reflected and/or emitted radiation from a ground scene to proportional electrical signals. These signals are then recorded on magnetic, optical and/or solid-state media and can be viewed as two-dimensional images on a computer or television monitor. Although electro-optical imaging systems have coarser spatial resolution than photographic cameras, they are capable of operating in numerous bands from more spectral regions of the electromagnetic spectrum, including near UV, visible, NIR, mid-infrared and thermal infrared.

The growing interest in airborne remote sensing was stimulated by research and development on multispectral video imaging systems and their applications in the 1980s and 1990s (Meisner and Lindstrom 1985; Nixon et al. 1985; Everitt et al. 1995). The increased use of this technology was also attributed to the immediate availability of the imagery for visual assessment, compatibility with computer processing systems, greater light sensitivity than film cameras, capability of capturing narrower spectral bands, and sensitivity further into the infrared spectrum than film cameras (Mausel et al. 1992).

Video imaging systems can have many configurations, including single-band black-and-white cameras, color video cameras, CIR video cameras, and multiple single-band cameras. Video cameras used in remote sensing can collect reflected radiation in the visible, NIR, mid-infrared regions of the spectrum. A video camera generates NTSC standard television signals. This enables the signals to be recorded on videotape or CD/DVD media. Many earlier video cameras use video tubes, with each tube having a particular spectral range. For a multispectral video imaging system, the simplest approach uses multiple black-and-white video cameras equipped with different lens filters to obtain multispectral images. For example, four video cameras with the visible to NIR sensitivity can be filtered to simultaneously obtain blue, green, red and NIR band images. The individual black-and-white band images can be viewed as normal color or CIR composite images. Normal color and CIR video cameras with single lenses are also available and these cameras can acquire color or CIR video images with single cameras.

As an alternative to video tubes, solid-state detectors are more commonly used in video and digital imaging systems. These sensors consist of a two dimensional array of detectors and readout electronics etched onto a silicon chip using integrated circuit technology. Depending on the microcircuit technology employed, several names are used for these detectors. The most common detector is called the charge-coupled device (CCD). Energy reaching the surface of the CCDs causes the generation of an electronic charge which is proportional in magnitude to reflectance of the ground area. A digital number for each spectral band is assigned to each pixel based on the magnitude of the electronic charge. CCD cameras typically have pixel arrays of 1024 by 1024 or 2048 by 2048, but some new cameras can have approximately a 9000 by 7000 array or bigger. The pixel depth or radiometric resolution varies from 8 to 16 bits (256 to 65536 gray levels). CCD cameras offer advantages of compactness, shock resistance, low power consumption, and near-perfect geometry.

Improvements and advancements in electronic imaging and computer technology in the 1990s resulted in the development of true digital imaging systems that provided imagery with higher pixel resolution (Pearson et al. 1994; Escobar et al. 1997). These systems, as with their immediate predecessors, became more popular and useful for assessing agricultural and other natural resources because of their higher image resolution. Most airborne digital imaging systems can provide multispectral image data at spatial resolutions ranging from less than 1 m to a few meters and at multiple narrow spectral bands in the visible to NIR regions of the electromagnetic spectrum. Airborne multispectral imagery has been widely used for assessing within-field crop growth and pest conditions (Moran et al. 1997; Pinter et al. 2003; Yang et al. 2010).

One limitation of CCD detectors is that they cannot operate at wavelengths longer than about 1.1 μm. Multispectral scanners are capable of operating simultaneously in the near UV, visible, NIR, mid-infrared, and thermal infrared regions of electromagnetic spectrum by using different types of detectors. Typical radiation detectors used in multispectral scanners and their wavelength sensitivities are: (1) photomultiplier tubes or CCDs, 0.3–0.9 μm; (2) silicon photodiodes, 0.9–2.5 μm; and (3) mixtures of certain metallic elements, 3–14 μm. This means that both reflected and emitted radiation can be collected by multispectral scanners. The ability to operate within this broad spectral region presents the possibility of identifying objects whose identifiable spectral signatures lie beyond the wavelength limits of the relative narrow confines of the visible spectrum. Multispectral scanners can be used on aircraft and spacecraft.

Hyperspectral imaging sensors or imaging spectrometers are a new generation of electro-optical sensors that can collect image data in tens to hundreds of very narrow, continuous spectral bands throughout the visible, NIR, mid-infrared and thermal infrared portions of the spectrum. These systems offer new opportunities for better differentiation and estimation of biophysical attributes for a variety of remote sensing applications. Many commercial airborne hyperspectral sensors such as AVIRIS, CASI, HYDICE, HyMap, AISA, and HySpex have been developed and used for various remote sensing applications. Advances in CCD cameras, frame grabber boards, and modular optical components have also led to developments of low-cost airborne hyperspectral imaging systems from off-the-shelf products (Mao 1999; Yang et al. 2003). Despite significant progress in airborne hyperspectral remote sensing, hyperspectral imagery has not been used as widely as multispectral imagery partially due to the high costs of data acquisition and the special needs for handling and processing vast volumes of data. Airborne hyperspectral imaging sensors employ an aircraft as a platform for pushbroom scanning. A raw hyperspectral image cube is formed by a collection of scanned lines with all the bands. Since the stability of the platform is affected by the surrounding turbulent atmosphere during image acquisition, variations in altitude, attitude, and velocity will cause geometric distortions on the hyperspectral image. However, inertia measurement units (IMUs) and/or gyro mounts are used in most hyperspectral imaging systems to damper and measure the variations for real-time or post-corrections.

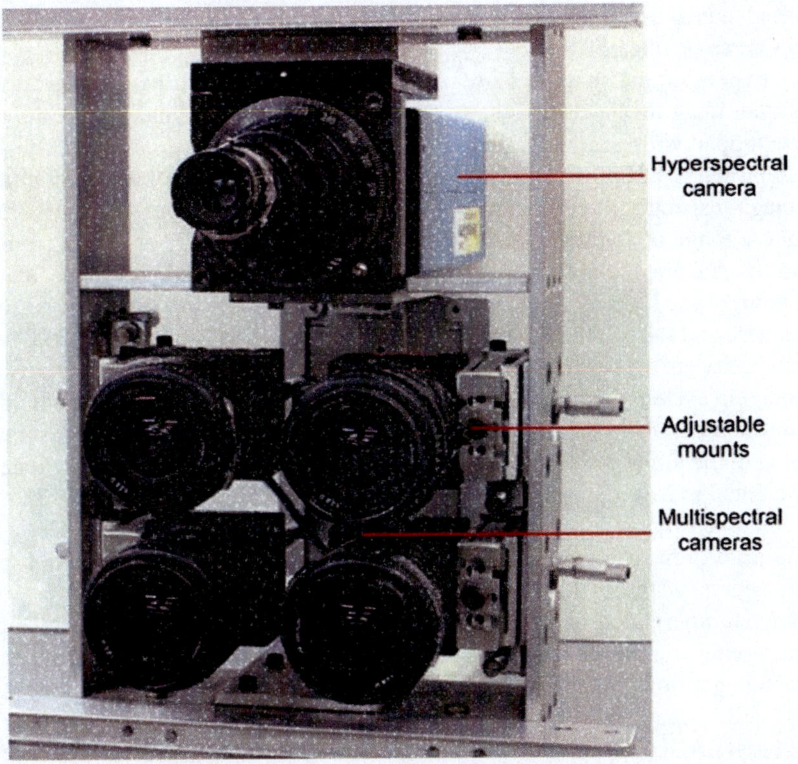

Fig. 13.3 A four-camera multispectral imaging system mounted along with a hyperspectral camera

Figure 13.3 shows a four-camera multispectral imaging system (Yang 2010) and a hyperspectral imaging camera (Yang et al. 2003) assembled at the USDA-ARS kika de la Garza Subtropical Agricultural Research Center in Weslaco, Texas. The two systems are mounted together and aligned to obtain both multispectral and hyperspectral imagery simultaneously. The multispectral system consists of four high resolution CCD digital cameras and a ruggedized PC equipped with a frame grabber and image acquisition software. The cameras are sensitive in the 400–1,000 nm spectral range and provide 2,048 × 2,048 active pixels with 12-bit data depth. The four cameras are equipped with blue (430–470 nm), green (530–570 nm), red (630–670 nm), and NIR (810–850 nm) bandpass interference filters, respectively, but have the flexibility to change filters for desired wavelengths and bandwidths. The cameras are arranged in a quad configuration and attached to adjustable mounts that facilitate aligning the cameras horizontally, vertically, and rotationally. The image acquisition software allows the synchronized black-and-white band images from the cameras to be viewed on the computer monitor in any one of the four modes: a quad, one band image at a time, a normal color composite, or a CIR composite. The band images are refreshed continuously to allow the operator to selectively save images with correct areas of interest.

The hyperspectral imaging system consists of a high performance digital CCD camera, an imaging spectrograph, an optional focal plane scanner, and a PC computer equipped with a frame grabbing board and camera utility software. The CCD camera provides 1,280(h) × 1024(v) pixel resolution and true 12-bit dynamic range. The imaging spectrograph is attached to the camera via an adapter to disperse radiation into a range of spectral bands. The effective spectral range resulting from this integration is from 467 to 932 nm. The optional focal plane scanner can be attached to the front of the spectrograph via another adapter for stationary image acquisition. The camera and the frame grabbing board are connected via a double coaxial cable, and the utility software allows for complete camera control and image acquisition. The imaging system captures one line image for all the bands at a time and an aircraft or the focal plane scanner serves as a mobile platform to carry out pushbroom scanning in the along-track direction. The horizontal and vertical binning capability of the camera makes it possible to obtain images with various spatial (160, 320, 640 and 1280 pixels in image width) and spectral (32, 64, 128, 256, 512 and 1024 bands) resolutions. For most applications, the hyperspectral sensor is configured to capture images with a swath of 640 pixels in 128 bands.

Many commercial airborne hyperspectral imaging sensors have been developed in recent years with improved spatial and spectral resolutions and high performance inertial navigation systems for increased position accuracy. For example, the AISA family of airborne hyperspectral sensors includes two sensors in the 0.4–0.97 μm range (AisaEAGLE and AisaEAGLET), one sensor in the 0.97–2.5 μm range (AisaHAWK), one sensor in the 0.4–2.5 μm range (AisaDUAL), and a thermal sensor in the 8–12 μm range (AisaOWL). The AisaEAGLE sensor can capture images with a swath of 1024 pixels and in up to 488 bands, while the AisaOWL can get a 384-pixel swath in up to 84 bands. All the sensors are equipped with a high performance, 3-axial inertial navigation sensor for monitoring the aircraft position and attitude. The sensor integrates solid state gyros and GPS with a real-time Kalman filter for increased accuracy.

13.8 Satellite Imaging Systems

Although remote sensing instruments are carried on ground-based and airborne platforms, remote sensors carried on man-made satellites provide large amounts of imagery commonly used today for a wide variety of applications. Satellite remote sensing systems not only cover large surface areas on the earth, but also view the same target area repeatedly. The sensor systems carried by satellites are primarily of the electro-optical type. Most of the satellite systems operate within the optical spectrum, which extends from 0.3 to 14 μm. Many other systems operate in the microwave portion of the spectrum, approximately 1 mm to 1 m wavelength. Remote sensing from space is rapidly changing with many countries and commercial firms developing and launching new systems on a regular basis. Therefore, it is important

Table 13.2 Wavelengths and spatial resolutions of the Landsat-7 ETM+ sensor

Band No.	Band Name	Wavelength (μm)	Spatial Resolution (m)
1	Blue-green	0.450–0.515	30
2	Green	0.525–0.605	30
3	Red	0.63–0.69	30
4	Near-infrared	0.75–0.90	30
5	Mid-infrared	1.55–1.75	30
6	Far-infrared	10.4–12.5	60
7	Mid-infrared	2.09–2.35	30
8	Panchromatic	0.52–0.90	15

to understand the basic characteristics of the major satellite systems available today so that appropriate satellite imagery can be selected for particular applications.

Landsat represents the first international satellite program designed specifically for collecting synoptic and repetitive multispectral image data of the earth's surface. Although the Landsat program is managed by the National Aeronautics and Space Administration (NASA), data from Landsat are collected and distributed by the U.S. Geological Survey (USGS). Since 1972, seven Landsat satellite systems have been launched with the launch failure of Landsat-6. Currently Landsat-5 and -7 are in operation. Landsat-5 was launched on March 1, 1984 and it carries the Thematic Mapper (TM) and the multispectral Scanner (MSS). The MSS sensor was powered off in 1995. Landsat-7 was launched on April 15, 1999 and is the latest satellite of the Landsat program. The instrument onboard this satellite is the Enhanced Thematic Mapper Plus (ETM+), which provides data continuity with Landsat-5. The main differences in data between Landsat-5 and -7 are the new 15-m panchromatic band and the enhanced thermal band with a spatial resolution of 60 m (versus 120 m for the TM sensor). The Landsat satellites have a 16-day repeat cycle and 185-km swath for imaging. The radiometric resolution is 8 bits. Table 13.2 gives the spectral wavebands and spatial resolution of the Landsat-7 ETM+ sensor.

SPOT is another successful international satellite remote sensing program initiated and designed by the French government, with the cooperation of other European organizations. Since 1986, five SPOT satellites have been launched with SPOT-4 and -5 currently in operation. SPOT-4 was launched in 1998 and provides 10-m panchromatic data over the range of 0.51–0.73 μm and 20-m multispectral data in the green (0.50–0.59 μm), red (0.61–0.68 μm), NIR (0.79–0.89 μm), and mid-infrared (1.58–1.75 μm) wavebands. With the successful launch of SPOT-5 on May 3, 2002, the SPOT program entered a new era by providing high resolution imagery with either 2.5- or 5-m resolution in the panchromatic band, 10-m resolution in the green, red, and NIR bands, and 20-m resolution in the mid-infrared band. The panchromatic band has a spectral range (0.48–0.71 μm) similar to that of SPOT-4, while the four multispectral bands remain the same. Each SPOT scene covers a 60 km by 60 km area and the pixel depth is 8 bits. For vertical observations, successful passes occur at 26-day intervals, but with its off-nadir viewing capability (+/–26°), successive imagery can be acquired every 2–3 days. The swath width of individual images varies from 60 to 80 km, depending on viewing angle.

In addition to the Landsat and SPOT satellite systems, there are many other earth observation satellite systems that have been in operation. India has launched several Indian Remote Sensing (IRS) satellite systems since 1988 and its satellite IRS-1D launched in 1997 provides 5.8-m panchromatic data, 23-m multispectral data in blue, green, red and NIR bands, and 70-m data in a mid-infrared band. China and Brazil have jointly launched several earth resource satellites (CBERS-1, -2 and -2B) since 1999. The CBERS satellites carry three sensors, including a high resolution CCD camera, a moderate multispectral scanner, and a low resolution wide-field imaging system. The CCD camera provides 20-m image data in a panchromatic band and four multispectral bands (blue, green, red, and NIR). Due to the success of these satellites, the two governments have decided to launch more CBERS satellites in the near future. Several other countries such as Japan and Russia have also developed and launched similar remote sensing satellites.

Since late 1990s, numerous high resolution satellite systems have been launched or are in development stage. These systems provide remote sensing data in much higher spatial and temporal resolutions than those described above. GeoEye, Inc. (Dulles, Virginia) first made history with the IKONOS satellite launch in 1999. IKONOS provides 1-m panchromatic images in the 0.45–0.90 μm spectral range and 4-m multispectral imagery in the blue (0.45–0.52 μm), green (0.51–0.60 μm), red (0.63–0.70 μm) and NIR (0.76–0.85 μm) bands. The panchromatic and multispectral imagery can be merged to create 1-m color imagery (pan-sharpened). The radiometric resolution is 11 bits, or 2048 gray levels. The image swath is 11.3 km at nadir and the revisit time is less than 3 days. Shortly after the successful launch and operation of IKONOS, DigitalGlobe, Inc. (Longmont, Colorado) launched the QuickBird satellite in 2001. QuickBird provides panchromatic and multispectral data in essentially the same spectral ranges as those of IKONOS, but at a higher spatial resolution. QuickBird acquires panchromatic data with 0.60-m resolution and four multispectral bands with 2.4-m resolution. Similarly, pixel depth is 11 bits. The image swath at nadir is 16.4 km and the sensor can tilt up to 45° off nadir.

GeoEye again made history with the launch of GeoEye-1 in 2008. It offers unprecedented spatial resolution by simultaneously acquiring 0.41-m panchromatic and 1.65-m four-band multispectral imagery. The spectral ranges are similar to those of IKONOS. The pixel dynamic range is also 11 bits. The image swath is increased to 15.2 km. On October 8, 2009, DigitalGlobe launched WorldView-2, the first high resolution 8-band multispectral satellite, to acquire panchromatic data at 0.46-m resolution and multispectral imagery at 1.84-m resolution. WorldView-2's unique combination of high spatial and spectral resolution provides new opportunities and potential for a variety of practical remote sensing applications. The imagery is distributed at either 0.5 or 0.6 m resolution for the panchromatic band and at either 2.0 or 2.4 m resolution for the multispectral bands, depending on the sensor's viewing angle. The image swath at nadir remains to be16.4 km and the average revisit time is about 1.1 days. Table 13.3 gives the spectral characteristics for the four high resolution satellite sensors.

These high resolution satellite imaging systems have significantly narrowed the gap in spatial resolution between satellite and airborne imagery. In addition to their high spatial resolution, these satellite sensors offer image data at 11 bits, 8 times as

Table 13.3 Spectral characteristics for WorldView-2, QuickBird, GeoEye-1 and IKONOS

	Spectral band (μm)			
Band name	WorldView-2	QuickBird	GeoEye-1	IKONOS
Panchromatic	0.450–0.800	0.450–0.900	0.450–0.800	0.450–0.900
Coastal	0.400–0.450			
Blue	0.450–0.510	0.450–0.520	0.450–0.510	0.445–0.516
Green	0.510–0.580	0.520–0.600	0.510–0.580	0.505–0.595
Yellow	0.585–0.625			
Red	0.630–0.690	0.630–0.690	0.655–0.690	0.632–0.698
Red Edge	0.705–0.745			
Near-IR1	0.770–0.895	0.760–0.900	0.780–0.920	0.757–0.853
Near-IR2	0.860–1.040			

many gray levels as the 8-bit resolution from traditional satellite sensors and many airborne imaging sensors. Moreover, the high revisit frequency and fast turnaround time of these high resolution satellites are certainly advantages over traditional satellites. These advantages combined with their relatively large area coverage and ability to take imagery over any geographic area make high resolution satellite imagery attractive for many applications, including crop pest detection and mapping.

13.9 Image Processing and Analysis

Image processing and analysis is an important component of remote sensing technology. Different imaging systems provide different types of imagery and therefore a variety of techniques need to be used to process and analyze the image data. These techniques are diverse, ranging from simple visual interpretation to sophisticated computer processing methods. Image processing and analysis generally involves image display and enhancements, image registration and rectification, image classification, accuracy assessment, and more advanced spectral analysis techniques. Because of limited space, the reader can refer to other textbooks for this special topic (Campbell 2002; Richards and Jia 2005; Lillesand et al. 2007; ERDAS 2008).

Two other spatial information technologies closely related to remote sensing are GPS and GIS. GPS data are often required to determine the geographic locations of airborne imagery and to geometrically correct and georeference the imagery. A GIS provides a platform for GPS data and remote sensing imagery to be displayed, analyzed, and integrated with other spatial data. At the same time, remote sensing imagery and GPS data have become primary data sources for modern GIS applications. Indeed, these technologies have been interrelated and the boundaries between them have become blurred. Although we emphasize the principles and systems of remote sensing in this chapter and will not provide further background information on GPS and GIS, we do emphasize and illustrate the interrelations among the three technologies in the next section. The reader is encouraged to consult available textbooks and references on these subjects (Gao 2002; Kennedy 2002; Madden 2009).

13.9.1 Detecting and Mapping Whitefly Infestations Using Remote Sensing, GPS and GIS – An Application Example

Damage caused by whiteflies of the *Bemisia* species complex involves a combination of factors, including the transmission of important plant pathogens (Brown and Bird 1992) and a general reduction in plant vigor as a result of feeding activities (Byrne et al. 1990). Whitefly-transmitted geminiviruses are a major constraint to the production of agricultural and vegetable crops in the tropical and subtropical regions of the world (Morales and Jones 2004). Moreover, feeding nymphs excrete copious quantities of honeydew which may contaminate cotton lint and commonly promote the growth of associated sooty mold fungi *(Capnodium* sp.) (Hendrix and Wei 1992). Heavy sooty mold deposits on the plant foliage are detrimental in the sense that they impede photosynthesis, but they are highly visible and can be used as an advantage for the detection of insect infestations in damage assessment surveys and remote sensing activities.

Any pest which supplies sufficient plant stress to significantly distort the reflectance signal is a candidate for detection by means of remote sensing. Remote sensing techniques employed for detection of insect activity are usually based on damage caused by the pest and not on detection of the actual organism. The types of damage that are usually detectable are sooty mold deposits that result from honeydew producing insects, defoliation (Harris et al. 1976), color changes (Hart and Ingle 1969), and geometric distortion of the shape of the plant (Payne et al. 1971). Past research has showed that CIR aerial photography can be used to remotely detect sooty mold deposits on citrus foliage caused by the honeydew producing insects, brown soft scale *(Coccus hesperidum* L.) and citrus blackfly *(Aleurocanthus woglumi* Ashby) (Hart and Myers 1968; Hart et al. 1973; Everitt et al. 1994). Although the CIR photography provided only indirect evidence of insect infestation, it was shown to be practical for use in regional surveys (Hart et al. 1973). Nuessly et al. (1987) evaluated CIR aerial photography for detecting sooty mold growing on honeydew secreted by *B. tabaci* on cotton *(Gossypium hirsutum* L.) plants in the Imperial Valley of California. Ground surveys indicated that cotton plants with no or low levels of sooty mold growth appeared bright red in CIR photographs, whereas leaves with sooty mold appeared dark red to black due to reduced reflectance. Since reflectance decreases as soot mold deposits increase, this verifies that remote sensing is useful to distinguish infestations by honeydew-producing insects. Several review articles or book chapters have been published to include the discussion of remote sensing for such insect pests affecting fruit trees and crops (Hart et al. 1971; Myers et al. 1983; Riley 1989; Hart 1990; Ryerson et al. 1997; Yang and Everitt 2005).

More recently, remote sensing has been integrated with GPS and GIS technologies to assist natural resource managers and agricultural consultants in developing sound management strategies. Aerial videography and GPS technology have been merged and shown to be useful tools for detecting and monitoring insect activity over forested areas (Myhre 1992). The latitude-longitude data provided by a GPS receiver were entered into a GIS to georeference forest pest problems. Richardson et al. (1993)

entered aerially obtained GPS coordinates into a GIS to map the distribution of undestroyed cotton fields in a regional management program for boll weevil (*Anthonomus grandis* Boheman) in south Texas. Everitt et al. (1994) integrated aerial videography with GPS and GIS technologies to map citrus blackfly (*Aleurocanthus woglumi* Ashby) infestations over a large agricultural area.

Everitt et al. (1996) described the application of airborne videography with GPS and GIS technologies for detecting and mapping silverleaf whitefly (*Bemisia argentijolii* Bellows and Perring) (also known as sweetpotato whitefly B-biotype) infestations in cotton in the Rio Grande Valley of Texas. This study is used as an example in this chapter to illustrate how remote sensing can be integrated with GPS and GIS for detecting and mapping whitefly infestations. The whitefly study area was a 62 km^2 (4.8 km × 12.9 km) agricultural area with a large number of cotton fields to the north of Weslaco in Hidalgo County. Aerial videography and ground truth observations were conducted for the study. Plant canopy reflectance measurements were made to help interpret the video imagery.

Video imagery was obtained with a three-camera multispectral digital video imaging system (Everitt et al. 1995). The system consisted of three CCD analog black-and-white video cameras, a computer equipped with an image digitizing board, a color encoder, and super(S)-VHS portable recorder. The cameras are visible/NIR (0.4–1.1 μm) light sensitive. Two of the cameras were equipped with visible green (0.555–0.565 μm) and red (0.623–0.635 μm) filters, respectively, while the third camera had a NIR (0.845–0.857 μm) filter. The image had a 640 × 480 pixel resolution. The NIR, red, and green image signals from the cameras were subjected to the RGB inputs of the computer digitizing board, giving a CIR composite digital image similar in color rendition to that of CIR film. In addition, the image signals were also subjected to a color encoder so that analog CIR video imagery could be recorded on the S-VHS recorder. Imagery of the whitefly experimental area was taken at altitudes ranging from 300 to 1,500 m on 11, 12 and 19 July 1995. The lower altitude imagery was taken to acquire more detail of some ground sites. All imagery was acquired with a fixed-wing aircraft between 1,130 and 1,530 h under sunny conditions.

A Trimble GPS receiver was integrated with the video system. The receiver continuously displayed the flight direction (bearing), altitude, time, ground speed, and latitude-longitude coordinates of the aircraft above the ground. Two modes were used simultaneously to obtain the GPS data for the whitefly survey. In one mode, the continuous GPS information was transferred and recorded onto the bottom of the red band image, which in turn was also superimposed on the composite image. The latitude-longitude coordinates on the video image corresponded to approximately the center of each scene. The second mode was the waypoint mode in which coordinates were manually entered and stored in the GPS memory. With this mode the coordinate data were directly transferred into the computer.

ATLAS GIS software was used to generate maps of Hidalgo County based on TIGER/Line files for the State of Texas. The TIGER/Line files were developed and trademarked by the U.S. Census Bureau and contained features such as roads, rivers as well as legal and statistical geographic areas. These files are available to the

Table 13.4 Canopy reflectance of non-infested and whitefly-infested cotton (as determined by deposits of sooty mold fungus on the foliage) at the green, red, and near-infrared wavelengths (Adapted from Everitt et al. 1996)

		Canopy reflectance (%)		
Site and date	Sooty mold level	Green	Red	Near-infrared
Field 1	None	4.7a	2.9a	36.5a
12 July 1995	Low-moderate	3.2b	2.2b	16.9b
	High	2.8b	2.5ab	9.0c
Field 2	None	5.6a	3.0a	34.5a
20 July 1995	Low-moderate	3.0b	2.1b	17.1b
	High	2.3c	2.0b	9.1c

Means within a column at each date followed by the same letter do not differ significantly at the 0.05 probability level according to Duncan's multiple range test

public for no charge and are designed for use with GIS software. The GPS data from the airborne survey were overlaid to the TIGER/Line map to geographically map the whitefly infestations.

Radiometric plant canopy reflectance measurements were made in two cotton fields near Weslaco, Texas during July 1995. Reflectance measurements were made on cotton plants with no sooty mold deposits on the leaves (control), plants with light to moderate sooty mold deposits, and those with heavy deposits. Control cotton plants had typical green leaves, plants with light to moderate sooty mold deposits had a light to moderate gray leaf appearance, and leaves from those with heavy deposits had a dark gray to black cast. Reflectance data in the visible green (0.52–0.62 μm), visible red (0.63–0.69 μm), and NIR (0.76–0.90 μm) spectral bands were taken on 12 randomly selected plant canopies for each sooty mold deposit level at each field with a multispectral radiometer. The sensor had a 15-degree field of view and was placed approximately 1 m above each plant canopy. Measurements were made between 1,100 and 1,500 h under sunny conditions. Ground truth surveys were made of the study sites to verify the presence of whitefly infestations. Observational data recorded were plant species, density, cover, and soil type. Ground photographs were taken to help interpret the aerial video imagery.

Mean canopy reflectance values of cotton plants with no sooty mold deposits on the foliage, those with low to moderate levels, and those with heavy levels for two cotton fields are given in Table 13.4. For both fields, cotton plants with low to moderate and high levels of sooty mold had lower reflectance values than those with no sooty mold, though the differences in reflectance were much more significant in the NIR band than in the visible bands.

The lower visible and NIR reflectance of cotton plants with sooty mold deposits on their leaves was attributed to the dark sooty mold fungus which absorbed a large percentage of the visible and NIR radiation. The more pronounced differences in the NIR spectral region between plants with no sooty mold deposits and those with sooty mold deposits agree with the findings of Gausman and Hart (1974).

Fig. 13.4 Color-infrared (CIR) composite and near-infrared (NIR), *red*, and *green black-and-white* band digital video images of a cotton field near Weslaco, Texas. The *arrow* on the CIR composite points to the high levels of sooty mold fungus on the cotton foliage caused by whitefly infestation. The infested area is also well delineated in the NIR image, but cannot be clearly distinguished in the *red* and *green* band images. The GPS data shown at the *bottom* of the CIR composite and the *red* band image include the bearing, ground speed (miles/h), altitude (feet), date, latitude, longitude, and time (Adapted from Everitt et al. (1996))

Figure 13.4 shows the CIR composite and NIR, red, and green band digital video images of a whitefly-infested cotton field near Weslaco, Texas. The latitude and longitude for the center of the field are 26°11.543′N and 97°59.560′W, respectively. The imagery was obtained at an altitude of approximately 910 m. The arrow on the CIR composite points to the dark gray to gray-black image response of high levels of sooty mold fungus on the cotton foliage caused by the whitefly infestation. Cotton plants with low to moderate levels of sooty mold fungus have a dull magenta to gray-brown response, while plants with no sooty mold deposits have a bright red tone. Sparsely vegetated and essentially bare soil areas have a green to light gray color. Most of the whitefly-infested plants could also be distinguished in the NIR black-and-white image. In contrast, the sooty mold deposits could not be clearly distinguished in the red and green band images, even though the green band image was better than the red band image. Analysis of the video imagery of the study area identified approximately 65 locations thought to be infested by whitefly. In several instances more than one infestation was recorded within large cotton fields. Ground reconnaissance of the study area verified the presence of whiteflies at all locations.

The superiority of the NIR image over the red and green images for distinguishing sooty mold deposits generally agrees with the canopy reflectance data, where the best differences among cotton plants with and without sooty mold deposits on the foliage occurred at the NIR wavelength interval (Table 13.4). These findings also concur with previous research on using videography for detecting citrus blackfly infestations (Everitt et al. 1994). Although the red and green canopy reflectance data showed potential for spectrally separating between cotton plants with and without sooty mold deposits, these differences apparently were not great enough to be distinguished in the red and green video images. Summy et al. (1997) used airborne digital video imagery to monitor damage caused by two honeydew-excreting insects, cotton aphid (*Aphis gossypii* Glover) and silverleaf whitefly, on cotton. They were able to identify the differences among treated and untreated plots using the NIR band images. Reisig and Godfrey (2006) explored airborne and QuickBird remote sensing imagery and ground reflectance data for their potential to distinguish cotton plants infested with aphid and spider mite (*Tetranychus spp.*) from non-infested cotton. Their results showed that cotton infested with cotton aphids above economic threshold levels was detected using NIR wavelengths from the remote sensing data.

The GPS data displayed at the bottom of the CIR and red band images of the whitefly-infested cotton fields are useful to locate infestations and could be entered into a GIS manually. The GPS latitude-longitude data obtained from the waypoint mode were used with the GIS to locate whitefly infestations in this study. Figure 13.5 (left) shows a GIS map of Hidalgo County with the study area denoted in the lower right portion of the map. The triangles depict GPS latitude-longitude coordinates for whitefly infestations within the area. Figure 13.5 (lower right) shows a more detailed GIS map of the study area depicting the 65 locations (triangles) where whitefly infestations occurred. With this map, one can associate the general road addresses and cannels with the marked location of each infested cotton field. The integration of the GPS with GIS technology enables the agricultural consultant to develop regional maps of a large agricultural area showing where insect infestations occur. The aerial imagery can provide a means of determining the severity and extent of infestations within each field. To identify and quantify whitefly-infested areas within each field, the imagery needs to be georeferenced and then classified using unsupervised and/or supervised classification techniques. These techniques have been successfully used on multispectral and hyperspectral imagery for quantifying the areas infested by insects or diseases within fields (Fitzgerald et al. 2004; Yang et al. 2010).

The results from this case study demonstrate that aerial multispectral videography is a useful tool for detecting whitefly infestations in cotton fields. The integration of videography, GPS, and GIS technologies is valuable for mapping the distribution of whitefly infestations over a large agricultural area. Maps can be produced that would aid agricultural consultants and farm managers to depict where insect infestations

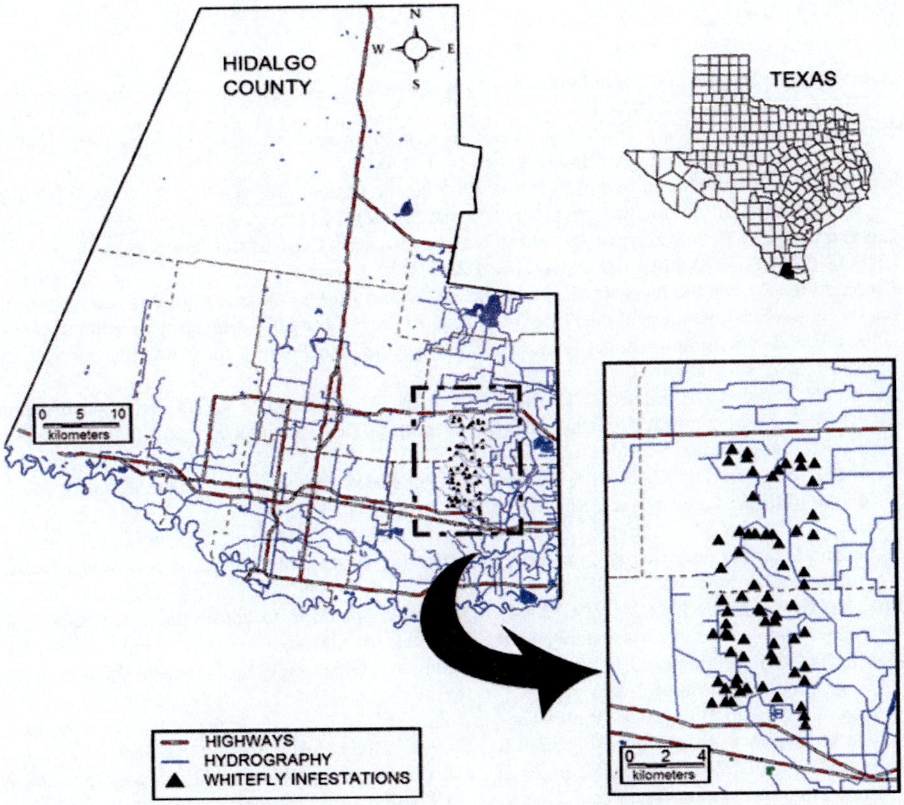

Fig. 13.5 A GIS map (*upper left*) of Hidalgo County with the whitefly study area denoted in the lower right portion of the map. The symbols (*triangles*) within the study area represent GPS latitude-longitude coordinates of whitefly infestations in cotton fields. A detailed GIS map (*lower right*) clearly depicts the locations of the whitefly infestations (Adapted from Everitt et al. (1996))

occur over a large geographic region. The video imagery can also serve as a permanent geographically located image database to monitor future contraction or spread of insect infestations over time.

There is no doubt that remote sensing has played an important role in the detection and management of many agricultural insect pests, including whiteflies. The integration of remote sensing with GPS and GIS has made it more valuable for practical applications. Nevertheless, early detection of most insect pests remains difficult to impossible with remote sensing. With the advances of these technologies, especially with the increases in spatial and spectral resolutions, revisit frequency, availability and affordability of these new-generation satellite imaging systems, remote sensing will play a more important role and become more accurate, practical and economical in the detection and management of agricultural insect pests at both field and regional levels.

References

Avery TE, Berlin GL (1992) Fundamentals of remote sensing and airphoto interpretation, 5th edn. Prentice Hall, Upper Saddle River

Brown JK, Bird J (1992) Whitefly-transmitted geminiviruses and associated disorders in the Americas and the Caribbean Basis. Plant Dis 76:220–225

Byrne DN, Bellows TS, Parrella MP (1990) Whiteflies in agricultural systems. In: Gerling D (ed.) Whiteflies: their bionomics, pest status and management. Intercept, Andover

Campbell JB (2002) Introduction to remote sensing, 3rd edn. The Guilford Press, New York

ERDAS (2008) ERDAS field guide, vol 1 and 2. ERDAS, Norcross

Escobar DE, Everitt JH, Noriega JR, Davis MR, Cavazos I (1997) A true digital imaging system for remote sensing applications. In: Proceedings of the16th biennial workshop on color photography and videography in resource assessment, American Society for Photogrammetry and Remote Sensing, Bethesda

Everitt JH, Escobar DE, Summy KR, Davis MR (1994) Using airborne video, global positioning system, and geographic information system technologies for detecting and mapping citrus blackfly infestations. Southwest Entomol 19:129–138

Everitt JH, Escobar DE, Cavazos I, Noriega JR, Davis MR (1995) A three-camera multispectral digital video imaging system. Remote Sens Environ 54:333–337

Everitt JH, Escobar DE, Summy KR, Alaniz MA, Davis MR (1996) Using spatial information technologies for detecting and mapping whitefly and harvester ant infestations in south Texas. Southwest Entomol 21:421–432

Fitzgerald GJ, Maas SJ, Detar WR (2004) Spider mite detection in cotton using hyperspectral imagery and spectral mixture analysis. Precision Agric 5:275–289

Gao J (2002) Integration of GPS with remote sensing and GIS: reality and prospect. Photogramm Eng Remote Sens 68:447–453

Gausman HW, Hart WG (1974) Reflectance of four levels of sooty mold deposits produced from the honeydew of three insect species. J Rio Grande Valley Hortic Soc 28:131–136

Harris MK, Hart WG, Davis MR, Ingle SJ, Van Cleave HW (1976) Aerial photographs show caterpillar infestation. The Pecan Quarterly 10:12–18

Hart WG (1990) Remote sensing. In: Rosen D (ed.) The armored scale insects, their biology, natural enemies and control, vol B. Elsevier Science, Amsterdam

Hart WG, Ingle SJ (1969) Detection of arthropod activity on citrus foliage with aerial infrared color film as a method of surveying for citrus blackfly. J Econ Entomol 66:190–194

Hart WG, Myers VI (1968) Infrared aerial photography for detection of populations of brown soft scale in citrus groves. J Econ Entomol 61:617–624

Hart WG, Ingle SJ, Davis MR, Mangum C, Higgins A, Boling JC (1971) Some uses of infrared aerial photography in entomology. In: Proceedings of the 3rd biennial workshop color aerial photography in the plant sciences, American Society of Photograrnmetry, Falls Church

Hart WG, Ingle SJ, Davis MR, Mangum C (1973) Aerial photography with infrared color film as a method of surveying for citrus blackfly. J Econ Entomol 66:190–194

Hendrix DL, Wei Y (1992) Detection and elimination of honeydew excreted by the sweetpotato whitefly feeding upon cotton. In: Proceedings of the beltwide cotton Conference, National Cotton Council, Memphis

Kennedy M (2002) The global positioning system and GIS: an introduction, 2nd edn. Taylor & Francis, New York

Lillesand TM, Kiefer RW, Chipman JW (2007) Remote sensing and image interpretation, 6th edn. Wiley, Hoboken

Madden M (2009) Manual of geographic information systems. American Society of Photogrammetry and Remote Sensing, Bethesda

Mao C (1999) Hyperspectral imaging systems with digital CCD cameras for both airborne and laboratory application. In: Proceedings of 17th biennial workshop on videography and color photography in resource assessment. American Society for Photogrammetry and Remote Sensing, Bethesda

Mausel PW, Everitt JH, Escobar DE, King DJ (1992) Airborne videography: current status and future perspectives. Photogramm Eng Remote Sens 58:1189–1195

Meisner DE, Lindstrom OM (1985) Design and operation of a color-infrared aerial video system. Photogramm Eng Remote Sens 51:555–560

Morales FJ, Jones PG (2004) The ecology and epidemiology of whitefly-transmitted viruses in Latin America. Virus Res 100:57–65

Moran MS, Inoue Y, Barnes EM (1997) Opportunities and limitations for image-based remote sensing in precision crop management. Remote Sens Environ 61:319–346

Myers VI, Bauer ME, Gausman HW, Hart WG, Heilman JL, McDonald RB, Park AB, Ryerson RA, Schmugge TJ, Westin FC (1983) Remote sensing in agriculture. In: Colwell RN (ed.) Manual of remote sensing. American Society of Photogrammetry, Falls Church

Myhre RJ (1992) Use of color airborne videography in the U.S. Forest Service. In: Proceedings of the Resource Technology 92. Symposium, American Society of Photogrammetry and Remote Sensing, Bethesda

Nixon PR, Escobar DE, Menges RM (1985) Use of a multi-band video system for quick assessment of vegetal condition and discrimination of plant species. Remote Sens Environ 17:203–208

Nuessly GS, Meyerdirk DE, Hart WG, Davis MR (1987) Evaluation of color-infrared aerial photography as a tool for the identification of sweetpotato whitefly induced fungal and viral infestations of cotton and lettuce. In: Proceedings of the 11th biennial workshop on color aerial photography and videography in the plant sciences and related fields, American Society of Photogrammetry and Remote Sensing, Falls Church

Payne JA, Hart WG, Davis MR, Jones LS, Weaver DJ, Horton BD (1971) Detection of peach and pecan pests and diseases with color infrared aerial photography. In: Proceedings of the 3rd biennial workshop on color aerial photography in the plant sciences, American Society of Photogrammetry, Falls Church

Pearson R, Mao C, Grace J (1994) Real-time airborne monitoring. Remote Sens Environ 49:304–310

Pinter PJ Jr, Hatfield JL, Schepers JS, Barnes EM, Moran MS, Daughtry CST, Upchurch DR (2003) Remote sensing for crop management. Photogramm Eng Remote Sens 69:647–664

Reisig D, Godfrey L (2006) Remote sensing for detection of cotton aphid- (Homoptera: Aphididae) and spider mite- (Acari: Tetranychidae) infested cotton in the San Joaquin Valley. Pest Manag 35:1635–1646

Richards JA, Jia X (2005) Remote sensing digital image analysis: an introduction, 4th edn. Springer, Berlin

Richardson AJ, Summy KR, Davis MR, Gomez A, Williams DW (1993) The use of 1990 Tiger/Line™ Census files for monitoring the Rio Grande Valley cotton stalk destruction program. In: Proceedings of the application advanced information technology Symposium, Stevens

Riley JR (1989) Remote sensing in entomology. Annu Rev Entomol 34:247–271

Ryerson RA, Curran PJ, Stephens PR (1997) Applications: agriculture. In: Philipson WR (ed.) Manual of photographic interpretation. American Society for Photogrammetry and Remote Sensing, Bethesda

Summy KR, Everitt JH, Escobar DE, Alaniz MA, Davis MR (1997) Use of airborne digital video imagery to monitor damage caused by two honeydew-excreting insects on cotton. In: Proceedings of the 16th biennial workshop on color photography and videography in resource assessment, American Society for Programmetry and Remote Sensing, Bethesda

Yang C (2010) A high resolution airborne four-camera imaging system for agricultural applications. ASABE paper no. 1008856, American Society of Agricultural and Biological Engineers, St. Joseph

Yang C, Everitt JH (2005) Remote sensing, GPS and GIS technologies for agricultural insect pest detection. In: Liu TX, Kang L (eds.) Entomological research: progress and prospects. Science Press, Beijing

Yang C, Everitt JH, Davis MR, Mao C (2003) A CCD camera-based hyperspectral imaging system for stationary and airborne applications. Geocarto Int J 18:71–80

Yang C, Fernandez CJ, Everitt JH (2010) Comparison of airborne multispectral and hyperspectral imagery for mapping cotton root rot. Biosyst Eng 107:131–139

Index

W.M.O. Thompson (ed.), *The Whitefly, Bemisia tabaci (Homoptera: Aleyrodidae) Interaction with Geminivirus-Infected Host Plants*,
DOI 10.1007/978-94-007-1524-0,

C

Q

T